Agricultural Waste Management

Agricultural Waste Management

Edited by **Thelma Bosso**

Callisto Reference
New York

Published by Callisto Reference,
106 Park Avenue, Suite 200,
New York, NY 10016, USA
www.callistoreference.com

Agricultural Waste Management
Edited by Thelma Bosso

International Standard Book Number: 978-1-63239-715-7 (Hardback)

Printed in China.

Contents

Preface

The purpose of the book is to provide a glimpse into the dynamics and to present opinions and studies of some of the scientists engaged in the development of new ideas in the field from very different standpoints. This book will prove useful to students and researchers owing to its high content quality.

Agricultural waste refers to the waste produced by various agricultural activities. It can include both organic waste, like unwanted produce, manure, etc; and inorganic waste like excess pesticides, agrochemicals, silage, waste from machineries like oil, grease, etc. Agricultural waste has become a growing concern for environmentalists. Thus, its management has become a priority. Agriculture waste management insures that the things used for agricultural activities are recycled, reused and the use of non-biodegradable waste is reduced. This book will elucidate the various concepts and theories to manage agricultural waste and provide case studies to comprehend the usefulness of their applications. The topics covered here in will offer readers new insights in this field. It is an essential guide for agriculturists, environmentalists, ecologists and those who wish to pursue this discipline further.

At the end, I would like to appreciate all the efforts made by the authors in completing their chapters professionally. I express my deepest gratitude to all of them for contributing to this book by sharing their valuable works. A special thanks to my family and friends for their constant support in this journey.

Editor

1

Studies on the reduction of organic load from palm oil mill effluent (POME) by bacterial strains

Jeremiah David Bala · Japareng Lalung · Norli Ismail

Abstract

Introduction Palm oil mill effluent (POME) contains large quantities of organic matter in the form of total suspended solids (TSS), volatile suspended solids (VSS), total solids (TS), oil and grease (O & G) that increase biochemical oxygen demand (BOD) and chemical oxygen demand (COD) of POME if left untreated. The main aim of the present study was to investigate the ability of bacterial strains either pure (individually) or mixed (combinations), to degrade and metabolize organic load from palm oil mill effluent.

Results Sequencing of the 16S rRNA of the isolates suggests that they were identified as *Micrococcus luteus* 101 PB, *Stenotrophomonas maltophilia* 102 PB, *Bacillus cereus* 103 PB, *Providencia vermicola* 104 PB, *Klebsiella pneumoniae* 105 PB and *Bacillus subtilis* 106 PB. The use of mixed cultures in the present study showed more extensive removal of organic load (COD and BOD) than pure single cultures. Mixed cultures were found to reduce the pollutant dynamically. Thus, the mixed cultures C_1 (*Bacillus cereus* 103 PB and *Bacillus subtilis* 106 PB) were the most effective bacterial combination for use in biological treatment technology of POME having the highest COD and BOD reduction rate. C_1 produced the highest degradative activity in reducing COD (90.64 %) and BOD (93.11 %).

Conclusion The indigenous microbial isolates from POME were observed to possess potential to degrade organic components whereas the use of mixed cultures resulted in more extensive degradation of COD and BOD than pure single cultures. This suggests that mixed culture of bacteria in the present study can be used for bioremediation of environment contaminated with polluted wastewaters. This study, however, indicates the prospect of isolating indigenous microorganisms in the POME for effective biotreatment of POME.

Keywords Bacteria · BOD · COD · Organic load · POME · Reduction · Treatment · Wastewater

Introduction

Palm oil industry is one of the leading agricultural industries in Malaysia with average crude palm oil production of more than 13 million tonne per year. However, production of such huge amount of crude palm oil has consequently resulted to even larger amount of palm oil mill effluent (POME). POME is a highly polluting wastewater with high chemical oxygen demand (COD) and biochemical oxygen demand (BOD) which can cause severe pollution to the environment, typically pollution to water resources (Lam and Lee 2011). The palm oil industry is identified as one of the agricultural industries in Malaysia that generates the highest pollution load into rivers throughout the country (Wu et al. 2007). Although POME is a non-toxic liquid waste with unpleasant smell, its COD and BOD values are high enough to cause serious pollution and environmental problem to the rivers. Chin et al. (1996) have reported that

J. D. Bala · J. Lalung · N. Ismail (✉)
Environmental Technology Division, School of Industrial Technology, Universiti Sains Malaysia (USM), 11800 Pulau, Pinang, Malaysia
e-mail: norlii@usm.my

J. D. Bala
e-mail: Jerrybrown316@yahoo.com

J. Lalung
e-mail: japareng@usm.my

POME contains a high concentration of organic matter, COD concentration in the range of 45,000–65,000 mg/L and BOD of 18,000–48,000 mg/L. COD in the range of 80,100–95,000 mg/L and BOD of 23,400–52,100 mg/L has also been encountered (Soleimaninanadegani and Manshad 2014).

The high COD value is responsible for distraction of aquatic life (Maygaonkar et al. 2012). Generally, COD and BOD are essential parameters used as indicators of composition and environmental impact of wastewaters (Jameel et al. 2011). Oil and grease are considered as hazardous pollutants particularly in the aquatic environments, since they are highly toxic to the aquatic organisms and can completely damage the ecology of the aquatic ecosystem. When discharged into the environment, it may have objectionable odour, cause undesirable appearance, burn on the surface of receiving water creating potential hazards and consume dissolved oxygen necessary to forms of life in water (Jameel and Olanrewaju 2011). Suspended solids can clog fish gills, either killing them or reducing their growth rate. They also reduce light penetration. This reduces the ability of algae to produce food and oxygen. POME contains higher concentrations of various pollutants and such may be categorized as a high strength wastewater which required effective treatment before being discharged into the environment.

Very few investigations have been conducted on aerobic digestion process for the treatment of organic pollutants present in POME (Wu et al. 2010). Therefore, this work represents one of the few studies on the aerobic biotreatment technology of POME-contaminated wastewater in Malaysia.

Biological treatment of POME by *Yarrowia lipolytica* NCIM 3589 has been studied (Oswal et al. 2002). Abass et al. (2012) and Soleimaninanadegani and Manshad (2014) have also reported aerobic digestion of POME by selected microorganisms. Treatment of POME has been variously reported by other workers (Sinnappa 1978; Agamuthu et al. 1986; Chin et al. 1996; Mohammed et al. 2014).

The anaerobic digestion treatment of POME using various types of bioreactors by researchers and the ponding systems in the mills uses unknown microbial populations (McHugh et al. 2003) to reduce the polluting power of wastes and wastewaters. This involves a consortium of unknown microorganisms catalyzing a complex series of biochemical reactions that mineralize organic matter producing methane and carbon dioxide. These microorganisms are not established and hence the substrate they degrade and utilize is not ascertained.

In the present study, known microorganisms were isolated from POME, which were further used to inoculate the POME. The efficiency for organic load reduction and the percent reduction was monitored. Therefore, so far to the best of our knowledge no work has been done on the isolation of different aerobic microorganisms in POME. This emphasizes the originality of the study.

The application of microorganisms such as *Trichoderma viride* spores, *T. viride* mycelium, *Yarrowia lipolytica* and *Saccharomyces cerevisiae* for the treatment of POME was not that effective in organic load reduction (Jameel and Olanrewaju 2011). This may be due to the fact that these microorganisms are not indigenous to POME (Jameel and Olanrewaju 2011). This gap offered researchers a greater opportunity to explore the organic load reduction of POME by their indigenous (autochthonous) isolates. These are organisms that exist/occur naturally in a particular biophysical environment.

Hence, serious measures have to be taken to prevent the growing pollution and ecological degradation related to POME. The aim of the present study was to investigate the ability of bacterial strains either pure (individually) or mixed (combinations), isolated from POME to degrade and metabolize organic load from POME.

Materials and methods

Sample collection

Raw palm oil mill effluent was collected from the site of a palm oil mill industry in a sterile container and brought back to the laboratory. The sample was transported to the laboratory in an ice box and analyzed for microbiological and physicochemical properties within 4 h of collection. The physicochemical characteristics of the sample were determined in accordance with the standard methods published by American Public Health Association (APHA 2005).

Sample preservation

The POME was preserved at a temperature less than 4 °C, to prevent the wastewater from undergoing biodegradation due to microbial action (APHA 2005).

Microorganisms

Four isolates (*Micrococcus luteus* 101 PB, *Stenotrophomonas maltophilia* 102 PB, *Bacillus cereus* 103 PB, and *Bacillus subtilis* 106 PB) were selected for POME inoculation and treatment based on the criteria that they were able to display good growth and degradation of oil and cellulose as sole source of carbon and energy in MSM from our previous work (Bala et al. 2014). The isolates were identified by sequence analysis of 16S rRNA genes. Sequences of primers used are:

27F: 5′-AGAGTTTGATCMTGGCTCAG-3′
1492R: 5′-GGGTTACCTTGTTACGACTT-3′

Plates 1 and 2 show genomic DNA and purified PCR product of bacteria isolated from POME inclusive of two others isolates which did not fulfil the criteria of the present study.

Analytical methods and characterization of POME

The following parameters BOD, COD, TSS, oil and grease were analyzed by standard methods (APHA 2005) for 5 days. These analyses were carried out according to the standard procedures (APHA 2005). Characterization of the POME was carried out before and after the treatment to determine the efficiency of the treatment. All the experiments were performed in triplicates.

Experimental procedure

Inoculation of POME with bacterial isolates

Using combination of bacterial strains The following bacterial combinations were investigated (in four laboratory batch-scale reactors inclusive of one control reactor) to determine which combination could produce the maximum organic load reduction. These include C_1 combination (consisting of *Bacillus cereus* 103 PB and *Bacillus subtilis* 106 PB) and C_2 combination (consisting of *Micrococcus luteus* 101 PB and *Stenotrophomonas maltophilia* 102 PB). Four percent inoculum containing 10^6 cells/mL with an optical density (OD) of 1.2 at 600 nm of each bacterial isolates was added into 250 mL of POME sample without addition of nutrients. They were incubated at 37 °C and at 150 rpm shaking speed. Bacterial cultures were incubated under aerobic conditions at 37 °C and agitated at 150 rev/min. Samples were then aseptically drawn from mixtures every 24 h for 5 days and analyzed for BOD, COD, TSS, oil and grease and pH from each combinations. Cell count was determined by serial dilution method of samples on nutrient agar plates and incubated at 37 °C for 24 h. Control reactor was not inoculated. Eight percent (8 %) of each inoculum containing 10^6 cells/mL with an optical density (OD) of 1.2 at 600 nm was used for pure/single bacterial strains. All the experiments were performed in triplicates. The efficiency for organic load reduction and the percent reduction was measured using the following equation (Piro et al. 2011). COD, BOD, TSS, oil and grease contents were analyzed with standard method (APHA 2005) and presented as percent reduction.

$$\text{Reduction}\,(\%) = 100 - \left[\frac{C_{\text{raw POME}} - C_{\text{f}}}{C_{\text{raw POME}}} \times 100\right]$$

where $C_{\text{raw POME}}$ is the concentration of COD, BOD_5, TSS and oil & grease of raw POME and C_f the concentration of these parameters after treatment. Each set of these experiments was done in triplicates.

Results and discussion

Palm oil mill effluent (POME) characteristics

Raw POME collected from the palm oil mill was thick, brownish in colour, colloidal suspension, dark, oily and viscous with an obnoxious odour. The raw sample consists of high COD concentration up to 75,900 mg/L, BOD 34,393 mg/L, TSS 14,467 mg/L, oil & grease 191 mg/L and pH 4.74 as depicted in Table 1. In comparison, other investigators have also reported values similar to the present study (Chin et al. 1996; Ahmad et al. 2003; Ahmad et al. 2005; Najafpour et al. 2006; Vijayaraghavan et al. 2007; Wanna and Pompan 2007; AbdulKarim et al. 2011; Lam and lee 2011).

Reduction efficiency (RE %) of COD using individual bacterial strains

Percent reduction efficiency of COD for all the isolates is presented in Fig. 1. Results revealed that reduction efficiency of COD for all the isolates is as follows: *Micrococcus luteus* 101 PB (67.19 %), *Stenotrophomonas maltophilia* 102 PB (61.92 %), *Bacillus cereus* 103 PB (78.60 %), *Bacillus subtilis* 106 PB (64.08 %) and control (15.63 %). The strain 103 PB with COD reduction rate of 78.60 % was the most effective followed by 101 PB

Table 1 Characteristics of raw palm oil mill effluent (POME) and discharge standard limits

Parameters	Raw POME (mg/L)	Discharge Effluent standard	
		Malaysia	Thailand
Chemical oxygen demand (COD)	75,900	–	<1,000
Biochemical oxygen demand (BOD)	34,393	100	<100
Total suspended solids (TSS)	14,467	400	<150
Volatile suspended solids (VSS)	13,033	–	–
Oil & grease (O&G)	191	50	<25
pH	4.74	5–9	5–9

All parameters are in mg/L except pH

Source: Malaysia: Pierzynski (2005), Thailand : Environmental Management Guideline (1997)

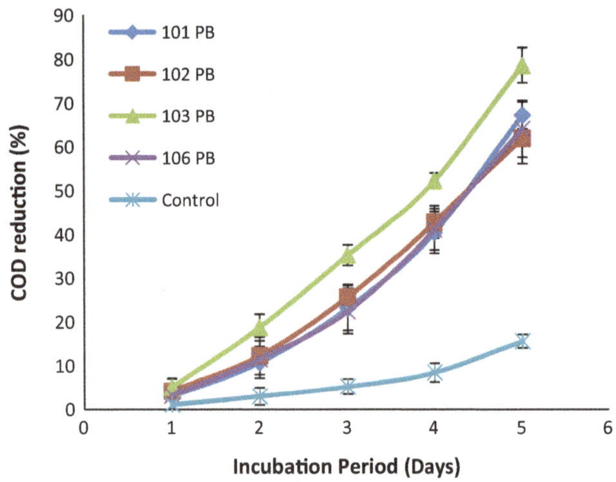

Fig. 1 Percentage reduction of COD in POME sample (pure culture)

(67.19 %), >106 PB (64.08 %) and >102 PB (61.92 %). Control reactor showed the lowest reduction efficiency compared to treatment reactors.

This suggests that our isolated strains from POME are effective for COD reduction. This study would help in understanding the role of bacteria in biological treatment of wastewaters such as those of oil processing from POME.

Comparisons of COD reduction efficiencies by bacteria isolated from POME revealed that percent reduction was similar with those results in previous publications (De Felice et al. 1997; Bhumibhamon et al. 2002; Oswal et al. 2002; Azbar and Yonar 2004; El-Bestawy et al. 2005; Takeno et al. 2005; Lan et al. 2009; AbdulKarim et al. 2011; Abass et al. 2012; Mohammed et al. 2014; Soleimaninanadegani and Manshad 2014) from oily industrial wastewaters. Thus, the findings of COD reduction appear useful for practical wastewater treatment as a compact treatment system for POME.

Reduction efficiency (RE %) of COD using combination of bacterial strains

Percent reduction efficiency of COD for bacteria combination is presented in Fig. 2. Results revealed that reduction efficiency of COD for the bacteria combination is as follows: C_1(*Bacillus cereus* 103 PB and *Bacillus subtilis* 106 PB) (90.64 %), >C_2 (*Micrococcus luteus* 101 PB and *Stenotrophomonas maltophilia* 102 PB)(71.84 %) and >control (14.35 %). The bacteria combination in reactor C_1 with COD reduction rate of 90.64 % was the most effective while reactor C_2 recorded COD reduction of 71.84 % Fig. 2. This study would help in understanding the role of mixed cultures in biological treatment of wastewaters such as those of oil processing like POME.

Comparison of the present study on bacteria combination with our pure individual bacterial culture in the previous study showed that bacteria combination recorded higher COD reduction as compare to single pure culture. This suggests that bacteria combination can be used for bioremediation of environment contaminated with polluted wastewaters. El-Masry et al. (2004) have reported bioremediation of oily polluted industrial wastewater with combination of bacterial isolates.

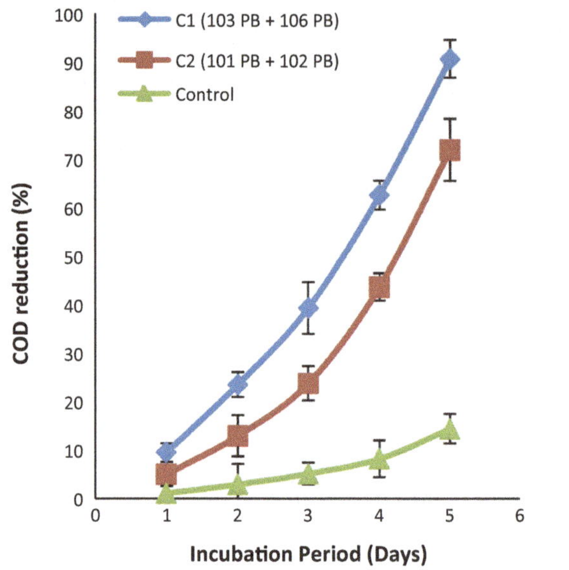

Fig. 2 Percentage reduction of COD in POME sample (mixed culture)

Results revealed that using bacteria as mixed cultures produce higher reduction of parameters than pure culture. Therefore, the mixed cultures of *Bacillus cereus* 103PB and *Bacillus subtilis* 106 PB can be considered as the best bacterial mixture that exhibited the best degradation compared to the other combination mixtures. This is in good agreement with previous workers who used mixed/consortium of organisms to achieve better results (Chigusa et al. 1996; Wakelin and Forster 1997; AbdulKarim et al. 2011). Oswal et al. (2002) reported that treatment of POME with *Yarrowia* and then with a consortium of mixed culture of bacteria developed from garden soil produced high degradative activity. The sequential treatments brought about a significant reduction in COD. El-Bestawy et al. (2005) also reported the combination of *Pseudomonas* sp. and *P. diminuta* as mixed cultures which produced the highest activity in reducing COD from contaminated industrial effluents. Other investigators have reported higher COD reduction with mixed cultures than single cultures of microorganisms. *Acinetobacter* sp. (KUL8), *Bacillus* sp. (KUL39), and *Pseudomonas* sp. (Bhumibhamon et al. 2002) and *Trichoderma harzianum* and *Penicillium* (AbdulKarim et al. 2011).

In consistence with the present study, biological treatment of oil-containing wastes significantly reduced organic load (COD) (EL-Gohary et al. 1987; Martine 1991;

Martirani et al. 1996; Odegaar et al. 1998; Raj and Murthy 1999; El-masry et al. 2004 and El-Bestawy et al. 2005). This biological treatment depends greatly on active microorganisms, which utilizes the organic substances present in the POME as nutrients and eventually degrades these organic matters into simple by-products such as methane, carbon dioxide, hydrogen sulphide and water (Jameel and Olanrewaju 2011; Jameel et al. 2011). Thus, the exploitation of these microorganisms isolated from POME for biodegradation and bioremediation purposes will offer a very efficient tool for purifying contaminated effluents water. Figures 3, 4, 5, 6 showed phylogenetic trees of the identified bacterial isolates (four bacterial isolates used for POME inoculation) from POME.

Reduction efficiency (RE %) of BOD using individual bacterial strains

Percent reduction efficiency of BOD for all the isolates is presented in Fig. 7. Results revealed that reduction efficiency of BOD for all the isolates are as follows: *Micrococcus luteus* 101 PB (55.21 %), *Stenotrophomonas maltophilia* 102 PB (64.73 %), *Bacillus cereus* 103 PB (90.98 %), *Bacillus subtilis* 106 PB (77.51 %) and control (28.95 %). The strain 103 PB with BOD removal rate of 90.98 % was the most effective followed by 106 PB (77.51 %), >102 PB (64.73 %) and >101 PB (55.21 %) compared to control reactor as illustrated in Fig. 7.

Results of BOD removal efficiencies by bacteria isolated from POME revealed that percent reduction was similar with those results in previous publications as reported by (Chin et al. 1996; El-Masry et al. 2004; El-Bestawy et al. 2005).

Reduction efficiency (RE %) of BOD using combination of bacterial strains

Percent reduction efficiency of BOD for bacteria combination was presented in Fig. 8. Results revealed that reduction efficiency of BOD for the bacteria combination is as follows: C_1 (*Bacillus cereus* 103 PB and *Bacillus subtilis* 106 PB) (93.11), >C_2 (*Micrococcus luteus* 101 PB and *Stenotrophomonas maltophilia* 102 PB) (67.33 %) and >control (20.81 %). The bacteria combination C_1 reactor that showed BOD removal rate of 93.11 % was the most effective while reactor C_2 recorded low BOD removal of 67.33 % (Fig. 8). The finding was in agreement with

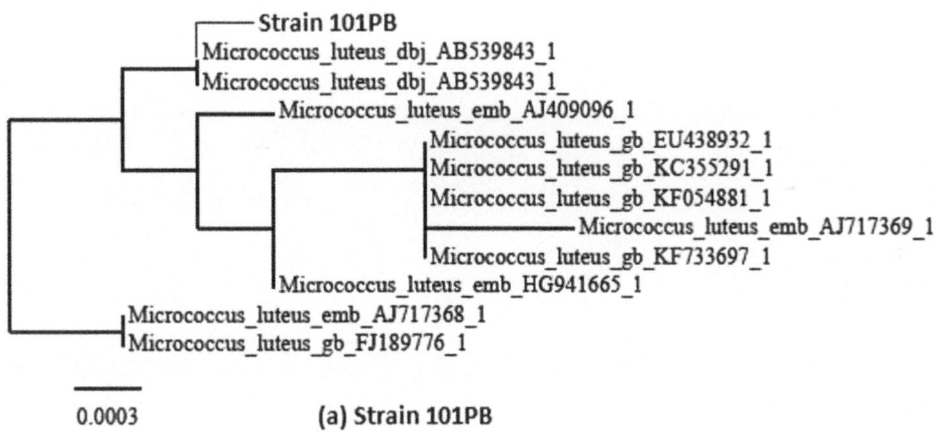

Fig. 3 Phylogenetic tree of *Micrococcus luteus* 101 PB based on 16S rRNA gene sequence comparisons

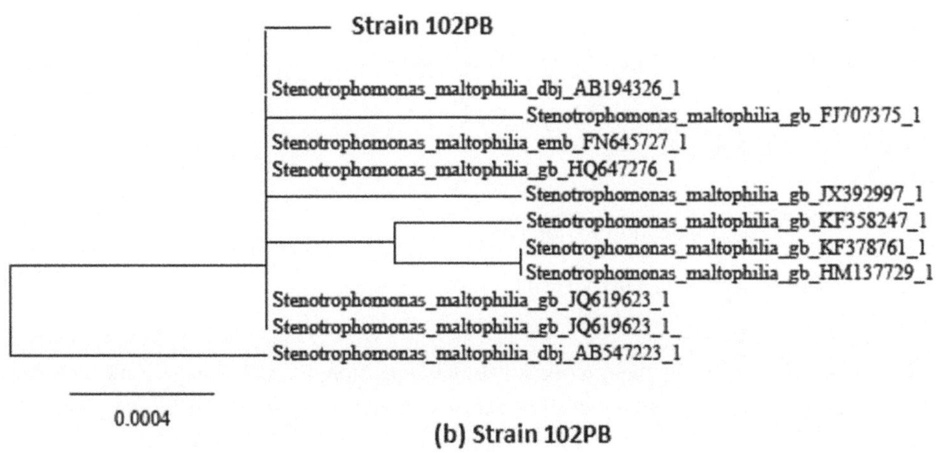

Fig. 4 Phylogenetic tree of *Stenotrophomonas maltophilia* 102 PB based on 16S rRNA gene sequence comparisons

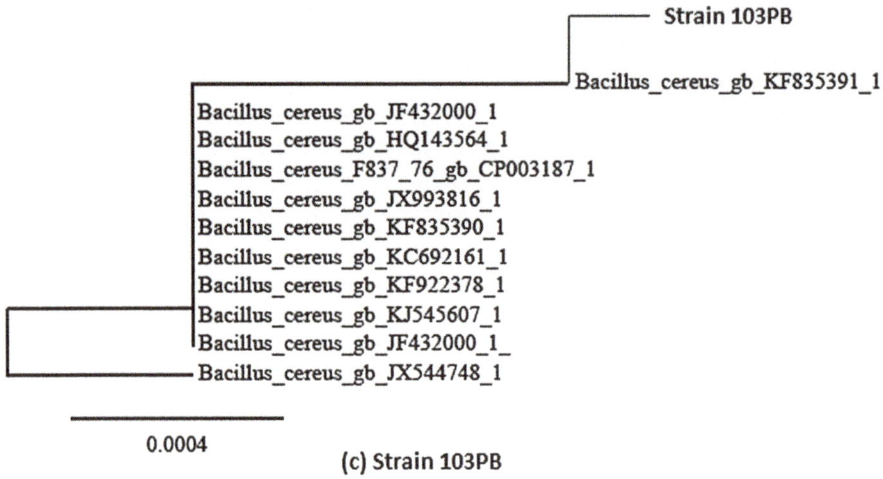

Fig. 5 Phylogenetic tree of *Bacillus cereus* 103 PB based on 16S rRNA gene sequence comparisons

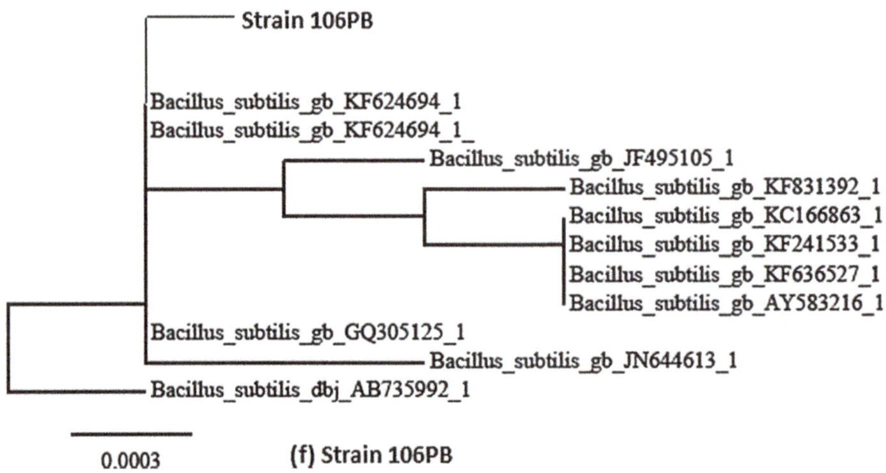

Fig. 6 Phylogenetic tree of *Bacillus subtilis* 106 PB based on 16S rRNA gene sequence comparisons

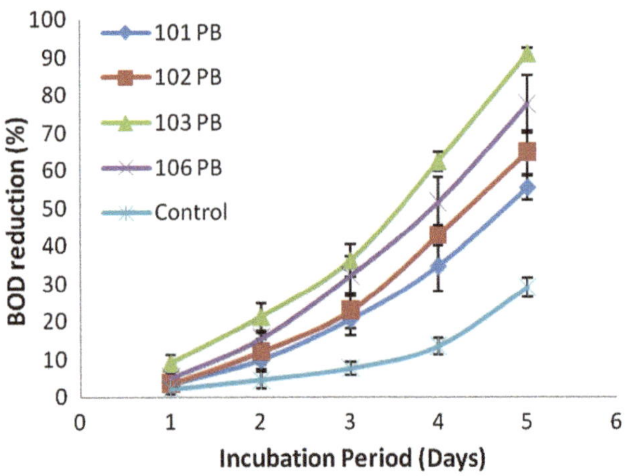

Fig. 7 Percentage reduction of BOD in POME sample (pure culture)

Qingwei et al. (1998) in their work for BOD reduction. Results of both studies revealed the effect of inoculation on organic matter biodegradation (Qingwei et al. 1998).

The synergistic effect of bacteria combination, for POME treatment brings about enhanced performance for effective biodegradation. However, synergism has been reported among mixed cultures of organisms which is attributed to the synergistic activities of the organisms Chigusa et al. (1996) and Benka-coker and Ekundayo (1997).

The composition of POME is mainly water, oil as well as cellulose wastes (Rashid et al. 2009; Ibrahim et al. 2012). This, therefore, necessitates the degradation of oil and cellulose in POME by lipase and cellulase producing bacterial. The strains (*Bacillus cereus* 103 PB and *Bacillus subtilis* 106 PB) possess the ability to produce extracellular lipase and cellulase which stimulates better waste treatment.

During degradation, oil and cellulose in the POME are broken down and mineralized (Singh et al. 2010). Generally, microbial oil degradation in POME is considered to occur as a result of hydrolysis of oil by secretion of lipase (oil degradation enzyme), which degrades the oil to organic acids and volatile fatty acids (VFAs) or reduces it to a low molecule via beta oxidation (fatty acid degradation pathway).

And finally, the oil is decomposed to CO_2 and H_2O (Koshimizu et al. 1997). Whereas cellulose degradation in POME is initiated primarily by microorganisms (fungi and bacteria) with the aid of extracellular enzymes. Cellulases are the enzymes that degraded cellulose. The enzymatic hydrolysis of cellulose requires the use of cellulase [1,4-(1,3:1,4)-β-D-glucan glucanohydrolase, EC 3.2.1.4], a multiple enzyme system consisting of endo-1,4,-β-D-glucanases [1,4-β-D-glucanases (CMCase, EC 3.2.1.4)] also known as endoglucanases, exo-1,4,-β-D-glucanases [1,4-β-D-glucan cellobiohydrolase, FPA, EC 3.2.1.91] also known as exoglucanases and β-glucosidase (cellobiase) (β-D-glucoside glucanohydrolase, EC 3.2.1.21); this act synergistically to convert cellulose into glucose (Lynd et al. 2002; Ray 2011). Endoglucanases hydrolyze the cellulose polymer internally at random in the amorphous site of cellulose polysaccharide chain, resulting in the formation of oligosaccharide of various lengths with a new chain ends. While exoglucanases act by removing cellobiose or glucose from both the reducing and non-reducing ends of the broken cellulose polymer, and on the other hand, β-glucosidase hydrolyses cellobiose and other cellodextrin into glucose (Lynd et al. 2002; Prado et al. 2011). Lynd et al. (2002) reported that cellulose hydrolytic enzymes are also active on hemicelluloses and hemicelluloses hydrolytic enzymes are commonly produced by cellulolytic microorganisms.

Degradation was more pronounced with mixed cultures of bacteria than with single cultures. The same findings

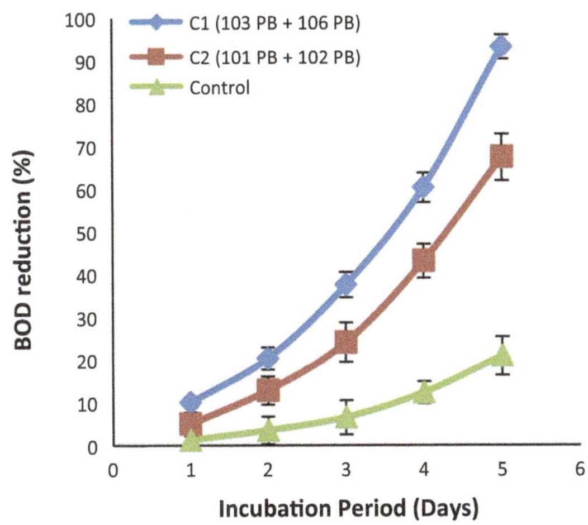

Fig. 8 Percentage reduction of BOD in POME sample (mixed culture)

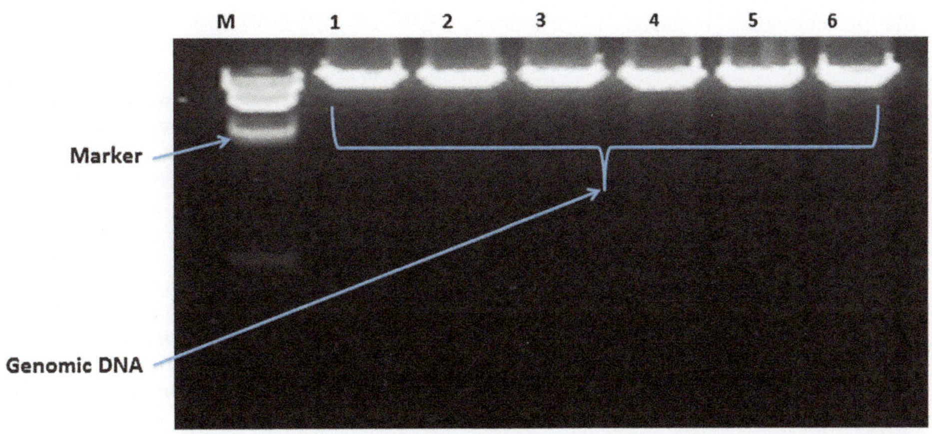

Plate 1 Gel Picture of genomic DNA: *Lane* 1: 101 PB; 2: 102 PB; 3: 103 PB; 4: 104 PB; 5: 105 PB; 6: 106 PB; M: Lambda/HindIII marker

Plate 2 Gel Picture of purified PCR product: *Lane* 1: 101 PB; 2: 102 PB; 3: 103 PB; 4: 104 PB; 5:105 PB; 6: 106 PB; M: 1 kb marker (Fermentas)

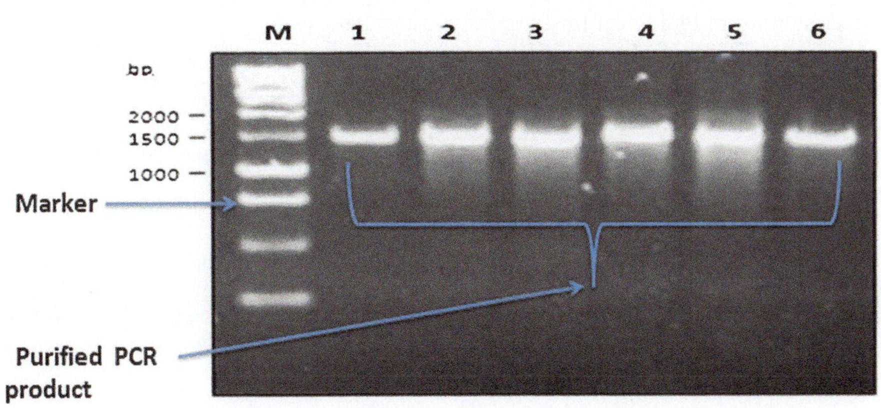

were confirmed by previous studies with mixed culture of organisms which revealed high degradative activity than with single cultures of microorganisms (Chigusa et al. 1996; Benka-coker and Ekundayo 1997 and Wakelin and Forster 1997).

In a similar study, characterization of the wastewater from food-processing plants indicated that the waste was highly contaminated with organic compounds as indicated by the high values of BOD_5 (EL-Gohary et al. 1987). However, the biological treatment of oil-containing wastes significantly removed the organic load (BOD_5) in the polluted wastewater as shown also by other authors (Martine 1991; Martirani et al. 1996; Raj and Murthy 1999). In the present study, using mixed culture was more effective in reducing BOD_5, which was previously indicated by Odegaar et al. (1998).

The indigenous microbial isolates from POME were observed to possess metabolic potential to degrade organic components whereas the use of mixed cultures resulted in more extensive degradation of COD and BOD than pure single cultures via one-step process under aerobic conditions.

The aerobic condition requires less hassle compared to anaerobic conditions in terms of microbial strains' sustainability (Fiestas 1984; Martine 1991). Degradation of organic load in the present study required no primary treatment (Valenzuela 1986; Tsonis 1993) and it is a one-step process. The present treatment proposal showed more advantages since no additional physical or chemical treatment was required and low energy was used. El-Bestawy et al.(2005) have reported that treatment of contaminated industrial effluents by mixed cultures of bacteria required no additional physical or chemical treatment.

Conclusion

It has been shown that the bacterial isolates from POME possess the ability to remove organic load (COD and BOD). This study would help in understanding the role of mixed culture in biological treatment of wastewaters such as those of oil processing like POME. However, a major finding of this study was that the performance of the mixed cultures could, therefore, be attributed to synergistic activities of the organisms. The mixed culture C_1 (*Bacillus cereus* 103 PB and *Bacillus subtilis* 106 PB) is the most effective bacteria and the best candidate to use in biological treatment technology of POME having the highest COD and BOD reduction rate thereby rendering the wastewater safe for effective reuse in the oil palm mills and eventual safe discharge and ultimately rendering the effluents more eco-friendly.

Authors' contributions JDB, JL & NI: conceived and planned the experiment. JDB: carried out the experiment and drafted the manuscript. JL and NI: formulated the objectives, analyzed the data, proofread and edited the manuscript, provided guidance and improved the quality of the final manuscript. All authors read and approved the final manuscript. (Jeremiah David Bala (JDB), Japareng Lalung (JL), Norli Ismail (NI).

Acknowledgments The authors would like to thank Universiti Sains Malaysia for the financial support under Research University (RUI) Grant, No.: 1001/PTEKIND/814147.

Conflict of interests The authors declared that they have no conflict of interest.

References

Abass AO, Jameel TA, Muyibi AS, Abdul Karim IM, Alam MdZ (2012) Investigation of the viability of selected microorganisms on the biodegradation of palm oil mill effluents (POME). Int J Chem Environ Eng 3(3):182–186

AbdulKarim MI, Daud NA, Alam MDZ (2011) Treatment of palm oil mill effluent using microorganisms. In: Alam MDZ, Jameel AT, Amid A (eds) Current research and development in biotechnology engineering at International Islamic University Malaysia (IIUM), 3rd edn. IIUM Press, Kuala Lumpur, pp 269–275 ISBN 9789674181444

Agamuthu P, Tan EL, Shaifal AA (1986) Effect of aeration and soil inoculum on the composition of palm oil effluent (POME). Agric Wastes 15:121–132

Ahmad AL, Ismail S, Bhatia S (2003) Water Recycling from palm oil mill effluent (POME) using membrane technology. Desalination 157:87–95. doi:10.1016/S0011-9164(03)00387-4

Ahmad AL, Sumathi S, Hameed BH (2005) Residual oil and suspended solid removal using natural adsorbents chitosan, bentonite and activated carbon: A comparative study. Chem Eng J 108(1–2):179–185. doi:10.1016/j.cej.2005.01.016

APHA (2005) Standard methods for the examination of water and wastewater in 21st edn. American Public Health Association (APHA), Washington

Azbar N, Yonar T (2004) Comparative evaluation of a laboratory and full-scale treatment alternatives for the vegetable oil refining industry wastewater. Process Biochem 39(7):869–875. doi:10.1016/S0032-9592(03)00193-6

Bala JD, Lalung J, Ismail N (2014) Biodegradation of palm oil mill effluent (POME) by bacterial. Int J Sci Res Publ 4(3):502–511

Benka-coker MO, Ekundayo JA (1997) Applicability of evaluating the ability of microbes isolated from an oil spill site to degrade oil. Environ Monit Assess 45:259–272. doi:10.1023/A:1005774602112

Bhumibhamon O, Koprasertsak A, Funthong S (2002) Biotreatment of high fat and oil wastewater by lipase producing microorganisms. Kasetsart J Nat Sci 36:261–267

Chigusa S, Hasegawa T, Yamamoto N, Watanabe Y (1996) Treatment of waste-Chappe water from oil manufacturing plant by yeasts. Wat Sci Tech 34(11):51–58

Chin KK, Lee SW, Mohammad HH (1996) A study of palm oil mill effluent treatment using a pond system. Water Sci Technol 34(11):119–123. doi:10.1016/S0273-1223(96)00828-1

De Felice B, Pontecorvo G, Carfagna M (1997) Degradation of waste waters from olive oil mills by *Yarrowia lipolytica* ATCC 20255 and *Pseudomonas putida*. Acta Biotechnol 17(3):231–239. doi:10.1002/abio.370170306

El-Bestawy E, El-Masry MH, El-Adl NE (2005) The potentiality of free Gram-negative bacteria for removing oil and grease from contaminated industrial effluents. World J Microbiol Biotechnol 21(6–7):815–822. doi:10.1007/s11274-004-2239-8

EL-Gohary FA, Aboelella SI, Ali HI (1987) Management of wastewater from soap and food industries: a case study. Sci Tot Environ 66:203–212. doi:10.1016/0048-9697(87)90088-X

El-Masry MH, El-Bestawy E, El-Adl NI (2004) Bioremediation of vegetable oil and grease from polluted wastewater using a sand biofilm system. World J Microbiol Biotechnol 20(6):551–557. doi:10.1023/B:WIBI.0000043162.17813.17

Environmental Management Guideline (1997) Palm oil industry. German Technical Cooperation, Thailand, p 5

Fiestas JA (1984) Directrices acutuales en la depuracion de aguas residuales de cartacter organico. Quim Ind 30:431–438

Ibrahim AH, Dahlan I, Adlan MN, Dasti AF (2012) Comparative study on characterization of Malaysian palm oil mill effluent. Res J Chem Sci 2(12):1–5

Jameel AT, Olanrewaju AA (2011) Aerobic biodegradation of oil and grease in palm oil mill effluent using consortium of microorganisms. In: Alam MDZ, Jameel AT, Amid A (eds) Current research and development in biotechnology engineering at International Islamic University Malaysia (IIUM), 3rd edn. IIUM Press, Kuala Lumpur, pp 43–51 ISBN 9789674181444

Jameel AT, Muyibi SA, Olanrewaju AA (2011) Comparative study of bioreactors used for palm oil mill effluent treatment based on chemical oxygen removal efficiencies. In: Alam MDZ, Jameel AT, Amid A (eds) Current research and development in biotechnology engineering at International Islamic University Malaysia (IIUM), 3rd edn. IIUM Press, Kuala Lumpur, pp 277–284 ISBN 9789674181444

Koshimizu S, Ohtake I, Yoshioka H, Saito Y, Taki H (1997) Development of kitchen wastewater treatment system using fats and oils degrading microorganisms. Kuuki-chouwa Eiseikougaku 71:999–1009 (in Japanese)

Lam MK, Lee KT (2011) Renewable and sustainable bioenergies production from palm oil mill effluent (POME): win–win strategies toward better environmental protection. Biotechnol Adv 29(1):124–141. doi:10.1016/j.biotechadv.2010.10.001

Lan WU, Gang GE, Jinbao WAN (2009) Biodegradation of oil wastewater by free and immobilized *Yarrowia lipolytica* W29. J Environ Sci 21(2):237–242. doi:10.1016/S1001-0742(08)62257-3

Lynd LR, Weimer PJ, Van Zyl WH, Pretorius IS (2002) Microbial cellulose utilization: fundamentals and biotechnology. Microbiol Mol Biol Rev 66(3):506–577. doi:10.1128/mmbr.66.3.506-577.2002

Martine AM (1991) Bioconversion of waste material to industrial products, 2nd edn. Kluwer Academic Publishers, Netherlands, p 576 ISBN 0-75140-423-3

Martirani L, Giardina P, Marzullo L, Sannia G (1996) Reduction of phenol content and toxicity in olive oil mill wastewater with the linolytic fungus *Pleurotus ostreatus*. Water Res 30(8):1914–1918. doi:10.1016/0043-1354(95)00330-4

Maygaonkar PA, Wagh PM, Permeswaran U (2012) Biodegradation of distillery effluent by fungi. Biosci Discov 3(2):251–258

McHugh S, O'Reilly C, Mahony T, Emer Colleran E, O'Flaherty V (2003) Anaerobic granular sludge bioreactor technology. Rev Environ Sci Biotechnol 2(2–4):225–245. doi:10.1023/B:RESB.0000040465.45300.97

Mohammed RR, Ketabachi MR, McKay G (2014) Combined magnetic field and adsorption process for treatment of biologically treated palm oil mill effluent (POME). Chem Eng J 243:31–42

Najafpour GD, Zinatizadeh A, Mohamed AR, Hasnain Isa M, Nasrollahzadeh H (2006) High-rate anaerobic digestion of palm oil mill effluent in an upflow anaerobic sludge-fixed film bioreactor. Process Biochem 41(2):370–379. doi:10.1016/j.procbio.2005.06.031

Odegaar H, Ruster B, Westrum T (1998) A new moving bed biofilm reactor—application and results. Water Sci Technol 19:157–165

Oswal N, Sarma PM, Zinjarde SS, Pant A (2002) Palm oil mill effluent treatment by a tropical marine yeast. Biores Technol 85(1):35–37. doi:10.1016/S0960-8524(02)00063-9

Pierzynski GM, Sims JT, Vance GF (2005) Soils and Environmental Quality, CRC Press

Piro P, Carbone M, Tomei G (2011) Assessing settleability of dry and wet weather flows in an urban area serviced by combined sewer. Water Air Soil Pollut 214(1–4):107–117. doi:10.1007/s11270-010-0408-y

Prado HFD, Leite RSR, Martins DB, Gomes E, Silva RD (2011) Cellulolytic enzymes isolated from Brazilian areas: production, characterization and applications. In: Golan AE (ed) Cellulase: types and action, mechanism, and uses. Nova Science Publishers Inc, New York

Qingwei L, Mancl KM, Tuovinen OH (1998) Effect of inoculation on the biodegradation of butterfat-detergent mixtures in fixedfilm sand columns. Biores Technol 64(1):27–32. doi:10.1016/S0960-8524(97)00155-7

Raj SA, Murthy DVS (1999) Synthetic dairy wastewater treatment using cross flow medium trickling filter. J Environ Sci Health, Part A 34(2):357–369. doi:10.1080/10934529909376841

Rashid SS, Alam MZ, Karim MIA, Salleh MH (2009) Management of palm oil mill effluent through production of cellulases by filamentous fungi. World J Microbiol Biotechnol 25(12):2219–2226. doi:10.1007/s11274-009-0129-9

Ray RC (2011) Solid-state fermentation for production of microbial cellulase: an overview. In: Golan AE (ed) Cellulase: types and action, mechanism, and uses. Nova Science Publishers Inc, New York

Singh R, Ibrahim MH, Esa N, Iliyana M (2010) Composting of waste from palm oil mill: a sustainable waste management practice. Rev Environ Sci Biotechnol 9(4):331–344. doi:10.1007/s11157-010-9199-2

Sinnappa S (1978) Studies of palm oil mill waste effluent. Malaysian Agricultural Journal 51:261–272

Soleimaninanadegani M, Manshad S (2014) Enhancement of Biodegradation of Palm Oil Mill Effluents by Local Isolated Microorganisms International Scholarly Research Notices 2014:1–8

Takeno K, Yamaoka Y, Sasaki K (2005) Treatment of oil-containing sewage wastewater using immobilized photosynthetic bacteria. World J Microbiol Biotechnol 21(8–9):1385–1391. doi:10.1007/s11274-005-5739-2

Tsonis SP (1993) Olive oil mill wastewater abatement by anaerobic digestion followed by total evaporation. Proceeding of the International Conference on Environmental Pollution, Sitges (Barcelona), vol 2. European Center for Pollution Research, UK

Valenzuela G (1986) Thermal concentration of vegetation water. In: Proceeding of the International Symposium on Olive By-products Polarization (FAO), Sevilla, Spain

Vijayaraghavan K, Ahmad D, Abdul Aziz ME (2007) Aerobic treatment of palm oil mill effluent. J Environ Manag 82(1):24–31. doi:10.1016/j.jenvman.2005.11.016

Wakelin NG, Forster CF (1997) An investigation into microbial removal of fats, oils and greases. Biores Technol 59(1):37–43. doi:10.1016/S0960-8524(96)00134-4

Wanna C, Pompan W (2007) Effect of temperature on the anaerobic digestion of palm oil mill effluent. Electron J Biotechnol 10(3):376–385

Wu TY, Mohammad AW, Jahim JM, Anuar N (2007) Palm oil mill effluent (POME) treatment and bioresources recovery using ultrafiltration membrane: effect of pressure on membrane fouling. Biochem Eng J 35(3):309–317. doi:10.1016/j.bej.2007.01.029

Wu TY, Mohammad AW, Jahim JM, Anuar N (2010) Pollution control technologies for the treatment of palm oil mill effluent (POME) through end-of-pipe processes. J Environ Manage 91:1467–1490

2

Hydrogen and syngas production from gasification of lignocellulosic biomass in supercritical water media

Farid Safari[1] · Ahmad Tavasoli[1] · Abtin Ataei[1] · Jun-Ki Choi[2]

Abstract

Purpose Novel biomass-processing technologies have been recently used for conversion of organic wastes into valuable biofuels like bio-hydrogen. Agricultural wastes are available and renewable energy resources to supply energy demand of the future. The purpose of this study is to investigate the production of hydrogen-rich syngas from wheat straw, walnut shell, and almond shell.

Methods Supercritical water gasification is a promising technology to convert biomass into useful fuels. Non-catalytic conversion of wheat straw, walnut shell, and almond shell into the hydrogen-rich gas in supercritical water media was performed using homemade batch microreactor system.

Results Hydrogen gas yields of 6.52, 4.26 and 4.1 mmol per 1 gram of wheat straw, walnut shell, and almond shell were observed, respectively. In addition, hydrogen and carbon gasification efficiencies equal to 42.6 and 46.9 % were calculated from gaseous products and elemental analysis of wheat straw, which were higher than other feedstocks' gasification efficiencies.

Conclusion Wheat straw had the highest and walnut shell had the lowest total gas and hydrogen gas yields. Taking into account the structural analysis, it was recognized that feedstocks with higher cellulose and hemicellulose and lower lignin contents were better gasified due to their easier hydrolysis and higher solubility in water.

✉ Abtin Ataei
a.ataei@srbiau.ac.ir

1 Department of Energy Engineering, Science and Research Branch, Islamic Azad University, Tehran, Iran

2 Department of Mechanical and Aerospace Engineering, Renewable and Clean Energy, University of Dayton, Dayton, OH, USA

Keywords Biomass · Hydrogen production · Gasification · Supercritical water media

Introduction

Sustainable development in the context of energy is one of the most important road maps for scientists and engineers in recent years. Using renewable energy resources, maximizing energy efficiency and improving energy conservation with novel methods are the strategies for achieving this aim. Research and development in methods of obtaining power from energy resources with less environmental effects and higher efficiencies would be a great leap forward in this way. Biomass is an available and renewable resource that easily can be obtained from nature. Wasting this resource results in harmful environmental effects and causes many problems, while it can be converted into useful products such as hydrogen by different conversional technologies. Conversion of biomass into these products is helping us to supply the energy demand of the future in a sustainable way. Many scientists regard hydrogen as a key energy carrier of the future. The amount of energy released during hydrogen combustion is higher than any other fossil fuels on a mass basis, with a lower heating value (LHV) 2.4, 2.8 and 4 times higher than that of methane, gasoline and coal, respectively. Generally, we have two main methods for hydrogen production from biomass. Thermochemical method, which is divided into three types of processes (pyrolysis, gasification and SCWG) and biological methods (Hepbasli et al. 2009).

In this study, we perform supercritical water gasification that is one of the most promising methods for thermochemical conversion of biomass into hydrogen-rich gas. This technology has many advantages compared with other thermochemical conversions. In this method, wet biomass can be converted into the hydrogen-rich syngas at lower temperatures with higher efficiencies due to the high dispersion and effective heat transfer of water in its supercritical condition (374 °C, 22.1 MPa) (Saxena 2008). Water in SCWG has a twofold role as reaction medium and reactant. The unique solvent properties allow supercritical water to solve the major part of biomass and reform it through a homogeneous mixture. Water–gas shift (producing H_2 and CO_2 from CO and H_2O) and methanation (producing CH_4 and H_2O from CO and H_2) reactions occur with biomass reforming in supercritical water, simultaneously (Calzavara et al. 2005).

Modell et al. (1978) reported the effect of temperature and concentration on the gasification of glucose and maple sawdust in water in the vicinity of its critical state. No solid residue or char was produced. Hydrogen gaseous yield up to 18 % was observed. After that, research and development on this method increased significantly. Guo et al. (2007), concluded that carbohydrates (i.e. cellulose and hemicelluloses) gasified much easier than xylen and lignin, and lignin was the most difficult to gasify. In addition, it was realized that lignin can interact with cellulose and change both the gas yield and product gas composition. Other researchers expressed that, in SCW gasification of real biomass feedstock, char/coke may originate from non-decomposed biomass by a solid–solid conversion; and coke formation in biomass gasification process due to the presence of lignin, not only depends on the lignin amount but also strongly depends on the structure of lignin and interactions between other components in the real biomass (Yanik et al. 2007). In the recent years, Tavasoli et al. (Barati et al. 2014) have studied the gasification of sugarcane bagasse in supercritical water condition. Hydrogen yield up to 37 % was observed in tubular batch microreactor, which was for the gasification in the temperature of 400 °C, with biomass loading of 0.05 g and water loading of 4 g in reaction time of 15 min. It was also shown that hydrogen yield was increased with the increment of reactor's temperature while it was decreased with the increment of feed concentration.

In this article, we investigate the gasification of wheat straw, walnut shell and almond shell in supercritical water media with tubular batch rector, which is designed for the efficient reacting condition, considering the latest experiments and research literature.

Materials and methods

Materials

The biomass particles used for the experiments were obtained from gardens and agriculture farms around Sanandaj, located in Kurdistan province, Iran. They were dried under atmospheric conditions for 48 h, and ground to particle sizes <150 micrometer in diameter. The elemental compositions of the whole biomass samples were analyzed in a CHNS analyzer.

Reaction setup and experimental outline

A batch microreactor made of 316 stainless steel tube with total volume of 23 mL has been used in this study (Fig. 1a). Variation of the reactor's pressure with time is shown in Fig. 1b. Feedstocks were mixed with a certain amount of deionized water and injected into the reactor using a syringe. The reactor was plunged in a molten salt bath that contained a mixture of potassium nitrate, sodium nitrate, and sodium nitrite. The molten salt bath temperature was controlled using an electrical heater and a PID temperature controller. Temperature and pressure were measured using a K-type thermocouple and a pressure gauge. Figure 1b, shows the typical changes in reactor pressure. After a given reaction

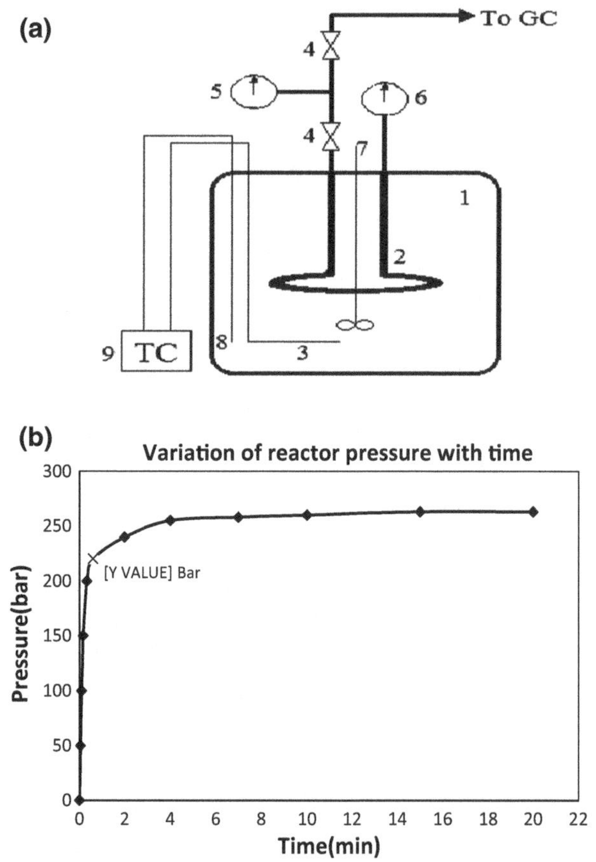

Fig. 1 **a** Scheme of the homemade tube batch microreactor: *1* molten salt bath *2* batch tube reactor, *3* electrical heater, *4* high-pressure valve, *5* low-pressure gauge *6* high-pressure gauge, *7* mixer *8* k-type thermocouple, *9* temperature controller. **b** A typical variation of reactor's pressure with time (*T* 440 °C, wheat L 0.05 g, water L 5 g)

time, the reactor was taken out of the molten salt bath and plunged in a water bath for rapid cooling to room temperature. All experiments were performed three times and the data reported here are the averages of repetitive runs.

At the end of each experiment, reactor free volume, final pressure and temperature were used to calculate the gas yield. The amounts of each product's gases were measured and analyzed using gas chromatographs (Varian 3400 and Teyfgostar-Compact). The carbon gasification efficiency (CGE), which is the ratio of the amount of carbon in the gaseous products to the amount of carbon in the biomass; and hydrogen gasification efficiency (HGE), which is the ratio of the amount of hydrogen in the gas phase to the amount of hydrogen in the biomass were measured after each experiment. Mathematically, CGE and HGE are defined as (Barati et al. 2014):

$$\mathrm{CGE} = \{Y_{CO} + Y_{CH_4} + Y_{CO_2} + 2Y_{C_2H_4} + 2Y_{C_2H_6}\} / \{\text{mmol Carbon/g biomass}\} \quad (1)$$

$$\mathrm{HGE} = \{Y_{H_2} + 2Y_{CH_4} + 2Y_{C_2H_4} + 3Y_{C_2H_6}\} / \{\text{mmol } H_2\text{/g biomass}\} \quad (2)$$

Results and discussion

Feedstock analysis

The feedstocks used in this study are lingo-cellulosic biomass. A lingo-cellulosic biomass typically has 30–60 % cellulose, 20–40 % hemicellulose and 15–25 % lignin (Nanda et al. 2013). The CHNS and structural analyses of the biomasses are given in Table 1. As indicated, the C and H content in the samples are 48.25 and 6.08 % for wheat straw, 50.2 and 5.67 % for walnut shell, and 49.70 and 5.43 % for almond shell, respectively. The balances are mostly oxygen. In addition, structural analysis shows that the lignin contents in their structure are 19.3, 35 and 38 %, respectively, for wheat straw, walnut shell and almond shell.

Reactions

Because of the lingo-cellulosic structure of biomasses used in this study, it is necessary to study the behavior of the intermediates and their degradation or reformation routes to gases. Prior to understanding the mechanisms, behavior and the breakdown of the complex lingo-cellulosic biomass in supercritical conditions, it is essential to investigate the SCWG of their model compounds such as cellulose, glucose, glycerol, lignin and phenolics. Some of the vital reactions that occur during the gasification of biomass in SCW are as given below (Reddy et al. 2014).

$$\text{Cellulose hydrolysis}: \; C_6H_{10}O_5 + nH_20 \rightarrow nC_6H_{12}O_6 \quad (3)$$

$$\text{Glucose reforming reaction}: \; C_6H_{12}O_6 \rightarrow 6CO + 6H_2 \quad (4)$$

$$\text{Lignin hydrolysis}: \; (C_{10}H_{10}O_3)_n + nH2O \rightarrow nC_{10}H_{12}O_4 + \text{Phenolics} \quad (5)$$

$$\text{Lignin steam reforming reaction}: \; \text{Phenolics} + H_2O \rightarrow CO + CO_2 + H_2 \quad (6)$$

Supercritical water gasification of biomass is a complex process but the overall conversion can be simplified as:

$$CH_xO_y + (2-y)\,H_2O \rightarrow CO_2 + (2 - y + x/2)\,H_2 \quad (7)$$

Thus x and y are the elemental molar ratios of H/C and O/C in biomass composition. This is an endothermic reaction and shows the reactance effect of water in addition to its solubility, and hydrogen in the water is released by the gasification reaction. In addition to reaction (7) some intermediate reaction can occur in order to complete the gasification process that is mentioned as follows (Rashidi and Tavasoli 2014; Ataei and Azimi 2012).

$$\text{Steam reforming}: \; CH_xO_y + (1-y)\,H_2O \rightarrow CO + (1 - y + x/2)\,H_2 \quad (8)$$

$$\text{Water-gas shift}: \; CO + H_2O \rightarrow CO_2 + H_2 \quad (9)$$

$$\text{Methanation}: \; CO + 3H_2 \rightarrow CH_4 + H_2O \quad (10)$$

The reforming reactions produce CO, CO_2 and H_2, whereas CH_4 is produced by methanation and hydrogenation reactions. Water–gas shift reaction (WGS) is another significant reaction of SCWG of biomass. It is obvious that water–gas shift should be accelerated and methanation should be avoided when hydrogen-rich gas is required (Reddy et al. 2014).

Table 1 Elemental and structural analysis of different feedstocks

Feedstock	Element analysis (CHNS) (%wt)					Structural analysis (%wt)		
	S	N	H	C	O	Lignin	Cellulose	Hemicellulose
Wheat straw	0.22	2.11	6.08	48.25	42.80	19.30	39.80	27.30
Walnut shell	0	0.84	5.67	50.2	42.64	38	36	25.43
Almond shell	0.06	1.10	5.43	49.70	41.60	35	30.70	32.60

Product analysis

Total gas yield and its composition is presented in Fig. 2, for the determined condition. The carbon and hydrogen gasification efficiency was also calculated and presented in Fig. 3. As shown, wheat straw has the highest hydrogen yield with 6.52 mol/g. Moreover, wheat straw has the highest hydrogen and carbon gasification efficiencies equal to 42.6 and 46.9 %, respectively. The total gas yields can

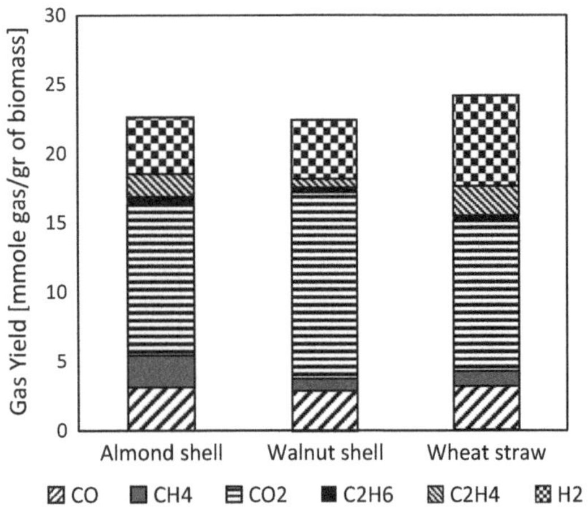

Fig. 2 Gasification yields for different feedstocks (*T* 440 °C, biomass loading: 0.05 g, water loading: 5 g)

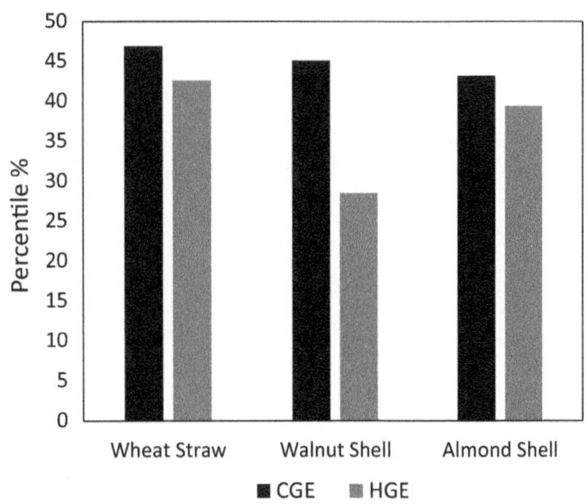

Fig. 3 Hydrogen and carbon gasification efficiencies of feedstocks (*T* 440 °C, biomass loading: 0.05 g, water loading: 5 g)

show that the structure of walnut shell with more lignin has the lowest total gas yield equal to 22.4 mol/g since the wheat straw has the highest yield with 24.2 mol gas per gram of it. This is because of the higher resistance of lignin during hydrolysis that can postpone the complete gasification. Lignin in the biomass acts as a binding agent that holds cellulose and hemicellulose together, providing firmness to the lingo-cellulosic network. This makes the biomass resistant to various chemicals and enzymes (Kumar et al. 2009). The presence of lignin in biomass makes it difficult to obtain cellulose and hemicellulose to produce fermentable sugars (Reddy et al. 2014). The results in Table 2 show the final composition of the syngas for different feedstocks. SCW gasification process is the result of combination of the series of complex and competing reactions mentioned earlier in this article.

In this study, in addition to comparison between different feedstocks in terms of total gas yield and hydrogen yield per mass unit, the relationship between biomass structure and gas yields was also investigated. There is no certain proved mechanism for supercritical water gasification. However, results indicate that the total gasification yields are approximately related to the total amount of cellulose and hemicellulose in the structure of biomass. This is because of higher solubility of cellulose and hemicellulose compared with lignin, but lignin structure can be dissolved in water in higher temperatures and resident times and specific conditions. Here, wheat straw with higher hydrogen content in its elemental analysis showed the highest hydrogen yield. Therefore, it can be said that the hydrogen content in CHNS analysis has an almost direct relationship with hydrogen gas yield.

Conclusion

The gasification of wheat straw, walnut shell and almond shell in supercritical water media was studied using a batch microreactor with volume 23 mL. These experiments were performed at the temperature 440 °C, 0.05 g biomass loading, and 0.05 g water loading with resident time of 15 min. The hydrogen gas yield of 6.52 mol per gram and the ratio of 27 % in total gas were observed for wheat straw as the highest hydrogen yield among experimented

Table 2 Gas composition and gasification efficiencies for supercritical water gasification of different feedstocks. (T: 440 °C, biomass loading: 0.05 g, water loading: 5 g)

Biomass	Gasification efficiency (%)		Gas composition (mmol gas/g of biomass)					
	HGE	CGE	CO	CH_4	CO_2	C_2H_6	C_2H_4	H_2
Wheat straw	42.6	46.9	3.1933	1.078	10.82	0.41	2.15	6.52
Walnut shell	28.5	45.1	2.86	0.883	13.47	0.362	0.607	4.26
Almond shell	39.4	43.2	3.12	2.31	10.80	0.647	1.65	4.1

feedstocks. Results indicated that wheat straw, walnut shell, and almond shell had the highest hydrogen yield, respectively, and wheat straw, almond shell, and walnut shell showed the highest hydrogen gasification efficiency, respectively. It was also recognized that cellulose and hemicellulose were better gasified than lignin because of their easier hydrolysis and higher solubility in water.

References

Ataei A, Azimi A, Behboodi Kalhori S, Foroughi Abari S, Radnezhad H (2012) Performance analysis of a co-gasifier for organic waste in agriculture. Int J Recycl Org Waste Agric 1:6

Barati M, Babatabar M, Tavasoli A, Dalai AK, Das U (2014) Hydrogen production via supercritical water gasification of bagasse using unpromoted and zinc promoted Ru/γ-Al2O3 nanocatalysts. Fuel Proc Tech 123:140–148

Calzavara Y, Dubien CJ, Boissonnet G, Sarrade S (2005) Evaluation of biomass gasification in supercritical water process for hydrogen production. Energy Conv Manage 46:615–631

Guo LJ, Lu YJ, Zhang XM, Ji CM, Guan Y, Pei AX (2007) Hydrogen production by biomass gasification in supercritical water: a systematic experimental and analytical study. Catal Today 129:275–286

Hepbasli A, Kalinci Y, Dincer I (2009) Biomass-based hydrogen production: a review and analysis. Int J Hydrog Energy 34:8799–8817

Kumar P, Barrett DM, Delwiche MJ, Stroeve P (2009) Methods for pretreatment of lignocellulosic biomass for efficient hydrolysis and biofuel production. Ind Eng Chem Res 48:3713–3729

Modell M, Reid RC, Amin S (1978) Gasification Process, US Patent 4,113,446, September 1978

Nanda S, Mohanty P, Pant KK, Naik S, Kozinski JA, Dalai AK (2013) Characterization of North American lignocellulosic biomass and biochars in terms of their candidacy for alternate renewable fuels. Bio Energ Res 6:663–677

Rashidi M, Tavasoli A (2014) Hydrogen Rich gas production via noncatalytic gasification of sugarcane bagasse in supercritical water media. Pet Coal 56(3):324–331

Reddy Sivamohan N, Nanda Sonil, Dalai Ajay K, Kozinski Janusz A (2014) Supercritical water gasification of biomass for hydrogen production. Int J Hydrogen Energy 39:6912–6926

Saxena RC, Seal D, Kumar S, Goyal HB (2008) Thermo-chemical routes for hydrogen rich gas from biomass: a review. Renew Sustain Energy Rev 12:1909–1927

Yanik J, Ebale S, Kruse A, Saglam M, Yüksel M (2007) Biomass gasification in supercritical water: part 1. Effect of the nature of biomass. Fuel 86:2410–2415

3

The effects of irrigation methods on some of soil and plant microbial indices using treated municipal wastewater

Payam Najafi · Jahangir Shams · Azita Shams

Abstract Shortage of irrigation water is a crucial problem in arid and semi-arid regions. Therefore, application of wastewater in such area seems to be an indispensable solution. The health aspect of the application should be considered by microbial evaluation of soil and plant irrigated with treated wastewater. In this study several irrigation methods were used. Results showed that the filtration reduced the microbial pollution of treated municipal wastewater. The maximum soil pollution was in the surface drip and furrow irrigation with treated municipal wastewater. The least level of total and fecal coliform were in sub-surface drip irrigation at the depth of 30 cm in condition of municipal wastewater reuse. The results of plants revealed that the highest coliforms were seen at the surface drip and furrow irrigation with treated municipal wastewater. The minimum coliforms numbers were observed in the sub-surface drip irrigation method. Therefore, the sub-surface drip irrigation can reduce the wastewater reuse problems at soil and plant. It is the best option to avoid the contamination in landscape.

Keywords Irrigation method · Microbial contamination · Municipal wastewater · Landscape

P. Najafi
Department of Water Engineering, Agricultural Faculty, Isfahan (Khorasgan) Branch, Islamic Azad University, Isfahan, Iran
e-mail: payam.najafi@gmail.com

J. Shams (✉) · A. Shams
Young Researchers Club, Isfahan (Khorasgan) Branch, Islamic Azad University, Isfahan, Iran
e-mail: shams.jahangir@yahoo.com

A. Shams
e-mail: shamsazita96@yahoo.com

Introduction

Nowadays, due to water scarcity it is very common to use the treated wastewater in agriculture. Risk analysis is an integrated framework for wastewater reclamation and reuse consisting of risk assessment and management (Ganoulis 2012). The reuse of treated effluent can decrease impacts of pig manure on the environment and water resources (Velho et al. 2012). Evaluation of soil quality and environmental observation are very important for the wastewater irrigation. Kalavrouziotis et al. (2008) reported that treated municipal effluent increased the concentrations of heavy metals and fecal coliform levels in the edible portion of the cabbages. The attentions of chemical constituents in soil layers were influenced by the water movement and chemical concentrations in irrigation water (Heidarpour et al. 2007). The most important concern with the sub-surface irrigation was the increased effluence in the top soil layer, since it might inhibit plant growth (Tabatabaei and Najafi 2009).

Methods

The experiment was performed at research center in Isfahan municipal wastewater treatment plant in 2008. The Table 1 presents the chemical analysis of fresh water and treated municipal wastewater (TMW). Soil texture was loam, EC (6.6 ds/m) and SAR (6.1).

Four biological parameters including the numbers of total coliform (TC), fecal coliform (FC), total plate count (TPC) and parasite zygote were measured in effluent of wastewater treaded plant after filtration. *Laurus nobilis* and *Buxus sempervirens* plants were cultivated in the plots with three replications. Experimental treatments included, T1

Table 1 Chemical analysis of the fresh water and treated municipal wastewater

Irrigation	EC (ds/m)	pH	SAR_{iw}	SAR_{adj}	BOD (mg/l)	COD (mg/l)	Na (meq/l)	NO_3 (mg/l)	NH_4 (mg/l)	SO_4 (meq/l)	HCO_3 (meq/l)	Mg (meq/l)
Fresh water	0.5	7.4	1	1.8	–	–	1.4	0.5	–	1.6	2	1.2
Treated municipal wastewater	1.4	7.5	3.3	8	34.4	85.3	6.3	1.28	29.2	2.2	6.8	2.4

Table 2 Level of microbial parameters of treated municipal wastewater

Parameter	Before filtration	After filtration	Removal efficiency (%)
Total plate count (N/ml)	8.6×10^6	9.3×10^4	98.9
Total coliform (MPN/100 ml)	9.3×10^7	4.3×10^3	99.0
Fecal coliform (MPN/100 ml)	4.3×10^7	2.3×10^3	99.0
Parasite zygote (N/L)	3.5	1.5	57.1

Table 3 Means of coliforms levels in 1 gr soil

Total coliform				Fecal coliform			
Treatment	First sampling	Secondary sampling	Average	Treatment	First sampling	Secondary sampling	Average
T1[c]	9×10^3	5×10^6	2×10^6	T1[d]	1.5×10^3	4.3×10^3	2.9×10^3
T2[b]	2×10^9	4×10^7	6×10^8	T2[a]	3×10^9	7×10^5	1.5×10^8
T3[c]	3×10^6	1×10^6	2×10^6	T3[b]	2×10^6	2×10^4	1×10^6
T4[c]	2×10^6	5×10^6	2×10^4	T4[c]	9.3×10^4	2×10^4	6×10^4
T5[a]	5×10^{10}	9×10^6	2×10^{10}	T5[a]	7.5×10^7	2.1×10^5	4×10^7

The mean for each experimental factor in each characteristic which has at least one letter in common are not statistically significant based on Duncan test at 5 % level

furrow irrigation with fresh water, T2 the surface drip irrigation (DI) with treated municipal wastewater, T3 the sub-surface drip irrigation (SDI) at the depth of 15 cm with TMW, T4 the SDI at the depth of 30 cm with TMW, T5 the furrow irrigation with TMW. The system outflow was regulated for a 2.5 $m^3/h/m^2$. The diameter of secondary pipe was 16 mm, the dripper distance was 30 cm, and the rate of drippers was 2.5 l/h. After irrigation, the samples were taken during of studies and then their TC and FC in soil and plant shoots were measured at three stages by MPN method.

Table 4 Comparison of the coliforms levels at plants shoots

Treatment	*Laurus*		*Buxus*	
	TC	FC	TC	FC
T1	ns	ns	3	ns
T2	35	ns	41	ns
T3	7	ns	2	ns
T4	6	ns	1	ns
T5	28	ns	41	35

The mean for each experimental factor in each characteristic which has at least one letter in common are not statistically significant based on Duncan test at 5 % level, *ns* not seen

Result and discussion

According to the results, biological parameters decreased by filtration of the wastewater effluent. Reduction of TPC, TC and FC parameters showed up to 99 % and level of parasite zygote up to 57 % after filtration (Table 2).

The highest levels of coliforms were related to the DI and furrow irrigation with treated municipal wastewater. There are no significant differences between the other treatments in total coliform. The minimum level of fecal coliform belonged to the T1, T4 and T3, respectively (Table 3).

It could be concluded that the use of SDI method, in comparison with furrow and DI, decreases the surface soil pollution. Of course in this condition, the pollution of wastewater relative to fresh water irrigation should have no significant difference. Results of *Laurus nobilis* shoots showed that, total coliform was not seen at furrow irrigation with fresh water treatment. The lowest level of TC was obtained in the sub-surface drip irrigation system with TMW at the depth of 30 and 15 cm, respectively. The maximum level was seen in the DI (35 MPN/100 ml) and furrow irrigation (28 MPN/100 ml) with treated municipal wastewater. Also, there was no fecal coliform in *Laurus*. In shoots of *Buxus sempervirens*, the highest level of TC indicated at DI and furrow irrigation with TMW (41 MPN/100 ml). The minimum level of TC was shown in the T4, T3 and T1, respectively. Also the fecal coliform parameter was not found in all of treatments except the furrow irrigation with treated municipal wastewater (Table 4).

Conclusion

Soil pollution decreased with the increased depth of installed drippers, as could be observed from the sub surface drip technique. The study showed that the furrow and drip irrigations had the highest soil microbial pollution. The lowest level of microbial contamination of soil was seen in the sub surface drip irrigation using the treated municipal wastewater. The study also revealed that shoots of *Buxus* and *Laurus* had high microbial contaminations with surface drip and furrow irrigations.

References

Ganoulis J (2012) Risk analysis of wastewater reuse in agriculture. Int J Recycl Organ Waste Agric 1:3

Heidarpour M, Mostafazadeh-Fard B, Koupai JA, Malekian R (2007) The effects of treated wastewater on soil chemical properties using subsurface and surface irrigation methods. Agric Water Manag 90:87–94

Kalavrouziotis IK, Robolas P, Koukoulakis PH, Papadopoulos AH (2008) Effects of municipal reclaimed wastewater on the macro-and-micro-elements status of soil and of *Brossica oleracea* Var. Italica, and *B. oleracea* Var. Gemmifera. J Agric Water Manag 95:419–426

Tabatabaei SH, Najafi P (2009) Effects of irrigation with treated municipal wastewater on soil properties in arid and semi-arid regions. Irrig Drain J 58:551–560. doi:10.1002/ird.449

Velho VF, Mohedano RA, Filho PB, Costa RHR (2012) The viability of treated piggery wastewater for reuse in agricultural irrigation. Int J Recycl Organ Waste Agric 1:10

4

Effect of macrophyte vermicompost on growth and productivity of brinjal (*Solanum melongena*) under field conditions

Ishtiyaq Ahmed Najar · Anisa B. Khan · Abdul Hai

Abstract

Background Macrophytes (fresh water plants) comprise a diverse group of the flora which play important roles in the maintenance of trophic food chains and biogeochemical processes, but are deleterious when present in excess. However, due to various anthropogenic activities, there is accumulation of nutrients in aquatic ecosystems resulting in massive macrophytic growth. These weeds do not possess any economic value and remain laid on lake shores after harvesting and become a source of odor problem, thus posing a challenge to the lake management authorities regarding their proper disposal. However, vermicomposting turns these macrophytes into materials useful in horticulture/agriculture for restoration of soil fertility, in addition to providing a solution to the nuisance of harvested weeds.

Results The study was conducted to investigate the effect of different rates (2, 4 and 6 t/ha) of macrophyte-based vermicompost on germination, growth and yield of *Solanum melongena* under field conditions. The data revealed that different rates of vermicompost produced varied and significant effect ($P < 0.05$) as compared to the control on germination, growth and yield parameters with maximum value recorded at 6 t/ha, followed by 4 t/ha and the least at 2 t/ha. The dose of 6 t/ha significantly ($P < 0.05$) increased germination (22.56 ± 2.5 %), number of fruits per plant (3.55 ± 0.07) mean fruit weight (73 ± 5.0 g), yield per plant (1.48 ± 0.05 kg) and marketable fruits (28.66 ± 3.0 %) when compared with the control. The study suggests that macrophyte-based vermicompost as a potential source of plant nutrients for sustainable crop production.

Conclusion Macrophyte-based vermicompost is an efficient quality yielder and economy enhancer for sustainable agriculture especially for the communities having vegetable gardens around lakes will benefit by using macrophyte vermicompost, a balanced and low-cost organic fertilizer.

Keywords Fruits · Germination · Macrophytes · *Solanum melongena* · Vermicompost · Yield

Introduction

Conventional farm systems have been characterized by a high input of chemical fertilizer leading to qualitative deterioration of soil as well as agricultural yield (Diacono and Montemurro 2010; Singh et al. 2007). However, a growing awareness of the adverse impacts of inorganic fertilizers on crop production as well as increasing environmental and ecological concerns has stimulated greater interest in the utilization of organic amendments for crop production. Organic manures act not only as a source of nutrients and organic matter, but also increase microbial diversity and activity in soil, which influence soil structure and nutrients turnover, in addition to improvement in other physical, chemical and biological properties of the soil (Albiach et al. 2000; Edmeades 2003). Thus organic amendments/manures are environmentally benign and help in maintaining soil

I. A. Najar (✉)
Hydrobiology Research Laboratory, S. P. College,
Srinagar 190001, Jammu and Kashmir, India
e-mail: ishtiyaq.env@gmail.com

A. B. Khan
Department of Ecology and Environmental Sciences,
Pondicherry Central University, Puducherry 605014, India

A. Hai
Department of Zoology, S. P. College,
Srinagar 190001, Jammu and Kashmir, India

fertility as well as agricultural productivity. Among various organic products, vermicompost has been recognized as potential soil amendment.

Vermicompost is a product of non-thermophilic biodegradation of organic material by combined action of earthworms and associated microbes (Pathma and Sakthivel 2013). It is a highly fertile, finely divided peat-like material with high porosity, aeration, water-holding capacity and low C:N ratios (Dominguez and Edwards 2004). Vermicompost is made up primarily of carbon (C), hydrogen (H) and oxygen (O) and possesses remarkable plant growth-promoting properties due to the presence of nutrients in plant-available forms such as nitrates, calcium, phosphorus and potassium (Edwards and Burrows 1988). The enhancement of plant growth by vermicompost may not only be nutritional, but also due to its content of biologically active plant growth-influencing substances (Warman and AngLopez 2010). The presence of plant growth regulators such as auxins, gibberellins, cytokinins of microbial origin (Tomati et al. 1988) and humic acids (Atiyeh et al. 2002) has been reported in vermicompost. Stimulation of root growth (initiation and proliferation of root hair), increased root biomass, enhanced plant growth and development have been reported with the application of vermicompost, because of the presence of humic acids (Tomati et al. 1988; Mylonas and Mccants 1980; Chen and Aviad 1990). Moreover, the positive influences of humic acids on plant growth and productivity, which seem to be concentration specific, could be mainly due to hormone-like activities of humic acids through their involvement in cell respiration, photosynthesis, oxidative phosphorylation, protein synthesis and various enzymatic reactions (Chen and Aviad 1990). Further, humic acids are molecules that regulate other processes of plant development, such as macro and micronutrient adsorption (Gutierrez-Miceli et al. 2008a) and metabolism, which influence protein synthesis. Significant increase in soil enzyme activities such as urease, phosphomonoesterase, phosphodiesterase and arylsulfatase has been associated with the application of vermicompost (Albiach et al. 2000). Vermicomposts also possess disease-suppressing potential on a wide range of phytopathogens (Sahni et al. 2008). Vermicomposts harbour a wide variety of efficient antagonistic bacteria aiding in suppression of diseases caused by devastating soil-borne phytopathogenic fungi (Singh et al. 2008; Pathma and Sakthivel 2012). Several studies also report that vermicompost application suppresses infection by insect pests, repel crop pests and induce biological resistance in plants against pests and diseases due the presence of antibiotics and actinomycetes (Munroe 2007).

There is accumulating scientific evidence that vermicomposts can influence the growth and productivity of plants significantly (Edwards 1998). Various greenhouse and field studies have examined the effects of a variety of vermicomposts on a wide range of crops including cereals and legumes (Souzaa et al. 2013; Chan and Griffiths 1988), vegetables (Doan et al. 2013; Edwards and Burrows 1988; Atiyeh et al. 2000a), ornamental and flowering plants (Edwards and Burrows 1988; Atiyeh et al. 2000b) and field crops (Najar and Khan 2013a; Bhattacharya et al. 2012; Wu et al. 2012; Valdez-Péreza et al. 2011). Most of these investigations have confirmed that vermicomposts usually have significant beneficial effects on plant growth and yield. Vermicomposts, whether used as soil additives or as components of greenhouse bedding plant container media, have improved seed germination and enhanced seedling growth and development with overall increase in plant productivity. The beneficial effects include, but are not limited to stimulation of seed germination (Atiyeh et al. 2000b; Arancon et al. 2007a; Lazcano et al. 2010), activation of growth (Lazcano et al. 2009; Uma and Malathi 2009), protection against pathogens (Singh et al. 2003; Rivera et al. 2004; Zaller 2006), nematodes (Serfoji et al. 2010) and herbivores (Yardim et al. 2006; Arancon et al. 2007b; Edwards et al. 2010), and increasing overall crop productivity (Nattudurai et al. 2013; Gutierrez-Miceli et al. 2007; Singh et al. 2008; Azarmi et al. 2009). Several studies have assessed the effect of vermicompost amendments on the growth of a wide range of marketable fruits cultivated in greenhouses (Arancon et al. 2003, 2004a; Atiyeh et al. 2000c), as well as in fields (Najar and Khan 2013a; Wang et al. 2010; Singh et al. 2008).

Large quantities of macrophytes are harvested from different lakes across the Kashmir Valley. These macrophytes could easily be used in horticulture/agriculture for the restoration of soil fertility or for the production of quality vegetables, as they have good nutrient value, mainly as nitrogen (N), phosphorous (P) and potassium (K). However, in the present form the nutrients are not readily available and the harvested weeds remain on lake shores and become a source of odor problem, thus posing a challenge to the lake management authorities regarding their proper disposal (Najar and Khan 2013b). However, vermicomposting turns these macrophytes into materials useful in horticulture and there would be great savings in primary plant nutrients and, in addition, a solution to the nuisance of harvested weeds. The objective of the study was to evaluate the response of different rates of macrophyte-based vermicompost on germination, growth and yield of *S. melongena* under field conditions.

Materials and methods

The experiment was conducted at the experimental site of Hydrobiology Research Laboratory, Srinagar, Jammu and

Kashmir, India (34°08′N, 74°50′E). The study area lies in the temperate zone, characterized by wet and cold winter and relatively dry and moderate hot summer with mean annual precipitation of about 1100 mm, mainly falling during winter and spring. The hottest months are July and August, when the maximum temperature rises above 30 °C. September has cooler nights and the severe winter sets in about the middle of December. The coldest month is January, with temperature falling below freezing point. The winters vary from year to year; some are severe with very heavy snowfall, while others are mild with moderate snowfall (Najar and Khan 2011, 2014).

Preparation of macrophyte vermicompost

Different macrophytes (fresh water weeds) were collected from Dal Lake, excess water was allowed to drain for 2 days under sunlight (Najar and Khan 2010) and the macrophytes mixed with cattle dung in a 5:1 ratio (6 kg of macrophytes:1.2 kg of cow dung). Healthy and adult individuals of earthworm (*Eisenia fetida*) were allowed to feed on macrophytes which were converted into vermicompost (macrophyte vermicompost) during 60 days' duration (Najar and Khan 2012, 2013b). The physiochemical characteristics of the macrophyte vermicompost are given in Table 1.

Experimental setup

Seeds and plants of *S. melongena* were germinated and grown during 2009 and 2010 to determine the effect of the vermicompost. Field plots were 4 m long and 4 m wide (16 m^2) and separated by 1 m from unplanted areas. The physiochemical characteristics of soil are given in Table 1. Macrophyte based vermicompost was applied at the rate of 0 t/ha (control), 2, 4 and 6 t/ha. The vermicompost was incorporated into the top 15 cm of soil in the whole experimental bed (Arancon et al. 2003; Najar and Khan 2013a). The plots were arranged in a randomized complete block design with four replications for each treatment. All the necessary cultural practices and plant protection measures were followed uniformly for all the treatments during the entire period of experimentation.

Table 1 Physicochemical characteristics of soil and macrophyte vermicompost (mean ± SE, $n = 3$)

Parameter	Soil	Vermicompost
pH	7.2 ± 0.08	7.73 ± 0.13
EC (mS/m)	0.08 ± 0.01	0.98 ± 0.12
Ca (g/kg)	2.64 ± 0.04	16.22 ± 1.37
Mg (g/kg)	0.51 ± 0.01	7.62 ± 1.03
Na (μg/g)	11 ± 0.23	32.03 ± 1.91
K (μg/g)	3.52 ± 0.11	35.31 ± 13.13
P (μg/g)	110 ± 11	586.23 ± 26.65
OC (g/kg)	22.8 ± 1.5	170.58 ± 7.65
ON (g/kg)	1.7 ± 0.2	8.76 ± 0.83

Germination (%)

Seeds of *S. melongena* were sown in the plotted fields and were considered as emerged or germinated when cotyledons project out through the surface of the soil. Germination rate was expressed as the cumulative number of seeds germinating during monitoring days relative to the total number of seeds sown for germination. Seedlings were transplanted at three true leaves stage. Seedlings were planted at a distance of 25 cm between two plants.

Growth parameters and yield parameters

Plants of *S. melongena* were harvested after every fortnight and assessed for growth (shoot length, root length and leaf area) and their respective dry weight. The roots and leaves were cleaned carefully to remove adhering soil and dust particles. Leaves per plant were counted and leaf area (cm^2) was measured by using a leaf area meter (Singh et al. 2007). Plant biomass was dried in an oven at 60 °C for 24 h and the dry weight obtained (Lazcano et al. 2009).

Marketable fruits

All the ripe fruits were harvested, weighed and graded as marketable and non-marketable. Fruits were classified as non-marketable when fruits showed sign of decay due to diseases or insect feeding or any other malformation on the fruit surface (Zaller 2007). Relative proportions of non-marketable fruits were expressed as percentages of the total number of fruits harvested. Marketable fruits do not show any sign of decay. However, non-marketable fruits were not evaluated for specific disease or type of insect feeding.

Data analyses

Data sets were subjected to multivariate statistical techniques: two-way analysis of variance (ANOVA) and Tukey's t test were used as a post hoc analysis to compare the means (Zar 2009). Pearson correlation coefficient (r) was employed to examine the relationship between the vermicomposting application and yield parameters. All the tests were achieved with a significance level $P < 0.05$. Statistical analyses and graphical presentations were performed using SPSS statistical software (Version 16.0).

Results

Application of different rates of vermicompost (2, 4 and 6 t/ha) in *S. melongena* significantly affected the germination of seeds, plant growth and development, marketable fruits and yield during the study period of 120 days.

Germination (%)

Germination was 44.33 ± 2.01 in 6 t/ha, 35.77 ± 2.02 in 4 t/ha and 21.77 ± 2.05 in the control (Fig. 1). The effect of vermicompost was significant ($P < 0.05$) on germination. Application of vermicompost at the rate of 2, 4 and 6 t/ha increased the germination in the respective treatments by 8.33 ± 0.9, 14 ± 1.1 and 22.56 ± 2.5 as compared to the control.

Plant growth parameters

The shoot and root length exhibited a maximum value of 69 ± 1.40 cm and 26.5 ± 1.28 cm in 6 t/ha and least values of 46 ± 1.17 cm and 19.5 ± 1.28 cm in the control, respectively (Fig. 2a, b). Shoot and root length varied significantly among the treatments ($F_3 = 11.59$; 20.01, $P < 0.05$) and during different fortnights ($F_7 = 101.20$; 73.57, $P < 0.05$). Dry weight showed significant variation ($F_3 = 5.80$; 5.88, $P < 0.05$) among the treatments and during different fortnights ($F_7 = 19.60$; 61.98, $P < 0.05$), with higher mean values of 8.62 ± 0.27 g (shoot) and 4.37 ± 0.23 g (root) in 6 t/ha and minimum of 3.42 ± 0.25 g (shoot) and 3.02 ± 0.06 g (root) in the control (Fig. 2c, d). Leaf area was 475.25 ± 7.40 cm^2 in 6 t/ha and 368.25 ± 10.65 cm^2 in the control (Fig. 2e). Application of vermicompost significantly increased the leaf area among the treatments ($F_3 = 8.28$, $P < 0.05$) and also exhibited significant variation during different fortnights ($F_7 = 129.26$, $P < 0.05$). Dry weight of leaves was less in the control with a mean value of 6.92 ± 0.24 g and

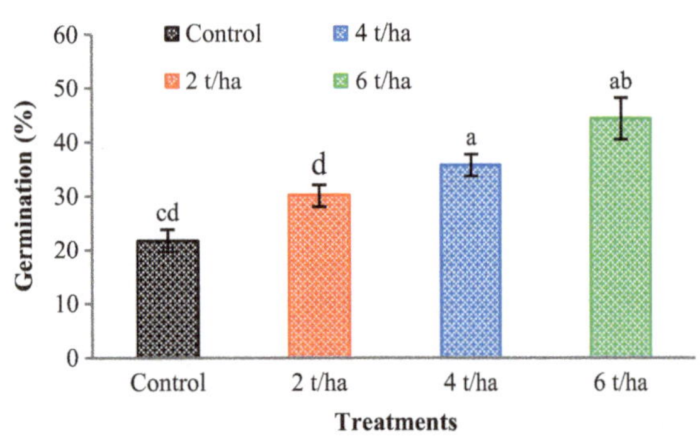

Fig. 1 Effect of macrophyte vermicompost on germination of *Solanum melongena* (mean ± SE, $n = 4$). Significant differences at $P < 0.05$ (Tukey's LSD test) are indicated by *different letters*

maximum in the treatment of 6 t/ha with a value of 12.47 ± 0.30 g (Fig. 2f), and differed significantly among the treatments ($F_3 = 6.09$, $P < 0.05$) and during different fortnights ($F_7 = 36.98$, $P < 0.05$).

Yield parameters

Vermicompost application showed positive correlation with clusters/plant ($r = 0.977$, $P < 0.05$), fruits/cluster ($r = 0.968$; $P < 0.05$), number of fruits/plant ($r = 0.965$; $P < 0.05$), mean fruit weight ($r = 0.988$; $P < 0.05$), yield/plant ($r = 0.965$; $P < 0.05$) and strong positive correlation with marketable fruits ($r = 0.997$; $P < 0.01$), whereas strong negative correlation with non-marketable fruits ($r = -0.997$; $P < 0.01$). Application of vermicompost resulted in 2.83 ± 0.16 clusters/plant in 6 t/ha and 2.5 ± 0.22 in 4 t/ha as compared to 2 ± 0.01 in the control. Thus, the application of vermicompost at the rate of 6 t/ha increased significantly ($P < 0.05$) the number of clusters per plant by 0.83 ± 0.02 and 0.5 ± 0.01 in 6 and 4 t/ha, respectively, when compared with the control (Fig. 3a). Fruits/cluster were observed to be significantly ($P < 0.05$) higher in vermicompost treatments, with a maximum of 4.5 ± 0.60 in 6 t/ha and 2.83 ± 0.30 in 2 t/ha, whereas it was 2.66 ± 0.21 in the control (Fig. 3b) with an overall increase of 1.84 ± 0.03, 1 ± 0.01 and 0.22 ± 0.01 fruits/cluster in the respective treatment. Fruits/plant recorded in the control were 6.83 ± 0.40 and varied significantly ($P < 0.01$) among treatments with a maximum value of 10.33 ± 0.66 in 6 t/ha. Vermicompost application enhanced the fruits/plant value by 0.78 ± 0.08 in 2 t/ha, 1.45 ± 0.09 in 4 t/ha and 3.55 ± 0.07 in 6 t/ha when compared with the control (Fig. 3c). Mean fruit weight (g) was 158.83 ± 5.41 in 6 t/ha, 126 ± 2.58 in 4 t/ha, 101.83 ± 4.09 in 2 t/ha and 85.83 ± 3.35 in the control. There was significant ($P < 0.05$) increase in mean fruit weight among the treatments, with a maximum increase of 73 ± 5.0 g recorded in the treatment of 6 t/ha and least of 16 in 2 t/ha (Fig. 3d). Yield/plant (kg) in the control was 0.49 ± 0.04, whereas among the treatments it was 1.97 ± 0.13 in 6 t/ha and 0.61 ± 0.07 in 2 t/ha (Fig. 3e) with a significant ($P < 0.01$) increase of 1.48 ± 0.05, 0.82 ± 0.01 and 0.12 ± 0.01 in the respective treatments when compared with the control. Marketable fruits (%) were significantly ($P < 0.01$) higher in vermicompost-amended plots: 92.49 ± 1.76 (6 t/ha), 85.13 ± 1.86 (4 t/ha) and 74.30 ± 3.47 (2 t/ha) against 63.83 ± 5.07 in the control (Fig. 4a). Among the amended plots, the maximum increase was recorded in 6 t/ha (28.66 ± 3.0). The effect of vermicompost application was significant ($P < 0.05$) on decrease in yield of non-marketable (%) fruits among the treatments and with the lowest value of 7.49 ± 1.76 (6 t/ha), 14.85 ± 1.86 (4 t/ha) and

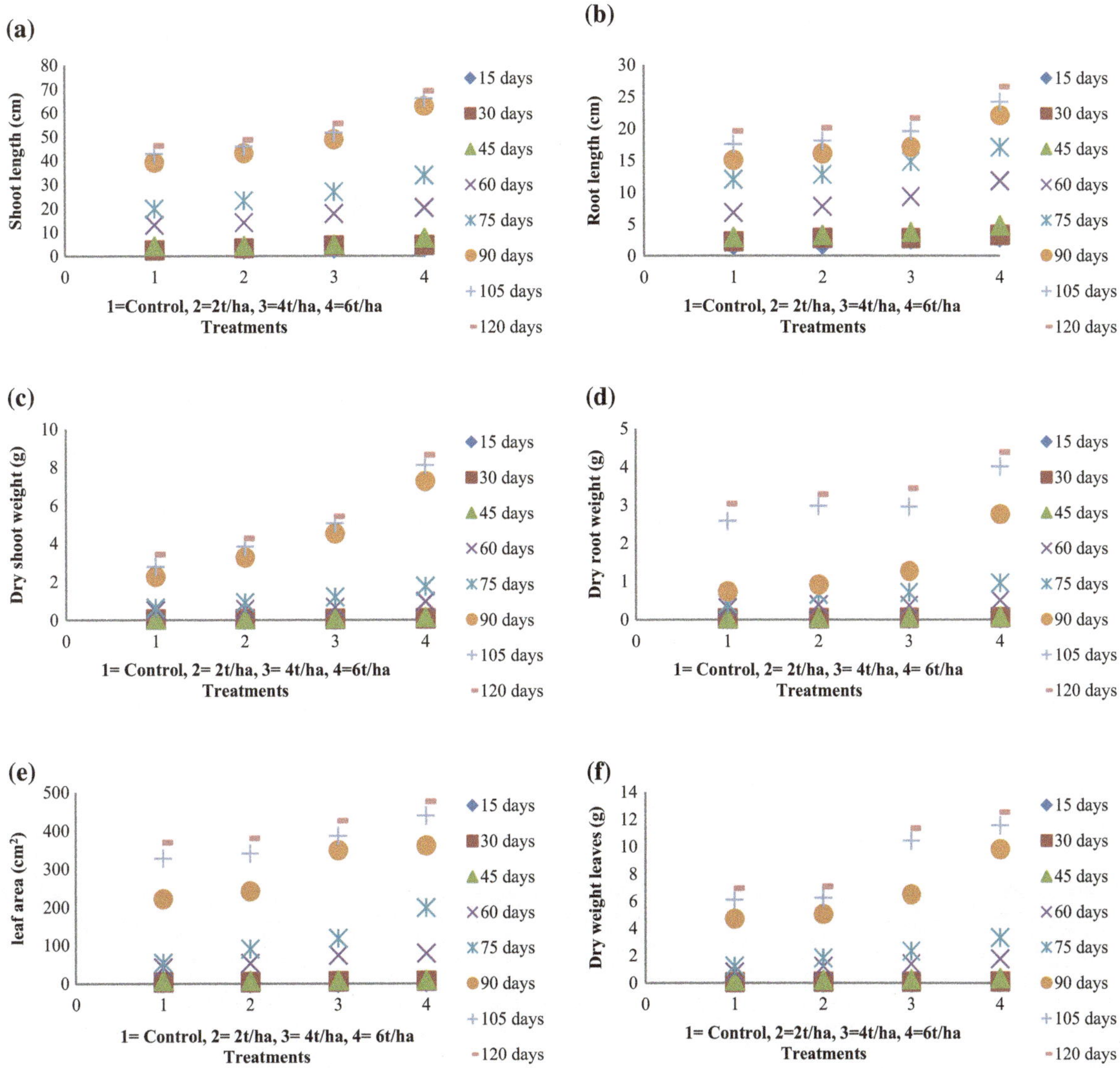

Fig. 2 Effect of macrophyte vermicompost on growth parameters of *Solanum melongena* (mean ± SE, $n = 4$)

25.69 ± 3.47 (2 t/ha). Among the vermicompost treatments, maximum decrease was recorded in 6 t/ha (28.67 ± 2.9) followed by 4 t/ha (21.47 ± 2.1) when compared with the control (Fig. 4b).

Discussion

Application of different rates of vermicompost in *S. melongena* resulted in varied response on germination, plant growth parameters, yield parameters, marketable and non-marketable fruits.

Germination

Vermicompost application resulted in differential effect on the germination of *S. melongena* among different treatments. Vermicompost has been reported to stimulate germination of several horticultural plant species such as green gram (*Phaseolus aureus*) Karmegam et al. (1999) and tomato plants (*L. esculentum*) Atiyeh et al. (2000b). Najar and Khan (2013a) also reported an increase in germination of *L. esculentum* by 10.33 % with the application of macrophyte-based vermicompost. According to Zaller (2007) and Lazcano et al.(2010), increase in seed germination of *Rumex obtusifolius* by 48 % and *Pinus pinaster* by 16 % has been reported with the application of cattle manure and rabbit manure vermicompost, respectively. Germination is an internally regulated process influenced mainly by genotype, although external factors such as light period, temperature, moisture and presence of certain chemical compounds can also alter this process either through promotion or inhibition (Kucera et al. 2005). When all these factors are integrated, it is mediated by signaling

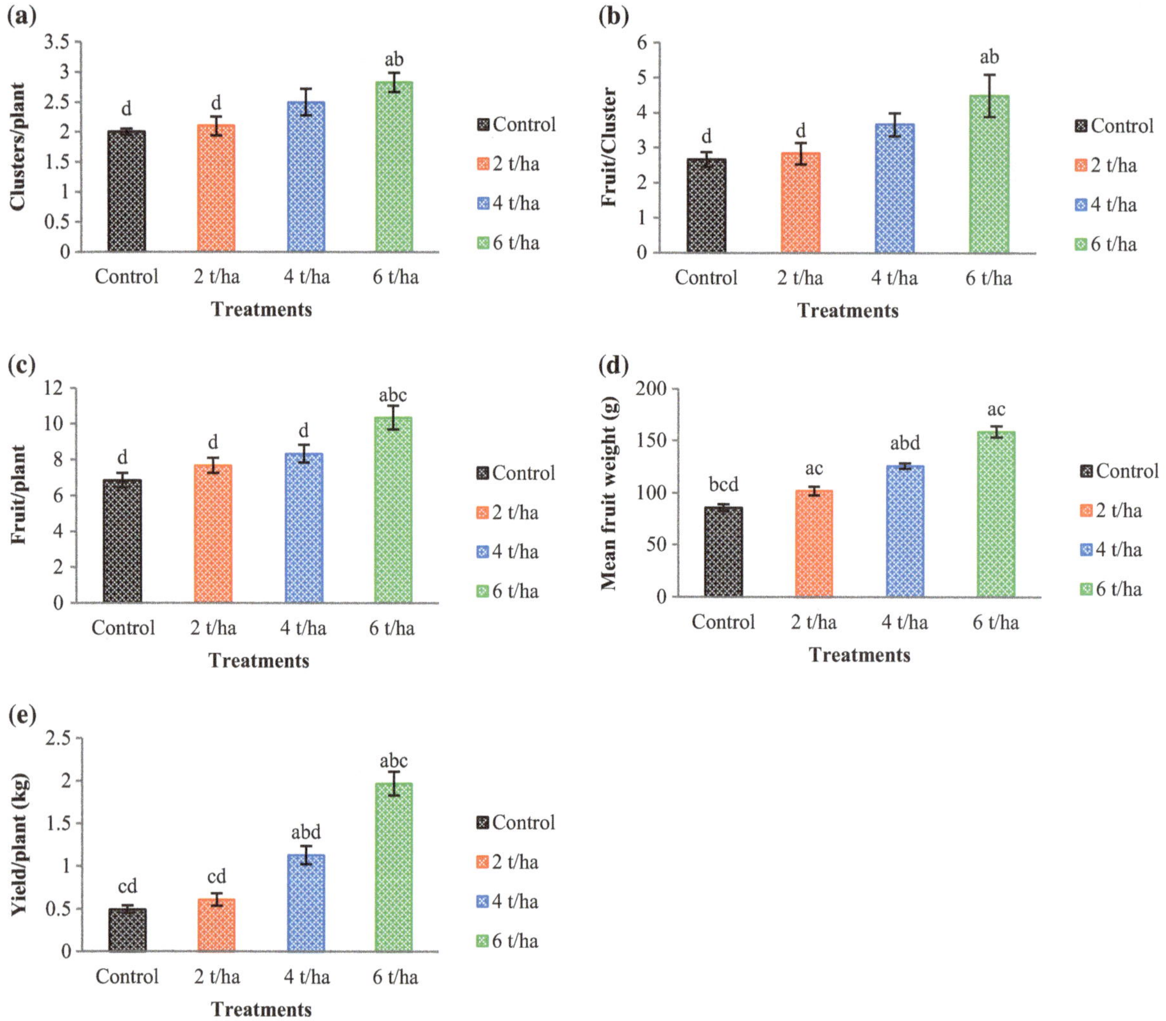

Fig. 3 Effect of macrophyte vermicompost on yield parameters of *Solanum melongena* (mean ± SE, $n = 4$). Significant differences at $P < 0.05$ (Tukey's LSD test) are indicated by *different letters*

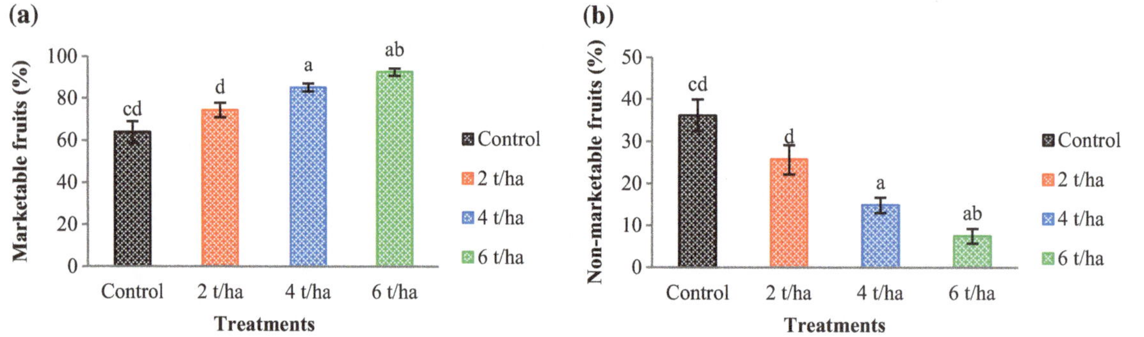

Fig. 4 Effect of macrophyte vermicompost on marketable and non-marketable fruits of *Solanum melongena* (mean ± SE, $n = 4$). Significant differences at $P < 0.05$ (Tukey's LSD test) are indicated by *different letters*

through multiple hormones that either promote or inhibit germination (Finkelstein 2004). The germination rate (%) in *S. melongena* increased with subsequent increase in vermicompost dose. The results corroborate those of Alves and Passoni (1997) that increasing doses of vermicompost application to the potting media of *Licania tomentosa* increased germination as compared to unamended soil. Lazcano et al. (2010) reported changes in the physical properties of the germination media with the incorporation of vermicompost that influence the moisture retention and aeration of the soil, thereby potentially affecting seed germination. Improvement in the soil physical properties with the addition of macrophyte-based vermicompost might have resulted in increased germination rate of *S. melongena*.

Plant growth parameters

Vermicompost has been demonstrated to be a valuable soil amendment that offers slow, but a balanced nutritional release pattern to plants, providing nutrients such as available N, K, Ca, Mg and P that can be taken up readily by plants (Edwards 1998; Edwards and Fletcher 1988). Plant growth parameters like plant length and dry weight; root length and dry weight; number of leaves, area and dry weight were significantly higher in plots amended with different doses of vermicompost. Increase in plant growth with vermicompost application has been reported in different studies. The incorporation of pig manure vermicompost enhanced shoot and root weight, leaf area and shoot:root ratios of tomato and French marigold (Bachman and Metzger 2008). Najar and Khan (2013a) reported significant increase in plant growth (shoot length, root length, leaf area and plant biomass) of *L. esculentum* in plots amended with 6 t/ha macrophyte vermicompost. Joshi and Vig (2010) also reported increase in plant growth parameters (plant height, number of leaves and plant dry biomass) with application of 45 % vermicompost (cattle dung) amended treatment in *L. esculentum*. According to Gupta et al. (2014), addition of cow dung and household-based vermicompost in appropriate quantities to the potting media resulted in increased growth and flowering of marigold seedlings including plant biomass, plant height, number of buds and flowers. Increase in length, biomass, number of seeds, number of shoots in *Vinca rosea* and tillers in *Oryza sativa* has been reported by Reddy (1986) in 50:50 soil to vermicompost mixtures. Tomati et al. (1983) showed positive effects of vermicompost on the growth of *Begonias* sp. and *Coleus* sp. (ornamental plants), especially a stimulation of rooting and time of flowering in plots amended with vermicompost. Plant height of maize increased significantly as compared to the control when grown in soil amended with vermicompost (Gutierrez-Miceli et al. 2008b). Azarmi et al. (2008) reported increase in leaf area and shoot dry weight by 43 and 27 %, respectively, in tomato with 15 t/ha sheep manure vermicompost applications, whereas Atiyeh et al. (2001) reported increase in shoot height of tomato plant with the amendment of 5 % pig manure vermicompost.

The different doses of vermicompost produced different responses in *S. melongena* and among these 6 and 4 t/ha showed maximum positive effect on growth parameters. The results corroborate the findings that different doses of vermicompost caused different responses in the growth parameters of *L. esculentum* plant (Azarmi et al. 2008). Pritam et al. (2010) reported higher number of flowers in treatments with vermicompost amendments, with reduced time taken to transform bud into flower that has been attributed to readily available nutrients and growth-regulating substances present in the vermicompost. According to Ravi et al. (2008) and Taleshi et al. (2011), availability of plant nutrients in vermicompost increases growth and leaf area index of plant, which in turn increases absorption of light leading to more dry matter and yield. The differential response of plants to different doses of vermicompost is due to production of lesser quantity of growth-promoting substances by lower doses of vermicompost than higher doses (Arancon et al. 2004a). Plant root morphology is known to be influenced by water and nutrient availability as well as by external application of hormones (Lopez-Bucio et al. 2003). Vermicomposts have hormone-like activity that aids in greater root initiation, increased root biomass and enhanced plant growth (Bachman and Metzger 2008). According to Forde and Lorenzo (2001), root growth and branching are favored in nutrient-rich environment and in the presence of hormones such as auxins that enable the plant to optimize the exploitation of the available resources which are in turn transformed into photoassimilates and transported again to the root, consequently influencing plant growth and morphology in a systemic manner.

Yield parameters

Vermicompost amendment has influenced the number of clusters/plant, number of fruits/cluster, mean fruit weight and total yield/plant when compared with the control. Atiyeh et al. (2000b) reported increase in tomato fruit yield with the amendments of 10–20 % vermicompost in potting media. Arancon et al. (2004a, 2006) also reported growth and yield improvement in different crops with vermicompost application. The results clearly indicated that the plants receiving vermicompost produced more fruits/cluster, clusters/plant and large-sized fruits with higher total yield than the control. Increase in yield of wheat in cattle dung-based vermicompost-amended soil has been reported by Joshi et al. (2013). According to Arancon et al. (2006), significant increase in plant growth and consequently lesser days to flowering with increase in fruit yield at higher doses of vermicompost amendment were observed. In other similar studies, increases in yield with vermicompost applications in okra, strawberry, eggplant, potato, cucumber cultivars, *Abelmoschus esculentus*, peppers, crossandra, lettuce and Amaranthus species were reported by Singh et al. (2008); Alam et al. (2007); Azarmi et al. (2009); Vijaya and Seethalakshmi (2011); Arancon et al. (2005); Papathanasiou et al. (2012) and Uma and Malathi (2009) respectively. A high level of phosphorous is necessary for plants to produce good yield (Orozco et al. 1996). Vermicompost contains macronutrients, beneficial microorganisms and hormones which influence the growth and yield of plants (Theunissen et al. 2010). Macronutrients play an important role in crop yield based on their role in activation of enzymes for chlorophyll synthesis, growth, fruit ripening and maintenance of the plant enzyme system (Grusak and Della Penna 1999). Vermicompost is known to provide a slow, balanced nutritional release pattern to plants, particularly in terms of release of plant-available N, soluble K, exchangeable Ca, Mg and P (Edwards and Fletcher 1988) which is subsequently used by plants efficiently. Vermicompost application generally improves the soil environment, particularly soil aeration, encouraging the proliferation of roots, which in turn draw more water and nutrient from distant areas and help to introduce and sustain beneficial micro-organisms into the rhizosphere (Padmavathiamma et al. 2008) in sufficient numbers for a longer period.

Marketable and non-marketable fruits

Optimum plant growth and development are essential for better quality yield (Theunissen et al. 2010). The application of vermicompost resulted in increase in marketable fruits and decrease in non-marketable fruits as compared to the control. Further, there was significant increase in marketable fruits with the macrophyte vermicompost when applied at 6 t/ha. The influence of macrophyte vermicompost on the marketable fruits of brinjal may be attributed to balanced plant nutrients present in vermicompost. Increase in vermicompost dose from 2 to 6 t/ha increased marketable fruits significantly, as increased dose provided more plant-available nutrients in the treatments. Singh et al. (2008) reported decreased occurrence of physiological disorders like albinism (4.6 %), fruit malformation (4.1 %) and gray mold (2.7 %) incidence with increase in marketable fruit yield (58.6 %) of Chandler strawberry in the plots amended with 7.5 t/ha vegetable waste-based vermicompost over the control. According to Arancon et al. (2003), there was a significant increase in the marketable fruits of tomato with the application of paper waste-based vermicompost when applied at 5 t/ha. Nath et al. (2011) observed decrease in the nematode population (*Meloidogyne incognita*) with the application of agro-based vermicompost in *S. melongena* and also the marketable fruit yield was consistently greater in vermicompost-treated plots as compared to the control. According to Chandrakumar et al. (2009), decrease in infection of *Leucinodes orbonalis* on *S. melongena* with the application of the vermicompost. Wang et al. (2010) recorded significant increase in marketable weight of Chinese cabbage (*Brassica campestris*) with the application of cow manure vermicompost. Plants treated with vermicompost receive nutrition in a balanced and sustained way than the control (Arancon et al. 2004b) and thus plants produce lesser number of non-marketable fruits.

Edwards et al. (2010) report statistically significant decrease in arthropods (aphids, buds, mealy bug and spider mite) populations and subsequent reduction in plant damage with vermicompost amendment. According to Ramesh (2000), plots amended with vermicompost exhibited decrease in the occurrence of leaf miner (*Aproaerema modicella*) on groundnuts. Further, Edwards et al. (2007) reported considerable suppression of root knot nematode (*Meloidogyne incognita*) and drastic suppression of spotted spider mites (*Tetranychus* spp.) and aphid (*Myzuspersicae*) in *L. esculentum* plants after application of vermicompost. Singhai et al. (2011) recorded appreciable suppression of common scab of potato through application of vermicompost. According to Arancon et al. (2005), vermicomposts provide some essential nutrient elements that are not available in inorganic fertilizers and these could either have increased the plants resistance to pests or made the plants less palatable to the pests, based on the observation that there was decrease in dry weight losses of peppers, tomatoes and cabbage grown with substitutions of different rates of vermicompost in response to the aphid, mealy bug and caterpillar infestations. According to Rao (2002), there were considerable decreases in the population of aphids, coccinellid beetles and spider mites in groundnuts grown on soils amended with vermicompost, compared to those grown on soils amended with inorganic fertilizer. Munroe (2007) reported the presence of chitinase enzyme in vermicompost which breaks down the chitin in the exoskeleton of insects and thus repels many different insect pests. Thus the significant increase in marketable fruit yield of *S. melongena* might be attributed to improved availability of nutrients from macrophyte based vermicompost, in addition to the protection from pests.

Conclusions

The present study reveals that application of macrophyte-based vermicompost is quite beneficial in field-grown *S. melongena* significant higher rate of germination, increased plant growth and yield parameters with higher marketable fruit. Vermicompost application of 6 t/ha resulted in relatively higher productivity and better quality of fruits. The study showed that the growth and yield of *S. melongena* was dose dependent and clearly indicated the advantages of macrophyte vermicompost in quality yield production. Such effects could be attributed to the nutritional status of vermicompost and improvement in soil properties. It encompasses economic aspect also, since faster growth with decreased days for crop maturation and enhanced better quality yield not only result in their early availability in market, but also with higher market values/returns. Thus, macrophyte based vermicompost is an efficient quality yielder and economy enhancer for sustainable agriculture, especially for people/communities having vegetable gardens around lakes. Further large-scale vermicomposting of macrophytes helps in controlling their menace considerably and keeps water bodies clean, along with esthetic improvement.

References

Alam MN, Jahan MS, Ali MK, Ashraf MA, Islam MK (2007) Effect of vermicompost and chemical fertilizers on growth, yield and yield components of potato in barind soils of Bangladesh. J ApplSci Res 3(12):1879–1888

Albiach R, Canet R, Pomares F, Ingelmo F (2000) Microbial biomass content and enzymatic activities after application of organic amendments to a horticultural soil. BioresourTechnol75:43-48

Alves WL, Passoni AA (1997) Compost and vermicompost of urban solid waste in *Licania tomentosa* (Benth) seedlings production to arborization. Pesqui Agropecu Brasil 32:1053–1058

Arancon NQ, Edwards CA, Bierman P, Metzger JD, Lee S, Welch C (2003) Effects of vermicomposts on growth and marketable fruits of field-grown tomatoes, peppers and strawberries. Pedobiologia 47:731–735

Arancon NQ, Edwards CA, Atiyeh R, Metzger JD (2004a) Effects of vermicomposts produced from food waste on the growth and yields of greenhouse peppers. Bioresour Technol 93:139–144

Arancon NQ, Edwards CA, Bierman P, Welch C, Metzger JD (2004b) Influences of vermicomposts on field strawberries: effects on growth and yields. Bioresour Technol 93:145–153

Arancon NQ, Edwards CA, Bierman P, Metzger JD, Lucht C (2005) Effects of vermicomposts produced from cattle manure, food waste and paper waste on the growth and yield of peppers in the field. Pedobiologia 49:297–306

Arancon NQ, Edwards CA, Bierman P (2006) Influences of vermicomposts on field strawberries: effects on soil microbial and chemical properties. Bioresour Technol 97:831–840

Arancon NQ, Edwards CA, Dick R, Dick L (2007a) Vermicompost tea production and plant growth impacts. Bio Cycle 48:51–52

Arancon NQ, Edwards CA, Yardim EN, Oliver TJ, Byrne RJ, Keeney G (2007b) Suppression of two-spotted spider mite (*Tetranychus urticae*), mealy bug (*Pseudococcus sp.*) and aphid (*Myzus persicae*) populations and damage by vermicomposts. Crop Prot 26:29–39

Atiyeh RM, Arancon NQ, Edwards CA, Metzger JD (2000a) Influence of earthworm-processed pig manure on the growth and yield of greenhouse tomatoes. Bioresour Technol 75:175–180

Atiyeh RM, Dominguez J, Subler S, Edwards CA (2000b) Changes in biochemical properties of cow manure during processing by earthworms (*Eisenia andrei*, Bouche) and the effects on seedling growth. Pedobiologia 44:709–724

Atiyeh RM, Edwards CA, Subler S, Metzger JD (2000c) Earthworm processed organic wastes as components of horticultural potting media for growing marigold and vegetable seedlings. Compost Sci Util 8:215–223

Atiyeh RM, Edwards CA, Subler S, Metzger JD (2001) Pig manure vermicompost as a component of a horticulture bedding plant medium: effect on physicochemical properties and plant growth. Bioresour Technol 78:11–20

Atiyeh RM, Lee S, Edwards CA, Arancon NQ, Metzger JD (2002) The influence of humic acids derived from earthworm-processed organic wastes on plant growth. Bioresour Technol 84:7–14

Azarmi R, Ziveh PS, Satari MR (2008) Effect of vermicompost on Growth and nutrient status of tomato (*Lycopersicumesculentum*). Pakistan J BiolSci

Azarmi R, Giglou MT, Hajieghrari B (2009) The effect of sheep manure vermicompost on quantitative and qualitative properties of cucumber (*Cucumissativus* L.) grown in the greenhouse. Afr J Biotechnol 8:4953–4957

Bachman GR, Metzger JD (2008) Growth of bedding plants in commercial potting substrate amended with vermicompost. Bioresour Technol 99:3155–3161

Bhattacharya SS, Iftikar W, Sahariah B, Chattopadhyay GN (2012) Vermicomposting converts fly ash to enrich soil fertility and sustain crop growth in red and lateritic soils. Resour Conserv Recy 65:100–106

Chan PLS, Griffiths DA (1988) The vermicomposting of pretreated pig manure. Biol Wastes 24:57–69

Chandrakumar HL, Kumar CTA, Chakravarthy AK, Kumar NG, Puttaraju TB (2009) Influence of organic materials against shoot and fruit borer, *Leucinodes orbonalis* Guen. (Lepidoptera: Pyralidae) on brinjal (*Solanum melongena* L.). Current Biotica 2(4):495–500

Chen Y, Aviad T (1990) Effects of humic substances on plant growth. In: MacCarthy P, Clapp CE, Malcolm RL, Bloom PR (eds) Humic Substances in Soil and Crop Sciences: Selected Readings. ASA and SSSA, Madison, pp 161–186

Diacono M, Montemurro F (2010) Long-term effects of organic amendments on soil fertility: a review. Agron Sustain Dev 30(2):401–422

Doan TT, Ngo PT, Rumpe C, Nguyen BN, Jouquet P (2013) Interactions between compost, vermicompost and earthworms influence plant growth and yield: a one-year greenhouse experiment. Sci Hortic 160:148–154

Dominguez J, Edwards CA (2004) Vermicomposting organic wastes: a review. In: Shakir HSH, Mikhail WZA (eds) Soil zoology for sustainable development in the 21st century. Self-Publisher, Cairo, pp 369–395

Edmeades DC (2003) The long-term effects of manures and fertilisers on soil productivity and quality: a review. Nutr Cyc Agroecosys 66:165–180

Edwards CA (1998) The use of earthworms in the breakdown and management of organic wastes. In: Edwards CA (ed) Earthworm ecology. CRC Press, Boca Raton

Edwards CA, Burrows I (1988) The potential of earthworm composts as plant growth media. In: Edwards CA, Neuhauser (eds) Earthworms in environmental and waste management. SPB Academic Publishers, The Netherlands, pp 211–220

Edwards CA, Fletcher KE (1988) Interaction between earthworms and microorganisms in organic matter breakdown. Agricul Ecosyst Environ 20(3):235–249

Edwards CA, Arancon NQ, Emerson E, Pulliam R (2007) Suppressing plant parasitic nematodes and arthropod pests with vermicompost teas. Bio Cycle 12:38–39

Edwards CA, Arancon NQ, Vasko-Bennett M, Askar A, Keeney G (2010) Effect of aqueous extracts from vermicomposts on attacks by cucumber beetles (*Acalymnavittatum*) (Fabr.) on cucumber and tobacco hornworm (*Manducasexta*) (L.) on tomatoes. Pedobiologia 53:141–148

Finkelstein RR (2004) Hormones in seed development and germination. In: Davies PJ (ed) Plant hormones: biosynthesis, signal transduction and action. Kluwer Academic Publishers, USA, pp 513–537

Forde B, Lorenzo H (2001) The nutritional control of root development. Plant Soil 232:51–68

Grusak MA, DellaPenna D (1999) Improving the nutrient composition of plants to enhance human nutrition and health. Annu Rev Plant phys Plant Mol Biol 50:133–161

Gupta R, Yadav A, Garg VK (2014) Influence of vermicompost application in potting media on growth and flowering of marigold crop. Int J Recycl Org Waste Agricult 3:47. doi:10.1007/s40093-014-0047-1

Gutierrez-Miceli FA, Santiago-Borraz J, Molina JAM, Nafate CC, Abud-Archila M, Llaven MAO, Rincon-Rosales R, Dendooven L (2007) Vermicompost as a soil supplement to improve growth, yield and fruit quality of tomato (*Lycopersicon esculentum*). Bioresour Technol 98:2781–2786

Gutierrez-Miceli FA, Gracia-Gomez RC, Rincon RR, Abud-Archila M, Maria- Angela OL, Gullin-Cruz MJ, Dendooven L (2008a) Formulation of liquid fertilizer for sorghum (*Sorghum bicolour* (L.) Moench) using vermicompost leachate. Bioresour Technol 99:6174–6180

Gutierrez-Miceli FA, Moguel-Zamudio B, Abud-Archila M, Gutierrez-Oliva VF, Dendooven L (2008b) Sheep manure vermicompost supplemented with a native diazotrophic bacteria and mycorrhizas for maize cultivation. Bioresour Technol 99:7020–7026

Joshi R, Vig AP (2010) Effect of vermicompost on growth, yield and quality of tomato (*Lycopersicum esculentum* L). Afr J Basic Appl Sci 2(3–4):117–123

Joshi R, Vig AP, Singh J (2013) Vermicompost as soil supplement to enhance growth, yield and quality of *Triticum aestivum* L.: a field study. Int J Recy Org Waste Agric 2:16. doi:10.1186/2251-7715-2-16

Karmegam N, Alagermalai K, Daniel T (1999) Effect of vermicompost on the growth and yield of greengram (*Phaseolus aureus* Rob.). Trop Agric 76(2):143–146

Kucera B, Cohn MA, Leubner-Metzger G (2005) Plant hormone interactions during seed dormancy release and germination. Seed Sci Res 15:281–307

Lazcano C, Arnold J, Tato A, Zaller JG, Dominguez J (2009) Compost and vermicompost a nursery pot components: effects on tomato plant growth and morphology. Span J Agric Res 7:944–951

Lazcano C, Sampedro L, Zas R, Dominguez J (2010) Vermicompost enhances germination of the maritime pine (*Pinus pinaster Ait.*). New For 39:387–400

Lopez-Bucio J, Cruz-Ramires A, Herrera-Estrella L (2003) The role of nutrient availability in regulating root architecture. Curr Option Plant Biol 6:280–287

Munroe G (2007) Manual of On-farm Vermicomposting and Vermiculture. Publication of Organic Agriculture Centre of Canada, Nova Scotia

Mylonas VA, Mccants CB (1980) Effects of humic and fulvic acids on growth of tobacco. I. Root initiation and elongation. Plant Soil 54:485–490

Najar IA, Khan AB (2012) Vermicomposting of fresh water weeds (macrophytes by *Eisenia fetida* (Savigny, 1826), *Aporrectodea caliginosa trapezoides* (Duges, 1828) and *Aporrectodea rosea rosea* (Savigny. Dyn Soil Dyn Plant 6(S1):73–77

Najar IA, Khan AB (2010) Vermicomposting of *Azolla pinnata* by using earthworm *Eisenia fetida*. The Bioscan 5(2):239–241

Najar IA, Khan AB (2011) Earthworm communities of Kashmir Valley, J&K, India. Trop Ecol 52(2):151–162

Najar IA, Khan AB (2013a) Effect of vermicompost on growth and productivity of tomato (*Lycopersicon esculentum*) under field conditions. Acta Biol Malaysiana 2(1):12–21

Najar IA, Khan AB (2013b) Management of fresh water weeds (macrophytes) by vermicomposting using *Eisenia fetida*. Environ Sci Pollut Res 20:6406–6417. doi:10.1007/s11356-013-1687-9

Najar IA, Khan AB (2014) Factors affecting distribution of earthworms in Kashmir Valley: a multivariate statistical approach. Proc Zool Soc 67(2):126–135. doi:10.1007/s12595-013-0081-4

Nath G, Singh DK, Singh K (2011) Productivity enhancement and nematode management through vermicompost and biopesticides in brinjal (*Solanum melongena* L.). World Appl Sci J 12(4):404–412

Nattudurai G, Vendan SE, Ramachandran PV, Lingathurai S (2013) Vermicomposting of coirpith with cow dung by *Eudrilus eugeniae* Kinberg and its efficacy on the growth of *Cyamopsis tetragonaloba* (L) Taub. J Saudi Soc Agric Sci. doi:10.1016/j.jssas.2012.12.003

Orozco SH, Cegarra J, Trujillo LM, Roig A (1996) Vermicomposting of coffee pulp using the earthworm *Eisenia fetida*: effects on C and N contents and the availability of nutrients. Biol Fertil Soils 22:162–166

Padmavathiamma PK, Li LY, Kumari UR (2008) An experimental study of vermibio waste composting for agricultural soil improvement. Bioresour Technol 99:1672–1681

Papathanasiou F, Papadopoulos I, Tsakiris I, Tamoutsidis E (2012) Vermicompost as a soil supplement to improve growth, yield and quality of lettuce. J Food Agric Environ 10(2):677–682

Pathma J, Sakthivel N (2012) Microbial diversity of vermicompost bacteria that exhibit useful agricultural traits and waste management potential. Springer Plus 1:26. doi:10.1186/2193-1801-1-26

Pathma J, Sakthivel N (2013) Molecular and functional characterization of bacteria isolated from straw and goat manure based vermicompost. Appl Soil Ecol 70:33–47

Pritam S, Garg VK, Kaushik CP (2010) Growth and yield response of marigold to potting media containing vermicompost produced from different wastes. Environmentalist 30:123–130

Ramesh P (2000) Effects of vermicomposts and vermicomposting on damage by sucking pests to ground nut (*Arachis hypogea*). Indian J Agric Sci 70(5):334

Rao KR (2002) Induce host plant resistance in the management of sucking insect pests of groundnut. Ann Plant Prot Sci 10(1):45–50

Ravi S, Channal HT, Hebsur NS, Patil BN, Dharmatti PR (2008) Effect of sulphur, zinc and iron nutrition on growth, yield, nutrient uptake and quality of safflower (*Carthamus tinctorius* L.). Karnataka. J Agric Sci 21:382–385

Reddy MV (1986) The effect of casts of *Pheretima alexandri* (Beddard) on the growth of *Vinca rosea* and *Oryza sativa* L. In: Edwards CA, Neuhauser EF (eds) Earthworms in waste and environmental management. SPB Academic Publishing, The Hague, pp 241–248

Rivera MC, Wright ER, Lopez MV, Garda D, Barrague MY (2004) Promotion of growth and control of damping-off (*Rhizoctonia solani*) of greenhouse tomatoes amended with vermicompost. Phyton 53:229–235

Sahni S, Sarma BK, Singh DP, Singh HB, Singh KP (2008) Vermicompost enhances performance of plant growth-promoting rhizobacteria in *Cicer arietinumrhizosphere* against *Sclerotium rolfsii*. Crop Protect 27:369–376

Serfoji P, Rajeshkumar S, Selvaraj T (2010) Management of root-knot nematode, Meloidogyne incognita on tomato cv Pusa Ruby by using vermicompost, AM fungus, *Glomus aggregatum* and mycorrhiza helper *bacterium, Bacillus* coagulans. J Agric Technol 6:37–45

Singh UP, Maurya S, Singh DP (2003) Antifungal activity and induced resistance in pea by aqueous extract of vermicompost and for control of powdery mildew of pea and balsam. J Plant Dis Protect 110:544–553

Singh R, Sharma RR, Tyagi SK (2007) Pre-harvest foliar application of calcium and boron influences physiological disorders, fruit yield and quality of strawberry (*Fragaria xananassa* Duch.). Sci Hortic 112:215–220

Singh R, Sharma RR, Kumar S, Gupta RK, Patil RT (2008) Vermicompost substitution influences growth, physiological disorders, fruit yield and quality of strawberry (*Fragaria ananassa* Duch.). BioresourTechnol 99:8507–8511

Singhai PK, Sarma BK, Srivastava JS (2011) Biological management of common scab of potato through *Pseudomonas* species and vermicompost Bio. Control 57:150–157

Souzaa MEP, Carvalhoa AMX, Deliberali DC, Juckscha I, Brownb GG, Mendonc ES, Cardosoa IM (2013) Vermicomposting with rock powder increases plant growth. Appl Soil Ecol 69:56–60

Taleshi K, Shokoh-far A, Rafiee M, Noormahamadi G, Sakinejhad T (2011) Effect of vermicompost and nitrogen levels on yield and yield component of safflower (*Carthamus tinctorius* L.) Under late season drought stress. Int J Agron Plant Prod 2(1):15–22

Theunissen J, Ndakidemi PA, Laubscher CP (2010) Potential of vermicompost produced from plant waste on the growth and nutrient status in vegetable production. Int J Phys Sci 5(13):1964–1973

Tomati U, Grappelli A, Galli E (1983) Fertility factors in earthworm humus. In: Proceeding of international symposium agric environ. Prospects in earthworm farming. Publication Ministerodella Ricerca Scientificae Technologia, Rome, pp 49–56

Tomati U, Grappelli A, Galli E (1988) The hormone like effect of earthworm casts on plant growth. Biol Fertil Soils 5:288–294

Uma B, Malathi M (2009) Vermicompost as a soil supplement to improve growth and yield of *Amaranthus species*. Res J Agric Biol Sci 5:1054–1060

Valdez-Péreza MA, Fernández-Luque F, Franco-Hernandezb O, Flores-Coteraa LB, Dendoovena L (2011) Cultivation of beans (*Phaseolus vulgaris* L.) in limed or unlimed wastewater sludge, vermicompost or inorganic amended soil. Sci Hortic 128:380–387

Vijaya KS, Seethalakshmi S (2011) Contribution of Parthenium vermicompost in altering growth, yield and quality of *Alelmoschus esculentus* (I) Moench. Advanced Biotech. 11(02):44–47

Wang D, Shi Q, Wang X, Wei M, Hu J, Liu J, Yang F (2010) Influence of cow manure vermicompost on the growth, metabolite contents, and antioxidant activities of Chinese cabbage (*Brassica campestriss sp. chinensis*) cow manure vermicompost. Biol Fertil Soils 46:689–696

Warman PR, AngLopez MJ (2010) Vermicompost derived from different feedstocks as a plant growth medium. Bioresour Technol 101:4479–4483

Wu Y, Zhang N, Wang J, Sun Z (2012) An integrated crop-vermiculture system for treating organic waste on fields. Eur J Soil Biol 51:8–14

Yardim EN, Arancon NQ, Edwards CA, Oliver TJ, Byrne RJ (2006) Suppression of tomato hornworm (*Manduca quinquemaculata*) and cucumber beetles (*Acalyma vittatum* and *Diabotrica undecimpunctata*) populations and damage by vermicomposts. Pedobiologia 50:23–29

Zaller JG (2006) Foliar spraying of vermicompost extracts: effects on fruit quality and indications of late-blight suppression of fieldgrown tomatoes. Biol Agric Hortic 24:165–180

Zaller JG (2007) Vermicompost as a substitute for peat in potting media: effects on germination, biomass allocation, yields and fruit quality of three tomato varieties. Sci Hortic 112:191–199

Zar JH (2009) Biostatistical analysis, 5th edn. Prentice Hall, Englewood Cliffs

5

Use of the co-composting time extract agar to evaluate the microbial community changes during the co-composting of activated sludge and date palm waste

Loubna El Fels[1] · **Yedir Ouhdouch**[2] · **Mohamed Hafidi**[1]

Abstract This study investigates changes in the microbiological community of two co-composting mixtures of activated sludge and date palm waste over a 6-month period. The use of the co-composting time extract medium to evaluate the abundance of cultivable indigenous microflora showed that the total microbial biomass was higher during the thermophilic phase. The mesophilic microflora was more abundant than the thermophilic bacteria, throughout the co-composting process. The proportion of mesophilic and thermophilic actinobacteria was high during the maturation phase especially for mixture A. However, thermotolerant fungal microflora increased during the thermophilic stage. Furthermore, correlating physico-chemical analysis of samples with bacterial diversity indicated that the bacterial communities underwent temperature changes. Enterococci and thermotolerant coliforms decreased significantly toward the end of the co-composting process which indicates the safety and sanitization of the end product.

Keywords Active sludge compost · Composting time extract agar · Indigenous microflora abundance and succession · Thermotolerant fungi · Coliforms

✉ Mohamed Hafidi
hafidi.ucam@gmail.com; hafidi@uca.ma

[1] Laboratory of Ecology and Environment (L2E) (Unit Associated with the CNRST, URAC32), Faculty of Science Semlalia, Cadi Ayyad University, BP: 2390, Marrakech, Morocco

[2] Laboratory of Biology and Biotechnology of Microorganisms, Faculty of Science Semlalia, Cadi Ayyad University, BP: 2390, Marrakech, Morocco

Introduction

It has been demonstrated that composts can be applied to agricultural fields as long-term fertiliser to improve soil structure (Jakobsen 1995; Barje et al. 2012). The evolution and the performance of the composting process are dependent on the microorganisms that populate composted substrates and are able to degrade the organic waste (Chroni et al. 2009). During the composting process, the microbial metabolic pathway influences the physico-chemical parameters, which in turn lead to changes of microbial community succession. In order to evaluate the process and the end product biotransformation, better knowledge of the functional microbial community dynamics is needed. Steger et al. (2007) showed that knowledge of the microorganism composition in composts, their community structure and the microbial succession should help to ensure high quality of the final compost product. It has been underlined that microbial activity during composting is mainly due to the mesophilic microbial community, such as (bacteria, fungi and actinobacteria) which play an important role in breaking down the organic compounds to humus-like material (Finstein and Morris 1975; Herrmann and Shann 1997; Ryckeboer et al. 2003).

The final composts should not contain pathogens or viable seeds and it should be stable and suitable for use as soil amendment (Epstein 1997; Tønner-Klank et al. 2007). Faecal matter is known to contain high numbers of naturally occurring enteric bacteria, and occasionally, disease-causing pathogens like bacteria, enteric viruses and parasites, which requires that the hygienic quality of treated organic faecal waste be ensured.

Sludge and date palm co-composting is an unexplored environment. In this context, the succession of indigenous microflora from this habitat has not been investigated. Current knowledge of the compost microbial community is based on different approaches such as direct analysis of phospholipid fatty acid patterns (PLFA) (Klammer and Baath 1998) or molecular techniques (Riddech et al. 2002). Various standard growth media are used for monitoring the cultivable microflora during composting such as malt extract agar (Wei et al. 2007). However, co-composting time extract agar was used as a selective media of only functional microflora able to degrade composted substrates (El Fels et al. 2014a). This plays a significant role to ensure that the microbial community composition is able to grow and transform the substrate available at given composting times, pH and temperatures.

The main objective of this study was to investigate the use of the co-composting time extract medium to evaluate the abundance and to assess the microbial community function in sewage sludge/date palm co-composting.

Materials and methods

Co-composting trials

Co-composting trials were conducted on a composting platform over a period of 6 months. Two trials with different proportions of mixed sewage sludge/palm waste were run:

- Mixture A: 1/3 sludge + date palm tree waste 2/3, total volume: 4 m^3
- Mixture B: 1/2 sludge + date palm tree waste 1/2, total volume: 4 m^3.

Each mixture was carefully homogenized, moisture was adjusted to 60 % (optimal value for composting) and then the mixtures were windrowed. Windrows were turned over by hand weekly to aerate the mixture. Homogeneous samples were taken at T_0 (first day of co-composting) and after each airing (aerating the mixture). Homogeneous samples (1 kg) were obtained by careful mixing of several sub samples taken at different points (height and length) of the windrow and quartering. The samples were kept at −20 °C before analysis.

Physico-chemical analysis

The temperature was measured every day at different levels (height and length) of the windrow using sensors with data memory (PH0700115 model 1.20, Ector Traceability software, ECTOR France). The samples were dried out at 105 °C. The pH was measured in an aqueous extract of the compost at room temperature (1 g/10 ml of distilled water) according to NF ISO 10390. Total organic carbon and Ash content were calculated after calcinations in a muffle furnace at 600 °C for 6 h. Total Kjeldahl nitrogen (TKN) was assayed in 0.5 g samples using classical Kjeldahl procedure and by steam distillation according to AFNOR T90-1110 standard. The physico-chemical parameters were analysed as mentioned in our previous study (El Fels et al. 2014a, b).

Biological analysis

Enumeration of cultivable indigenous microflora

Samples of each co-composting time (0, 15, 22, 30, 60, 90 and 180 days) were first mixed, suspended in sterile distilled water (10 g in 100 ml) homogenized by vortexing and finally treated 10 to 15 min by sonication according to Ouhdouch et al. (2001); El Fels et al. (2014a) All treated samples were serially diluted up to 10^{-6} and functional microflora enumerated by plating and spreading 0.1 ml of 10^{-4}, 10^{-5} and 10^{-6} dilutions over the surface of the co-composting time extracts agar (CTEA) prepared according to El Fels et al. (2014a) as follows: One litre of distilled water and 35 g of co-composting time sample were mixed overnight. Agar (15 g) was added to the filtrate collected after it had been filtered and sterilized at 120 °C for 15 min. The pH was adjusted to that of the co-composting time before sterilization. For each co-composting time, the plates were incubated at 28 °C for mesophilic and 45 °C for thermophilic microflora enumeration.

For actinobacteria enumeration, the media were supplemented with 40 μg/ml of actidione to inhibit the development of fungi (Olson 1968) and 10 μg/ml of nalidixic acid to inhibit the Gram-negative bacteria capable of swarming without affecting the growth of actinobacteria (Bulina et al. 1997; Barakate et al. 2002). For fungus evaluation, the media were supplemented with 5 μg/ml of chloramphenicol to inhibit the development of bacteria. For other microorganisms, the media were supplemented with 40 μg/ml of actidione to inhibit the development of fungi. After 7 days of incubation, the actinobacteria were recognized on the basis of morphological and microscopic features following directions given by International *Streptomyces* Project (ISP) (Shirling and Gottlieb 1966).

For the initial and final co-composting time, total coliform and thermotolerant coliform counts were determined on desoxycholate (0.1 %) lactose agar (Biokar diagnostics) after incubation for 24 h at 37 and 44 °C, respectively. Microbial analyses were performed in triplicate.

Statistical analysis

The results are presented in the form of averages ± SEM (standard error of the mean). Comparison of the averages was made by ANOVA. The differences are considered significant at $p < 5$ %.

Principal components analysis (PCA) was applied to the correlation matrix between microflora succession and physical–chemical variables. The statistical treatments were carried out using the software SPSS win version 10.

Results and discussion

Physical–chemical characterisation

The species composition changed markedly as composting progressed, and particular correlations could be drawn between the biological and physico-chemical states of the material (El Fels et al. 2014a, b). In the present example (Fig. 1), the temperature indicates a typical co-composting pattern characterised by two major phases: A thermophilic (stabilization) phase is characterised by a rise in temperature, which peaked at 65 °C. This is the result of intense microbial activity resulting from the degradation of the simple molecules present in the substrate (Barje et al. 2012; El Fels et al. 2014a, b). The maturation phase is characterised by a decrease of temperature due to the exhaustion of easily metabolisable organic compounds from the medium, with compounds that are only resistant to degradation (e.g. lignin, cellulose) (Amir et al. 2005).

The principal characteristics of sludge using of composting are C/N around 20.8, moisture content 46.46 ± 0.26 and TKN 1.5 ± 0.2.

The mean features of co-composting products are presented in Table 1; the final products presented a C/N ratio of around 10, $NH_4^+/NO_3^- \leq 1$, which provides information on the importance of organic matter oxidation and a significant degree of stability of substrate. As indicated in the literature these values confirm the maturity of the compost (Barje et al. 2012; El Fels et al. 2014b).

Coliforms versus time of co-composting process

Most published work has used enterococci and thermotolerant coliforms as faecal contamination indicator organisms (Redlinger et al. 2001). Their presence indicates poor hygiene. During co-composting, we noted a significant decrease of the order of 90 % toward the end of co-composting (Table 1). For compost hygiene, the recommended Faecal Coliform (FC) and Faecal Streptococcus (FS) densities are of 5×10^2 bacteria/g and 5×10^3 bacteria/g (Vuorinen and Saharinen 1997; Bhatia et al. 2013), respectively.

The removal of coliforms is presumably linked to the physical co-composting parameters, especially the high temperature and unfavourable growth conditions established during the thermophilic phase. Several authors (Gaspard and Schwartzbrod 2003; Cofie et al. 2006; Koné et al. 2007; El Fels et al. 2014b) showed that the rise of temperature during the thermopilic stage leads to hygeinization. Furthermore, Redlinger et al. (2001); Pietronave et al. (2004) and Bhatia et al. (2013) noted that a significant decrease of faecal coliforms and faecal Streptococci during composting. Paniel et al. (2010) demonstrated that faecal *Enterococcus* was not observed during the stabilization phase. Palmisano and Barlaz (1996); Lung et al. (2001); Tønner-Klank et al. (2007) showed that proper composting can kill many human pathogens in different types of compost, and the temperatures achieved during the treatment (50–70 °C) are able to kill the enteric pathogens. This stage of the process is defined as disinfection or sanitization.

In spite of the abundance of faecal coliforms in sewage sludge, the results indicate the good hygienic quality of the end products of co-composted sewage sludge and date palm waste.

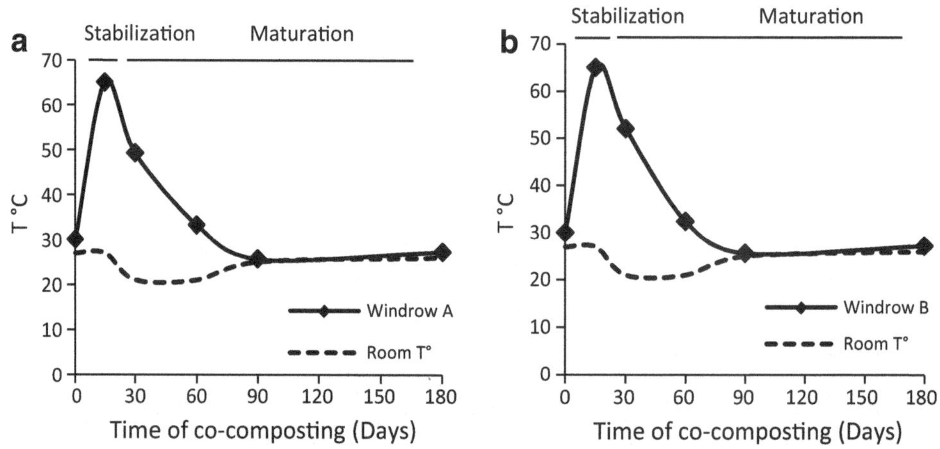

Fig. 1 Temperature versus time during the co-composting process for two mixtures A (**a**) and B (**b**), according to El Fels et al. (2014b)

Table 1 Physico-chemical and biological parameters during co-composting for mixtures A and B

Mixture	Time of co-composting (day)	%TKN[a]	Moisture (%)	NH_4^+/NO_3^-	C/N	DEC (%)[a]	pH	FC UFC/g	TC UFC/g
A	T_0	1.31 ± 1.2	58.83	13.75	26.2	–	6.34 ± 0.03	12×10^4	345×10^4
	T_{180}	2.18 ± 1.1	66	0.12	10.09	40.07 ± 1.1	6.79 ± 0.06	≤85	≤125
B	T_0	1.28 ± 1.09	60.97	15.6	27.4	–	6.04 ± 0.28	2×10^4	456×10^4
	T_{180}	2.28 ± 1.21	66	0.14	10.08	40 ± 1.49	7.03 ± 0.08	≤90	≤175

TKN total Kjeldahl nitrogen, *DEC* decomposition rate, *FC* faecal coliforms, *TC* total coliforms

[a] Results expressed per unit weight dry matter

Indigenous microflora abundances during co-composting

In this study, both mesophilic and thermophilic microflora evolution showed a peak during the thermophilic stage for mixtures A and B. The total mesophilic microflora increased from 43×10^8 to 137×10^8 CFU/g on day 22 (thermophilic phase), and from 63×10^8 to 294×10^8 CFU/g (Fig. 2); the evolution of thermophilic microflora in CTEA medium varied in the range from 5.4×10^8 to 40×10^8 CFU/g, and from 7.7×10^8 to 114×10^8 CFU/g for mixtures A and B, respectively (Fig. 2) (El Fels et al. 2014a).

These high values during the thermophilic phase could result from the high levels of mesophilic microflora and especially that of thermotolerant microflora in the co-composted substrates (El Fels et al. 2014a). However, the mesophilic microflora were more abundant than thermophilic microflora. It was also noted that mixture B microflora abundance was greater than that of mixture A. These results could be related to the higher proportion of sewage sludge in mixture B and resulting abundance of mesophilic microflora rate (El Fels et al. 2014a). Thereafter all mesophilic and thermophilic microflora abundances declined at maturation stage. Sneath (1986), demonstrating that temperatures from 26–48 °C lead to favourable conditions for the growth of actinomycetes and of other heat-tolerant bacteria. Hassen et al. (2001); Marshall et al. (2004); Bhatia et al. (2013), showed that the initial phase of the composting process is characterised by the intense activity and growth of mesophilic microflora which reach abundances two to four orders of magnitude higher than the thermophilic group. Chroni et al. (2009) demonstrated that relatively low temperatures (around 50 °C) seem to facilitate the survival of total mesophilic bacteria.

At the beginning of co-composting, the mesophilic microflora start vigorous oxidation of easily biodegradable compounds. This intense microbial activity leads to a rapid increase in temperature which improves the development and proliferation of total thermotolerant followed by thermophilic microflora. This group becomes responsible for the degradation process during the stabilization phase. Total microflora abundances declined after only the first month of co-composting. This could be related to the restricted conditions, particularly the loss of easily metabolisable substrates. The decrease of temperature affects thermophilic microflora development and promotes the relocation of the new mesophilic microflora. This result

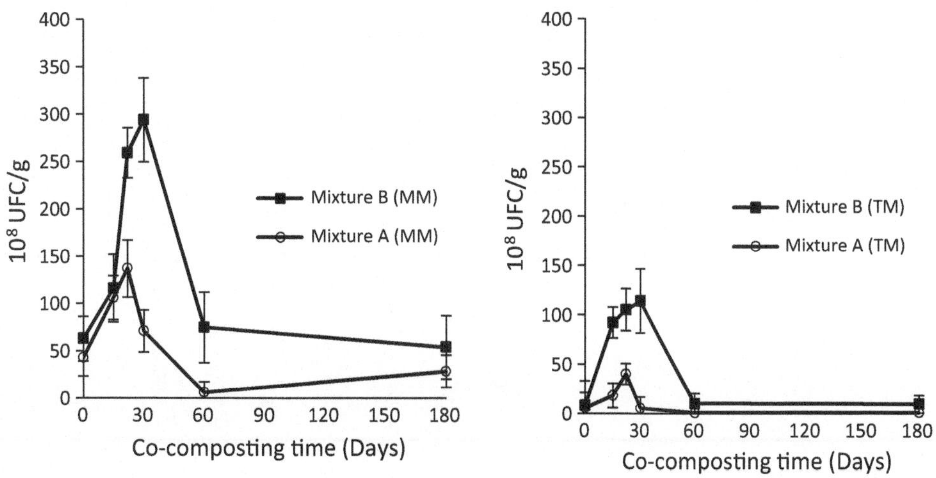

Fig. 2 Mesophilic and thermophilic microflora evolution during co-composting of mixture A and B on CTEA medium, according to El Fels et al. (2014a). *TM* thermophilic microflora, *MM* mesophilic microflora, *CTEA* co-composting time extract agar

is in line with data reported by Tønner-Klank et al. (2007). Chroni et al. (2009) showed that most of the microbial groups examined exhibited their highest counts towards the end of the thermophilic phase.

Many factors such as oxygen content, moisture, composition of the raw composted substrate, pH and temperature affect the succession of microbial communities during composting. Paul and Clark (1996), showed that the rapid changes of physicochemical conditions in composting processes are likely to select for a succession of different microbial communities. It could be expected that temperature and the availability of substrates are the main factors. Sidhu et al. (2000) showed that the overall population of bacteria declined with compost maturity.

A difference was noted between mixtures A and B. The thermophilic phase determined by the indigenous microflora was limited to 22 days for mixture A but reached 30 days for mixture B (El Fels et al. 2014a). In this study, the actinobacteria, which proliferate during the maturation phase, show a significant development at 60 and 90 days, for mixtures A and B, respectively (Fig. 3a, b). This observation is confirmed by a significant correlation between actinobacteria and maturation phase according to PCA (Fig. 4a, b). This difference could be attributed to the higher concentration of lignocellulosic substrate in mixture A. Filippi et al. (2002) identified that some products cause a temporary lag-phase in the decomposition of the organic fraction during composting. Adani et al. (2001) showed that the oxygen level, organic matter stabilization and moisture limitation generate a biological activity lag-phase.

Indigenous microflora dynamics allowed a distinction between two composting main phases, thermophilic and maturation, which characterise of the progression of the composting process. This suggests that the indigenous microflora play a significant role in the degradation of matter and in the maturity of the co-composting process of sewage sludge and date palm.

Enumeration of mesophilic and thermophilic indigenous microflora groups

Three main aerobic microorganisms were monitored in the microbiological analysis of co-composting stages: fungi, actinobacteria and other bacteria. These microorganisms belong to the dominating taxa and functional groups present during co-composting. Finstein and Morris (1975); De Bertoldi et al. (1983) and Marshall et al. (2004) showed that microbiological activity as composting advances can be attributed to different groups of microorganisms such as the total aerobic mesophilic microflora (TAM).

Functional mesophilic and thermophilic actinobacteria show a peak of about 80 and 70 % for the mesophilc, and

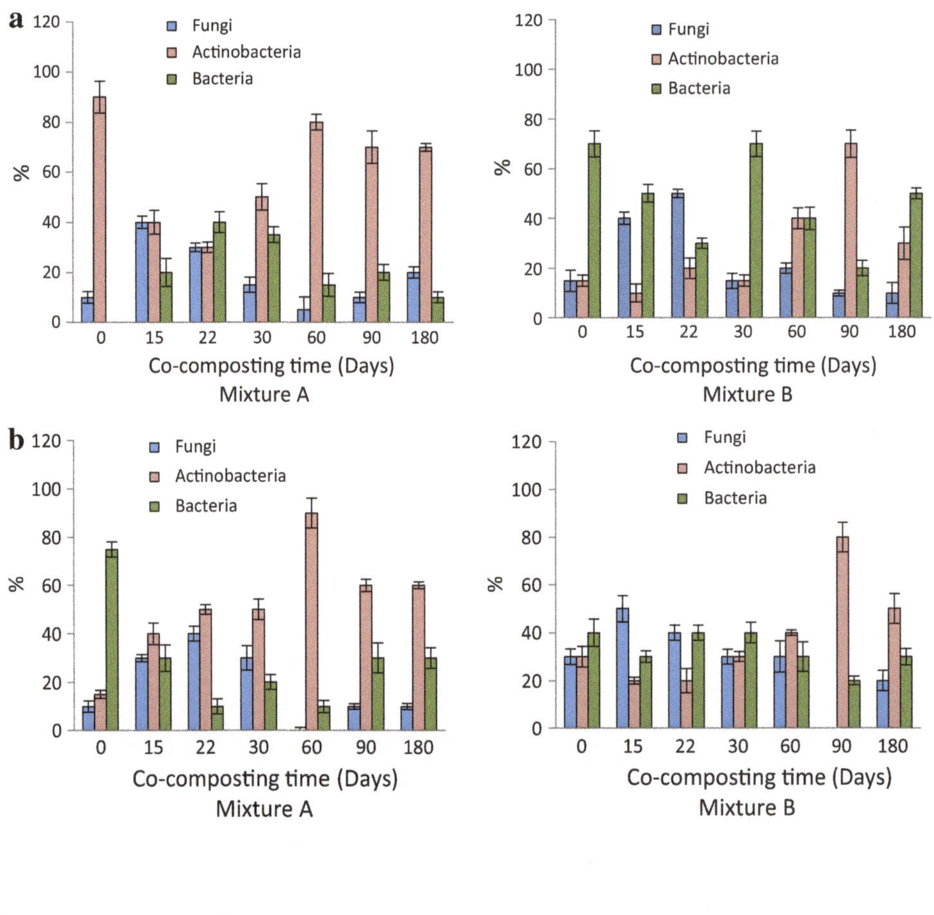

Fig. 3 **a** Mesophilic indigenous microflora groups during co-composting process Fungi = Moulds and yeasts. **b** Thermophilic microflora groups versus time of co-composting Fungi = Moulds and yeasts

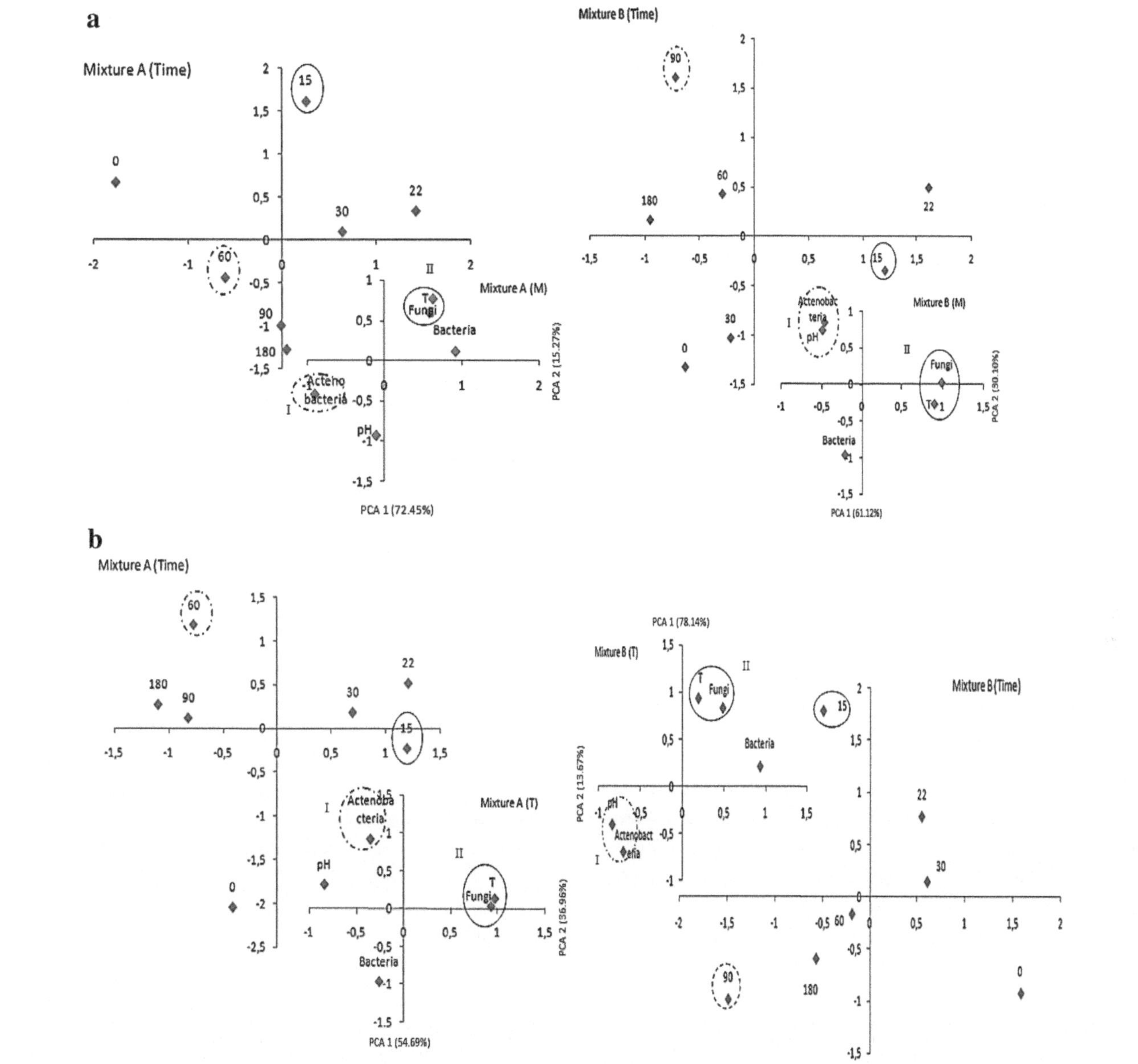

Fig. 4 **a** Principal components analysis (PCA) between mesophilic microflora evolution and physico-chemical parameters versus time for mixtures A and B. *M* mesophilic. **b** Principal components analysis (PCA) between thermophilic microflora and physicochemical parameters versus time for mixtures A and B. *T* thermophilic

90 and 80 % for thermophilic microflora, respectively, for mixtures A and B during the maturation phase of co-composting (Fig. 3a, b). Partanen et al. (2010); Kurola et al. (2011) demonstrated that high proportions of actinobacteria indicate good composting conditions.

The results of the (PCA) run between the all identified mesophilic and thermophic microflora profile and physico-chemical parameters, during co-composting, show that the actinobacteria (domain I) are inversely correlated with temperature and highly correlated with pH (case of mixture B) (Fig. 4a, b), which can explain their high evolution, especially during the maturation phase when pH increases and temperature decreases significantly. However, some actinobacteria isolates in this study were thermotolerant (results not shown). Partanen et al. (2010) and Sundberg et al. (2011) showed a correlation between a rising pH with an increase in actinobacteria. However, Goodfellow and Williams (1983) reported that various actinobacteria are involved in all compost stages which demonstrates a wide temperature range for their growth, often with an optimum between 25 and 30 °C for mesophiles and 45–55 °C for thermophiles.

As reported by Tuomela et al. (2000), the enzymatic capacity of actinobacteria to attack recalcitrant molecules explains their activity and their proliferation during the maturation stage. This could explain the increase of actinobacteria groups during the co-composting of lignocellulose-based waste (Fig. 3a, b). However, towards the end of the process the actinobacteria group shows a drop in mixture B and slightly decreases in mixture A. This could be related to the richness of lignocellulose substrate in mixture A. Pérez et al. (2002) and Nakasaki et al. (2009) showed that actinobacteria produce an active enzyme to degrade lignocellulotic compounds. Tuomela et al. (2000); Steger et al. (2007) and Xiao et al. (2011) have shown that actinomycetes are activated later and contribute to the degradation of recalcitrant compounds and the formation of the stable compounds of "humic substance".

The fungal microflora increases during the thermophilic stage of co-composting. They reached 40 and 50 % for the mesophilic and thermophilic microflora of mixtures A and B, respectively (Fig. 3a, b). The PCA shows that fungi, temperature and day 15 (when temperature is around 65 °C) (Fig. 4a, b) are highly correlated for both mesophilic and thermophilic microflora. This can explain that the fungal microflora of co-composted substrates is especially the thermotolerant type.

During the intermediate stage the fungal microflora decreased, and this is due to the decrease of temperature. Nevertheless, toward the end a reinstatement of fungal microflora was noted. Chroni et al. (2009) explained that the thermophilic fungi exhibited a small increase as temperature rose in the first month of composting, stabilizing thereafter. Steger et al. (2007), showed that fungi and actinobacteria commonly proliferate during periods with comparatively low temperatures (<45 °C). Several authors show that the final phase, which includes both cooling-down and maturation stages, is characterised by the development of a new mesophilic fungal community (Finstein and Morris 1975; Ishii et al. 2000). Ait Baddi et al. (2004) reported that during the maturation phase, the biodegradation is limited to complex compounds such as high-molecular-weight lignocelluloses. Mouchacca (1997), identified a fungus which has been studied in detail owing to its capacity to produce cellulase and the enzyme able to degrade the lignin in the range of 36–40 °C with a maximum for 40–46 °C. The fungi with their lignolytic enzymatic capacity can release small aromatic subunits by depolymerization as shown by Kirk and Farrell (1987), which can explain their development on lignocellulosic substrates at the end of co-composting and their contribution to degradation of substance and stabilization of the end products.

The change in the mesophilic microbial population of co-composting, generally, bacteria and yeasts, showed a rise–fall–rise trend; this pattern is due to the nature and life styles of these microorganisms determined by physicochemical conditions. Tuomela et al. (2000) showed that during the co-composting process, the microbial community must adapt to gradually reduced nutrients, oxygen and water with compost maturity. Principal components analysis (PCA) shows that bacteria were never correlated with temperature (Fig. 4a, b).

Bhatia et al. (2013) demonstrated that mesophilic microflora are critical for the degradation of organic substrates which in turn leads to a rapid increase in temperature. The easily metabolisable substrate exhaustion and impoverishment lead to cooling of the windrow; this new condition encourages the installation of the second co-composting mesophilic microflora (bacteria-yeast; fungus; and actinomycetes) (Fig. 3a). Chroni et al. (2009) show that high numbers of proteolytic bacteria can be expected to appear after the end of the thermophilic stage, as a result of a large amount of proteins that are released when microorganisms die.

The various indigenous microflora such as bacteria, actinobacteria and fungi in composted sewage sludge and date palm waste have the ability to breakdown organic matter. On the other hand, the decrease of temperature and nutrient limitation with compost maturity causes a significant change in the indigenous microflora. Furthermore the temperature is one of the key indicators of co-composting and determines the rate at which many of the biological processes take place and plays a selective role on the succession of the microbiological communities.

Conclusion

After 6 months of co-composting, the results concluded as to the growth of significant functional microbial communities, by using the CTEA medium adjusted to composting time and recorded pH. Three cultivable and functional indigenous groups were enumerated: actinobacteria, bacteria and fungi. The functional actinobacteria group presented a peak during the maturation stage, especially for mixture A. The bacterial diversity revealed in the co-composting process underwent the changes linked to variations in temperature and availability of new metabolic substrates. In addition, the co-composted substrates are characterised by the abundance of thermotolerant fungi.

Enterococci and the thermotolerant coliforms decreased significantly during the co-composting process which indicates the safety and sanitization of the end product.

The results presented in this study give new information on how the development of the compost microbial community structure is linked to the co-composting time.

References

Adani F, Lozzi P, Genevini P (2001) Determination of biological stability by oxygen uptake on municipal solid waste and derived products. Compost Sci Util 9:163–178

Ait Baddi G, Alburquerque JA, Gonzálvez J, Cegarra J, Hafidi M (2004) Chemical and spectroscopic analyses of organic matter transformations during composting of olive mill wastes. Int Biodeterior Biodegrad 54:39–44

Amir S, Hafidi M, Merlina G, Hamdi H, Jouraiphy A, El Gharous M, Revel JC (2005) Fate of polycyclic aromatic hydrocarbons during composting of lagooning sewage sludges. Chemosphere 58:449–458

Barakate M, Ouhdouch Y, Oufdou K, Beaulieu C (2002) Characterization of rhizospheric soil streptomycetes from Moroccan habitats and their antimicrobial activities. World J of Microbiol Biotechnol 18:49–54

Barje F, El Fels L, El Hajjouji H, Amir S, Winterton P, Hafidi M (2012) Molecular behaviour of humic acid-like substances during co-composting of olive mill waste and the organic part of municipal solid waste. Int Biodeterior Biodegrad 74:17–23

Bhatia A, Madan S, Sahoo J, Ali M, Pathania R, Kazmi AA (2013) Diversity of bacterial isolates during full scale rotary drum composting. Waste Manag 33:1595–1601

Bulina TI, Alferova IV, Terekhova LP (1997) A novel approach to isolation of actinomycetes involving irradiation of soil samples with microwaves. Microbiology 66:231–234

Chroni C, Kyriacou A, Georgaki I, Manios T, Kotsou M, Lasaridi K (2009) Microbial characterization during composting of biowaste. Waste Manag 29:1520–1525

Cofie OO, Agbottah S, Strauss M, Esseku H, Montangero A, Awuah E, Kone D (2006) Solid–liquid separation of faecal sludge using drying beds in Ghana: implications for nutrient recycling in urban agriculture. Water Res 40:75–82

De Bertoldi M, Vallini G, Pera A (1983) The biology of composting: a review. Waste Manag Res 1:167–175

El Fels L, El Ouaqoudi F-Z, Barje F, Hafidi M, Ouhdouch Y (2014a) Two culture approaches used to determine the co-composting stages by assess of the total microflora changes during sewage sludge and date palm waste co-composting. J Environ Health Sci Eng 12:132

El Fels L, Zamama M, El Asli A, Hafidi M (2014b) Assessment of biotransformation of organic matter during co-composting of sewage sludge-lignocelullosic waste by chemical FTIR analyses and phytotoxicity tests. Int Biodeterior Biodegrad 87:128–137

Epstein E (1997) The science of composting Lancaster. Technomic Publishing Company, Lancaster

Filippi C, Bedini S, Levi-Minzi R, Cardelli R, Saviozzi A (2002) Cocomposting of olive oil mill by-products: chemical and microbiological evaluations. Compos Sci Util 10(1):63–71

Finstein MS, Morris ML (1975) Microbiology of municipal solid waste composting. Adv Appl Microbiol 19:113–151

Gaspard PG, Schwartzbrod J (2003) Parasite contamination (helminth eggs) in sludge treatment plants: definition of a sampling strategy. Int J Hygiene Environ Health 206:117–122

Goodfellow M, Williams ST (1983) Ecology of actinomycetes. Annu Rev Microbiol 37:189–216

Hassen A, Belguith K, Jedidi N, Cherif A, Cherif M, Boudabous A (2001) Microbial characterization during composting of municipal solid waste. Bioresour Technol 80:217–225

Herrmann RF, Shann JF (1997) Microbial community changes duringthe composting of municipal solid waste. Microb Ecol 33:78–85

Ishii K, Fukui M, Takii S (2000) Microbial succession during a composting process as evaluated by denaturing gradient gel electrophoresis analysis. J Appl Microbiol 89:768–777

Jakobsen ST (1995) Aerobic decomposition of organic wastes 2. Value of compost as fertilizer. Resour Conserv Recycl 13:57–71

Kirk TK, Farrell RL (1987) Enzymatic "combustion": the microbial degradation of lignin. Annu Rev Microbiol 41:465–505

Klammer M, Baath E (1998) Microbial community dynamics during composting of straw material studied using phospholipid fatty acid analysis. FEMS Microbiol Ecol 27:9–20

Koné D, Cofie O, Zurbrugg C, Gallizzi K, Moser D, Drescher S, Strauss M (2007) Helminth eggs inactivation efficiency by faecal sludge dewatering and co-composting in tropical climates. Water Res 41:4397–4402

Kurola JM, Arnold M, Kontro MH, Talves M, Romantschuk M (2011) Wood ash for application in municipal biowaste composting. Bioresour Technol 102:5214–5220

Lung AJ, Lim CM, Kim JM, Marshall MR, Nordstedt R, Thompson NP, Wei CI (2001) Destruction of *Escherichia coli* O157:h7 and *Salmonella enteritidis* in cow manure composting. J Food Prot 64:1309–1314

Marshall MN, Reddy AP, Vander Gheynst JS (2004) Microbial ecology of compost. In: Lens P, Hamelers B, Hoitink H, Bidlingmaier W (eds), Resource Recovery and Reuse in Organic Solid Waste Management. IWA Publishing, London, pp 193–224

Mouchacca J (1997) Thermophilic fungi: biodiversity and taxonomic status. Cryptogamie Mycol 18:19–69

Nakasaki K, Tran LTH, Idemoto Y, Abe M, Rollon AP (2009) Comparison of organic matter degradation and microbial community during thermophilic composting of two different types of anaerobic sludge. Bioresour Technol 100:676–682

Olson EH (1968) Actinomycetes isolation agar. In: Difco: supplementary literature. Difco Lab Detroit, Michigan

Ouhdouch Y, Barakate M, Finance C (2001) Actinomycetes from Maroccan habitats: screening for antifungal activities. Eur J Soil Biol 37:1–6

Palmisano AC, Barlaz MA (1996) Microbiology of solid waste. CRC Press, Boca Raton

Paniel N, Rousseaux S, Gourland P, Poitrenaud M, Guzzo J (2010) Assessment of survival of *Listeria monocytogenes*, *Salmonella infantis* and *Enterococcus faecalis* artificially inoculated into experimental waste or compost. J Appl Microbiol 108:1797–1809

Partanen P, Hultman J, Paulin L, Auvinen P, Romantschuk M (2010) Bacterial diversity at different stages of the composting process. BMC Microbiol 10:94

Paul EA, Clark FE (1996) In: 2nd edn. Soil microbiology and biochemistry. Academic Press, San Diego

Pietronave S, Fracchia L, Rinaldi M, Martinotti MG (2004) Influence of biotic and abiotic factors on human pathogens in a finished compost. Water Res 38:1963–1970

Pérez J, Munoz-Dorado J, de la Rubia TDLR, Martinez J (2002) Biodegradation and biological treatments of cellulose, hemicellulose and lignin: an overview. Int Microbiol 5(2):53–63

Redlinger T, Graham J, Corella-Barud V, Avitia R (2001) Survival of faecal coliforms in dry-composting toilets. Appl Environ Microbiol 67:4036–4040

Riddech N, Klammer S, Insam H (2002) Characterisation of microbial communities during composting of organic wastes. Springer, Berlin, Heidelberg, pp 43–51

Ryckeboer J, Mergaert J, Coosemans J, Deprins K, Swings J (2003) Microbial aspects of biowaste during composting in monitored compost bin. J Appl Microbiol 94:127–137

Shirling EB, Gottlieb D (1966) Methods for characterization of *Streptomyces* species. Int J Syst Bacteriol 16:313–340

Sidhu J, Gibbs RA, Ho GE, Unkovich I (2000) The role of indigenous microorganisms in suppression of Salmonellare-growth in composted biosolids. Water Res 35:913–920

Sneath PHA, (1986) Endospore-forming gram-positive rods and cocci. In: Bergey's Manual of Determinative Bacteriology, vol 2. William & Wilkins, USA, pp 1104–1207

Steger K, Sjogren AM, Jarvis A, Jansson JK, Sundh I (2007) Development of compost maturity and Actinobacteria populations during full-scale composting of organic household waste. J Appl Microbiol 103:487–498

Sundberg C, Franke-Whittle I, Kauppi S, Yu D, Romantschuk M, Insam H, Jönsson H (2011) Characterisation of source-sorted household waste intended for composting. Bioresour Technol 102:2859–2867

Tønner-Klank L, Møller J, Forslund A, Dalsgaard A (2007) Microbiological assessments of compost toilets: in situ measurements and laboratory studies on the survival of faecal microbial indicators using sentinel chambers. Waste Manag 27:1144–1154

Tuomela M, Vikman M, Hatakka A, Itavaara M (2000) Biodegradation of lignin in a compost environment: a review. Bioresour Technol 72:169–183

Vuorinen AH, Saharinen MH (1997) Evolution of microbiological and chemical parameters during manure and straw co-composting in a drum composting system. J Agric Ecosyst Environ 66:19–29

Wei Z, Xi B, Zhao Y, Wang S, Liu H, Jiang Y (2007) Effect of inoculating microbes in municipal solid waste composting on characteristics of humic acid. Chemosphere 68:368–374

Xiao Y, Zeng GM, Yang ZH, Ma YH, Huang C, Xu ZY, Huang J, Fan CZ (2011) Change in actinomycetal communities during continuous thermophilic composting as revealed by denaturing gradient gel electrophoresis. Bioresour Technol 102:1383–1388

6

Effect of gamma-irradiated textile effluent on plant growth

Fahmida Parvin · Zannatul Ferdaus · Shafi M. Tareq · Tasrina R. Choudhury · Jahid M. M. Islam · Mubarak A. Khan

Abstract

Introduction Textile industry produces large quantities of highly colored effluents, which are generally toxic and resistant to biodegradation. The present study was carried out to find out the efficiency of using gamma radiation to degrade or decontaminate combined textile effluent and its potential application as irrigation water on spleen amaranth.

Results The change of absorption spectra, physicochemical parameters of gamma-irradiated textile effluent revealed that all these properties have improved after irradiation at (10–15) kGy. Analysis of ammonium and total nitrogen concentration of the effluent also showed that both of them are increased after irradiation and can be used for irrigation purposes. Application of irradiated textile effluents on spleen amaranth have increased dry mass (10.77 g), plant height (10.53/week), root length (19.00 cm), number of leaves (6/week) compared to the plants which were nourished with raw textile effluent and only water. Elemental analysis showed trace amount of heavy metal uptake by the plant. Moreover, the plants contain mineral nutrients essential for plant growth.

Conclusions The treatment of textile effluents by gamma irradiation (10–15 kGy) is a promising tool as it produces zero waste and has the possibility of alternative use for irrigation water with fertilizing properties.

Keywords Textile effluent · Gamma radiation · Plant · Irrigation water · Fertilizing effect

F. Parvin (✉) · Z. Ferdaus
Department of Environmental Sciences, Jahangirnagar University, Savar, Dhaka, Bangladesh
e-mail: fahmidaprvn497@gmail.com

S. M. Tareq
School of Biosciences, The University of Nottingham, Jalan Broga, 43500 Semenyih, Selangor Darul Ehsan, Malaysia

T. R. Choudhury · J. M. M. Islam · M. A. Khan (✉)
Institute of Radiation and Polymer Technology, Bangladesh Atomic Energy Commission, Dhaka, Bangladesh
e-mail: makhan.inst@gmail.com

Introduction

Textile effluents contain substantial pollution loads in terms of high alkalinity, high biological oxygen demand (BOD), high chemical oxygen demand (COD) and high suspended solids (SS). It also contains dye particle at a high concentration (Selambakkannu et al. 2011) and most of these are non-biodegradable. Existing conventional treatment using physical and chemical treatment is not capable of decomposing the recalcitrant and toxic organic pollutants rather it transfers the pollutants from a liquid medium to a solid medium via coagulation/chemical process (Rabby 2011). A number of research have been going on adsorption processes (Panda et al. 2009; Islam et al. 2013), but those are costly, time consuming and cannot produce zero waste. Biological methods are simple in applications, generally cheap and often used to remove organics and color of textile effluent. However, dye containing effluent cannot be readily degraded by conventional biological processes, e.g., activated sludge process, because the structure of most commercial dye compounds are generally complex and many dyes are non-biodegradable due to the chemical nature and molecular size (Kim et al. 2002). And this process is also time consuming. So it is required to look for a most effective, less time consuming, zero waste producing wastewater treatment technology.

Ionizing radiation (known as electron beam and gamma radiation) is promising for the treatment of textile dye solution because the effect of radiation can be intensified in aqueous solution, in which the dye molecules may be degraded effectively by the primary products formed through radiolysis of water (Dahlan 2001; Wang et al. 2006; Takács et al. 2007; Selambakkannu et al. 2011). However, to the best of our knowledge, radiation-induced decoloration and degradation of the true textile mill effluent has not performed yet.

Industrial textile effluent contains azo dyes. So after irradiation it may convert into amide, which can be converted into ammonium by hydrolysis (Bagyo et al. 1997). So, this decontaminated wastewater containing nitrogen compound could be used for irrigation. Spleen amaranth is rejuvenator (rasayana) and useful in the treatment of fevers, cough, rheumatism, paralysis, tremors, kapha and vata disorders (Grubben and Denton 2004). Because of this importance, spleen amaranth was chosen to evaluate the fertilizing effect of irradiated textile effluent. Keeping in view the above-cited facts, the present study was carried out to find out the efficiency of using gamma radiation to degrade or decontaminate combined textile effluent and its potential application as irrigation water on spleen amaranth.

Experimental

Sampling and irradiation

Effluents used in this research were obtained from a textile industry which is located at Ganda, Savar, Bangladesh. The sample was a mixture of wastewater produced from several processes such as washing, dying, waxing and rinsing. The samples were irradiated by gamma rays from the cobalt-60 gamma radiation source. The dose was determined by Amber Perpex dosimeter; type 3042F. In this study 10–25 kGy dose have been chosen for irradiation as lower doses were not sufficient for complete mineralization of textile effluent (Takács et al. 2007).

Physicochemical analysis

Samples were analyzed before and after irradiation. pH, EC, TDS and DO of both irradiated and raw samples were measured using portable multiparameter (Model no-sension^{TM} 156, Supplier-Hach Company USA, 2000) within half an hour of the collection of sample. The turbidity of the samples is also measured by turbidity meter HANNA H193703 microprocessor turbidity meter, range (0.00–1000) FTU (Formazin Turbidity Unit). The BOD of the samples was determined by 5-day BOD test at 20 °C temperature of over an elapsed period of 5 days in the dark. The COD of the samples was determined by closed reflux, titrimetric method. The concentration of nitrogen and ammonium was determined by the Kjeldahl and Kjeldahl distillation methods, respectively (Mulvaney 1996). UV–Visible spectra of irradiated and raw textile effluent were obtained by T-60 UV–visible spectrophotometer using quartz cell (size 10 × 10 mm) in UV-win 5 software.

Analysis of fertilizing effects

The irradiated and raw textile effluents were irrigated into spleen amaranth to observe the fertility effect of gamma ray-irradiated textile effluents using pot experiments. Six pots of spleen amaranth (three replications) have prepared by adding garden soil with the appropriate moisture content where they were successively nourished by water, raw and the irradiated (10–25) kGy textile effluent. In every week, the plants were monitored prior to measuring the plant height, maximum leaf area, number of leaves to compare the effect of irradiated textile effluent with the raw and control one.

The plant heights were measured from the border of the pot to the top of the main plant stem. It was not measured from the top of the soil; as the soil was condense with watering over time. To measure the number of leaves, every visible leaf of each plant was counted including the tips of new leaves just beginning to emerge. For measuring the leaf area, the leaves were traced on graph paper and the squares covered were counted to get an estimate of the surface area for each leaf (Parvin et al. 2013). At 55 DAS (days after sown) the plants were harvested and lengths of roots were measured. The plants were dried in oven at 105 °C for 12 h (Mollah et al. 2009) and after half an hour interval the weight was taken to measure the dry mass.

Elemental analysis of samples

After irradiation the textile effluent was analyzed by atomic absorption spectrophotometer (AAS) prior to determine the changes in metal concentration. Cu, Cd, Zn, Co, Cr, Mn, Ni, Pb, Na, K, Cu, Fe, Zn, Mg and Mn were analyzed using flame AAS (Varian AA240FS), As by hydride generation AAS (Varian AA240) and Hg by cold vapor AAS (novAA350, Analytik jena). The calibration curves were prepared for each element individually applying linear correlation by least square method.

The samples of spleen amaranth and the soil that was used for the planting of spleen amaranth were also analyzed to determine the concentration of trace metals. Dried samples of soil and vegetable were ground into a fine powder (80 mesh size) using a mortar and duplicate samples (0.3 g) of each vegetable and soil samples were weighted in digestion flask and treated with 3 ml of concentrated HNO_3. Digestions of solid sample (soil and plant) were carried out in microwave-accelerated reaction system (MARS); model no—MARS 5.

Results and discussions

Changes of physicochemical parameter of the textile effluent after irradiation

The pH of the textile wastewater which was collected from the industry was 10.33. But after irradiation, at 10 kGy the pH of the effluent is reduced to 8.14 which comply with the limit of Bangladesh industrial effluent standards for discharge in irrigated land (Table 1). It may be deduced that the oxidation of aromatic compounds induced by ionizing radiation gives some lower molecular weight aliphatic compounds such as aldehyde and organic acids and that further oxidation of the organic acid formed leads to the formation of carbon dioxide (Spinks and Woods 1964; Suzuki et al. 1975) and lowering of the pH of the water. But with the augment of absorbed doses the pH remains almost similar. 10 kGy may be enough to oxidize all the aromatic compounds that remain in wastewater and no more aromatic compound remain to oxidize in elevated absorbed doses.

High BOD_5 (135 ppm) and COD (670 ppm) value of the textile wastewater has been decreased dramatically by irradiation and the rate of reduction has increased with increasing absorbed dose. The reduction of BOD_5 and COD implies that oxidizable materials found in the sample have been reduced due to complete decomposition of organic pollutants by ionizing radiation (Tchobanoglous and Schroeder 1985; Miyata 1993).

When the textile effluent was exposed to gamma irradiation, the water molecule undergoes radiolysis process to produce ionized and excited water molecules and free electrons (reactive species). Reactions between pollutants and primary products of water radiolysis ($\cdot OH$, e^-_{aq}, H^+) and secondary short-lived species formed from the pollutants causes the removal of pollutants from the wastewater (Anbar and Neta 1967; Buxton et al. 1988). And these reduced values of BOD_5 and COD comply with the limit of Bangladesh industrial effluent standards for discharge in irrigated land (BOD_5 100 and COD 400 ppm) (GoB, ECR 1997). After gamma irradiation, the DO value of the effluent becomes improved and approaches to the standard value (4.5–8 ppm) for irrigation water (GoB, ECR 1997). This occurs because of decreasing the BOD level, increasing of transparency of the textile effluent and owing to the presence of hydrolysis product of water (O_2, H_2O_2, etc.,) by the effect of gamma radiation (Miyata 1993).

At 10 kGy irradiation, the turbidity of the effluent is reduced from 161.65–120.95 FTU. The turbidity values decreased with increasing absorbed dose that indicates the reduction of suspended particulate matter and the breaking down of organic particles by the action of gamma radiation. This process also revealed by the reduction of COD (Soutsas et al. 2010; Selambakkannu et al. 2011). Total suspended solids of the textile effluent were also reduced from 440–210 ppm at 10 kGy of gamma irradiation, as they were converted to precipitates resulting from the degradation of organic substances and suspended matter in wastewater (Compton 1971).

TDS value (3,410 mg/L) of the raw effluent (Table 1) exceeded the standard value of water quality for irrigation (GoB, ECR 1997). But after gamma ray irradiation it becomes lower to 1,410 at the absorbed dose of 15 kGy which indicates the degradation of pollutant in the textile wastewater. There are two probable cause of TDS reduction; one is degradation of dye molecules induced by the reaction with oxidative species from water hydrolysis (Somasiri et al. 2006) and the other reason is the breakdown of larger organic chain into smaller chain adjacent to the main azo or anthraquinone groups (N=N bonds) due to radiation (Nickelsen et al. 1992), the TDS value shows an increasing trend at 20 and 25 kGy absorbed doses, but it still remains lower than the standard value of water quality for irrigation. EC is correlated closely with TDS (Rouse 1979) and similar trend of reduction also found in case of EC.

Table 1 Physicochemical parameters of the irradiated and raw sample of textile effluent

Parameter	Unit	Raw effluent	10 kGy	15 kGy	20 kGy	25 kGy
TSS	mg/L	440	210	240	230	235
TDS	mg/L	3,410	1,620	1,410	1,850	1,730
Turbidity	FTU	161.65	120.95	101.22	95.65	93.12
DO	mg/L	0.5	5.50	6.0	6.5	6.5
pH	–	10.33	8.14	8.13	8.17	8.16
EC	µS/cm	4,140	2,210	1,900	2,700	2,890
BOD	mg/L	135	55	40	45	35
COD	mg/L	670	255	262	210	222

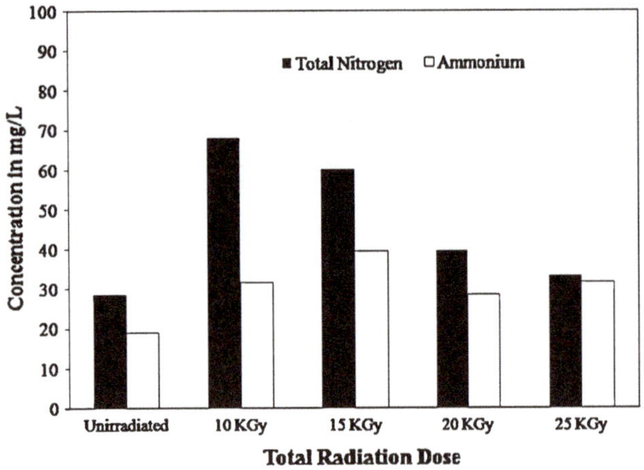

Fig. 1 Comparison of total N and ammonium content in both irradiated and raw (unirradiated) textile effluent

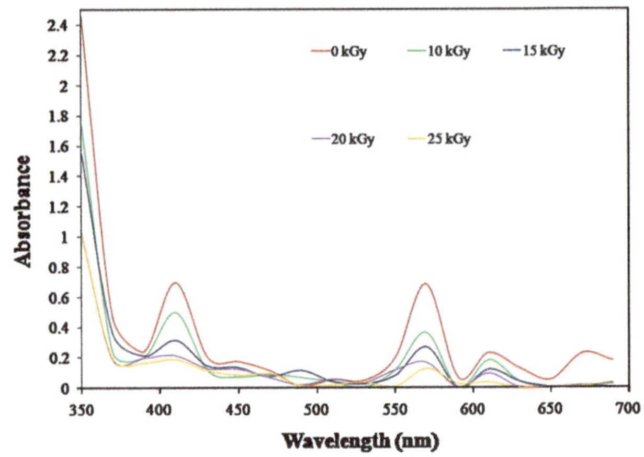

Fig. 2 Changes of absorption spectra for irradiated and raw (0 kGy) textile effluent

Changes of total nitrogen and ammonium concentration in textile effluent after gamma irradiation

After gamma irradiation both the nitrogen (N) and ammonia content of the textile effluent has increased. Figure 1 shows that raw textile effluent contains 28.5 mg/L total N but after irradiation at 10 kGy the total N content increased to more than double and amounts to 68 mg/L. Owing to applying radiation, the refractory fractions of the azo dyes have broken down easily and nitrogen molecules come out into the solution rapidly after digestion. Nickelsen et al. (1992) and Somasiri et al. (2006) also revealed that radiation causes the degradation of azo dyes and releases the nitrogen (N_2) into effluent. Besides this, textile effluent contains azo dye, so after gamma irradiation it was converted into amide, which then converted into ammonium by hydrolysis (Bagyo et al. 1997) and maximum concentration of ammonium (39.50 mg/L) was found when the effluent is irradiated at 15 kGy. One reason is the degradation of dye molecules induced by the reaction with oxidative species from water hydrolysis (Somasiri et al. 2006) and the other reason is the breakdown of larger organic chain into smaller chain adjacent to the main azo or anthraquinone groups (N=N bonds) due to radiation (Nickelsen et al. 1992).

Analysis of absorption spectra for irradiated and raw textile effluent

The raw textile effluent has very strong absorption bands in visible region at (400–430) and (550–600) nm (Fig. 2). After gamma irradiation the intensity of the main absorption decreased gradually as the dose increased. The changes in absorbance are found in these two regions are due to the presence of green (visible range band 550–650 nm) and blue (visible range band 300–400 nm) dyes present in the combined raw effluent collected from the textile industry. These intensity decreases reflects the destruction of conjugated system of the dye molecules by the attack of hydroxyl radicals (·OH) formed by the water hydrolysis with the action of gamma ray radiation. Solphan in (2001) and Sumartono in (2006) has also found that commercial azo dyes and commercial anthraquinone dyes show the sharp changes of absorption peak after irradiation in oxygen-saturated condition.

Elemental analysis of the irradiated and raw textile effluent

The analytical data of various elements concentration in textile effluents and their standard are represented in Table 2. Among the heavy metals lead, chromium, mercury, cobalt, nickel and cadmium have prime importance because they are subject to bioaccumulation and may pose a risk to human health when transferred to the food chain (Itanna 2002). But, analytical results revealed that the concentration of lead, chromium, mercury, arsenic, cobalt, nickel and cadmium in effluents of textile industry for both raw and gamma-irradiated remains within the acceptable limit for the drinking water set by the World Health Organization (WHO 2000) and irrigation water set by DoE (GoB, ECR 1997).

On the contrary, Table 2 also revealed that the nutrient contents like Na, K, Cu, Fe, Zn, Mg and Mn are considerably higher in irradiated effluents. These are essential micro and macroelements for supporting the plant growth when it will apply in the purpose of irrigation. So these gamma ray-irradiated textile effluent containing Na, K, Cu, Fe, Zn, Mg and Mn has the potentiality as the plant fertilizer sources.

Table 2 Elements concentration (in mg/L) for raw and gamma-irradiated textile effluent

Metal name	Raw effluent	10 kGy	15 kGy	20 kGy	25 kGy	WHO standard for drinking water	DoE standard for irrigation water
Chromium (Cr)	0.003	0.008	0.008	0.003	0.003	0.05	1
Lead (Pb)	<0.02	0.038	0.038	0.027	0.027	0.01	0.1
Cadmium (Cd)	<0.004	<0.004	<0.004	<0.004	<0.004	0.003	0.05
Cobalt (Co)	<0.004	0.013	0.013	0.013	0.013	–	
Nickel (Ni)	<0.004	0.015	0.016	0.015	0.013	0.02	1
Copper (Cu)	<0.002	0.092	0.056	0.072	0.057	2	3
Zinc (Zn)	0.038	0.013	0.024	0.010	0.013	3	10
Arsenic (As)	<0.0003	<0.0003	<0.0003	<0.0003	<0.0003	0.01	0.2
Mercury (Hg)	0.001	0.001	<0.0003	<0.0003	<0.0003	0.001	0.01
Iron (Fe)	<0.004	0.212	0.326	0.218	0.313	0.3	2
Manganese (Mn)	0.140	0.326	0.224	0.238	0.295	0.1	5
Magnesium (Mg)	0.004	0.004	0.004	0.004	0.004	–	
Sodium (Na)	0.123	0.723	0.775	0.718	0.686	200	
Calcium (Ca)	0.024	0.033	0.035	0.033	0.034	–	
Potassium (K)	0.001	0.02	0.022	0.019	0.019	–	

Table 3 Morphological characters of spleen amaranth

Sample ID	Plant height average (cm)						No. of leaf average						Maximum leaf area (cm^2)					
	20 DAS	27 DAS	34 DAS	41 DAS	48 DAS	55 DAS	20 DAS	27 DAS	34 DAS	41 DAS	48 DAS	55 DAS	20 DAS	27 DAS	34 DAS	41 DAS	48 DAS	55 DAS
Controlled	12	21.3	22	27.2	35.5	43.7	7	11	16	22	25	29	14	22	36	48	62	80
0 KGy	12.5	16.9	19.5	21.6	27.2	32.9	8	10	13	16	20	24	15	18	26	38	51	65
10 KGy	11.6	25.2	34.3	46	59.4	74.8	7	13	20	29	34	42	14	25	47	68	89	110
15 KGy	12	23	32.3	44.2	53	68.3	7	12	18	23	28	35	15	24	42	56	79	98
20 KGy	11.3	22.6	29	40.2	49	66.6	7	12	19	22	26	33	13	24	40	52	73	93
25 KGy	11.8	21.1	25.5	31.5	39	52.4	8	11	17	20	24	31	14	22	41	48	67	88

Fertilizing effect of textile effluent

Application of irradiated textile effluent showed a marked effect on the plant morphology. Plants nourished with textile effluent irradiated at 10 and 15 kGy showed best performance (Table 3) among the plants nourished by effluent irradiated at four different radiation doses, raw textile effluent and control sample, in case of these three-growth parameters (highest plant height, maximum leaf area and number of leaf average).

Increase of root length indicates the growth of plant. Gamma-irradiated textile effluent has increased the root length of spleen amaranth compared to the controlled sample and sample nourished by raw textile effluent (Fig. 3). Highest root length at 55 DAS was found 19.00 cm with the treatment of textile effluent irradiated at 10 kGy, whereas 14.00 cm found for controlled sample and 10.33 cm for raw textile effluent. This indicates that gamma ray-irradiated textile effluent has marked effect on root length. Similar results are also found in case of dry weight. The results suggest that the application of irradiated textile effluent containing the higher concentration of ammonium and nitrogen causes an increase in the biochemical and physiological functions of plants that lead to an increase in dry matter. So for the best production of vegetables total dose of (10–15) kGy at dose rate of 13.0 kGy/h of gamma radiation is required and the gamma ray-irradiated textile effluent should be applied at three times per week to get the optimal growth results.

Bioaccumulation of elements in spleen amaranth

Heavy metal content as well as the micro and macronutrient content were analyzed for plant irrigated by gamma ray-irradiated textile effluent. The results of metal analysis (Table 4) demonstrate that the concentrations of heavy metal in the plant sample are present in an acceptable limit,

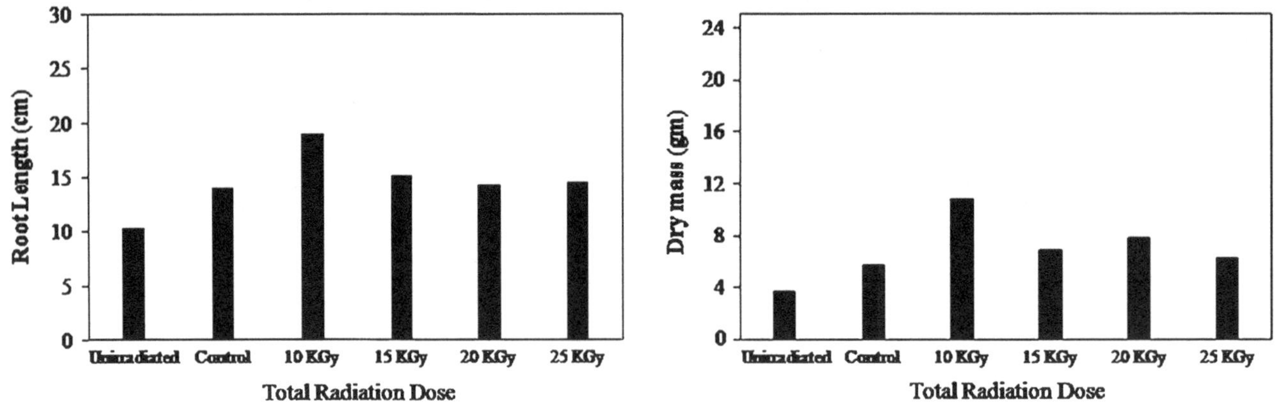

Fig. 3 Comparison of root length and dry mass of spleen amaranth after irrigation with raw (unirradiated) and gamma-irradiated textile effluent at different absorbed doses

Table 4 Elements concentration (mg/kg) in spleen amaranth nourished by gamma-irradiated and raw textile effluent

Metal name	Raw effluent	Control	10 kGy	15 kGy	20 kGy	25 kGy
Chromium (Cr)	0.76	0.55	0.60	0.40	0.48	0.63
Lead (Pb)	3.96	1.86	2.9	1.6	1.6	2.1
Cadmium (Cd)	0.12	0.12	0.44	0.5	0.6	0.3
Cobalt (Co)	1.86	1.56	1.63	1.53	1.6	2.03
Nickel (Ni)	1.01	<0.004	<0.004	<0.004	<0.004	<0.004
Copper (Cu)	20.59	15.64	30.76	25.72	32.66	23.80
Zinc (Zn)	56.62	55.2	66.35	62.56	59.93	69.94
Arsenic (As)	<0.1	0.24	0.15	0.19	0.15	0.22
Mercury (Hg)	<0.3	<0.3	<0.3	<0.3	<0.3	<0.3
Iron (Fe)	365	456.9	422.9	522.6	380.7	407
Manganese (Mn)	0.33	0.48	0.56	0.43	0.47	0.36
Magnesium (Mg)	183	184	244	217	204	194
Sodium (Na)	4.24	14.02	62.9	6.98	8.7	11.2
Calcium (Ca)	101	98	287	119	141	163
Potassium (K)	130	111	182	183	182	156

excepting Pb and Cd (Itanna 2002; Naser et al. 2011). But their concentrations remain in slightly higher concentration than the acceptable limit.

Elemental analysis of the soil in which the plants were grown were also performed (Table 5). And by analyzing the metal concentration in irradiated textile effluent (Table 2) and soil (Table 5), it has been found that the source of heavy metal in the plant largely come from the soil, not from the effluent.

The concentration of Cu, Na, Ca, K and Mg is comparatively higher in the plants nourished by gamma ray-irradiated textile effluent than the plants nourished with raw textile effluent and only water. This is due to the bioavailability of the nutrient from the textile effluent after gamma ray irradiation. In the raw effluent the nutrient element remains as a ligand that binds to the central metal atom in a complex form. But after irradiation the complex form has broken down and the nutrient element becomes available as a free element. And plants are then able to uptake the nutrient from water. Depending upon the exposure route, all these elements (Cu, Na, Ca, K and Mg) are least toxic trace elements if ingested (USEPA 1995). Finally it can be easily concluded that the hazardous and toxic elements in the plant are present in negligible amounts but the nutrients essential for plant growth and human health are found in higher rate. So the gamma ray-irradiated textile effluents have proved as the fertilizer-containing irrigation water.

Conclusions

Decoloration and degradation of textile dyes and organic waste in aqueous solutions by gamma irradiation has been accomplished successfully. All the physicochemical

Table 5 Metal concentration (mg/kg) in soil used for plantation

Metal	Concentration
Chromium (Cr)	59.94 ± 1.1
Lead (Pb)	33.96 ± 2.42
Cadmium (Cd)	0.76 ± 0.01
Cobalt (Co)	10.9 ± 0.8
Nickel (Ni)	34.05 ± 2.4
Copper (Cu)	8.84 ± 0.4
Mercury (Hg)	<0.003
Zinc (Zn)	122.2 ± 0.1
Arsenic (As)	1.12 ± 0.08
Potassium (K)	45 ± 0.1
Iron (Fe)	80 ± 1.4
Manganese (Mn)	124 ± 1.6
Magnesium (Mg)	139 ± 2.42
Sodium (Na)	223 ± 2.8

properties were improved after irradiation and optimum result was found at (10–15) kGy gamma-radiation doses. Total nitrogen and ammonium content in the textile effluent has been increased after irradiation, which indicated the possibility of the textile effluent to be used as fertilizer-containing irrigation water. And it is also proved by the enhanced growth rate of spleen amaranth irrigated by gamma ray-irradiated textile effluent compared to that of the plant nourished with the raw textile effluent and only water. The elemental analysis revealed that the plants uptake trace amount of heavy metal and moreover contain satisfactory amount of mineral nutrients. At the dose rate of (10–15) kGy the psychochemical parameter as well as the plant production is satisfactory. All the results indentified that the treatment of textile effluents by gamma irradiation (10–15 kGy) is a promising tool as it produces zero waste and has the possibility of alternative use for irrigation water with fertilizing properties.

References

Anbar M, Neta P (1967) Radiation chemical destruction of phenol in oxygenated aqueous solutions. Int J Appl Radiat Isot 18:493

Bagyo ANM, Arai H, Miyata T (1997) Radiation-induced decoloration and sedimentation of colloidal disperse dyes in water. App Radiat Isot 48:175–181

Buxton GV, Greenstock CL, Helman WP, Ross AB (1988) Critical review of rate constants for reactions of hydrated electrons, hydrogen atoms and hydroxyl radicals in aqueous solution. J Phys Chem Ref Data 17(2):513–886

Compton DMJ (1971) Destruction of organic substances in waste water by ionizing radiation. Isot Radiat Technol 18:453

Dahlan KZM (2001) Radiation sciences. Advances in tissue banking, vol 5. World Scientific Publishing Co. Pte. Ltd

GoB (Government of Bangladesh) (1997) Environment conservation rules, ECR-Schedule 3, Standards for water

Grubben GJH, Denton OA (2004) Plant resources of tropical Africa 2. Vegetables. PROTA Foundation, Wageningen; Backhuys, Leiden; CTA, Wageningen

Islam JMM, Habib SMA, Parvin F, Rahman MF, Saadat AHM, Khan MA (2013) Removal of industrial dye effluent (drimarene yellow) by renewable natural resources. Am Acad Scholar Res J 5:144–150

Itanna F (2002) Metals in leafy vegetables grown in Addis Ababa and toxicological implications. Ethiop J Health Dev 16(3):295–302

Kim TH, Park C, Lee J, Shin EB, Kim S (2002) Comparison of disperse and reactive dye removals by chemical coagulation and Fenton oxidation. J Hazard Matter 112:95–103. doi:10.1016/j.jhazmat.2004.04.008

Miyata T (1993) Liquid wastes chemistry, UNDP/IAEA/RCA regional training course on radiation technology for environmental conservation, JAERI

Mollah MZI, Khan MA, Khan RA (2009) Effect of gamma irradiated sodium alginate on red amaranth (*Amaranthus cruentus* L.) as growth promoter. J Radiat Phys Chem 78:61–64. doi:10.1016/j.radphyschem.2008.08.002

Mulvaney RL (1996) Nitrogen—inorganic forms. Methods of soil analysis, part 3—chemical methods. Soil Science Society of America, Madison, pp 1129–1139

Naser HM, Sultana S, Mahmud NU, Gomes R, Noor S (2011) Heavy metal levels in vegetables with growth stage and plant species variations. Bangladesh J Agric Res 36(4):563–574. doi:10.3329/bjar.v36i4.11743

Nickelsen MG, Cooper WJ, Waite TD, Kurucz CN (1992) Removal of benzene and selected alkyl substituted benzenes from aqueous solution utilizing continuous high-energy electron irradiation. Environ Sci Technol 26:144–152

Panda GC, Das SK, Guha AK (2009) Jute stick powder as a potential biomass for the removal of congo red and rhodamine B from their aqueous solution. J Hazard Mater 164:374–379. doi:10.1016/j.jhazmat.2008.08.015

Parvin F, Yeasmin F, Islam JMM, Molla E, Khan MA (2013) Effect of gamma irradiated sodium alginate on Malabar spinach (*Basella alba*) and spinach (*Spinacia oleracea*) as plant growth promoter. Am Acad Sch Res J 5(5):63–71

Rabby F (2011) Radiation induced waste water treatment in textile industry. Bangladesh Textile Today, Dhaka

Rouse RD (1979) Water quality management in pond fish culture. In: Research and development series no. 22, Project: AID/DSAN-G 0039, Auburn University, Alabama

Selambakkannu S, Bakar KA, Ming TT, Sharif J (2011) Effect of gamma and electron beam irradiation on textile waste water. J Sains Nukl Malays 23(2):67–73

Somasiri W, Ruan W, Xiufen L, Jian C (2006) Decolorization of textile wastewater containing acid dyes in UASB reactor system under mixed anaerobic granular sludge. Electron J Environ Agric Food Chem 5(1):1224–1234

Soutsus K, Karyannis V, Poulios I, Riga A (2010) Decoloration and degradation of reactive Azo dyes via heterogeneous photo catalytic processes. Desalination 250:345–350. doi:10.1016/j.desal.2009.09.054

Spinks JWT, Woods RJ (1964) An introduction to radiation chemistry, vol 8. Wiley, New York

Sumartono A (2006) Degradation of standard dyes and textile wastewater as a pollutant model using gamma radiation. Indones J Chem 6(3):225–230

Suzuki N, Nagai T, Hotta H, Washino M (1975) The radiation-induced degradation of azo dyes in aqueous solutions. Int J Appl Radiat Isot 26(12):726–730. doi:10.1016/0020-708X(75)90129-5

Takács E, Wojnárovits L, Pálfi T (2007) Azo dye degradation by high-energy irradiation: kinetics and mechanism of destruction. Nukleonika 52(2):69–75

Tchobanoglous G, Schroeder ED (1985) Water quality: characteristics, modeling and modification. Addison-Wesley Publishing Company, Canada

United States Environmental Protection Agency (1995) Water quality standards revision of metals criteria. Federal Regist 60:22229–22237

Wang M, Yang R, Wang W, Shen Z, Bian S, Zhu Z (2006) Radiation-induced decomposition and decoloration of reactive dyes in the presence of H_2O_2. Radiat Phys Chem 75(2):286–291. doi:10.1016/j.radphyschem.2005.08.012

WHO (World Health Organization) (2000) WHO water quality guidelines, 2nd edn. World Health Organization, Europe Regional Office, Copenhagen

Comparison of ammonia emissions from animal wastes and chemical fertilizers after application in the soil

Majid Rostami[1] · Stefano Monaco[2] · Dario Sacco[2] · Carlo Grignani[2] · Elio Dinuccio[3]

Abstract

Background Application of different chemical fertilizers and manures is a major source of ammonia (NH_3) emission. The rate and total amount of NH_3 emission are related to different parameters such as climatic conditions, soil characteristics and kind of fertilizer. The current study has indicated the NH_3 emission from bovine slurry, pig slurry and ammonium nitrate fertilizer after application on two soils. Two different methods were used to measure NH_3 emissions: the method that use acid traps and the method that use photoacoustic infrared gas analyzer.

Results In both soils the rate of NH_3 emission was the greatest from the denser bovine slurry, declined in the pig slurry followed by the ammonium nitrate treatment and the control. The rate of soil infiltration could be the main factor that explained these differences. For all treatments the amount of total NH_3 losses reduced in the more acidic soil. For all fertilizers the highest NH_3 fluxes were measured in the first hours after spreading. A good agreement observed between the two methods is used for determining of NH_3 emission. The use of a multi-gas monitor (MGM) is simple and accurate and produces a continuous series of NH_3 concentration in time.

Conclusion The rate and amount of NH_3 emission was related to the kind of fertilizers and interaction of these treatments with soils. The results of current study confirmed that comparison of chemical fertilizers and slurry for NH_3 emission is difficult because the reaction of these two groups of fertilizer is totally different.

Keywords Ammonia emission · Laboratory methods · Mineral fertilizer · Slurry

✉ Majid Rostami
Majidrostami7@yahoo.com

Stefano Monaco
Stefano.Monaco@unito.it

Dario Sacco
Dario.Sacco@unito.it

Carlo Grignani
Carlo.Grignani@unito.it

Elio Dinuccio
Elio.Dinuccio@unito.it

[1] Department of Agronomy, University of Malayer, 657719-95863 Malayer, Iran

[2] Department of Agricultural, Forest and Food Sciences, University of Turin, L.go Braccini 2, 10095 Grugliasco, Italy

[3] Mechanics Section, Department of Agricultural, Forest and Food Sciences, University of Turin, L.go Braccini 2, 10095 Grugliasco, Italy

Introduction

Ammonia (NH_3) may be released into the atmosphere from basically all ammonium-containing products. Livestock and especially animal manures are the most important sources of NH_3 emissions in Europe, followed by the application of mineral nitrogen fertilizers (Leip et al. 2011). The increase in animal stocking and in the price of chemical fertilizers encourage farmers to use animal manure and slurry as an option to reduce the use of commercial fertilizers. However, the handling and spreading of these fertilizers may pose an agronomic and environmental risk, not only because of leakage of nitrate to ground waters but also because of gaseous losses of NH_3 (Asman 1992). Ammonia can form secondary particulate matter in the atmosphere that may have adverse effects on human health (Moldanová et al. 2011).

Ammonia emissions from manure applied to the soil are produced primarily by physical and chemical processes and secondarily by biological ones (Monaco et al. 2012). Ammonia losses from manure are harmful from the agronomic point of view, because they decrease the amount of manure N available for the crop (Smil 1999). Olivier et al. (1998) estimated that about 70 % of global NH_3 emission is related to food production and, in particular, to manure management.

A comprehensive understanding of the post application fate of fertilizers is essential for the development of best management practices that aim to minimize off-site transport and maximize nutrient use efficiency. Different variables affect both the rate and extent of emissions following soil application of manures (Meisinger et al. 2001). The dominant factors influencing losses can be categorized as: manure characteristics (dry matter content, pH, NH_4–N content), application methods (incorporation, time of application), soil properties (soil moisture, soil texture, soil pH), and environmental factors (temperature, wind speed, rainfall, relative humidity).

A number of techniques have been developed to quantify NH_3 emission. Accuracy and mechanism of these methods are very different. Such techniques fall generally in two groups: micro-meteorological methods (usually used for large scale areas) and enclosure methods (commonly used on small plots for comparative experiments). Chamber methods that belong to second category are usually used for measuring emission at the small scale both in the field and in the laboratory. Three measurement schemes are commonly used for the chamber methods: the open chamber, the closed static chamber and the closed dynamic chamber. All methods employ an inverted chamber covering a small area of soil. The lower edge of the chamber usually is inserted into the soil to a shallow depth.

The analysis of NH_3 emission have been carried out using different methods as acid traps or direct measurement through a multi-gas monitor (MGM). Between them, using MGM is easier because this system provides a real-time analysis of NH_3 concentration (Dinuccio et al. 2008). Experiments for measuring NH_3 emissions are usually carried out in the field where soil and moisture conditions and other environmental factors are variable and hard to control. Unfortunately, little effort has been made to standardize the laboratory methods for NH_3 emission measurement. The objectives of this experiment were to measure NH_3 emissions from different slurries using closed dynamic chambers with both acid traps and MGM method under controlled environmental conditions and assessing the influence of two different soils on modifying NH_3 emissions.

Materials and methods

The amount of volatilized NH_3 after surface application of different fertilizers in two soil types was assessed under constant and controlled environmental conditions using dynamic chamber technique (Roelle and Aneja 2002). The soils used in the experiment presented different physical and chemical characteristic (Table 1) and were representatives of arable soil types of the western Po river plain (Northern Italy). In particular, Poirino soil came from a farm field cultivated mainly with maize for grain in rotation with wheat receiving chemical fertilizations, while Tetto Frati soil (TF) was collected from an experimental field continuously cultivated with maize for grain fertilized with cattle slurry and urea. The soil samples were collected from the tilled top 20 cm layer on, air-dried and sieved using 2 mm. Cylindrical glass jars (3.1 l) were filled with 1.1 l of each soil type moistened to reach the field capacity (FC). In particular, 1595 and 1475 g of dry soil was moistened with 367 and 428 ml of deionized water for Poirino and TF soil, respectively. Soil moisture content at FC (−33 kPa) was measured on 4 replicates for each soil type using pressure plates and was equal to 0.23 and 0.29 g of H_2O per g of dry soil for Poirino and TF, respectively. Bulk density of soils at FC were measured by drying 100 cm^3 of soil (four replicates) at 105 °C for 3 days, and were equal to 1.45 and 1.34 g of dry soil per cm^3.

Organic fertilizers used in the experiment were pig slurry (PS) and bovine slurry (BS). Slurries were collected from storage facilities of two farms few days before the beginning of experiment, stored at 4 °C and analyzed for the main characteristics (Table 2). Fertilizers were applied at a rate of 85 kg ha^{-1} of total N. In particular, 34.8 and 67.4 g of PS and BS were gently distributed on the surface of the jars (154 cm^2), providing 131 mg jar^{-1} of total N and 95 and 65 mg jar^{-1} of ammoniacal N for PS and BS, respectively. For each soil type, a treatment fertilized with ammonium nitrate at the rate of 85 kg ha^{-1} of total N was also prepared. Ammonium nitrate was solubilised in 15 ml deionized water and distributed on the soil surface. In addition, one extra jar in which BS was

Table 1 Physical and chemical properties of soils

Soil	Sand (%)	Silt (%)	Clay (%)	CEC Meq/100 g	pH	OM (%)	N (%)	C/N ratio
Poirino	15.8	75.5	8.7	16.7	6.2	2.4	0.14	9.6
TF	48.4	43.1	8.5	12.4	7.9	2	0.17	7

Table 2 Physical and chemical properties of the slurries

Slurry type	Dry matter (%)	pH	N (%)	TAN (%)
Bovine	5.8	8.5	19	10
Pig	2.3	8.0	38	27

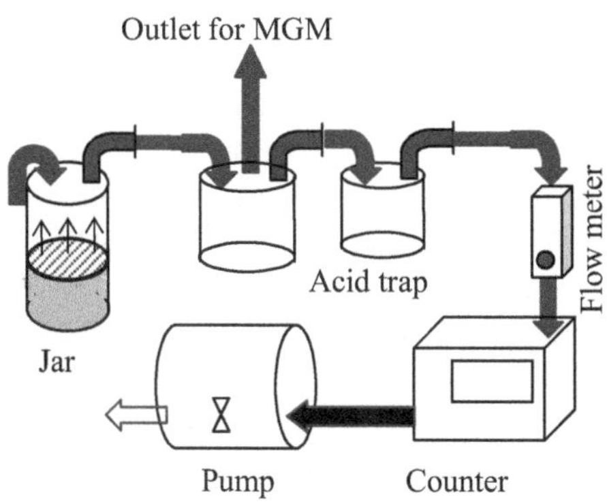

Fig. 1 Schematic diagram of the measuring system

incorporated in Poirino soil was also analyzed. For each soil type, an unfertilized soil was used as control treatment. One replication for each treatment was measured in three different sessions of measurement in a growth chamber properly equipped for NH_3 emission assessment using open dynamic chamber method. Each session was conducted at 25 °C under controlled environmental conditions. Ammonia concentration in the incoming air (background) was always measured and found negligible (0.08 ppm).

Immediately after fertilization, jars were closed with air-tight lids prepared with one input and one output port directing air into the headspace and connecting the jars to the measurement system (Fig. 1). In particular, each jar was connected first to an expansion bottle and then connected in an air-tight way with a Drechsel bottle, flow meter, gas meter and pump. The flow rate (2 l min^{-1}) was chosen to ensure a minimal exchange rate, but also to prevent rapid drying of the soil surface. It corresponded to an air renewal rate of one headspace volumes min^{-1} with an average air flow rate of 0.12 m^3 h^{-1}. The air flow rate was monitored over each jar using one volumetric air flow meter per jar. From the expansion bottles another outlet tube was inserted for measuring of NH_3 concentration using photoacoustic infrared gas analyzer (Innova 1412).

In the acid trap method, the air stream passing through the jar and the expansion bottle arrived to a NH_3 scrubber containing 100 ml of 0.1N H_2SO_4, in which absorbed NH_3 was converted to ammonium sulfate. All of the acid samples were stored at 4 °C until the analysis. The amount of NH_3 absorbed in each scrubber was determined using an ionometer (Ion lab, WTW). The NH_3–N concentration in the outgoing air was calculated from the data of ionometer multiplied by the volume of acid trap.

Measurements with acid traps were carried out in four intervals of 5 h during the first 4 days after slurry application. Measuring for 4 days was considered enough to reveal the possible differences in NH_3 emission, because according to earlier studies NH_3 emission is at its highest rate on the day of application and declines sharply in the following few days (Yang et al. 2003). The total amount of emitted NH_3 during the 4 days was derived by interpolation of the integral form of a power curve of the amount of NH_3 trapped in the Drechsel bottle during the four measurements.

Simultaneously, measurements with MGM were carried out during the first 2 days after fertilization, when the NH_3 emissions were expected to be high for fertilized treatments. Gas sample suction for MGM started approximately 30 min after the closing of chamber to allow stabilization of NH_3 concentration in the expansion bottle and was repeated three times for treatment during each interval of measurement. Before each measurement, Teflon tubes connecting the system to MGM were cleaned with background air flush.

The NH_3 readings by MGM (mg/l) were multiplied by a correction factor for the atmospheric pressure. NH_3 concentrations (mg/l) were converted into the flux of NH_3 leaving each jar as follows:

$$\text{FNg} = \text{air volume} \times |\text{conc}| \times 10^{-6} \times \rho\text{NH}_3 \times (14/17)$$

where FNg is the NH_3–N flux in mg N per jar; air volume is the throughput of air during one measurement; |conc| is the value of the corrected volume concentration (vol. mg/l) of NH_3; ρNH_3 is the density of NH_3 in mg l^{-1} (ρNH_3 = 696 mg l^{-1}; 25 °C, 1013 hPa).

The total amount of NH_3 emitted during each interval of measurement was calculated separately for each treatment by the integration on time basis of the measurements carried out with MGM, using linear, exponential and power function. The results were then compared with total amount of NH_3 emitted as measured with acid traps. Calculations and drawing the trend lines was done with using Microsoft Excel-2003, and SPSS.

Results

Comparison of manures using the acid trap method

Immediately after the application of fertilizers, NH_3 emission started and continued up to the end of the experiment. Based on Kruskal–Wallis test ($P \leq 0.05$) differences among treatments were significant. Until the third day of experiment, the ranking of treatments did not change and the rate of NH_3 emissions were highest in BS, followed by PS and ammonium nitrate, and were very low in control for both soils. Incorporation of BS strongly reduced emission to the level of control treatment. On day 4 the ranking of the treatments was different, but emissions were very low (Fig. 2).

One day after application of fertilizers, NH_3 fluxes declined sharply by about sevenfold in slurry treatments and threefold in mineral fertilizer (Fig. 2). Ammonia fluxes had fallen to the lowest levels by the third or fourth day depending on the treatments. Based on results the most NH_3 losses occurred during the first day after application of fertilizer.

With respect to their NH_3 emissions both the slurry treatments showed systematic time course patterns over the 4 days. The soils fertilized with pure mineral fertilizer, also showed similar patterns, but at much lower levels. The emission rate did not change in time from the manure incorporated treatment and the control treatment, that showed a low and steady flux during the whole experiment.

The total emissions of NH_3 during the experimental period were calculated using integral of measured emission rate with acid trap method (Fig. 3a) and reported as a percentage of ammoniacal nitrogen applied (Fig. 3b). In both the cases Kruskal–Wallis test reported significant differences among treatments.

The rate and total amount of NH_3 emission was related to kind of fertilizers. In this experiment the range of NH_3 emission from BS was 39–82 % of total ammoniacal nitrogen (TAN) applied and the amount of emitted NH_3 in PS and ammonium nitrate varied from 13–38 to 2–18 % of TAN, respectively. This shows that the higher dry matter content of the slurry the higher emission rate.

Figure 3 shows that there was a similar ranking of treatments for total amount of emitted NH_3 and of the ratio between emitted NH_3 and TAN. In other words, in our experiment the amount of TAN in the different fertilizers did not affect the final NH_3 emissions. The amount of total

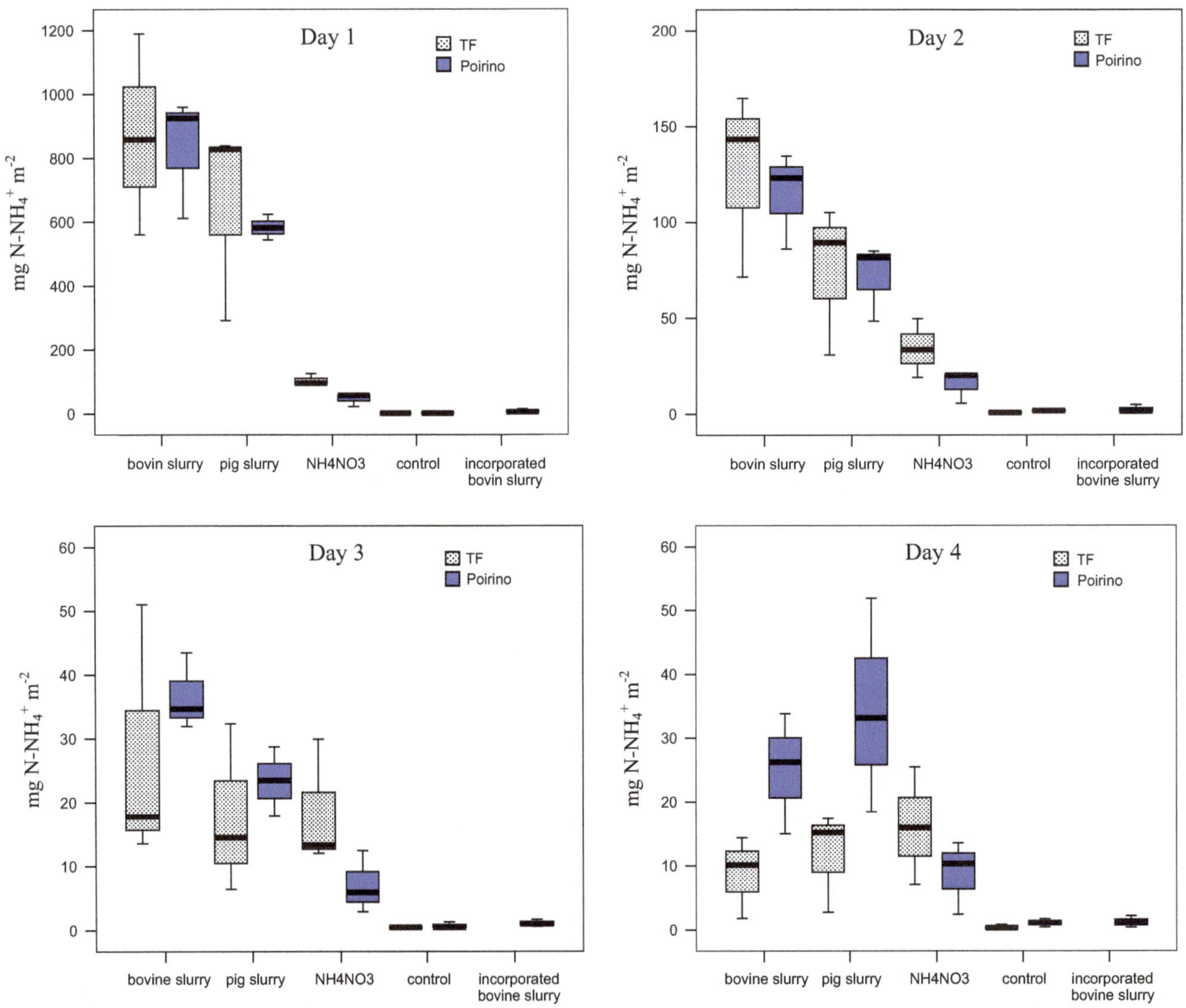

Fig. 2 Ammonia emission from different treatments during measurement times in each day (measured with acid traps)

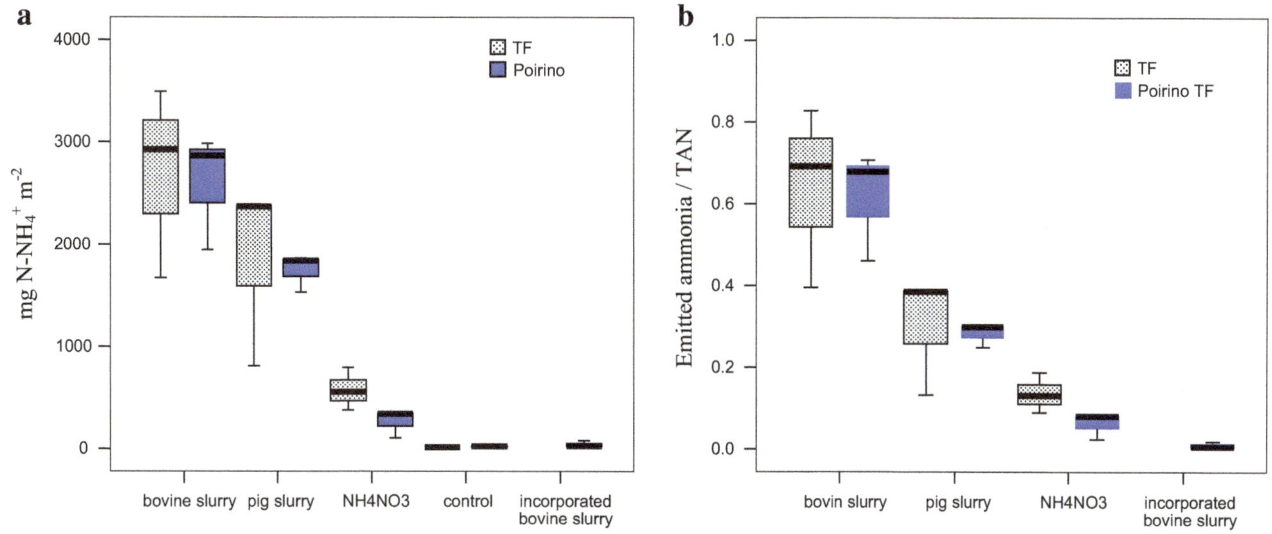

Fig. 3 Total amount of emitted ammonia from different treatments during the experiment (**a**), and ratio between emitted ammonia and total ammoniacal nitrogen (TAN) in different treatments (**b**) (measured with acid traps)

NH_3 losses in the calcareous TF soil was higher than in the more acidic Poirino soil for PS and ammonium nitrate treatment, the two fertilizers that are expected to better infiltrate into soil because of lower dry matter. Instead no differences observed for the BS, which was the fertilizer characterized by higher dry matter.

Comparison of the two methods for ammonia measurements

Using MGM the highest NH_3 fluxes were measured during the first hour following the application of fertilizer in all the fertilized treatments (Fig. 4). But it is likely that the fluxes immediately after application were still greater, due to the delay (about 30 min) in the beginning of measurements. The time variation patterns of NH_3 concentration in the air fluxes were similar in both the methods. In the second day after application of fertilizers, NH_3 fluxes declined sharply by about sixfold in slurry treatments and less than twofold in mineral fertilizer (Fig. 4).

For the interpolation of the amount of emitted NH_3 measured with MGM in each interval, three different mathematical models were used. A strong linear relationship was observed between the NH_3 concentration by acid trap and MGM methods (Fig. 5), but in general, the amount of emitted NH_3 measured with MGM was higher than that measured by acid trap. Comparison of the three methods used for calculation of the total amount of emission showed that they are all acceptable, but the best method for representing data of MGM in short times (5 h) is using the

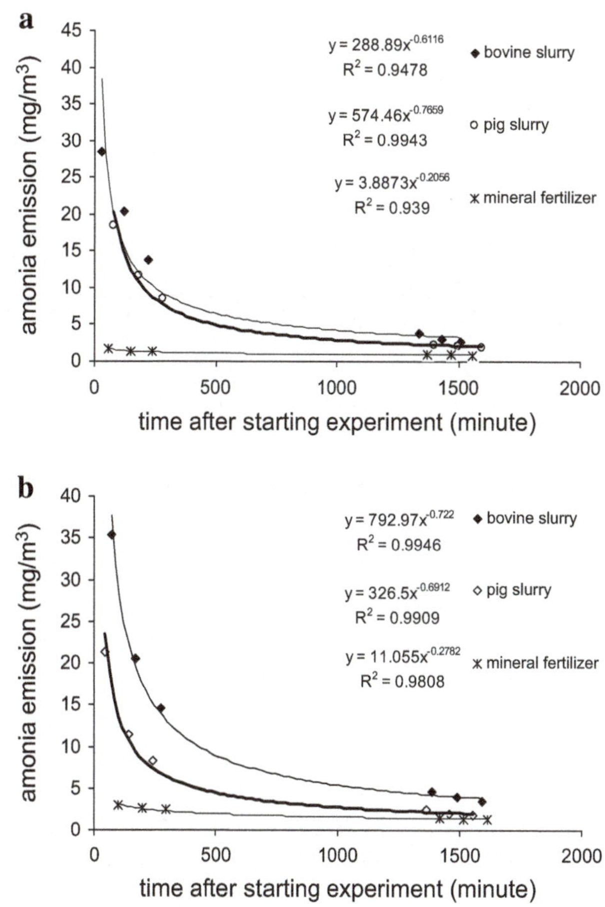

Fig. 4 Rate of ammonia emission measured with MGM in **a** Poirino and **b** TF soil

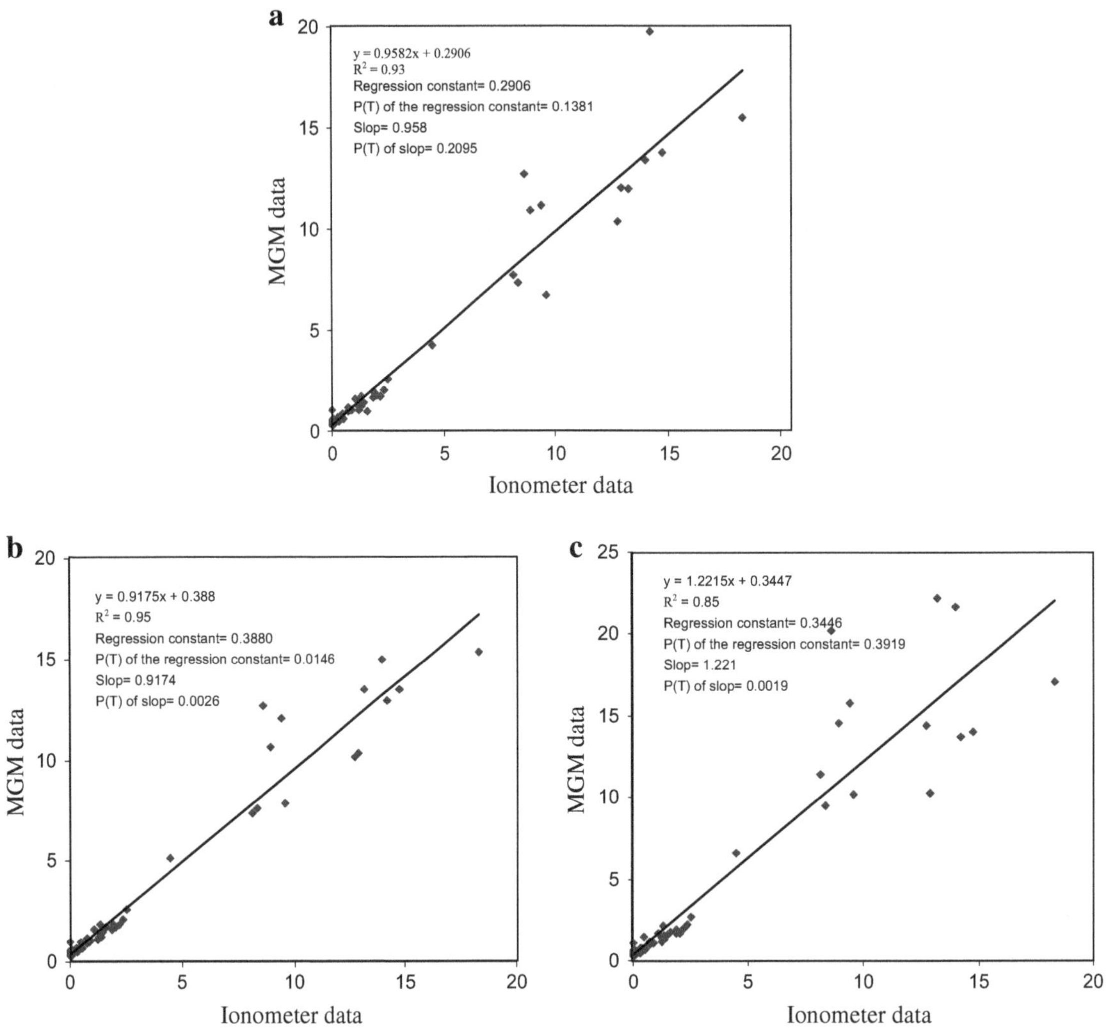

Fig. 5 Regression between amounts of ammonia emission measured by ionometer and derived from MGM measurements interpolated with **a** linear, **b** exponential and **c** power function

linear function, whereas the use of a linear function for representing the trend of NH_3 emission in longer duration (day) is not justified.

Discussion

The highest rate of NH_3 fluxes was measured using MGM in the first day and in the second day after application of fertilizers NH_3 fluxes declined sharply. Huijsmans et al. (2001) reported that the rate of NH_3 emission decreased with time. NH_3 emission also decreases with a reduction in slurry water content because the water content initially decreases due to infiltration of the slurry liquid into the soil.

Meisinger et al. (2001) reported that losses of NH_3 through emission are very rapid during the first 6–12 h after application. Results of previous study showed that 30–70 % of the total NH_3 loss from cattle slurry occurred in the first 4–6 h, and 50–90 % in the first day (Stevens and Laughlin 1997); however, this result was related to the application of slurries in grassland and is not totally comparable with our results obtained under controlled conditions.

The TAN concentration of pig slurry was higher and the dry matter content was lower than that of the bovine slurry, as it is often the case (Dinuccio et al. 2008), but in this experiment the higher amount of NH_3 emission was observed when BS was applied. This result is probably related to interaction between soil and slurry.

Higher dry matter content of the bovine slurry decreased infiltration into the soil, and it increased the chances of NH_3 losses, as it was also indicated by Stevens et al. (1992). The slurry dry matter content has been shown to be an important factor in determining the NH_3 emission potential. Sommer and Olesen (2000) showed a linear relationship between cattle slurry dry matter content and NH_3 emission in the range 4–12 % DM, but outside this range dry matter had a little effect on NH_3 emission. Smith et al. (2000) reported a similar linear relationship but in a shorter range (2–5 % of slurry dry matter) and concluded that for every 1 % increase in slurry dry matter, NH_3 losses increased about 6 %. In the MANNER model that was developed for predicting NH_3 emission from manure, it was suggested that for every 1 % increase in manure DM, NH_3 loss increases about 5 % (Chambers et al. 1999). In the Poirino soil the difference between total amount of emitted NH_3 from BS and PS was higher than in TF soil (Fig. 2), and this difference probably was related to the soil texture because the importance of slurry dry matter (for NH_3 emission) in soils with a poor infiltration rate is higher (Jarvis and Pain 1990).

Soil pH and cation exchange capacity (CEC) also influenced NH_3 emission. When the fertilizer could infiltrate in the soil, the higher pH of the TF soil enhanced NH_3 emission in comparison with the soil of Poirino. This confirms that lower pH leads to a lower proportion of aqueous NH_3 and therefore decreases NH_3 emission (Li 2000). Huijsmans et al. (2003) and FAO (2001) reported similar results, in which the increase of pH and decrease of CEC could raise NH_3 emission of 10 times. Soil surface pH greatly changes when slurry is applied to soil, due to the different buffering capacities of the soil and slurry. The highest reaction of NH_3 emission to pH was observed when pH ranges from 7 to 10, and NH_3 emission below 7 decreases dramatically (Hartung and Phillips 1994). Using a mechanistic model for estimating the NH_3 emission from slurry after application in soil Genermont and Cellier (1997) pH was the main factor that influenced NH_3 emission. They reported that the reduction of the pH from 7 to 6 resulted in 19 % decrease of NH_3 emission.

The total amount of emitted NH_3 measured in our experiment was close to the values reported in other experiments. Stevense and Laughlin (1997) showed that NH_3 losses from surface applied liquid cattle manure in grassland was in the range 40–70 %, whereas NH_3 losses from poultry litter in pasture ranged from 28 to 46 % of TAN (Marshall et al. 1998).

This research work confirmed the general idea that the amount of emitted NH_3 from mineral fertilizer is less than from slurry. Whitehead and Raistrick (1990) estimated that only 3.4 % of the applied N in chemical fertilizer (five nitrogen compounds) was lost as NH_3. In reporting emissions from fertilizers for the UK, Lee and Dollard (1994) used emission factors of 3 % for ammonium nitrate.

Incorporation of slurry almost suppressed NH_3 emission and in this treatment total amount of emitted NH_3 was negligible. Other researchers also mentioned that incorporation of manure was one of the best methods for decreasing NH_3 emission. Rodhe et al. (2006) reported that by injection of slurry into the soil the emission decreased by 39 % in comparison with the band spreading of slurry. By incorporating and mixing slurries with soil, contact area between the slurry and the air was reduced, and therefore NH_3 emission in comparison with surface application was decreased. Sommer et al. (1991) reported that evaporation of water from the manure lead to an increase of the aqueous NH_3 concentration in the manure and to an increase in NH_3 emission.

Results of this experiment showed that for calculating total amount of emitted NH_3 by MGM, different functions should be used because the trend of NH_3 emission based on the time scale of measurement is different. Regarding the NH_3 emission rate measured with MGM in the first 2 days, power function can represent the data in better comparison to linear functions. These results are in agreement with results of other experiment (Bussink et al. 1996) that showed that the trend of NH_3 emission rate from applied manure was not linear. The results of this experiment are also in contrast with results from Chambers et al. (1997), who observed linear rates of emission from poultry litter for up to 3 weeks after application.

Conclusion

The rate and amount of NH_3 emission were related to the kind of fertilizers and interaction of these treatments with soils. Comparison of chemical fertilizers and slurry for NH_3 emission is difficult because the reaction of these two groups of fertilizer is totally different. The amount of emitted NH_3 from different fertilizers was related to the amount of water added to the soil with the application and consequently to the infiltration of the slurry into the soil. It seems that 4 days are enough for the estimation of the NH_3 emission from slurry, but longer time is needed for chemical fertilizers. The strong linear correlations between the results of the two measurement methods demonstrated that MGM was also utilizable for determination of the total amount of emitted NH_3 in different treatments even though a slight overestimation existed.

References

Asman WAH (1992) Ammonia emission in Europe: updated emission and emission variations. RIVM, Bilthoven, p 88

Bussink DW, Lowry A, Harper A, Corre WJ (1996) Ammonia transport in a temper- temperate grassland, ll: Diurnal

fluctuations in response to weather and management conditions. Agron J 88:621–626
Chambers BJ, Smith K, Van der Weerden T (1997) Ammonia emissions following the land spreading of solid manures. In: Jarvis S, Pain B (eds) Gaseous nitrogen emissions from grasslands. CAB International, Wallingford
Chambers BJ, Lord EI, Nicholson FA, Smith KA (1999) Predicting nitrogen availability and losses following application of organic manure to arable land: MANNER. Soil Use Manag 15:137–143
Dinuccio E, Berg W, Balsari P (2008) Gaseous emissions from the storage of untreated slurries and the fractions obtained after mechanical separation. Atmos Env 42:2448–2459
FAO (2001) Global inventory of NH_3 emissions from mineral fertilizers and animal manure applied to croplands and grasslands. Bouwman, Rome
Genermont S, Cellier P (1997) A mechanistic model for estimating ammonia volatilization from slurry applied to bare soil. Agric Forest Meteo 88:145–167
Hartung J, Phillips VR (1994) Control of gaseous emissions from livestock buildings and manure stores. J Agric Eng Res 57:173–189
Huijsmans JFM, Hol JMG, Hendriks MMWB (2001) Effect of application technique, manure characteristics, weather and field conditions on ammonia volatilization from manure applied to grassland. Neth J Agric Sci 49:323–342
Huijsmans JFM, Hol JMG, Vermeulen GD (2003) Effect of application method, manure characteristics, weather and field conditions on ammonia volatilization from manure applied to arable land. Atmos Env 37:3669–3680
Jarvis SC, Pain BF (1990) Ammonia volatilisation from agricultural land. In: Proceedings the fertilisers society no. 298. Greenhill House, Peterborough, p 35
Lee DS, Dollard GJ (1994) Uncertainties in current estimates of emissions of ammonia in the United Kingdom. Eviron Pollut 86:267–277
Leip A, Achermann B, Billen G, Bleeker A, Bouwman AF, de Vries W, Dragosits U, Döring U, Fernall D, Geupel M, Heldstab J, Johnes P, Le Gall AC, Monni S, Nevečeřal R, Orlandini L, Prud'homme M, Reuter HI, Simpson D, Seufert G, Spranger T, Sutton MA, van Aardenne J, Voß M, Winiwarter W (2011) In: Sutton et al. (Eds.), The European nitrogen assessment. Sources, effects and policy perspectives. Cambridge University Press, Cambridge, pp 345–376
Li CS (2000) Modeling trace gas emissions from agricultural ecosystems. Nut Cycl Agroecosys 58:259–276
Marshall SB, Wood CW, Braun LC, Cabrera ML, Mullin MD, Guertal EA (1998) Ammonia volatilization from tall fescue pastures fertilized with broiler litter. J Environ Qual 27:1125–1129
Meisinger JJ, Lefcourt AM, Thompson RB (2001) Construction and validation of small mobile wind tunnels for studying ammonia volatilization. App Eng in Agr 17:375–381
Moldanová J, Grennfelt P, Jonsson Å, Simpson D, Spranger T, Aas W, Munthe J, Rabl A (2011) Nitrogen as a threat to European air quality. In: Sutton et al. (Eds.), The European nitrogen assessment. Sources, effects and policy perspectives. Cambridge University Press, Cambridge, pp 405–433
Monaco S, Sacco D, Pelissetti S, Dinuccio E, Balsari P, Rostami M, Grignani C (2012) Laboratory assessment of ammonia emission after soil application of treated and untreated manures. J Agric Sci 150:65–73
Olivier JGJ, Bouwman AF, Van der Hoek KW, Berdowski JJM (1998) Global air emission inventories for anthropogenic sources of NO_x, NH_3 and N_2O in 1990. Env Poll 102:135–148
Rodhe L, Pall M, Yamulki S (2006) Nitrous oxide, methane and ammonia emissions following slurry spreading on grassland. Soil Use Manag 22:229–237
Roelle PA, Aneja VP (2002) Characterization of ammonia emissions from soils in the upper coastal plain, North Carolina. Atmos Env 36:1087–1097
Smil V (1999) Nitrogen in crop production: an account of global flows. Glob Biogeochem Cyc 13:647–662
Smith KA, Jackson DR, Misselbrook TH, Pain BF, Johnson RA (2000) Reduction of ammonia emission by slurry application technique. J agric Engng Res 77:277–287
Sommer SG, Olesen JE (2000) Modelling ammonia volatilization from animal slurry applied with trail hoses to cereals. Atmos Env 34:2361–2372
Sommer SG, Olesen JE, Christensen BT (1991) Effects of temperature, wind speed and air humidity on ammonia volatilization from surface applied cattle slurry. J Agric Sci 117:91–100
Stevens RJ, Laughlin RJ (1997) The impact of cattle slurries and their management on ammonia and nitrous oxide emissions from grassland. In: Jarvis SC, Pain BF (eds) Gaseous nitrogen emissions from grasslands, CAB International, Wallingford, p 233
Stevens RJ, Laughlin RJ, Frost JP (1992) Effects of separation, dilution, washing and acidification on ammonia volatilization from surface-applied cattle slurry. J Agric Sci 119:383–389
Whitehead DC, Raistrick N (1990) Ammonia volatilization from five nitrogen compounds used as fertilizers following surface application to soils. J Soil Sci 47:387–394
Yang Z, Nimbi H, Kanda K, Suga Y (2003) Measurment of ammonia volatilization from a field, in upland Japan, spread with cattle slurry. Env Poll 121:463467

8

The reuse of peanut organic wastes as a growth medium for ornamental plants

A. Mohammadi Torkashvand[1] · M. Alidoust[1] · A. Mahboub Khomami[2]

Abstract

Background The reuse of organic wastes in agriculture is an appropriate method of environment management. Most plants with ornamental leaves are cultivated in soilless media, in which peat is the basic medium; but the use of peat is questionable due to ecological damage to the environment and economic disadvantages to ornamental plant producers. Peanut shells as remained wastes of cultivating peanut has a considerable volume, compost of which can be used as available sources of ornamental plants medium. A pot experiment was conducted to investigate the possibility of using peanut shell compost in the appropriate bed in the cultivation of ornamental plants. *Dracaena marginata* was selected as the test plant. The control treatment consisted of a 2:1 ratio of peat to perlite, and peat was replaced by 15, 30, 45, 60 and 100 % v/v of peanut shell compost. The study was done based on a complete randomized design in three repetitions in the ornamental plants and flowers research station, Lahijan, Iran.

Results The results showed that the bulk density of the planting beds decreased with the addition of peanut shell compost in comparison to the control. Peanut shell compost increased total nitrogen and available phosphorus and potassium of the media. Peanut shell compost increased plant growth indices, but the growth was reduced in 100 % in the compost treatment. The compost led to increase in nutrient uptake by plants.

Conclusion In conclusion, it became known that peanut shell compost is an appropriate alternative to peat as the growth medium of ornamental plants. Because of the favorable properties and high porosity, peanut shell compost can be mixed with low-porosity substrates and applied as a substitute to peat in the cultivation media. It is not suggested to increase peanut compost level by 60 and 100 %.

Keywords Compost · *Dracaena* · Environment · Nutrient · Peat · Perlite · Soilless

Introduction

The cultivation of peanut is common in several populous countries, in which it is a major food crop. According to the World Food and Agriculture Organization, the major peanut producing countries are India, China, USA, Indonesia and Myanmar. Peanut cultivation worldwide covers about 22.2 million hectares, comprising 16.3 million hectares in Asia, 7.39 million hectares in Africa and 0.7 million hectares in South and Central America (Maiti and Ebeling 2002; Murata 2003). The average yield of pods on a global scale increased from 1980 to 2003. Although the highest yield, 9.6 tons per hectare, has been reported in many countries, the yields are usually 3–4 ton/ha. Technology, proper crop management, soil conditions and climate fluctuations are the factors that directly affect the yield of peanut (Maiti and Ebeling 2002).

The cultivated area of peanut in Iran is about 3218, 2718 ha of which is located in the Guilan Province. About 1980–3500 kg/ha pods is harvested and 89327.6 tons produced in Guilan. The average shell wastes of 1 kg yield is

✉ A. Mohammadi Torkashvand
m.torkashvand54@yahoo.com; torkashvand@iaurasht.ac.ir

[1] Department of Horticulture, Faculty of Agriculture, Islamic Azad University, Rasht Branch, Rasht, Iran

[2] Ornamental Plants and Flowers Research Station, Lahijan, Iran

35–40 %; therefore, about 3388–3873 tons of peanut shells wastes are annually produced in Iran (Agriculture Organization of Guilan Province 2010). As a result, it has been decided to evaluate the possibility of using this organic waste as a growth medium for *Dracaena marginata* as a test plant.

Today, most plants with ornamental leaves are cultivated in soilless media, among which peat is a basic medium (Atieyh et al. 2000). However, the use of peat causes ecological damage to the environment and economic disadvantages to ornamental plant producers. These factors caused researchers to use beds of high quality and low expense, instead of peat (Krumfolz et al. 2000).

Millions of tons of different agricultural wastes are produced annually across the country, but unfortunately a major fraction is burned or left unattended, leading to environmental pollution (Mohammadi Torkashvand 2010). With increasing awareness of environmental problems of wastes and to decrease the use of non-renewable sources such as peat, further use of composted biosolids has been suggested in farming (Bugbee 2002; Papafotiou et al. 2005). However, in recent years, the wastes have been chemically or biologically treated to obtain useful products before the final disposal (Rajesh Banu et al. 2001, 2007).

Some studies show that peat can be replaced by organic wastes such as municipal wastes, sewage sludge, livestock manure, paper, waste from pruning, and fungi beds and other organic waste after composting (Gayasinghe et al. 2010). Investigations of *Ficus benjamina* variety Starlight in a growth medium containing a ratio of 1:1 of peat to olive waste (as volume) showed the plant growth to be very high during 10 months (Chen et al. 1989). Papafotiou et al. (2005) used olive waste compost as an alternative to peat to cultivate some ornamental plants and suggested that 25, 75 and 75 % v/v of this compost could be used for cultivating *Ficus benjamina*, *Cordyline* and *Syngonium podophyllum*, respectively.

Applications of manure, wood chips and paper wastes mixed with volcanic material in the cultivation bed of croton (*Codiaeum variegatum*) and chrysanthemum showed that these materials can be used as planting beds (Cull 1981). The bark of broad leaves and conifer trees, sewage sludge, sawdust, mushroom compost and municipal waste compost are other materials that can be used as planting beds (Fred et al. 1997; Scharpenseel and Kunth 1987; Verdonck and Gabriels 1992). The aim of this study is to evaluate the possibility of using cellulose waste compost of peanuts (shells) as a growth medium for *Dracaena*.

Methods

This experiment was conducted at the Ornamental Plants and Flower Research Center, Lahijan, Iran, to evaluate the impacts of composted peanut shell wastes as an alternative to peat on the growth medium for ornamental plants. The peanut shells were supplied by a peeling peanut factory located in Astane Ashrafie Township, Guilan, Iran, and transferred to the research station. Peanut shells were placed in 1 M^3 wooden boxes (Fig. 1) that had pores to provide aerobic conditions and ensure the activity of microorganisms. The temperatures and aerations were recorded during 4 months and the prepared compost was applied as the growth medium. Some physical and chemical characteristics of peanut shells (total nitrogen, phosphorus, potassium, organic carbon, C/N ratio, EC and pH in 1:5 extract peanut shells to water) before and after composting were measured (Table 1).

D. marginata was selected as the test plant. It is a perennial plant with evergreen, shiny and smooth leaves and often has a cluster makeup at the end. It has green leaves with a dominant red color in the periphery and ivory-colored stripes at the central parts. These species have lower chlorophyll than typical green species (Edwards 1999). The peanut shell compost was produced and passed through a 20 mm sieve. Some physical and chemical characteristics (total nitrogen, phosphorus, potassium, organic carbon, C/N ratio, EC and pH in 1:5 extract peanut shells to water) were measured (Table 1). A factorial completely randomized design was used with two factors including (a) growth media with different levels of peanut shell compost and (b) nutrient solution.

Factor A six treatments of the growth medium are as follows:

1. Control: medium with a ratio 2:1 v/v of peat to perlite,
2. 15 % compost: peat was replaced by 15 v/v of peanut shell compost in the medium,
3. 30 % compost: peat was replaced by 30 v/v of peanut shell compost in the medium,
4. 45 % compost: peat was replaced by 45 v/v of peanut shell compost in the medium,
5. 60 % compost: peat was replaced by 60 v/v of peanut shell compost in the medium,

Fig. 1 The applied *box* for providing peanut shell compost

Table 1 Some properties of composted peanut shells and peat used in the experiment

Property	Peanut shell wastes	Composted peanut shells	Peat
Total nitrogen (%)	0.87	2.76	0.63
Total phosphorus (%)	1.87	0.67	0.03
Total potassium (%)	1.19	1.48	0.03
Organic carbon (%)	30.0	27.1	55.7
C/N ratio	34.5	9.8	88.5
pH (1:5)	5.89	5.06	4.62
EC (dS/m)	1.38	4.30	0.32

6. 100 % compost: peat was replaced by 100 v/v of peanut shell compost in the medium.

Factor B

1. With the use of nutrient solution,
2. Without the use of nutrient solution.

Therefore, 12 treatments in three replications were used in this study. The applied peat was purchased from the SAB Company in Germany. After preparing the media, the rooted cutting of *D. marginata* was transferred to pots. For this aim, first after preparing the beds, the roots of *D. marginata* were removed from the transplant pot and completely washed. Then for each pot with a new bed, one plant was cultivated and then transferred to the greenhouse to pass the plant growth period. Plant height and crown diameter, dry weights of the leaves and roots and fresh weights of the leaves and roots were measured at the end of the experiment.

The total Kjeldahl nitrogen (TKN) and the total organic carbon (TOC) of the samples were estimated using a micro-Kjeldahl method (Singh and Pradhan 1981) and Walkey and Blacks (1934) rapid titration method, respectively. The pH and EC were determined on an extract of 1:5 ratio of compost to water by weight. Phosphorus and potassium were determined by spectrophotometric and flame photometric methods, respectively. The physical properties of beds were measured by Gabriels et al. (1993).

During the growth period, the pots were irrigated with distilled water. The leaves were harvested after 4 months and the dry matter yield was determined after drying of the harvested leaves at 70 °C for 48 h. Subsamples of the dry leaves were ground and then the dry ash was extracted with 2 M HCl from the furnace at 550 °C. The concentrations of Ca, Mg, Fe, Mn and Zn were measured in the extracts by atomic absorption spectrophotometry, K by flame photometry, and P by spectrophotometry. The experiment was a completely randomized design in three replications and MSTATC software was used for the variance analysis of data by least significant difference (LSD) test.

Results

Physical and chemical characteristics of the cultivation media

Some physical and chemical characteristics of the planting media are provided in Table 2. The use of peanut shell compost tends to decrease the bulk density of the growth media. The highest decrease was observed in 100 % compost; consequently, the porosity of this medium increased by 2 % more than that of the control. The C/N ratios in the peanut shell compost treatments were much lower than in the control due to the high organic carbon and low total nitrogen in peat compounds. The use of peanut shell compost caused a decrease in the available phosphorus of cultivation beds. A noticeable aspect was an increase in potassium in the cultivation beds in proportion to the used peanut shell compost. Potassium in 100 % perlite treatment was 40 times more than that of the control treatment, but it should be noted that the salinity of the bed increased from 0.21 dS/m in the control to 4.38 dS/m in the 100 % perlite treatment. Acidity (pH) increased in proportion to the compost used, so pH increased by 1.18 units in the 100 % compost treatment compared with that in the control.

Impacts of treatments on plant growth

ANOVA results (Table 3) showed that the effect of compost (A) on all indices of growth and the impact of nutrient solution (B) on the growth indices of plants (except the dry weight of root) were significant at a 5 % level. The interaction impact A × B was significant at a 5 % level in terms of leaf numbers, height, crown diameter and the fresh and dry weight of leaves.

Table 4 indicates the impact of treatments on the growth of plants. The highest recorded leaf number was observed at 15, 30 and 60 % compost, respectively. The highest plant height (23.61 cm) was obtained for the 15 % treatment of peanut shell compost, which had no significant difference from the 30 % treatment with peanut shell compost (22.67 cm). However, there was a significant difference in the control (19.11 cm). The lowest height (11.67 cm) was related to 100 % peanut shell compost. The largest crown diameter (2.36 cm) was found for the 30 % peanut shell compost, a more significant difference than that found for the control. All peanut shell compost treatments caused an increase in crown diameter compared to the control. The largest dry weight of root (8.29 g) was obtained in 45 % peanut shell compost, which differed significantly from that of the control (2.25 g), but showed no difference from 15, 30 and 60 % compost. The lowest dry weight of root (5.77 g) was related to 100 % peanut shell compost. The treatments of peanut shell compost

Table 2 The effect of treatments on some physicochemical properties of the growth media

Treatment	Bulk density (g/cm^3)	Porosity (%)	Total nitrogen (%)	Available P (mg/kg)	Available K (mg/kg)	Available Na (mg/kg)	Available Ca (mg/kg)	Available Mg (mg/kg)	Organic Carbon (%)	C/N ratio	pH (1:5)	EC (dS/m)
Control	0.23	92.0	0.80	26.3	50.0	306.4	7.6	81.3	47.6	47.0	5.32	0.21
15 % peanut shell compost	0.22	92.0	2.90	18.7	248.6	310.0	6.0	81.2	33.7	11.6	5.51	1.16
30 % peanut shell compost	0.21	92.0	2.90	24.3	450.0	315.4	5.6	59.0	32.1	11.6	5.99	2.26
45 % peanut shell compost	0.20	92.0	3.02	23.3	747.2	350.2	8.0	82.8	35.1	11.6	6.12	2.47
60 % peanut shell compost	0.20	93.0	2.74	16.0	994.4	360.9	7.4	49.3	31.8	11.6	6.32	4.11
100 % peanut shell compost	0.16	94.0	2.20	7.1	2000.0	404.3	7.6	74.9	24.6	11.1	6.50	4.38

significantly increased the leaf dry weight, more than in the control; however, the lowest leaf dry weight (4.02 g) obtained in 100 % peanut shell compost was significantly lower than that of the control.

Table 5 shows that the nutrient solutions had a further impact on the growth of plants in all treatments, compared to plants receiving no nutrient solution. Based on Table 6, the highest plant height (26.0 cm) was obtained for 15 % peanut shell compost with a nutrient solution. This differed significantly from the control (13.0 cm) without a nutrient solution. The comparison between the control treatments, the nutrient solution and compost treatments without a nutrient solution indicated that the growth indices of the control treatment were approximately the same with compost treatment and without a nutrient solution.

Impacts of treatments on nutrient concentration in leaves

Based on Table 7, the effect of compost (A) on the concentration of N, P, Ca, Mg, Fe, Zn and Mn in leaves and the impact of nutrient solution (B) on P, K, Fe and Mn were significant at a 5 % level. The interaction impact A × B was significant at all nutrient concentrations (excluding N) at a 5 % level. The concentration of P, Mg and Mn in leaves decreased in treatments with peanut shell compost (Table 8). Nitrogen concentration increased in 15 % compost treatments and then decreased in other treatments of compost, so they showed no significant difference from the control.

Nutrient solution significantly increased the phosphorus and potassium and decreased the zinc and manganese concentrations in leaves, compared to leaves without a nutrient solution (Table 5). The interaction impact A × B on the nutrient concentration in *Dracaena* leaves is provided in Table 9. The results show that the highest leaf nitrogen (2.74 %) was related to a 15 % peanut shell compost treatment with a nutrient solution, which differed significantly from 0, 30, 45, 60 and 100 % treatments in both nutrition conditions. The control treatment had a higher concentration of phosphorus (2.13 %) than the 15, 30, 45 and 60 % treatments of peanut shell compost. The lowest phosphorus concentration was observed for the 45 % peanut shell compost without a nutrient solution. The highest K concentration was observed for the 30 % shell compost with a nutrient solution. The impact of peanut shell compost was significant on leaf calcium. The highest calcium (8.74 %) was observed for the 30 % peanut shell compost with a nutrient solution, a significant difference from the other treatments.

Based on Table 7, the effect of compost (A) on the uptake of all nutrients by plants and the impact of nutrient solution (B) on N, P, K, Ca, Mg, Fe and Mn were

Table 3 The ANOVA results of the treatment effect on growth of *Dracaena marginata*

Variation Resources	Freedom Degree	Mean squared								
		Leaf number	Height	Crown diameter	Fresh weight of root	Dry weight of root	Fresh weight of stem	Dry weight of stem	Fresh weight of leaf	Dry weight of leaf
A: compost	5	86.4*	108.3*	1.10*	723.0*	7.63*	98.0*	3.87*	347.6*	16.5*
B: nutrient solution	1	2131.0*	285.4*	3.82*	887.6*	11.3^{ns}	500.5*	14.2*	3108.6*	65.0*
A × B	5	77.6*	19.5*	0.87*	245.9^{ns}	2.19^{ns}	19.3^{ns}	0.15^{ns}	82.9*	6.34*
Error	24	296.7	2.59	0.28	104.3	2.57	198.8	13.5	28.3	2.30

* significant at 5 % level

^{ns} not significant at 5 % level

Table 4 The effect of treatments on the growth of *Dracaena marginata*

Treatment	Leaf Number	Height (cm)	Crown diameter (cm)	Fresh weight of leaf (g)	Dry weight of leaf (g)	Fresh weight of stem (g)	Dry weight of stem (g)	Fresh weight of root (g)	Dry weight of root (g)
Control	21 cd	19.1 bc	1.33 c	33.0 b	4.75 cd	13.1 bc	2.25 cd	47.2 c	5.80 b
15 % peanut shell compost	26 ab	23.6 a	2.33 a	41.8 a	6.62 b	18.0 a	3.66 a	62.9 ab	7.42 ab
30 % peanut shell compost	27 ab	22.7 a	2.36 a	42.4 a	7.01 ab	16.4 ab	3.36 ab	63.5 a	8.27 a
45 % peanut shell compost	25 bc	20.6 b	2.25 a	41.3 a	8.67 a	15.0 bc	2.91 bc	67.4 a	8.29 a
60 % peanut shell compost	29 a	18.7 c	2.00 ab	37.7 ab	6.21 bc	12.0 c	2.63 bc	50.8 bc	6.94 ab
100 % peanut shell compost	19 d	11.7 d	1.58 bc	22.7 c	4.02 d	6.5 d	1.43 d	39.6 c	5.77 b

Values followed by the same letters in each row and column are not significantly different at the 0.05 level (least significant difference)

Table 5 The effect of nutrient solution on plant growth, nutrient concentration in leaf and nutrient uptake (mg/pot) by plant

Nutrient solution	The growth factors of *dracaena marginata*								
	Leaf number	Height (cm)	Crown diameter (cm)	Fresh weight of leaf (g)	Dry weight of leaf (g)	Fresh weight of stem (g)	Dry weight of stem (g)	Fresh weight of root (g)	Dry weight of root (g)
Without NS*	17 b	16.6 b	1.65 b	50.3 b	6.52 a	9.77 b	20.1 b	27.2 b	4.87 b
With NS	32 a	22.2 a	2.30 a	60.2 a	7.64 a	17.23 a	3.3 a	45.8 a	7.56 a

Nutrient solution	Nutrient concentration in leaf							
	Nitrogen (%)	Phosphorus (%)	Potassium (%)	Calcium (%)	Magnesium (%)	Iron (mg/kg)	Zinc (mg/kg)	Manganese (mg/kg)
Without NS	1.49 a	0.67 b	1.86 b	6.06 b	1.25 a	278.6 a	44.50 a	121.9 a
With NS	1.79 a	0.90 a	3.21 a	6.60 b	1.42 a	278.4 a	27.89 b	96.75 b

Nutrient solution	Nutrient uptake (mg/pot) by plant							
	Nitrogen	Phosphorus	Potassium	Calcium	Magnesium	Iron	Zinc	Manganese
Without NS	64.4 b	29.6 b	89.1 b	290.4 b	55.3 b	1.34 b	0.19 a	0.52 b
With NS	133.9 a	57.4 a	236.1 a	491.9 a	96.7 a	1.94 a	0.20 a	0.72 a

* Nutrient solution

significant at a 5 % level. The interaction impact A × B was not significant only on N uptake at a 5 % level. The results of Table 10 show that the highest nitrogen uptake (162.1 mg/pot) was obtained for the 15 % peanut shell compost, which showed a significant difference from the 0, 30, 45, 60 and 100 % treatments. The lowest uptake of nitrogen was observed for the 100 % peanut shell compost (53.27 mg/pot). The control treatment had the highest phosphorus uptake (70.28 mg/pot), which did not show a significant difference from the 15 % compost; however,

Table 6 The effect of treatments and nutrient solution on the growth of *Dracaena marginata*

Treatments	Nutrient solution	Leaf number	Height (cm)	Crown diameter (cm)	Fresh weight of leaf (g)	Dry weight of leaf (g)	Fresh weight of stem (g)	Dry weight of stem (g)	Fresh weight of root (g)	Dry weight of root (g)
2 Peat + 1 perlite	Without	13 e*	13.0 f	0.67 e	44.4 cd	5.50 cd	7.58 gh	1.59 ef	20.18 ef	3.42 de
1.7 Peat + 0.3 compost + 1 perlite		12 e	21.2 bc	2.11 bcd	55.6 bc	6.71 abcd	12.51 def	3.14 abc	27.93 de	4.83 cde
1.4 Peat + 0.6 compost + 1 perlite		24 d	19.8 cde	2.08 bcd	68.7 ab	8.37 ab	13.91 cde	2.79 bcde	36.70 cd	6.96 bc
1.1 Peat + 0.9 compost + 1 perlite		16 e	17.5 de	1.35 de	53.4 bc	7.21 abc	10.03 defg	2.02 cdef	30.42 d	5.77 cd
0.8 Peat + 1.2 compost + 1 perlite		25 cd	17.2 e	2.03 bcd	47.7 cd	6.98 abcd	9.51 efgh	1.93 cdef	32.16 d	5.82 cd
2 Compost + 1 perlite		11 e	10.8 f	1.65 cd	31.9 d	4.38 d	5.08 h	0.99 f	15.74 f	2.42 e
2 Peat + 1 perlite	With	30 bcd	25.2 a	2.00 bcd	49.9 c	6.11 bcd	18.66 abc	2.90 bcd	45.94 b	6.09 bc
1.7 Peat + 0.3 compost + 1 perlite		39 a	26.0 a	2.55 abc	70.3 ab	8.13 abc	23.42 a	4.17 a	55.70 a	8.41 d
1.4 Peat + 0.6 compost + 1 perlite		31 bc	25.5 a	2.65 ab	58.4 bc	8.18 abc	18.93 ab	3.93 ab	48.09 ab	7.05 bc
1.1 Peat + 0.9 compost + 1 perlite		34 ab	23.7 ab	3.14 a	81.3 a	9.37 a	19.96 a	3.80 ab	52.09 ab	11.58 a
0.8 Peat + 1.2 compost + 1 perlite		33 ab	20.1 cd	1.96 bcd	54.0 bc	6.91 abcd	14.44 bcd	3.32 ab	43.19 bc	6.60 bc
2 Compost + 1 perlite		28 bcd	12.7 f	1.51 de	47.4 cd	7.16 abc	7.97 fgh	1.87 def	29.63 d	5.63 cd

Values followed by the same letters in each column are not significantly different at the 0.05 level (least significant difference)

* LSD (least significant difference) shows the significant difference ($\rho = 0.05$) among the different treatments

there was a remarkable difference with other peanut shell compost treatments. The highest and the lowest uptakes of potassium were, respectively, related to the 45 and 100 % peanut shell compost. Potassium uptake in 15 and 30 % of peanut shell compost treatments (182.6 and 185.0 mg/pot, respectively) showed no significant difference from the control. The lowest calcium uptake was related to the 100 % peanut shell compost, which showed no significant difference from the control. Manganese uptake increased significantly for the 15 % compost compared to the control, but decreased significantly in other peanut shell compost treatments when compared with the control.

A nutrient solution significantly increased the uptake of N, P, K, Ca, Mg, Fe and Mn compared to no nutrient solution (Table 5). Table 11 shows the interaction impact of A × B on the nutrient uptake by plants. The 15 % compost treatment with a nutrient solution showed the best uptake of nutrients compared to the control. The control treatment with a nutrient solution had a better nitrogen, phosphorus potassium and magnesium uptake than the compost treatment without a nutrient solution.

Discussion

The total nitrogen increased in proportion to the peanut shell compost used. This may have been due to the increased nitrogen provided by composted peanut shells rather than peat. Grigatti et al. (2007) reported increasing nitrogen in pot cultivation beds by adding manure compost to the soil rather than peat. The nitrogen of cultivated beds was at a sufficient level for plant growth according to Nappi and Barberis (1993). The phosphorus decreased in proportion to the compost used (Grigatti et al. 2007). The highest and the lowest phosphorus for growth media was related to the control and the 100 % treatment with peanut shell compost, respectively. The phosphorus levels for composted peanut shells were higher than for peat, but the available P of compost treatments decreased. It appeared that microorganisms had converted mineral P to organic P during organic matter mineralization, thereby temporarily decreasing the availability of phosphorus (Mohammadi Tarkashvand et al. 2005). Grigatti et al. (2007) and Prez-Murcia et al. (2006) also reported a decrease in phosphorus of growth media containing green waste compost and sewage sludge. Meanwhile, there occurred a nutrient concentration and C/N ratio change during the growth of plants and decomposition of organic matter in beds. Logakanthi et al. (2006) reported that the C/N ratio of vegetable waste was reduced by 69 % during composting, along with the consumption of 50 % phosphorus by fungal species. Peanut shell compost increased the potassium levels of media in proportion to the compost used

Table 7 The ANOVA results of the treatment effects on nutrient concentration in leaf and nutrient uptake by plant

Variation in resources	Freedom degree	Mean squared							
		Nutrient concentration in leaf							
		Nitrogen	Phosphorus	Potassium	Calcium	Magnesium	Iron	Zinc	Manganese
A: compost	5	1.21*	1.98*	0.023^{ns}	2.83*	20.5*	31683.5*	1153.3*	56607.6*
B: nutrient solution	1	0.81^{ns}	0.45*	16.4*	2.67^{ns}	0.26^{ns}	0.38^{ns}	2483.3*	5680.6*
A × B	5	0.16^{ns}	1.67*	0.51*	2.66*	3.02*	40086.0*	1122.9*	9268.3*
Error	24	0.23	1.21	0.14	0.29	5.26	2273.2	93.9	484.1
Variation in resources	Freedom degree	Mean squared							
		Nutrient uptake by plant							
		Nitrogen	Phosphorus	Potassium	Calcium	Magnesium	Iron	Zinc	Manganese
A: Compost	5	7518.2*	3868.2*	11587.7*	51094.7*	5298.0*	0.936*	0.054*	1.34*
B: Nutrient solution	1	38527.1*	6983.3*	194378.1*	365396.0*	15387.9*	3.22*	0.001 ^{ns}	0.34*
A × B	5	1685.5 ^{ns}	3153.6*	2749.1*	19605.2*	5272.2*	1.44*	0.059*	0.067*
Error	24	782.1	209.7	1849.5	9476.7	896.7	0.197	0.005	0.019

* significant at 5 % level

^{ns} not significant at 5 % level

Table 8 The effect of treatments on the nutrient concentration of leaf

Treatment	Nitrogen (%)	Phosphorus (%)	Potassium (%)	Calcium (%)	Magnesium (%)	Iron (mg/kg)	Zinc (mg/kg)	Manganese (mg/kg)
Control	1.83 b	1.61 a	2.65 a	6.39 b	1.83 b	341.7 a	27.8 b	258.1 a
15 % peanut shell compost	2.46 a	1.02 b	2.57 a	6.29 b	0.81 cd	185.5 b	48.9 a	204.2 b
30 % peanut shell compost	1.36 b	0.49 d	2.60 a	7.28 a	0.25 cd	314.4 a	24.7 b	74.2 c
45 % peanut shell compost	1.27 b	0.80 c	2.45 a	5.15 c	0.70 cd	185.7 b	28.4 b	53.7 cd
60 % peanut shell compost	1.55 b	0.33 e	2.46 a	6.29 b	0.76 cd	329.6 a	58.2 a	32.2 d
100 % peanut shell compost	1.35 b	0.32 e	2.37 a	6.57 b	2.76 a	313.9 a	29.2 b	32.1 d

Values followed by the same letters in each row and column are not significantly different at the 0.05 level (least significant difference)

compared to the control (Grigatti et al. 2007). The amount of potassium in peanut shell compost was 50 times more than in peat, leading to an increase in the medium's potassium level.

The higher nitrogen and lower carbon levels of peanut shell compost compared to peat caused a decrease in the C/N ratio of compost treatments (Gayasinghe et al. 2010). This ratio was lower than the allowed level, which is suitable for growing ornamental plants. Davidson et al. (1994) reported that the C/N ratio of ideal compost was lower than 20 for growth of plants. A C/N ratio of more than 30 may cause problems for plant growth (Zucconi et al. 1981). The pH values of the cultivation beds were at the optimum range for culturing ornamental plants, so according to Abad et al. (2001) the appropriate pH for desirable growth was determined as 5.3–6.5.

Some favorable factors in terms of beauty such as the size and appearance of the plant are the main criteria for determining ornamental plants' tolerance to salinity. The maximum allowed salinity level of saturation extract for *Dracaena* is 6–8 dS/m (Alizadeh 1999). Therefore, it is possible that salinity may cause some problems in the growth of plants with higher compost level. Grigatti et al. (2007) found that the pH and EC of cultivation beds increased by adding 25–100 % v/v green waste compost instead of peat to cultivation beds.

A comparison of the control treatment with a nutrient solution with compost treatments, but without a nutrient solution indicated that growth indices in the control treatment were approximately the same with compost treatments and without a nutrient solution. The growth of *Dracaena* includes height, leaf number, dry leaf weight and dry stem weight, which in 15, 30, 45 and 60 % compost were higher than in the control and in 100 % peanut shell compost. It seems that the impact of peanut shell compost occurs due to the presence of humic materials; thus, Chen et al. (1989) claimed that the impact of compost on *Ficus benjamina* growth may be similar to the role of growth

Table 9 The effect of treatments and nutrient solution on the nutrient concentration of leaf

Treatments	Nutrient solution	Total nitrogen	Phosphorus	Potassium	Calcium	Magnesium	Iron	Zinc	Manganese
2 Peat + 1 perlite	Without	1.70 bc	2.13 a	1.80 d	6.58 bcde	1.73 bc	442.4 a	25.01 c	345.4 a
1.7 Peat + 0.3 compost + 1 perlite		2.18 ab	1.13 b	1.73 d	5.71 ef	0.73 de	195.2 bc	69.50 b	225.7 b
1.4 peat + 0.6 compost + 1 perlite		1.04 c	0.52 d	1.53 d	5.83 dfe	1.08 cd	374.7 c	26.86 c	53.10 e
1.1 Peat + 0.9 compost + 1 perlite		1.17 c	0.07 f	2.14 cd	5.17 f	0.67 de	208.7 bc	21.80 c	53.14 e
0.8 Peat + 1.2 compost + 1 perlite		1.36 bc	0.10 ef	1.76 d	6.67 bcd	1.09 cd	246.6 b	86.62 a	32.62 e
2 Compost + 1 perlite		1.50 bc	0.10 ef	2.18 cd	6.38 bcde	2.20 b	203.9 bc	37.23 c	21.27 e
2 Peat + 1 perlite	With	1.97 abc	1.10 b	3.33 ab	6.21 bcde	1.93 b	241.0 bc	30.54 c	170.8 c
1.7 Peat + 0.3 compost + 1 perlite		2.74 a	0.91 bc	3.41 ab	6.86 b	0.90 de	175.8 bc	28.41 c	182.7 c
1.4 Peat + 0.6 compost + 1 perlite		1.69 bc	0.47 de	3.68 a	8.74 a	1.41 bcd	254.1 b	22.46 c	95.23 d
1.1 Peat + 0.9 compost + 1 perlite		1.38 bc	0.12 ef	2.67 bc	5.14 f	0.73 de	162.7 c	34.98 c	54.30 e
0.8 Peat + 1.2 compost + 1 perlite		1.75 bc	0.56 cd	3.16 ab	5.90 cdef	0.23 e	412.6 a	29.80 c	31.84 e
2 Compost + 1 perlite		1.20 c	2.23 a	2.90 b	6.76 bc	3.32 a	424.0 a	21.16 c	45.56 e

Table 10 The effect of treatments on the nutrient uptake (mg/pot) by plant

Treatment	Nitrogen	Phosphorus	Potassium	Calcium	Magnesium	Iron	Zinc	Manganese
Control	89.4 b	70.3 a	130.8 cd	300.9 cd	90.5 ab	1.49 b	0.138 b	1.08 b
15 % peanut shell compost	162.1 a	65.6 a	185.0 ab	426.8 ab	55.1 cd	1.22 b	0.286 a	1.30 a
30 % peanut shell compost	95.6 b	35.6 b	182.6 ab	509.6 a	87.5 abc	2.19 a	0.172 b	0.52 c
45 % peanut shell compost	109.8 b	69.3 c	222.9 a	453.3 ab	61.8 bcd	1.51 b	0.226 a	0.47 c
60 % peanut shell compost	96.9 b	21.3 bc	185.7 bc	388.9 bc	39.4 d	2.07 a	0.348 a	0.20 d
100 % peanut shell compost	53.3 c	24.5 bc	98.9 d	267.6 d	121.8 a	1.37 b	0.105 b	0.15 d

Values followed by the same letters in each row and column are not significantly different at the 0.05 level (least significant difference)

regulators in plant. The growth of plants decreased significantly in the 100 % peanut shell compost treatment due to the large number of pores and a decrease in water-holding capacity. Pool and Conover (1991) also reported the weak growth of dracaena grown in organic beds with a high pore presence and low water-holding capacity. The growth of dracaena in the control bed was low, so indexes such as leaf number and dry stem and leaf weight in this treatment showed no significant difference compared to 100 % treatment with peanut shell compost. This may be due to the large C/N ratio in the control bed and the decreased need for nitrogen of the plant, compared to the 15, 30, 45 and 60 % treatments with peanut shell compost. Gayasinghe et al. (2010) used manure compost (CMC) and synthetic compounds (SA) as an alternative to peat in cultivating *Tagetes paluta* and concluded that plant height, the number of flowers for each plant, the dry and fresh weights of stems, root length and dry and fresh weights of roots increased in the combined treatment using 40 % SA and 60 % v/v CMC.

The nutrient concentration in dracaena leaves was within the presented range as shown by Denis et al. (2003), denoting the effectiveness of nutrient range for plant growth. Lack of significant potassium level changes in treatments using peanut shell compost may be due to a higher yield of dry leaf matter in these treatments, compared to the control. In fact, potassium uptake occurred in plants, but the higher yield of plants and the dilution effect indicated no observable difference from the control. Additionally, in most cases, concentration variations did not follow from the values of these elements in the cultivation beds. In terms of nutrient concentration in plant organs, these depended on different factors such as plant growth, ionic competition and deposition; therefore, sometimes it is impossible to use nutrient concentrations in plants as a reliable parameter for assessing plant growth. The impact of nutrient dilution resulting in additional yields can also give rise to confusion. In this regard, nutrient uptake by plants from growth media is considered a more reliable parameter.

Increasing the amount of peanut shell compost (less than 60 %) caused an increase in nitrogen uptake by plants when compared with the control; this may have been due to the decomposed organic material in the substrate medium and as a result of increasing the amount of available nitrogen to plants. Increasing peanut shell

Table 11 The effect of treatments and nutrient solution on the nutrient uptake (mg/pot) by plant

Treatments	Nutrient solution	Total nitrogen	Phosphorus	Potassium	Calcium	Magnesium	Iron	Zinc	Manganese
2 Peat + 1 perlite	Without	59.0 ef	66.7 a	59.9 fg	223.9 cd	64.0 cde	1.46 cd	0.086 e	0.69 b
1.7 Peat + 0.3 compost + 1 perlite		100.9 cde	54.6 ab	83.8 efg	276.0 bcd	35.6 de	0.95 de	0.24 cd	1.08 b
1.4 Peat + 0.6 compost + 1 perlite		72.3 def	33.5 bc	105.2 efg	402.5 b	74.9 bcd	1.79 bc	0.16 de	0.37 d
1.1 Peat + 0.9 compost + 1 perlite		66.1 ef	4.1 d	130.9 def	298.9 bcd	40.6 de	1.19 cd	0.123 de	0.31 d
0.8 Peat + 1.2 compost + 1 perlite		78.0 cdef	6.0 d	102.7 efg	388.9 b	15.2 e	1.43 cd	0.20 ab	0.19 de
2 Compost + 1 perlite		34.5 f	1.3 d	52.5 g	152.5 d	53.3 cde	0.38 e	0.09 e	0.05 e
2 Peat + 1 perlite	With	119.8 bc	73.9 a	201.7 bcd	377.9 bc	117.0 b	1.52 cd	0.19 de	1.14 b
1.7 Peat + 0.3 compost + 1 perlite		223.2 a	76.6 a	286.2 a	577.7 a	74.6 bcd	1.49 cd	0.33 bc	1.53 a
1.4 Peat + 0.6 compost + 1 perlite		118.9 bc	37.8 b	259.9 ab	616.8 a	100.1 bc	2.60 a	0.187 de	0.67 c
1.1 Peat + 0.9 compost + 1 Perlite		153.3 b	13.5 cd	315.0 a	607.7 a	83.1 bcd	1.83 bc	0.41 ab	0.63 c
0.8 Peat + 1.2 compost + 1 perlite		116.0 bcd	36.6 bc	208.6 bc	490.9 b	63.6 cde	2.72 a	0.50 a	0.21 de
2 Compost + 1 perlite		72.0 def	17.8 cd	145.3 cde	382.7 bc	140.2 a	2.36 ab	0.12 e	0.25 de

compost caused a decrease in the C/N ratio when compared with the control. This process was also reported by Oworu et al. (2010), so an increase in the uptake of nitrogen by the plant was observed when compost was added to the growth medium of an Amaranthus ornamental plant.

The increasing amount of peanut shell compost decreased phosphorus uptake in leaves in comparison to the control (compost more than 15 %). It appears that microorganisms converted mineral P to organic P during organic matter mineralization, thereby temporarily decreasing the availability of phosphorus (Mohammadi Tarkashvand et al. 2005). The lowest phosphorus uptake was obtained in the 45 % compost treatment. This may have been due to the dilution effect, because of the larger size of the plant and a decreasing P concentration. Grigatti et al. (2007) also reported a decrease of phosphorus uptake by mimulus and salvia plants in beds containing green waste compost and sewage sludge when compared with controls (white peat).

Increasing K uptake in compost treatments is due to an increase in the availability of potassium in media by addition of peanut shell compost. Calcium uptake increased in compost treatments due to the higher yield of these treatments, compared to the control. The decrease in Ca uptake in the 100 % compost treatment was due to the lower yield of plant than in the control. Manganese uptake increased in 15 % compost and then decreased in other treatments of peanut shell compost in proportion to the applied compost. This increase was due to the higher yield of plants when compared with the control, while the subsequent decrease was due to the decreasing Mn concentration in leaves (Mn concentration in leaves was 258.1, 204.2, 74.2, 53.7, 32.2 and 32.1 mg/kg, respectively, in the control and 15, 30, 45, 60 and 100 % of plants with peanut shell compost). Magnesium uptake decreased in compost treatments compared to the control, whereas it increased in the 100 % compost treatment. This may have been due to ionic competition and the interaction effect between Ca and Mg. The increase of Fe uptake in 30 and 60 % compost treatments, compared to the control, was due to the higher yield of plants with these treatments. The highest zinc uptake was observed for the 60 % compost treatment, which may have been due to the higher concentration of Zn in the shoots of plants compared to other treatments (58.21 mg/kg).

Conclusion

Peanut shells as wastes of peanut have a considerable volume in Guilan Province, Iran, the compost of which can be used as a cheap and accessible resource in the growth medium of ornamental plants. Peanut shell compost because of its favorable properties and high porosity can be mixed with low-porosity substrates and applied as a substitute to peat in the cultivation media. Increase in peanut compost level as 60 and 100 % replacement is not suggested. Investigation of the effects of peanut shell compost on the growth of other varieties of ornamental plants has been proposed.

Acknowledgments The authors would like to thank the Islamic Azad University, Rasht Branch, for the use of equipment and also Mr. Mahboob Khomami for his assistance.

References

Abad M, Noguera P, Bures S (2001) National inventory of organic wastes for use as growing media for ornamental potted plant production: case study in Spain. Bioresour Technol 77:197–200

Agriculture Organization of Guilan Province (2010) Crops data of Guilan province, Rasht, Iran

Alizadeh A (1999) Soil–water–plant relationships. Astan-e- Qods Razavi Press, 355 p (In Persian)

Atieyh RM, Arancon N, Edwars CA, Metzger JD (2000) Influence of earthworm processed pig manure on the growth and yield of greenhouse tomatoes. Bioresour Technol 75:175–180

Bugbee JG (2002) Growth of ornamental plants in container media amended with biosolids compost. Compos Sci Utilz 10:92–98

Chen Y, Inbar Y, Hadar Y (1989) Composted agricultural wastes as potting media for ornamental plants. Soil Sci 145:298–303

Cull DC (1981) Alternatives to peat as container media: organic resources in UK. Acta Hort 126:69–81

Davidson H, Mecklenburg R, Peterson C (1994) Nursery management: administration and culture, 3rd edn. Prentice Hall, Englewood Cliffs

Dennis B, Chen J, Richard J, Kelly C (2003) Cultural guidelines for commercial production of interiorscape *Dracaena*. University of Florida. URL at http://edis.ifas.ufl.edu. Accessed 24 May 2012

Edward F (1999) Dracaena Marginata. Environmental horticulture department, Institute of Agricultural Sciences. University of Florida, Gainesville, 32611

Fred DR, Hariss HM, Watanabe R, Stanley RW (1997) Plant growth in potting media using compost. Horticulture Research Note. University of Hawaii. pp 4

Gabriels R, Kerrsbulkand WV, Engels H (1993) A rapid method for the determination of physical properties of growing media. Acta Hort 342:243–247

Gayasinghe GY, Liyana Arachchi ID, Tokashiki Y (2010) Evaluation of containerized substrates developed from cattle manure compost and synthetic aggregates for ornamental plant production as a peat alternative. Resour Conserv Recycl 54:1412–1418

Grigatti M, Giorgioni ME, Ciavatta C (2007) Compost-based growing media: influence on growth and nutrient use of bedding plants. Bioresour Technol 98:3526–3534

Krumfolz LA, Wilsonand SB, Stoffella PJ (2000) Use of compost as a media amendment for containerized production of perennial cat whiskers. SNA Res Conf 45:69–72

Logakanthi S, Rajesh Banu J, Esakkiizaj JS, Vijayalakshmi GS (2006) Fungal composting- a novel method for green waste composting. Asian J Microbiol Biotechnol Environ Sci Paper 8(2):205–208

Maiti R, Ebeling W (2002) The peanut (*Arachis hypogaea*) crop. Science Pup., Inc., p 376

Mohammadi Tarkashvand A, Kalbasi M, Shariatmadari H (2005) Effects of converter slag on some chemical characteristics of acid soils. J Crop Prod Process 8(4):47–62

Mohammadi Torkashvand A (2010) Improvement of compost quality by addition of some amendments. Aust J Crop Sci 4(4):252–257

Murata M R (2003) The impact of soil acidity amelioration on groundnut production on sandy soils of Zimbabwe. Ph.D. Thesi, Uuniversity of Pretoria, Zimbabwe

Nappi P, Barberis R (1993) Compost as growing medium: chemical, physical and biological aspects. Acta Hort 342:249–256

Oworu OO, Dada OA, Majekodunmi OE (2010) Influence of compost on growth, nutrient uptake and dry matter partitioning of grain amaranths (Amaranthus hypochondriacus L.). Libyan Agr Res Center J Int 1(6):375–383

Papafotiou M, Phsyhalou M, Kargas G, Chatzipavlidis I, Chronopoulos J (2005) Olive-mill waste compost as growth medium component for the production of poinsettia. Hort Sci 102:167–175

Perez-Murcia MD, Moral R, Moreno-Caselles J, Perez-Espinosa A, Paredes C (2006) Use of composted sewage sludge in growth media for broccoli. Bioresour Technol 97:123–130

Pool RT, Conover CA (1991) Potential for eucalyptus mulch used as a component of potting mixes for foliage plant production. University of Florida, IFAS, Central Florida Research and Education Center – Apopka, CFREC-Apopka Research Report, RH-91-13

Rajesh Banu J, Logakanthi S, Vijayalakshmi GS (2001) Biomanagement of paper mill sludge using indigenous (Lampito mauritii) and two exotic (*Edrilus eugineae* and *Eisenia foetida*) earthworms. J Environ Biol 22(3):181–185

Rajesh Banu J, Raj E, Kaliappan S, Beck D, Yeom IT (2007) Solid state biomethanation of fruit wastes. J Environ Biol 28(4):741–745

Scharpenseel HW, Kunth K (1987) Use and importance of Azolla Anabaena in industrial countries, Azolla utilization (IRRI): 241–245

Singh R, Pradhan K (1981) Determination of nitrogen and protein by Kjeldahl method. In: Forage Evaluation Science. Pvt. Publishers Ltd., New Delhi, p 23

Verdonck O, Gabriels R (1992) Part I reference method for the determination of physical properties of plant substrates. Part II reference method for the determination of chemical properties of plant substrates. Acta Hort 302:169–179

WalkeyJA Black JA (1934) Estimation of organic carbon by the chromic acid titration method. Soil Sci 37:29–31

Zucconi F, Pera A, Forte M, Bertoldi M (1981) Evaluating toxicity of immature compost. Biol Cycle 22:54–57

9

Effect of soil type and vermicompost applications on tomato growth

Marc A. Zucco[1] · S. Alan Walters[2] · She-Kong Chong[2] · Brian P. Klubek[2] · Joseph G. Masabni[3]

Abstract

Background Vermicomposts (VC) improve plant growth and development beyond that normally observed from just soil nutrient transformation and availability. These increases in plant productivity have been attributed to improved soil structure and soil microbial populations that have higher levels of activity and greater production of biological metabolites, such as plant growth regulators. Although there have been many studies on the benefits of VC as a fertilizer source, little research has focused on the effects and/or interactions of soil type and VC application rates on vegetable crop productivity. This paper identifies optimum application rate(s) of VC on tomato growth responses for three different textural classes of soils (loamy sand, silt loam, and silty clay).

Results Soils with high VC rates (0.4 and 0.8 g/g) produced taller plants with more leaf and flower numbers, higher leaf chlorophyll content, greater plant biomass, and more total leaf area compared to soils with low VC rates (0.05, 0.1, and 0.2 g/g). Tomato growth increases were also observed at the low VC soil amendment rates compared to the nontreated control. Tomatoes grown in the sandy soil amended with VC generally had the greatest growth responses (plant height, leaf and flower number, and leaf chlorophyll content) compared to the clay or silt loam soils, with the silt loam soil generally providing the least response.

Conclusions This research indicated that VC is a suitable alternative fertilizer for tomato, with approximately 0.5–0.6 g/g VC added to soil resulting in optimal tomato plant growth. Moreover, this rate provided tomato growth results similar to the standard inorganic fertility program. The sandy soil with VC amendments generally increased tomato plant growth parameters the most compared to the clay and loam soils, with the loam soil generally providing the least.

Keywords Earthworms · Organic wastes · Recycling · *Solanum lycopersicon*

✉ Joseph G. Masabni
jmasabni@ag.tamu.edu

Marc A. Zucco
m_zucco@yahoo.com

S. Alan Walters
awalters@siu.edu

She-Kong Chong
skchong@siu.edu

Brian P. Klubek
bklubek@siu.edu

1 USDA, Natural Resources Conservation Service, Grayling, MI 49738, USA

2 Department of Plant, Soil, and Agricultural Systems, Southern Illinois University, MC 4415, 1205 Lincoln Dr., Carbondale, IL 62901, USA

3 Department of Horticultural Sciences, Texas A&M University, 2134 TAMU, College Station, TX 77843-2134, USA

Introduction

Vermicomposts (VC) are finely divided, peat-like materials with high porosity, aeration, drainage, water-holding capacity, and microbial activity (Edwards 2004; Pandya et al. 2014). These composts result from a non-thermophilic biodegradation and stabilization of organic materials through interactions of earthworms and microorganisms (Edwards 2004; Arancon et al. 2003; Agnieszka et al.

2013). A wide range of organic residues, including sewage sludge, animal wastes, crop residues, and industrial refuse are increasingly being converted by earthworms to form VC (Edwards 2004; Roberts et al. 2007; Pascal et al. 2010). The earthworms breakdown the organic residues, which stimulate greater microbial activity, increase nutrient mineralization rates, and rapidly convert the wastes into a humus-like substance that has a finer structure than ordinary composts while possessing greater and more diverse microbial populations (Atiyeh et al. 2000; Yadav and Garg 2011).

Various types of composts, including VC, are often used in sustainable farming systems to improve soil physical properties, provide plant nutrients, and recycle organic wastes. Vermicomposts have been shown to increase plant growth and crop yields in managed and natural ecosystems (Edwards 2004; Arancon et al. 2003; Gutiérrez-Miceli et al. 2007; Pascal et al. 2010). Organic wastes converted to VC have beneficial effects on plant growth and development that are unrelated to increases observed only from soil nutrient transformation and availability. Vermicomposts improve seed germination, seedling vigor, and plant productivity more than what would have been possible from inorganic mineral nutrient sources, while using as little as 10–40 % of the total plant rooting volume (Subler et al. 1998; Gopalakrishnan et al. 2012; Alsina et al. 2013). These increases in plant productivity have been attributed to improved soil structure and soil microbial populations with higher levels of activity and greater production of biological metabolites, such as plant growth regulators (Pascual et al. 1997; Canellas et al. 2002; Atiyeh et al. 2002; Roberts et al. 2007).

Although there have been many studies relating to the benefits of using VC as a fertilizer source (Arancon et al. 2003, 2004a, b; Alsina et al. 2013), little research has focused on the effects and/or interactions of soil type and VC applications on vegetable crop productivity. Therefore, the objective of this research was to identify the optimum application rate(s) of VC on tomato growth responses for three different textural classes of soils (loamy sand, silt loam, and silty clay).

Methods

A study was conducted at the Southern Illinois University Horticulture Research Center Greenhouse in Carbondale, Illinois, USA to determine the effects and/or interactions of soil type and VC applications on tomato growth. The experiment was a 3 × 7 factorial in a randomized complete block design with four replications and was repeated twice in the greenhouse. Three different textural classes of soils (sandy, loamy, and clayey) were amended with seven VC treatments: (1) untreated control (0 g/g VC), (2) standard fertilizer treatment (SFT) with 0 g/g VC and 2 g per pot of an inorganic 12-12-12 fertilizer [N–P–K; nitrogen, phosphorus from P_2O_5, potassium from K_2O]; (3) 0.05 g/g VC; (4) 0.1 g/g VC; (5) 0.2 g/g VC; (6) 0.4 g/g VC; and (7) 0.8 g/g VC. The VC was obtained from New Horizon Organics (Jerseyville, IL, USA). The sandy soil came from an inclusion of sandy orthents within a Medway silty clay loam (Fine-loamy, mixed, mesic Fluvaquentic Hapludolls) in Union county, Illinois, USA (Miles 1979; Fehrenbacher et al. 1984). The other soils were collected in Jackson county, Illinois, USA with the loamy soil collected from a Hosmer silt loam (Fine-silty, mixed, mesic Typic, Fragiudalf) and the clayey soil collected from a Darwin silty clay (Fine, smectitic, mesic Vertic Haplaquolls) (Herman 1979; Fehrenbacher et al. 1984). All soil samples were collected (surface to 15 cm depths) from fields historically used to produce corn (*Zea mays*) and soybean (*Glycine max*). Particle size distribution of each soil was analyzed by the hydrometer method to confirm soil textural class (Table 1; Sabey et al. 2003). All soils were steam pasteurized at approximately 70 °C for 6 h to reduce the presence of soilborne pathogens and then sieved to 4 mm. The moisture content of the sandy soil was adjusted to 10 % and the loamy and clayey soils were adjusted to 20 % to allow for optimal mixing. After mixing, the soil mix was packed into 15-cm-diameter plastic pots to provide 1.7 L of soil volume in each pot.

The nutrient and organic matter content and pH for the VC used in the experiment were determined by analyzing three 20 g VC bulk dried samples (Brookside Laboratories Inc., New Knoxville, OH, USA). The VC had a pH of 6.1, organic matter content of 0.52 g/g, and N, P, and K content of 20,710, 3610, and 4580 mg/kg, respectively. The levels of Ca, Mg, and S were 33,330, 5040, and 2540 mg/kg, respectively; and the micronutrients B, Fe, Cu, Zn, Mn, and

Table 1 Results of textural analysis of Illinois soils (Sabey et al. 2003) used for the tomato vermicompost greenhouse study

Soil separate content (%)			
Soil series[a]	Sand	Silt	Clay
Hosmer silt loam	5.7	70.0	24.3
Darwin silty clay	7.7	48.0	44.3
Sandy orthents (loamy sand)	81.0	8.7	10.3

[a] Hosmer silt loam (fine-silty, mixed, mesic Typic, Fragiudalf) and Darwin silty clay (fine, smectitic, mesic Vertic Haplaquolls) were both collected in Jackson county, Illinois, USA (Fehrenbacher et al. 1984; Herman 1979); and an inclusion of sandy orthents was collected within a Medway silty clay loam (Fine-loamy, mixed, mesic Fluvaquentic Hapludolls) in Union county, Illinois, USA (Miles 1979; Fehrenbacher et al. 1984)

Na were 31, 5750, 55, 192, 690, and 1290 mg/kg, respectively.

'Sun Chief' tomato seeds (Seedway Inc., Elizabethtown, PA, USA) were germinated in a greenhouse and transplanted into pots at the one-leaf stage. Plants were watered 2–3 times per week for the first month after planting and then daily after that to prevent wilting. Tomato suckers were removed and plants were pruned to two fruiting stems at 1 month after transplanting, and each tomato plant was then tied to a 0.5–0.7-m-long bamboo stake. The tomato was chosen for this study for two primary reasons: (1) it is a widely grown vegetable in both greenhouse and field production systems throughout the world, and growers of this crop have a great interest in alternative nutrient systems, and (2) it was also selected since the tomato plant is a high nutrient feeder, which would determine if VC could provide sufficient nutrient resources for a crop that requires high amounts of nitrogen and other nutrients for maximum growth and development.

Plant height (cm), leaf and flower number per plant, leaf chlorophyll content, total leaf area per plant, and dry foliage and root weight (g) were measured at the termination of each experiment at 2 months after transplanting. The height of each plant was measured from the soil line to the tip of the tomato main plant stem. The leaf chlorophyll content was measured using a Minolta SPAD-502 chlorophyll meter (Special Products Analysis Division, Konica Minolta Sensing Inc., Osaka, Japan) on ten randomly selected mature leaves from the mid-portion of tomato plants within each pot. Shoots were cut at the soil surface with all leaves removed from each plant and total leaf area (cm^3) measured with a portable leaf area meter (Model LI-3000, LI-COR, Lincoln, NE, USA). Roots were washed free of soil using tap water. Root and foliage samples were oven dried at 80 °C for 48 h and weighed.

Data were subjected to analysis of variance procedures using the general linear models procedure of SAS (SAS Inst., Cary, NC, USA) appropriate for a factorial experimental design to determine the effects of soil type and VC rate on tomato growth response variables. Fisher's Least Significance Difference (LSD) test at $P \leq 0.05$ was used to make comparisons among soil types. Orthogonal contrasts were used to make comparisons between the various combinations of low and high VC rates, nontreated control, and standard inorganic fertilizer treatments. Furthermore, data were also analyzed using regression analysis to determine the influence of VC rate on tomato growth responses.

Results

No interactions ($P > 0.05$) were detected between the two greenhouse experiments with either soil type or VC rate, and between soil type and VC rate for most tomato growth responses (data not presented). Thus, tomato growth data are presented by soil type and VC rate main effects (Tables 2, 3).

Soil type

Tomatoes grown in the sandy soil had the greatest plant heights, leaf and flower numbers per plant, and leaf chlorophyll content compared to the clay and loam soils (Table 2), with the loam soil generally providing the least. Although the clay and sandy soils provided similar high dry shoot foliage weights, the clay soil produced the greatest root dry weights compared to the other soils. The three soils differed ($P \leq 0.05$) for root-to-shoot ratios and leaf area with the loam soil having the highest, followed by the clay and sandy soils, respectively. Although total leaf area per plant was not influenced by soil type, the amount of leaf area per leaf was highest for the loam soil, followed by the clay soil, with the sand soil type providing the least.

Vermicompost rate

The low VC rates (0.05, 0.1, and 0.2 g/g) did not improve the tomato plant height, leaf and flower numbers per plant, leaf chlorophyll content, and dry root biomass compared to that of the control (0 g/g) (Table 3). Although no differences ($P > 0.05$) were detected between the high VC rates (0.4 and 0.8 g/g) and the SFT for most tomato growth responses (except leaf and flower number per plant), differences ($P \leq 0.05$) were generally observed between the low and high VC rates for most growth responses (except leaf SPAD value). Tomato leaf and flower number per plant responded to VC applications, as differences were detected among most contrasts evaluated. However, no VC rate provided tomato leaf and flower numbers on a per plant basis that were similar to SFT. Higher leaf SPAD values were also obtained for SFT compared to either 0 g/g control or low VC rates. Although differences were not detected ($P > 0.05$) for leaf chlorophyll content between the low and high VC rates, the low rate provided lower SPAD values compared to SFT. The highest VC rate evaluated (0.8 g/g VC) also had high leaf chlorophyll content, which was similar to SFT. Most contrasts were significant ($P \leq 0.05$) for dry shoot and root weight, except the high VC rates provided tomato shoot and root biomass that was similar to SFT and the 0.0 g/g control and low VC rates had similar dry root weight. For tomato root-to-shoot ratio, most contrasts were again significant ($P \leq 0.05$), except the high VC rates were similar to SFT. Tomato leaf area was also influenced by VC rate, as differences were detected ($P \leq 0.05$) between the 0 g/g control with both low and high VC rates as well as SFT; the high VC rates also differed from the low VC rates for tomato leaf area.

Table 2 Influence of soil type on 'Sunchief' tomato growth responses combined for the vermicompost rates and for two greenhouse experiments

Soil type[A]	Plant	Flower	Leaf	Leaf SPAD	Dry Wt (g)[D]		Root-to-shoot	Leaf area (cm^3)[F]	
	Ht (cm)[B]	No./plant	No./plant	Value[C]	Shoot	Root	Ratio[E]	Leaf	Total
Clay	86.0b	10.1b	6.7b	40.4b	20.6a	13.9a	0.67b	36.0b	241.3a
Loam	80.4c	9.1c	4.3c	40.0b	15.5b	11.6b	0.75a	51.4a	221.2a
Sand	90.7a	11.2a	8.6a	43.5a	21.7a	11.7b	0.54c	26.2c	225.0a

All measurements were taken at the termination of the experiments at 2 months after transplanting. Tomato growth response means followed by the same letter within a column do not differ significantly at $P \leq 0.05$

[A] Clay was Darwin silty clay (Fine, smectitic, mesic Vertic Haplaquolls), loam was a Hosmer silt loam (Fine-silty, mixed, mesic Typic, Fragiudalf), and sand came from an inclusion of sandy orthents within a Medway silty clay loam (Fine-loamy, mixed, mesic Fluvaquentic Hapludolls)

[B] Plant heights were measured for each plant from the soil line to the tip of the tomato main plant stem

[C] Leaf SPAD values were measured using a Minolta SPAD-502 chlorophyll meter (Special Products Analysis Division, Konica Minolta Sensing Inc., Osaka, Japan) on 10 randomly collected leaves from the mid-portion of plants in each plot

[D] Roots and foliage samples were oven dried at 80 °C for 48 h

[E] Root-to-shoot ratio is the root weight divided by shoot weight

[F] Leaf area was measured using a portable leaf area meter (Model LI-3000, LI-COR, Lincoln, NE, USA)

Tomato plant growth responses generally increased with the increasing VC application rates (Table 3). Tomato plant height increased in a quadratic manner with the increasing VC rate. Approximately 0.66 g/g VC provided the optimal rate to maximize tomato plant height [$y = 78.42 + 0.39$ (VC rate) $- 0.003$ (VC rate)^2, $R^2 = 0.95, P = 0.0097$]. Quadratic relationships also explained the increase in leaf and flower number per plant with increasing VC rate, although there was only a 2.6 leaf and 4.0 flower increase per plant from the 0 g/g control to the highest VC rate (0.8 g/g). The optimum VC application rates to achieve maximum leaf [$y = 8.92 + 0.04$ (VC rate) $- 0.0002$ (VC rate)^2, $R^2 = 0.96, P = 0.0077$] and flower numbers [$y = 3.77 + 0.15$ (VC rate) $- 0.001$ (VC rate)^2, $R^2 = 0.95, P = 0.0103$] per plant were about 0.80 and 0.59 g/g, respectively. Tomato leaf chlorophyll content increased in a linear manner with the increasing VC application rate [$y = 39.67 + 0.05$ (VC rate), $R^2 = 0.70, P = 0.0381$]. Tomato dry shoot and root biomass increased in a quadratic [$y = 13.67 + 0.24$ (VC rate) $- 0.0015$ (VC rate)^2, $R^2 = 0.99, P = 0.0005$] and linear [$y = 11.74 + 0.02$ (VC rate), $R^2 = 0.64, P = 0.0576$] manner, respectively, with the increasing VC application rate; and approximately, 0.70 and 0.80 g/g VC provided the highest tomato shoot and root biomass, respectively, which was similar to the SFT in both cases. The decrease in the root-to-shoot ratio with the increasing VC rate was best described by a quadratic model [$y = 0.82 + 0.007$ (VC rate) $- 0.00005$ (VC rate)^2, $R^2 = 0.97, P = 0.0064$], with about 0.60 g/g VC rate providing the lowest root-to-shoot ratio. Lastly, total tomato leaf area increased in a quadratic manner as VC application rate increased [$y = 181.74 + 2.66$ (VC rate) $- 0.025$ (VC rate)^2, $R^2 = 0.87, P = 0.0459$], with leaf area optimized at the 0.53 g/g VC rate.

Discussion

The use of VC as a fertilizer source for tomato growth provided no advantage compared to a standard inorganic fertilizer, although the higher amounts of VC applied improved tomato growth over the untreated control. However, other benefits provided by VC such as soil structure improvement and increase in soil microbial populations (Edwards and Burrows 1988; Canellas et al. 2002) compared to petroleum-based synthetic fertilizers should be considered and may be more ideal for certain situations. Results from this study indicated: (1) VC enhanced tomato growth particularly for the sandy soil type; (2) VC application rate affected tomato plant growth response, and soils with higher amounts of VC (0.4 and 0.8 g/g) generally resulted in taller tomato plants with greater leaf and root biomass, more leaves and flowers, higher chlorophyll content and increased leaf area than the non-treated control or low VC rates; and (3) the soil VC application rate to achieve optimal tomato growth parameter responses was between 0.5 and 0.6 g/g.

Although VC additions will improve soil structure primarily through increases in organic matter, tomato growth differed among soil types that were amended with VC. The sandy soil with VC amendments generally produced greater tomato plant heights, leaf and flower numbers per plant, and leaf chlorophyll content compared to the clay and loam soils, with the loam soil generally providing the least (Table 2). The influence of soil type on the tomato growth parameters was most likely due to VC improvements in soil structure allowing for greater water retention and aeration, especially in the sandy soil. However, in this experiment, the loamy and clayey soils appeared to become

Table 3 Influence of vermicompost (VC) rate on 'Sunchief' tomato plant growth responses combined over the three soil types evaluated

Vermicompost Rate (g/g)[a]	Plant Ht (cm)[b]	Leaf No./plant	Flower No./plant	Leaf SPAD Value[c]	Dry Wt (g)[d] Shoot	Dry Wt (g)[d] Root	Root-to-shoot Ratio[e]	Leaf area[f] (cm^3)
0.0	77.9	8.8	3.8	38.8	14.1	11.9	0.84	168.7
Low								
0.05	80.1	9.0	4.1	39.1	14.4	11.2	0.77	201.4
0.1	82.0	9.7	5.3	41.9	16.0	11.7	0.73	211.5
0.2	87.0	9.7	6.9	41.0	17.8	12.9	0.72	234.6
High								
0.4	88.0	10.3	7.4	41.3	21.1	12.8	0.61	237.0
0.8	91.1	11.4	7.8	43.5	23.2	13.2	0.57	237.4
SFT	93.9	12.2	10.5	43.7	24.6	13.5	0.55	232.8
Contrasts[g]								
None vs. low	NS	NS	NS	NS	**	NS	**	**
None vs. high	*	***	***	*	***	**	***	***
None vs. SFT	**	***	***	**	***	*	***	***
Low vs. high	*	***	**	NS	***	*	***	**
Low vs. SFT	**	***	***	*	***	**	***	NS
High vs. SFT	NS	**	**	NS	NS	NS	NS	NS
Significance of VC rate trend[h]								
Linear	**	**	*	*	**	*	**	NS
Quadratic	**	**	**	NS	**	NS	**	*

All measurements were taken at the termination of the experiments at 2 months after transplanting. The three soil types evaluated were clay (silty clay), loam (silt loam), and sand (silty clay loam)

NS, *, **, *** Nonsignificant or significant at $P \leq 0.05$, $P \leq 0.01$, or $P \leq 0.0001$, respectively

[a] SFT is the standard fertilizer treatment and was an inorganic 12-12-12 (N–P–K) complete fertilizer applied at 2 g per 1.7 L pot

[b] Plant heights were measured for each plant from the soil line to the tip of the tomato main plant stem

[c] The leaf SPAD value was measured using a Minolta SPAD-502 chlorophyll meter (Special Products Analysis Division, Konica Minolta Sensing Inc., Osaka, Japan) on 10 randomly collected leaves from the mid-portion of each plant

[d] Root and foliage samples were oven dried at 80 °C, with dry weights taken after 48 h

[e] Root-to-shoot ratio was root dry weight divided by shoot dry weight

[f] Leaf area was measured using a portable leaf area meter (Model LI-3000, LI-COR, Lincoln, NE, USA)

[g] Low is the combination of 0.05, 0.1 and 0.2 g/g vermicompost rates and High is the combination of 0.4 and 0.8 g/g vermicompost rates, while None = 0 g/g vermicompost

[h] SFT is used only in contrast comparisons and is not included in the calculation of linear and quadratic models

more compacted over the duration of the experiment, which would definitely have an effect on plant growth (Brady and Weil 2008).

Vermicompost application rate affected all tomato plant growth characteristics in this study and provided further evidence of VC as a suitable alternative-type fertilizer for tomato which is comparable to other studies including Arancon et al. (2003), Gutiérrez-Miceli et al. (2007), and Roberts et al. (2007). This study indicated that addition of 0.5 to 0.6 g/g VC to soils provided optimal tomato plant growth, since this rate provided tomato growth results similar to the standard inorganic fertility program and application rates >0.6 g/g tended to result in excessive amounts of shoot growth as well as low root-to-shoot ratios (Table 3). Arancon et al. (2003) indicated that the improvements in tomato plant growth due to VC application may be due to increases in microbial populations in soils provided by these materials. However, other research has indicated that increased plant growth and yields with VC may be due to the production of plant growth regulators by microorganisms or to the effects of humates (Canellas et al. 2002). Regardless, the addition of organic composts or VC to soil tends to improve soil structure, increase microbial population and activity, and increase water retention (Edwards and Burrows 1988; Edwards 2004). Therefore, the use of organic composts, such as VC, can provide various benefits to soils while, at the same time, providing a fertility source for selected vegetable crops.

Vermicomposts can be used as a source of nutrients for vegetable crops, as well as increasing soil organic matter, improving soil structure, and providing increases in soil microbial populations which all have beneficial effects on plant growth and development (Arancon et al. 2003, 2004a; Gutiérrez-Miceli et al. 2007). Furthermore, the recycling of organic wastes into composts for use as alternative fertilizers can reduce the need and dependence on synthetic, inorganic fertilizers by both small- and large-scale vegetable producers. Many amateur home gardeners as well as commercial vegetable growers are interested in using environment-friendly alternative type fertilizers, such as VC, that can supply sufficient amounts of nutrients for crop growth and development while at the same time improving the physical properties of soils. Thus, there is great potential to increase the use of these fertilizers by both commercial vegetable growers and home vegetable gardeners.

Conclusions

This research provided further evidence for VC as a suitable alternative fertilizer for tomato, with approximately 0.5–0.6 g/g VC added to soil resulting in optimal tomato plant growth. Moreover, this rate provided tomato growth similar to the standard inorganic fertility program. The sandy soil with VC amendments generally increased tomato plant growth parameters the most compared to the clay and loam soils, with the loam soil generally providing the least.

Acknowledgments The investigators would like to express their sincere thanks to Illinois Dept of Agriculture Sustainable 2000 Grant Program and Illinois Department of Commerce and Economic Opportunity for funding this research. Special thanks are also extended to Mr. Chad Hurley at New Horizon Organics Inc. in Jerseyville, IL and all personnel at the Horticulture Research Center, SIUC for their assistance in conducting this research. The tomato seed used in this study was generously provided by Rupp Seeds Inc., Wauseon, OH.

Conflict of interest The authors declare that they have no conflict of interest.

Author contribution This paper is based on a thesis submitted by Mr. Marc Zucco, who is a former graduate student at the Department of Plant, Soil, and Agricultural Systems at SIUC, Carbondale, Illinois, USA. He is now a soil scientist with the Natural Resources Conservation Service (NRCS) in Grayling, Michigan, USA. Dr. Alan Walters is Professor of Vegetable Science and Director of the Southern Illinois University Sustainable Farming Center. This center includes a vermicomposting facility in which University food wastes are recycled and used as a fertilizer source and soil enhancer for the Organic Vegetable Farm located next to the facility. Dr. Chong is an Emeritus Professor of Soil Science, who was involved in vermicompost research prior to his retirement and is now involved in various international activities, especially in China. Dr. Brian Klubek is Emeritus Professor of Soil Microbiology and former Chair of the Dept of Plant, Soil, and Agricultural Systems at Southern Illinois University, Carbondale, Illinois, USA. Dr. Joseph Masabni is an Assistant Professor and Extension Vegetable Specialist at Texas A&M University and collaborated with the authors in writing this manuscript.

References

Agnieszka R, Kacprzak M, Vandenbulcke F, Paytycz B (2013) Soil amendment with municipal sewage sludge affects the immune system of earthworms *Dendrobaena veneta*. Appl Soil Ecol 64:237–244

Alsina I, Dubova L, Steinberga V, Gmizo G (2013) The effect of vermicompost on the growth of radish. Acta Horticulturae 1013:359–365

Arancon NQ, Edwards CA, Bierman P, Metzger JD, Lee S, Welch C (2003) Effects of vermicomposts on growth and marketable fruits of field-grown tomatoes, peppers and strawberries. Pedobiolagia 47:731–735

Arancon NQ, Edwards CA, Atiyeh R, Metzer JD (2004a) Effects of vermicomposts produced from food waste on the growth and yields of greenhouse peppers. Bioresource Technol 93:139–144

Arancon NQ, Edwards CA, Bierman P, Welch C, Metzger JD (2004b) Influences of vermicomposts on field strawberries: 1. Effects on growth and yields. Bioresour Technol 93:145–153

Atiyeh RM, Dominguez J, Subler S, Edwards CA (2000) Changes in biochemical properties of cow manure during processing by earthworms (*Eisenia andrei*) and the effects on seedling growth. Pedobiologia 44:709–724

Atiyeh RM, Lee S, Edwards CA, Arancon NQ, Metzger JD (2002) The influence of humic acids derived from earthworm-processed organic wastes on plant growth. Bioresour Technol 84:7–14

Brady NC, Weil RR (2008) The nature and properties of soils, 14th edn. Pearson/Prentice Hall, Upper Saddle River

Canellas LP, Olivares FL, Okorokova-Façanha AL, Façanha AR (2002) Humic acids isolated from earthworm compost enhance root elongation, lateral root emergence, and plasma H+-ATPase activity in maize roots. Plant Physiol 130:1951–1957

Edwards CA (2004) Earthworm ecology, 2nd edn. CRC Press, Boca Raton

Edwards CA, Burrows I (1988) The potential of earthworm composts as plant growth media. In: Edwards CA, Neuhauser EF (eds) Earthworms in Environmental and Waste Management. SPB Academic Publ., The Hague, pp 211–220

Fehrenbacher JB, Alexander JD, Jensen IJ, Darmody RG, Pope RA, Flock MA, Vass EE, Scott JW, Andrews WF, Bushue JJ (1984) Soils of Illinois. Bul 778. Illinois Agr Expt Stat and USDA Soil Conserv Serv. Champaign-Urbana, IL

Gopalakrishnan S, Humayun P, Vadlamudi S, Vijayabharathi R, Bhimineni RK, Rupela O (2012) Plant growth-promoting traits of streptomyces with biocontrol potential isolated from herbal vermicompost. Biocontrol Sci Tech 22(10):1199–1210

Gutiérrez-Miceli FA, Santiago-Borraz J, Molina JAM, Nafate CC, Abud-Archila M, Llaven MAO, Rincón-Rosales R, Dendooven L (2007) Vermicompost as a soil supplement to improve growth, yield and fruit quality of tomato (*Lycopersicon esculentum*). Bioresour Tech 98:2781–2786

Herman RJ (1979) Soil survey of Jackson county, IL. Illinois Agr Expt Stat Soil Rpt 106. Champaign-Urbana, IL

Miles CC (1979) Soil survey of Union county, IL. Illinois Agr Expt Stat Soil Rpt 110. Champaign-Urbana, IL

Pandya U, Maheshwari DK, Saraf M (2014) Assessment of ecological diversity of rhizobacterial communities in vermicompost and analysis of their potential to improve plant growth. Biologia 69(8):968–976

Pascal J, Plumere T, Thu TD, Rumpel C, Duc TT, Orange D (2010) The rehabilitation of tropical soils using compost and vermicompost is affected by the presence of endogenic earthworms. Appl Soil Ecol 46(1):125–133

Pascual JA, Garcia C, Hernandez T, Ayuso M (1997) Changes in the microbial activity of an arid soil amended with urban organic wastes. Biol Fertil Soils 24:429–434

Roberts P, Jones DL, Edward-Jones G (2007) Yield and vitamin C content of tomatoes grown in vermicompost wastes. J Sci Food Agric 87:1957–1963

Sabey BR, Klubek BP, Chong S-K, Varsa EC (2003) Introductory experimental soil science, 2nd edn. Stipes Publ, Champaign

Subler S, Edwards CA, Metzger JD (1998) Comparing vermicomposts and composts. Biocycle 39:63–66

Yadav A, Garg VK (2011) Recycling of organic wastes by employing *Eisenia fetida*. Bioresource Tech 102(3):2874–2880

10

Simultaneous biodegradation of organic (chlorophenols) and inorganic compounds from secondary sludge of pulp and paper mill by *Eisenia fetida*

Santosh Kumar Karn · Swapan Kumar Chakrabarti

Abstract

Background Present research communicates the role of *Eisenia fetida* in converting pulp and paper mill sludge into valuable products by removing the high concentration of different chlorophenols and metals present in the sludge.

Results Pulp and paper mill sludge was toxic to *Eisenia fetida* and therefore, it was mixed with composted material in different ratio of experimental set up such as Set up: 1, (95:5), Set up: 2 (80:20) and Set up: 3 (50:50) on dry weight basis. We found very promising results by the application *Eisenia fetida* in the sludge. *Eisenia fetida* removed all the twelve different chlorophenols more than 95 % from the sludge which is declared as carcinogenic compounds by United States Environmental Protection Agency (US-EPA) in 90 days of incubation. Heavy metals concentration was decreased significantly during the incubation period whereas some metal concentration was also increased in the sludge due to mineralization. Carbon (C) and Nitrogen (N) were increased in the composted material, C:N ratio for Set 1 9:4 in Set 2 13:4 and Set 3 18:4 was observed which acts as soil conditioner. Whereas some other metals like (Ca, Mg, Na, K, Fe and Zn) was increased that are essential for the plant growth. There was great reduction in adsorbable organic halogen (AOX) 95 % and extractable organic halogen (EOX) 90 % as compare to control. *Eisenia fetida* also removed pentachlorophenol (PCP) from the sludge with artificially increased concentration about 100 mg/kg.

Conclusions By observing chlorophenols concentration, metals and carbon–nitrogen ratio, *Eisenia fetida* is acting as a potential candidate for the reclamation of industrial sludge. The result indicated that vermicomposting with *Eisenia fetida* is better option to manage the sludge or convert the sludge into nutrient-rich composted material in a short span of time.

Keywords *Eisenia fetida* · Chlorophenols · Vermicompost · Sludge · Pulp and paper mill · AOX · EOX · Metals

Introduction

The rapid increase in population and the increasing demand for industrial establishments overexploitation of available resources to meet human requirements have created problems such as pollution of the land, air and water environments. Pulp and paper mills generate significant amount of biodegradable sludge (34 and 105 kg/t product in large and small paper mills, respectively) during the papermaking and pulp making stages (Pokhrel and Viraraghavan 2004). The disposal of industrial sludge from effluent treatment plant is a global concern to the industries not a unique problem for the pulp and paper industry. Among the four major modes of sludge disposal; sea discharge, landfill, incineration and land application, now a days most of industry disposing the sludge by only land application. According to the recent estimate carried out by the US-EPA (1991) half of the 6,500 municipal landfill sites have been

S. K. Karn (✉)
Key Laboratory of Marine Environmental Corrosion and Biofouling, Institute of Oceanology, Chinese Academy of Sciences, No. 7 Nanhai Road, Qingdao 266071, Shandong, China
e-mail: santoshkarn@gmail.com

S. K. Chakrabarti
Environment Management Division, Thapar Centre for Industrial Research and Development, Yamunanagar 135001, India

closed by the end of the last century and another 54 % of the remaining landfills will be closed within next 5 years. The final disposal route for excess sludge generated by wastewater treatment is becoming a serious issue mainly due to the growth of population and sludge accumulation in large cities and growth in the amount and complexity of the related industrial activities. For this reason, it is necessary to reduce the amount of sludge generation by improving the efficiency of the treatment methodologies and implementing new technologies able to use sludge as raw material for compost, generate biomolecules and energy.

To cope up with this problem United States undertook a long-term research and demonstration program involving all the segments of society from regulators to farmers for the utilization of biosolids for gainful purposes and to find a long-term solution. Organochlorines are found in the sludge produced at pulp and paper mills. Accounting for as much as four percent of the total weight of the material contaminated sludge is spread on the land, buried in landfills or incinerated releasing chlorinated by-products into the air including polychlorinated phenols and dioxins (Mantykoski et al. 1989). In forests where pulp mill sludge has been dumped chlorinated phenols/dioxins have accumulated in the tissues of field animals and caused biochemical effects in birds (ERT 1987). Vermicomposting are well-known processes for solid organic waste reclamation: the final product vermicompost can be used as sources of organic matter for soil amendment, as sources of nutrients for soil fertilization or as growing media constituents for soilless cultivation (González et al. 2010). Vermicomposting is a bio-oxidative process which engages earthworms and microorganisms. The microorganisms both in the earthworm guts and in the feedstock are responsible for the biochemical degradation of the organic matter while the earthworms are responsible for the fragmentation of the substrate which increases the surface area exposed to the microorganisms. Hence the earthworms directly modify the physical properties of the material and indirectly modify its chemical properties. It is well-established that a large number of organic wastes can be ingested by earthworms and egested as peat-like material termed as vermicompost. It is much more fragmented, porous and microbially active than parent material (Edwards 1988; Edwards and Bohlen 1996) due to humification and increased decomposition. Recent study conducted by Huang et al. (2014) found that changes physiochemical properties and microbial profiles during vermicomposting to make the reliable material and Hanc and Chadimova (2014) found nutrient recovery after vermicomposting. Due to the certain limitations of the other processes, vermicomposting receiving more attention for stabilization of various wastes including for decreased duration of treatment process, increased pathogen reduction and better product quality (Hait and Tare 2011). Butt (1993) showed that solid paper mill sludge was a suitable feed for *Lumbricus terrestis* under laboratory conditions. Elvira et al. (1998) have also reported vermicomposting of paper mill sludge using *Eisenia andrei* under laboratory as well as field conditions. Gajalakshmi et al. (2002) studied the vermicomposting of paper waste using anecic earthworm *Lampito mauriti*. Today this is needful to treat the industrial sludge for the removal of organic and inorganic contaminants and manage it properly for the useful application; therefore, present study deals with the treatment of chlorophenols and inorganic constituents from the sludge of pulp and paper mill using *Eisenia fetida* further it can be applied as a compost material.

Materials and methods

Collection of sludge sample

Dewatered secondary sludge samples of pulp and paper mill were collected from the effluent treatment plant of BILT, Unit-Shree Gopal, Yamunanagar, Haryana, India. This mill adopts kraft process for pulping of raw materials mainly eucalyptus, poplar and bamboo. The effluent and sludge generated by the process is treated in activated sludge process for the biological removal of organics.

Vermicomposting set up

Dewatered secondary sludge was collected in large-sized plastic containers and then brought to the laboratory for further processing. Sludge was mixed with composted material in different proportions Set up: 1 (95:5) Set up: 2 (80:20) and Set up: 3 50:50) to maintain the moisture content and reduce the toxicity level of the secondary sludge for the adaptation of worms. Vermicomposting experiments were performed in plastic trays. No external addition of Nitrogen (N) and Phosphorus (P) was added during vermicomposting. pH, temperature and moisture were maintained in the range of 7.5–8.5; 27–30 °C and 60–75 %, respectively throughout the period of study (for 90 days). The trays were covered with a jute mat and were kept in the shed. The composted material with earthworm 1 kg was spread and mixed at the top layer. Normally (1 kg worms present in 10 kg dry material). Earthworm species *Eisenia fetida* were obtained from a stock culture of Krishi Vigyan Kendra (KVK) Tepla Ambala, Haryana an ICAR unit of Govt. of India, with decomposed cow dung spiked with plant litters. Wheat straws (WS) were used to prepare as bedding materials for the experiments.

Further in another set chlorophenols degradation ability *E. fetida* was observed in sludge (proportion like set up: 2) with artificially spiked concentration of PCP in the sludge at the rate of 100 mg/kg and mixed thoroughly in plastic trays mentioned above and inoculated *E. fetida* and incubated under the same condition for 3 months. The tray were covered with jute mate to maintain the pH; temperature, moisture as mentioned as before.

Sludge characterization

The sludge sample were analysed for physiochemical characteristic such as moisture, pH, carbon, hydrogen, nitrogen, sulphur, zeta potential, absorbable organic halogen (AOX) and extractable organic halogen (EOX). Two grams of sludge sample was taken and dried into oven at 105 °C for 24 h. The sample was then grinded in mortar pestle to get particles having size of about 0.1 mm. The dried sludge sample (10-100 mg) was taken for CHNS analysis using Elemental analyzer (Thermo Scientific, USA). Zeta potential was analysed by maintaining pH of the sample to neutral and placing the sample in zeta potential analyzer, Muteck SZP06 (BTG Mutek GmbH, Germany). The AOX and EOX concentrations were determined using Euroglas Netherlands instrument ECS-2000 according to the manufacturer's recommended procedure.

Determination of chlorophenol

Chlorophenols extraction and analysis was done by (NCASI (1986) method CP-85.01) method with a little modification. Secondary sludge samples were taken in ice bath and sonicated using a Sonicator (SG-25) Roop Telesonic Ultrasonics Ltd. India with variable power intensity between 3 and 7.4 W/cm^2 and a frequency of 24 kHz. The sample was immersed in ice and sonicated for total of 10 min with 1.5 min burst followed by 5 min rest in ice. Next added 1.3 ml K_2CO_3 (pH 11.5) again sonicated for 2 min and added 1.5 ml of acetic anhydride and sonicated for 2 min again added 0.5 ml acetic anhydride further added 5 ml of hexane and again sonicated and centrifuged and further extracted with hexane. After centrifugation at 10,000 rpm for 15 min, the supernatant was used to analyse the chlorophenols.

The GC analysis were performed in electron capture detection mode with a gas chromatograph Nucon GC-5765 (Centurion Scientific, India) capillary column DB-5 (30 meter length 0.025 mm i.d. 0.25 μm film thickness) was used at a temperature program of 50 °C (2 min) then raised to 10 °C/min to 280 °C where it was held for 10 min. Helium was used as the carrier gas at a constant flow of 1.2 ml/min. The samples were analysed in splitless mode at an injection temperature of 250 °C and detector temperature 280 °C. The injected volume was 0.1 μl.

Determination of metals

Metals was determined as per (APHA 1995) first sludge sample was digested with nitric and perchloric acid (3:1) and filtered through Whatman no. 42 paper. The filtrate was used for characterization of phosphorus content and the trace elements were analysed on furnace atomic absorption spectroscopy (AAS) and rest were analysed on flame AAS. The hydride forming elements were analysed using hydride generator.

Statistical analysis

Data were statistically analysed by analysis of variance (ANOVA) and the mean differences were compared by Tukey–Kramer Multiple Comparison Test at $p < 0.05$. All experiments were performed in three replicates and analyses were performed using GraphPad Prism (v 4.03) software. (CA, USA).

Results and discussion

Characterizations of sludge

In recent time, interest for vermicomposting (using earthworms to breakdown organic materials) has increased (Hand 1988; Edwards 1988; Edwards and Bohlen 1996) due to its potential. In its basic form, this is a low-cost technology system that primarily uses earthworms in the processing or treatment of organic wastes (Hand 1988). Certain species of earthworm can consume organic material residuals very rapidly and fragment them into much finer particles and reducing the pollutants by passing them through a grinding gizzard. The earthworms derive their nourishment from microorganisms that grow upon these materials. At the same time, they promote further microbial activity since the fecal material or casts that they produce is much more fragmented and microbially active than what they consume (Edwards 1988; Edwards and Bohlen 1996). During this process, the important plant nutrients in the material (particularly the N, K, P and Ca) are released and converted through microbial action into forms that are much more soluble and available to plants than those in the parent compounds. Therefore, first, collected secondary sludge samples were characterized for various physical and chemical characteristics such as pH, organic matter, CHNS, AOX, and EOX described in (Table 1). All the twelve different chlorophenols were also analysed and described in (Table 2) and metals are described in (Table 3). Sludge sample were rich in organic C, H, N and S content. Vermicomposted material significantly modified the physical and chemical properties of all sets mixtures.

Table 1 Chemical characteristic of secondary sludge from pulp and paper mill and material containing *Eisenia fetida*

Parameter	Secondary sludge	Material containing worms
Moisture (%)	89.5 ± 2.1	51.34 ± 1.8
Organic (%)	72 ± 1	78 ± 2.4
Carbon (%)	24.2 ± 1.3	27.1 ± 2.1
Hydrogen (%)	5 ± 0.5	2.4 ± 0.7
Nitrogen (%)	3.2 ± 0.6	1.5 ± 0.05
Sulphar (%)	0.7 ± 0.01	0.5 ± 0.01
Kjeldhal Nitrogen (%)	5.8 ± 1.1	2.2 ± 0.6
AOX (mg/Kg)	2,400 ± 67	580 ± 11
EOX (mg/kg)	812 ± 16	132 ± 6
C/N ratio	7.5 ± 0.8	8.4 ± 0.3
Zeta potential (mv)	−15.2 ± 0.05	−12.1 ± 0.04

Values are given as (Mean ± SE)

ND not detected

Table 2 Chlorophenols concentration in secondary sludge from pulp and paper mill and also material containing *Eisenia fetida*

Chlorophenolic compounds	Secondary sludge (mg/kg)	Material containing worms (mg/kg)
2,4,6-TCP	0.351 ± 0.02	<0.011
2,4,5-TCP	0.191 ± 0.01	ND
2,3,4,5-Tetrachlorophenol	0.154 ± 0.06	ND
3,4,6-Trichloroguacol	0.153 ± 0.03	ND
3,4,5-Trichloroguacol	0.123 ± 0.02	ND
4,5,6-Trichloroguacol	0.143 ± 0.04	ND
3,4,6-Trichlorocatecol	0.322 ± 0.07	ND
3,4,5-Trichlorocatecol	0.373 ± 0.01	ND
Tetrachloroguaiacol	0.362 ± 0.01	ND
Tetrachlorocatecol	0.312 ± 0.02	ND
Trichlorosyringol	0.299 ± 0.01	0.05
Pentachlorophenol	0.345 ± 0.01	ND

Values are given as (Mean ± SE)

ND not detected

Table 3 Metallic characterizations of secondary sludge and material containing *Eisenia fetida*

Metals	Secondary sludge (mg/kg)	Material containing worms (mg/kg)
Na	1,300 ± 19	2,110 ± 43
K	2,600 ± 26	5,033 ± 13
Mg	5,423 ± 12	3,555 ± 19
Ca	1,400 ± 23	4,929 ± 22
Al	2,900 ± 27	943 ± 12
P	416 ± 11	368 ± 11
Fe	2,453 ± 41	1,791 ± 33
Cr	36 ± 7	2.8 ± 0.7
Mn	133 ± 11	127 ± 2
Co	2.0 ± 0.3	0.7 ± 0.02
Zn	19 ± 3	67 ± 2
Ni	24 ± 1	1.8 ± 0.2
Cu	87 ± 4	18 ± 2
Cd	2 ± 0.05	0.1 ± 0.02
Pb	28 ± 3	3.2 ± 0.2
Hg	ND	ND
Se	ND	ND
As	ND	ND

Values are given as (Mean ± SE)

ND not detected

The vermicompost was much darker in color than before and has been processed more homogeneous mixture after 90 days of earthworm activity, the pH observed normal about 7.2 in the final vermicompost. Organic percentages were increased after composting for Set 1 it increased 36 % for Set 2; 35 % whereas for Set 3, it increased up to 50 %. The C:N ratio, one of the most widely used indicating for maturity of organic waste. In our experiments C:N ratio was between 9.4 and 18.4 after 90 days of worms activity. In Set 1, we observed 9:4; in Set 2, 13:4 and Set 3, 18:4. In initial stage, C:N ratio was 8:4 for Set 1, 9:3 for Set 2 and was 14:3 for Set 3. According to Senesi (1989) a decline of C:N ratio to less than 20 indicates an advanced degree of organic matter stabilization and reflects a satisfactory degree of maturity of organic waste. Total K and N were also increased by the end of vermicomposting due to mineralization of organic matter. Previously Benitez et al. (2000) has also reported that decomposition of organic materials by earthworm accelerates the N mineralization process and subsequently changes the N profile of the substrate. Kaushik and Garg (2004) have reported a 2.0–3.2 fold increase in TKN during vermicomposting of textile mill sludge mixed with cow dung and wheat straw. Therefore, vermicomposting is a considerable technology for the waste or specially sludge to increase the soil nutrient condition. Different organochlorine contaminants like AOX and EOX in the sludge were analysed (described in Table 4). AOX values ranged from 451 to 5,140 mg/kg dry solids. However, 93–95 % decreases in the AOX and 90–92 % decrease in EOX level, after composting which makes it less toxic more stable and reliable material for use. Earthworms accumulate many lipophilic organic pollutants from the surrounding soil environment not only through passive absorption through the body wall of the dissolved fraction in the interstitial water but also by intestinal uptake during the passage of soil through the gut. The accumulation increases as the concentration of the pollutant in the soil environment (Belfroid et al. 1995a, b).

Table 4 Chlorophenols concentration in three different sets before and after treatment with *Eisenia fetida*, results shown of every month analysed chlorophenols concentration and also compared with control (untreated)

Chlorophenolic compounds (mg/kg)	Mixed Secondary sludge + composted material 5 % (untreated)	After 1 month	After 2 month	After 3 month	Control (without worms)	Mixed secondary sludge + Composted material 20 %	After 1 month	After 2 month
2,4,6-TCP	0.335 ± 0.022cA	0.303 ± 0.06cA	0.142 ± 0.01cC	0.012 ± 0.01dE	0.345 ± 0.03cA	0.280 ± 0.03cB	0.220 ± 0.03dC	0.101 ± 0.02dD
2,4,5-TCP	0.182 ± 0.071eA	0.152 ± 0.022eB	0.075 ± 0.06eD	ND	0.095 ± 0.001fC	0.152 ± 0.02eB	0.110 ± 0.03eC	0.099 ± 0.07dC
2,3,4,5-Tetrachlorophenol	0.148 ± 0.02eA	0.110 ± 0.011eB	0.043 ± 0.05fC	ND	0.146 ± 0.03eA	0.124 ± 0.01eB	0.086 ± 0.002fB	ND
3,4,6-Trichloroguacol	0.150 ± 0.062eA	0.105 ± 0.04eB	0.039 ± 0.03fD	ND	0.057 ± 0.002gC	0.122 ± 0.03eA	0.099 ± 0.01eC	ND
3,4,5-Trichloroguacol	0.117 ± 0.03fA	0.101 ± 0.06eA	0.042 ± 0.001fC	ND	0.119 ± 0.01gA	0.098 ± 0.001fA	0.067 ± 0.02fB	ND
4,5,6-Trichloroguacol	0.136 ± 0.02eA	0.107 ± 0.04eB	0.038 ± 0.001fD	ND	0.039 ± 0.001fD	0.115 ± 0.001A	0.076 ± 0.01fC	ND
3,4,6-Trichlorocatecol	0.305 ± 0.08dA	0.276 ± 0.02dA	0.150 ± 0.02cC	0.12 ± 0.01dDE	0.307 ± 0.03dA	0.258 ± 0.03dA	0.218 ± 0.03 dB	0.086 ± 0.01dD
3,4,5-Trichlorocatecol	0.354 ± 0.04cA	0.310 ± 0.03cB	0.120 ± 0.02dC	0.021 ± 0.01cD	0.349 ± 0.02cA	0.298 ± 0.01cB	0.252 ± 0.06cB	0.103 ± 0.02dC
Tetrachloroguaiacol	0.343 ± 0.04cA	0.323 ± 0.01cA	0.134 ± 0.01cD	0.014 ± 0.001cE	0.310 ± 0.04dA	0.290 ± 0.02cC	0.262 ± 0.07cC	0.125 ± 0.06cD
Tetrachlorocatecol	0.312 ± 0.022dA	0.287 ± 0.02dA	0.136 ± 0.02cB	0.019 ± 0.001cD	0.320 ± 0.04dA	0.250 ± 0.04 dB	0.221 ± 0.02 dB	0.089 ± 0.01dC
Trichlorosyringol	0.296 ± 0.056dA	0.254 ± 0.01dA	0.118 ± 0.02dC	0.021 ± 0.001cD	0.278 ± 0.02dA	0.240 ± 0.02 dB	0.219 ± 0.02 dB	0.097 ± 0.01dC
Pentachlorophenol	0.327 ± 0.043cA	0.296 ± 0.03cA	0.143 ± 0.02cB	0.022 ± 0.001cCD	0.333 ± 0.03cA	0.276 ± 0.02cA	0.232 ± 0.04dA	0.122 ± 0.07cB
AOX	2,674 ± 19aA	1,989 ± 23aC	739 ± 41aE	198 ± 12aF	2,497 ± 16aA	2,413 ± 22aB	1,786 ± 23aC	678 ± 41aE
EOX	957 ± 26bA	764 ± 27bB	278 ± 29bD	73 ± 16bE	873 ± 26bA	765 ± 12bB	513 ± 11bC	219 ± 13bD

Chlorophenolic compounds (mg/kg)	After 3 month	Control (without worms)	Mixed secondary sludge + composted material 50 %	After 1 month	After 2 month	After 3 month	Control (without worms)
2,4,6-TCP	0.012 ± 0.01dE	0.243 ± 0.03 dB	0.177 ± 0.01cC	0.100 ± 0.003cD	0.025 ± 0.002dE	ND	0.113 ± 0.02eD
2,4,5-TCP	ND	0.171 ± 0.05eA	0.098 ± 0.002 dC	0.029 ± 0.001dE	ND	ND	0.086 ± 0.01eC
2,3,4,5-Tetrachlorophenol	ND	0.132 ± 0.02eA	0.078 ± 0.001 dB	0.022 ± 0.001dC	ND	ND	0.56 ± 0.01fB
3,4,6-Trichloroguacol	ND	0.063 ± 0.001fC	0.077 ± 0.001dC	0.019 ± 0.001dD	ND	ND	0.072 ± 0.01fC
3,4,5-Trichloroguacol	ND	0.086 ± 0.002fB	0.062 ± 0.001 dB	0.031 ± 0.001dC	ND	ND	0.071 ± 0.03fB
4,5,6-Trichloroguacol	ND	0.122 ± 0.02eA	0.072 ± 0.003dC	0.034 ± 0.001dD	ND	ND	0.063 ± 0.01fC
3,4,6-Trichlorocatecol	ND	0.289 ± 0.01dA	0.167 ± 0.02cC	0.122 ± 0.06cC	0.026 ± 0.001dD	ND	0.153 ± 0.06dC
3,4,5-Trichlorocatecol	0.026 ± 0.001cD	0.267 ± 0.02 dB	0.188 ± 0.05cC	0.113 ± 0.07cC	0.034 ± 0.003cD	ND	0.185 ± 0.04cC
Tetrachloroguaiacol	0.020 ± 0.01cE	0.286 ± 0.03dC	0.182 ± 0.03cD	0.099 ± 0.02cD	0.019 ± 0.001dE	ND	0.178 ± 0.06cD
Tetrachlorocatecol	ND	0.341 ± 0.02cA	0.157 ± 0.04cB	0.124 ± 0.03cB	0.043 ± 0.001cD	ND	0.163 ± 0.01cB
Trichlorosyringol	ND	0.289 ± 0.03dA	0.152 ± 0.01cB	0.122 ± 0.04cB	0.049 ± 0.001cD	ND	0.153 ± 0.02 dB

Table 4 continued

Chlorophenolic compounds (mg/kg)	After 3 month	Control (without worms)	Mixed secondary sludge + composted material 50 %	After 1 month	After 2 month	After 3 month	Control (without worms)
Pentachlorophenol	0.012 ± 0.01dD	0.255 ± 0.01dA	0.175 ± 0.03cB	0.119 ± 0.03cB	0.056 ± 0.003cC	ND	0.187 ± 0.02cB
AOX	134 ± 17aF	2,363 ± 28aB	2,367 ± 43aB	1,532 ± 21aD	697 ± 24aE	128aF	2,287 ± 29aB
EOX	45 ± 11bE	713 ± 17bB	612 ± 21bC	456 ± 15bC	198 ± 13bD	52bE	567 ± 17bC

Values sharing common lowercase letter within column and uppercase letter within row are not significant at $P < 0.05$ value. Data are mean and standard deviation of triplicates. All the concentration represented in the table in mg/kg

ND not detected

The secondary sludge sample was also characterized for 12 chlorophenolic compounds identified by USEPA as carcinogenic compound described in later section.

Chlorophenol concentration

Chlorophenols are important biocides and as by-product of bleaching in the pulp and paper industry. Their widespread use has resulted in broad distribution of these compounds in the environment. Environmental contamination with chlorophenols is widespread due to the importance of these chemicals as industrial intermediate, pesticides and solvents. *Eisenia fetida* can be a significant fate for the chlorophenol removal because physical and chemical methods are not feasible due to high cost and generate secondary pollutants and a single microorganism is unable to mineralize a wide range of different chlorophenols. Field and Sierra-Alvarez (2008) wrote a comprehensive review on the aerobic and anaerobic biotransformation of chlorophenols by microorganism. From the present result, we found that all the 12 different chlorophenols decreased significantly as compared to control having very high concentration (Table 4). Chlorophenols were decreased gradually with time; at 90 days from all the three sets, we did not observe any chlorophenols from the sample whereas in control chlorophenols, concentration was remaining the same (Table 4). From the 2nd experiment having spiked concentration of PCP about (100 mg/kg) after 90 days, we found that there is also significant decrease in the concentration of PCP and left about 0.02 mg/kg, which shows that vermicomposting having high potential for the removal of chlorophenol from the pulp and paper mill sludge. It is well-established that a large number of organic wastes can be ingested by earthworm and egested as peat-like material termed as vermicompost. Edwards (1988); Kaushik and Garg (2003, 2004) have reported the vermicomposting of textile mill sludge using *Eisenia fetida*. Butt (1993) showed that solid paper mill sludge was a suitable feed for *Lumbricus terrestis* under laboratory conditions. Elvira et al. (1998) have reported vermicomposting of paper mill sludge using *Eisenia andrei* under laboratory as well as field conditions and found suitable for mineralization and compost formation.

Some report also indicates that other annelids, such as aquatic Polychaetes, can metabolize benzopyrene, because they possess cytrochrome P450 enzymes capable of degrading this compound (Driscoll and McElroy 1997). The same enzymatic activity was found in terrestrial earthworms such as *Eisenia fetida* (Achazi et al. 1998). This may be a reason for *Eisenia fetida* to remove organic compounds and metabolize successfully. Autochthonous microorganisms degrade hydrocarbons (Johnsen et al. 2005), but if earthworms are added to soil, they will

Table 5 Metallic concentrations in all the three sets before and after treatment. Metal concentration was observed at every month shown in table composted by *Eisenia fetida*

Metals (mg/kg)	Mixed secondary sludge + composted material 5 %	After 1 month	After 2 month	After 3 month	Control (without worms)	Mixed secondary sludge + composted material 20 %	After 1 month	After 2 month
Na	1,346 ± 20eD	1480 ± 43eC	1,568 ± 22eB	1,594 ± 12eB	1,332 ± 21eD	1,693 ± 4 dB	1,680 ± 16eB	1,850 ± 2eA
Mg	3,046 ± 26cB	3,287 ± 23cB	3,690 ± 43cA	3,619 ± 13cA	3,012 ± 22cB	2,843 ± 12bC	3,300 ± 76bB	3,520 ± 22cA
Al	4,894 ± 33aC	5,097 ± 21aB	5,476 ± 44aA	5,505 ± 32aA	4,767 ± 12aC	4,729 ± 21aC	5,096 ± 67aB	5,597 ± 21aA
K	1,850 ± 13cD	2,100 ± 22dC	2,320 ± 56 dB	2,378 ± 13 dB	1,798 ± 31dD	2,037 ± 12cC	2,300 ± 31cB	2,450 ± 12 dB
Ca	3,561 ± 32bCB	3,898 ± 35bB	4,400 ± 23bA	4,526 ± 31bA	3,572 ± 12bB	2,854 ± 13bD	3,200 ± 21bC	3,789 ± 13bB
Fe	2,025 ± 14dA	2,100 ± 17dA	2,150 ± 87dA	2,156 ± 12dA	2,017 ± 13dA	1,902 ± 4cA	1,995 ± 2dA	2,012 ± 12eA
Cr	12.4 ± 2 hB	13.5 ± 13iA	13.9 ± 2hA	14.1 ± 2hA	11.9 ± 2iB	11.5 ± 2hB	12 ± 1iB	12.1 ± 2iB
Mn	124 ± 21fB	128 ± 12fAA	130 ± 41eA	130 ± 21eA	129 ± 11fA	112 ± 2eC	115 ± 21fC	124 ± 6fB
Co	1.2 ± 1iA	1.1 ± 0.2jA	1.2 ± 0.4iA	1.2 ± 0.5jA	1.0 ± 0.2kB	1.2 ± 0.2jA	1.2 ± 0.1A	1.1 ± 0.1A
Ni	15.5 ± 3hA	16.1 ± 2iA	16.5 ± 2hA	16.7 ± 2hA	15.3 ± 2iA	13.3 ± 2hB	13.5 ± 1iB	13.7 ± 1iB
Cu	22.5 ± 2hB	24.8 ± 2hB	27.5 ± 3gA	28.2 ± 10gA	21.87 ± 3hB	19.9 ± 3gB	19.5 ± 2hB	19.0 ± 2hB
Zn	74.5 ± 7gB	77.8 ± 12gA	80.9 ± 11fA	82.2 ± 13fA	74.1 ± 12gB	65.8 ± 4fC	66.0 ± 4gC	66.5 ± 6gC
Cd	1.0 ± 02iA	0.8 ± 0.1jB	0.0kE	1.0 ± 0.2jA	0.8 ± 0.1kB	0.7 ± 0.1jB	0.5 ± 0.1kC	0.5 ± 0.1kC
Pb	3.7 ± 1jA	2.2 ± 0.5jB	0.9 ± 0.05iC	4.0 ± 1iA	3.3 ± 1j	3.4 ± 1iA	3.0 ± 1jA	3.0 ± 1jA
Hg	ND	ND	ND	ND	ND	ND	ND	ND
Se	ND	ND	ND	ND	ND	ND	ND	ND
As	ND	ND	ND	ND	ND	ND	ND	ND

Metals (mg/kg)	After 3 month	Control (without worms)	Mixed secondary sludge + composted material 50 %	After 1 month	After 2 month	After 3 month	Control (without worms)
Na	1,898 ± 6eA	1,634 ± 45 dB	1,214 ± 23dD	1,350 ± 12dD	1,455 ± 17dC	1,466 ± 23dC	1,321 ± 33dD
Mg	3,575 ± 19cA	2,794 ± 21bC	2,422 ± 24bD	2,600 ± 15bC	3,078 ± 19bB	3,106 ± 13bB	2,403 ± 12bD
Al	5,739 ± 14aA	4,716 ± 17aC	3,055 ± 45aE	3,340 ± 23a	3,600 ± 12aD	3,724 ± 19aD	3,032 ± 23aE
K	2,469 ± 12 dB	2,002 ± 9cC	2,102 ± 16bC	2,466 ± 11cB	2,780 ± 8bA	2,818 ± 12cA	2,023 ± 11cC
Ca	3,963 ± 12bB	2,817 ± 11bD	2,154 ± 12bE	2,300 ± 6cE	2,542 ± 9cD	2,662 ± 11cD	2,107 ± 16cE
Fe	2,015 ± 13eA	1,876 ± 21cB	1,412 ± 3dC	1,450 ± 9dC	1,452 ± 11dC	1,446 ± 15dC	1,398 ± 17dC
Cr	12.3 ± 2iB	10.9 ± 1hC	7.7 ± 2D	7.5 ± 2h	7.5 ± 2hD	5.7 ± 2hE	7.1 ± 2hD
Mn	124 ± 5fB	113 ± 4eC	95 ± 2eE	105 ± 11eC	115 ± 2eC	117 ± 11eC	87 ± 13eE
Co	1.3 ± 0.3A	0.97 ± 0.1B	0.8 ± 0.2j	1.2 ± 0.2iA	1.1 ± 0.2kA	1.1 ± 0.3jA	0.7 ± 0.2jC
Ni	13.9 ± 2iB	13.8 ± 2hB	9.6 ± 2hD	10.2 ± 2hD	10.7 ± 2hC	10.7 ± 1gC	9.4 ± 2hD
Cu	18.5 ± 2hB	19.1 ± 1gB	14.9 ± 2gD	15.9 ± 1gD	16.2 ± 3gD	16.2 ± 2gD	14.7 ± 3gD
Zn	67.0 ± 4gC	64.23 ± 2fC	55.2 ± 3fD	56.5 ± 6fD	57.2 ± 4fD	57.2 ± 6fD	54 ± 5fE
Cd	0.6 ± 0.1kC	0.6 ± 0.1jC	0.5 ± 0.1jC	0.6 ± 0.1C	0.7 ± 0.2kB	0.7 ± 0.1jB	0.4 ± 0.1jD
Pb	2.8 ± 1jB	3.3 ± 1iA	2.7 ± 1iB	2.5 ± 1iB	2.4 ± 1jB	2.4 ± 1iB	3.1 ± 1iA

Table 5 continued

Metals (mg/kg)	After 3 month	Control (without worms)	Mixed secondary sludge + composted material 50 %	After 1 month	After 2 month	After 3 month	Control (without worms)
Hg	ND	ND	ND	ND	ND	ND	ND
Se	ND	ND	ND	ND	ND	ND	ND
As	ND	ND	ND	ND	ND	ND	ND

Values sharing common lowercase letter within column and uppercase letter within row are not significant at $P < 0.05$ value. Data are mean and standard deviation of triplicates. Metals described in the table in mg/kg

ND not detected

improve aeration, and stimulate microbial activity, thus increasing biodegradation. Eijsackers et al. (2001) reported that there was a steady decrease in the concentrations of Phenenthrene in soil when they added worms and only very low concentrations of Phenanthrene were detected after 40 days.

Metal concentration

Metals concentration in present sludge having higher concentration is described in (Table 5) because of various source such as metal in the raw material and process involved in pulping and paper-making stage. So, the vermicompost made from pulp and paper mill sludge may have higher metal concentrations. Many metal in the sludge mentioned in table is essential and plays significant role for the plant growth and nutrition, but sometime in higher concentration may have detrimental effect on the plant growth. Therefore, before application of the composted material, determining the heavy metals concentration is needed. From the present study, we found that after vermicomposting metals concentration was increased in the composted material as compared to before, in as such sludge. Essential elements such as Na, K, Mg, Ca which is necessary for the plant growth, metabolism were increased after composting these metals also need for external supply for the plant growth. Previously, Suthar and Singh (2008) have reported that an earthworm processed waste material contains a higher concentration of K due to enhanced microbial activity during the vermicomposting therefore, enhances the rate of mineralization. Garg and Kaushik (2005) have also reported an increase in Ca content during the vermicomposting of industrial waste. Result showed that heavy metals such as Fe, Cu, Zn, Cr, Ni concentration in the final vermicompost in all three sets were slightly increased as compare to the initial concentration described in (Table 5). High concentration of Cu and Cr may be harmful for the plant but Zn having beneficial effect to plants. Pb and Cd concentration in final vermicomposted material shown decreased which is more toxic to the plants. Wang et al. (2013) also evaluated the role of earthworm in the sewage sludge and observed the reduction of Pb and Cd concentration after vermicomposting. Al concentration was also increased in the composted material. Singh and Kalamdhad (2012) reported that, the contents of total metals (Fe, Cu, Zn Cr, Ni) concentration increased during the water hyacinth (*Eichhornia crassipes*) composting process. Likewise, similar results were obtained by Singh and Kalamdhad (2013) concerning the increase of these heavy metals during the vermicomposting of water hyacinth employing *E. fetida*. Hait and Tare (2012) reported in their experiments that vermicomposting caused a significant increase in total heavy metals (Cu, Co, Fe, Mn, Zn,

Cr) contents and a significant decrease in water-soluble heavy metals contents as compared to the compost material. Heavy metal like Hg, Se, As was not found in the present sludge. Metal such as Co, Mn was also increased which plays a major role as micronutrients and this result is supported by the finding of Hait and Tare (2012). Elvira et al. (1998) also reported an increase in heavy metals concentrations in vermicompost of paper mill sludge. Deolalikar et al. (2005) suggested that weight and volume reduction due to breakdown of organic matter during vermicomposting may be the reason for increase in heavy metal concentrations in vermicompost. In present results, we also observed weight loss which was about 13, 17 and 20 % gradually for Set 1, Set 2 and Set 3.

Conclusions

We would like to recommend that earthworms (*Eisenia fetida*) have a great potential to remove chlorophenol from the sludge, even mineralize many metals that are resistant to degradation. Earthworm is extremely resistant to toxic chlorophenols and able to tolerate the high concentrations, normally not present in the soil. Applying *Eisenia fetida* to a contaminated sludge/site might be an environmentally friendly way to remove the chlorophenols.

Author contribution All the lab work was performed by SKK and suggestion was provided by SKC. Manuscript was prepared by SKK and checked by SKC. Both the authors have read and approved the final manuscript.

Acknowledgments The authors are thankful to KVK Tepla, for providing *Eisenia fetida* sp. and TCIRD Yamunanagar, for providing the industrial sludge and Biotechnology Engineering Department, Ambala College of Engineering and Applied Research for providing the facilities for the work.

Conflict of interest The authors declare that they have no competing interest.

References

Achazi RK, Flenner C, Livingstone DR, Peters LD, Schaub K, Schiwe E (1998) Cytochrome P450 and dependent activity in unexposed and PAH-exposed terrestrial annelids. Comp Biochem Physiol C 121:339–350

APHA (1995) Standard methods for the examination of water and wastewater, 19th edn. American Public Health Association, American Water Works Association, Water Environment Federation, Washington, DC

Belfroid A, Meiling J, Drenth HJ, Hermens J, Seinen W, Gestel KV (1995a) Dietary uptake of super lipophilic compounds by earthworms (*Eisenia andrei*). Ecotoxicol Environ Saf 31:185–191

Belfroid A, Meiling J, Drenth HJ, Hermens J, Seinen W, Gestel KV (1995b) Dietary uptake of super lipophilic compounds by earthworms (*Eisenia andrei*). Ecotoxicol Environ Saf 31:185–191

Benitez E, Nogales R, Masciandro G, Ceccanti B (2000) Isolation by isoelectric focusing of humic urease complexes from earthworm (*Eisenia fetida*) processed sewage sludges. Biol Fert Soils 31:489–493

Butt KR (1993) Utilization of solid paper mill sludge and spent brewery yeast as a feed for soil-dwelling earthworms. Biores Technol 44:105–107

Deolalikar AV, Mitra A, Bhattacharyee S, Chakraborty S (2005) Effect of vermicomposting process on metal content of paper mill solid waste. J Env Sci Eng 47:81–84

Driscoll SBK, McElroy AE (1997) Elimination of sediment-associated benzopyrene and its metabolites by polychaete worms exposed to 3-methylcholanthrene. Aqu Toxicol 39:77–91

Edwards CA (1988) Breakdown of animal, vegetable and industrial organic wastes by earthworms. In: Edwards CA, Neuhauser EF (eds) Earthworms in waste and environmental management. SPB Academic Publishing, The Hague, pp 21–31

Edwards CA, Bohlen PJ (1996) Biology and ecology of earthworms. Chapman & Hall, London

Eijsackers H, Van Gestel CAM, DeJonge S, Muijs B, Slijkerman D (2001) Polycyclic aromatic hydrocarbons-polluted dredged peat sediments and earthworms: a mutual interference. Ecotoxicology 10:35–50

Elvira C, Sampedro L, Benitez E, Nogales R (1998) Vermicomposting of sludge from paper mill and dairy industries with *Eisenia andrei*: a pilot scale study. Biores Technol 64:205–211

ERT (1987) A resource engineering company land treatment effects on wildlife populations in red pine plantations. Nekoosa Papers, Inc, Wisconsin, US

Field AJ, Sierra-Alvarez R (2008) Microbial degradation of chlorinated phenols. Rev Env Sci Biotech 7:211–241

Gajalakshmi S, Ramasamy EV, Abbasi SA (2002) Vermicomposting of paper waste with the anecic earthworm *Lampito mauriti* Kingburg. Int J Chem Technol 9:306–311

Garg VK, Kaushik P (2005) Vermistablisation of solid textile mill sludge spiked with poultry droppings by an epigeic earthworm *Eisenia fetida*. Biores Technol 96:1063–1071

González M, Gómez E, Comese R, Quesada M, Conti M (2010) Influence of organic amendments on soil quality potential indicators in an urban horticultural system. Biores Technol 101:8897–8901

Hait S, Tare V (2011) Optimizing vermistabilization of waste activated sludge using vermicompost as bulking material. Waste Manage 31:502–511

Hait S, Tare V (2012) Transformation and availability of nutrients and heavy metals during integrated composting vermicomposting of sewage sludges. Ecotoxicol Env Saf 79:214–224

Hanc A, Chadimova Z (2014) Nutrient recovery from apple pomace waste by vermicomposting technology. Biores Technol 168:240–244

Hand P (1988) Earthworm biotechnology (vermicomposting). In: Greenshields R (ed) Resources and applications of biotechnology. The Macmillan Press Ltd., London, pp 49–58

Huang K, Li Fusheng, Yongfen W, Xiaoyong F, Xuemin C (2014) Effects of earthworms on physicochemical properties and microbial profiles during vermicomposting of fresh fruit and vegetable wastes. Biores Technol 170:45–52

Johnsen AR, Wick LY, Harms H (2005) Principles of microbial PAH-degradation in soil. Env Pollut 133:71–84

Kaushik P, Garg VK (2003) Vermicomposting of mixed solid textile mill sludge and cow dung with the epigeic earthworm *Eisenia foetida*. Biores Technol 90:311–316

Kaushik P, Garg VK (2004) Dynamics of biological and chemical parameters during vermicomposting of solid textile mill sludge mixed with cow dung and agricultural residues. Biores Technol 94:203–209

Mantykoski K, Paasivirta J, Mannila E (1989) Combustion products of biosludge from pulp mill. Chemosphere 19:413–416

NCASI method CP-85.01 (1986) NCASI methods for the analysis of chlorinated phenolics in pulp industry wastewater. National Council of the Paper Industry for Air and Stream Improvement, New York, USA

Pokhrel D, Viraraghavan T (2004) Treatment of pulp and paper mill wastewater a review. Sci Total Env 333:37–58

Senesi N (1989) Composted materials as organic fertilizers. Sci Total Env 81(82):521–524

Singh J, Kalamdhad AS (2012) Concentration and speciation of heavy metals during water hyacinth composting. Biores Technol 124:169–179

Singh J, Kalamdhad AS (2013) Effect of *Eisenia fetida* on speciation of heavy metals during vermicomposting of water hyacinth. Ecol Eng 60:214–223

Suthar S, Singh S (2008) Feasibility of vermicomposting in biostabilization of sludge from a distillery industry. Sci Total Environ 394:237–243

US Environmental Protection Agency (1991) Proposed regulation of land application of sludge from pulp and paper mills using chlorine and chlorine derivative bleaching processes. Fed. Reg. 56(91), 40 CFR. Part 744, OPTS-62100; FRL 3873

Wang L, Zheng Z, Zhang Y, Chao J, Gao Y, Luo X, Zhang J (2013) Biostabilization enhancement of heavy metals during the vermiremediation of sewage sludge with passivant. J Hazard Mater (244/245):1–9

11

Aggregation stability and organic carbon fraction in a soil amended with some plant residues, nanozeolite, and natural zeolite

Milad Mirzaei Aminiyan · Ali Akbar Safari Sinegani · Mohsen Sheklabadi

Abstract This research is aimed to study comparison of the effects of applied nanozeolite and zeolite on aggregation and organic carbon (OC) in each aggregate size fraction in treated soil with different levels of some plant residues and incubation for 90 days. The analysis of variance showed that the effects of applied nanozeolite, zeolite, plant residues, and their interaction and incubation times on MWDw and OC in aggregate size fractions were statistically significant. The results showed that MWDw and OC contents in each aggregate size fraction increased with the additions of nanozeolite, zeolite, and plant residues. This study also revealed that nanozeolite and alfalfa straw were more effective for increasing the MWDw and OC contents in aggregate size fractions than zeolite and wheat straw, respectively. In other words, higher percentage nanozeolite, zeolite, and plant residues resulted in increasing the MWDw and OC in aggregate size fractions. The highest amounts of OC were observed in A and B compounds compared with the control and other treatments. However, the comparison of OC contents in all the treatments showed that proportion of nanozeolite for increasing the OC content was greater than that of zeolite. Also the highest OC contents were observed in larger aggregate size fractions. The MWDw and OC contents increased with the increasing days of incubation in treated soil. Thus, it was concluded that compound A can be more effective to improve the soil carbon sequestration.

Keywords Nanozeolite · Zeolite · Plant residue · Organic carbon · Aggregation · Incubation

M. Mirzaei Aminiyan (✉) · A. A. Safari Sinegani · M. Sheklabadi
Soil Science Department, College of Agriculture, Bu-Ali Sina University, Azadegan Street, 6517833131 Hamedan, Iran
e-mail: miladmir67@yahoo.com

A. A. Safari Sinegani
e-mail: safari_sinegani@yahoo.com

M. Sheklabadi
e-mail: sheklabadi@yahoo.com

Introduction

Aggregation has a major effect on carbon cycling in soil. Aggregates are composed of primary mineral particles and organic binding agents (Haynes and Swift 1990; Tisdall and Oades 1982). Soil aggregation and soil structure are important aspects of soil fertility in influencing the root distribution and the uptakes of water and nutrients (Pachepsky and Rawls 2003; Bronick and Lal 2005). The initial unit of aggregation is called microaggregate. According to Tisdall and Oades (1982), the microaggregates (<0.25 mm) are bounded together by organic compounds of different origins to form macroaggregates (>0.25 mm).

The macroaggregates affect soil carbon storage by occluding organic residues, making them less accessible to degrading organisms and their enzymes (Six et al. 2000). The soil organic carbon (SOC) protection by aggregates has been extensively studied (Tisdall and Oades 1982; Elliott 1986; Gupta and Germida 1988; Angers 1992). Christensen (2001) considered the structural control of soil aggregates as the accessibility of decomposers for the SOC, diffusion of oxygen, degradation of products, transport of moisture, etc. Beare et al. (1994) showed that macroaggregates (>0.25 mm) in nontilled soils provide an important mechanism for the SOC protection. Six et al. (2002) asserted that SOC protection is greater within free microaggregates (<0.25 mm) than within macroaggregates.

Carter et al. (2003) concluded that macroaggregates play an important role in the SOC storage, although protection by macroaggregates is labile because macroaggregates provide an environment conducive for interactions between clay particles and the SOC. Even though the effect of the SOC on soil aggregation seems to be more evident in the wet-sieved aggregates (Puget et al. 2000), the aggregates obtained from dry sieving might have more information on the processes occurring in situ. Previous studies of these sites between 1995 and 2002 showed that tillage practices influenced the accumulation of the SOC fractions (Wander and Bidart 2000; Yoo and Wander 2008).

Soil aggregation, more specifically of macroaggregates (250–2,000 μm) and microaggregates (53–250 μm), has direct influence on soil properties such as, but not limited to, bulk density, pore size distribution (Hillel 1982), microbial community structure (Paul and Clark 1996), and soil resistance to erosion (Valmis et al. 2005). Soil organic matter (SOM) indirectly contributes to soil structure by serving as a nucleus for aggregate formation (Six et al. 1998). SOM has been found to increase with reclamation age, but there is not always a corresponding increase in soil macroaggregation (Malik and Scullion 1998).

The fractionation of water-stable aggregates and density fractionation may thus be helpful for an improved understanding of carbon dynamics affected by soil management, since aggregate and density fractions are more sensitive to changes in soil management than total OC (Oades 1988; Pikul et al. 2007; Puget et al. 2000; Von Lützow et al. 2006). For example, Puget et al. (2000) concluded that water-stable macroaggregates were enriched in younger organic materials and have faster turnover times than microaggregates. Meanwhile, it is currently ascertained that the quality of organic matter is involved in both the accumulation of OC and the stabilization of soil particle sizes and aggregates (Piccolo et al. 2004; Spaccini et al. 2002; Yamashita et al. 2006).

Organic matter inputs through organic amendment, in addition to supplying nutrients, improve soil aggregation, and stimulate microbial diversity and activity (Shiralipour et al. 1992; Carpenter-Boggs et al. 2000). Soil aggregate stability is important for carbon (C) sequestration (Chivenge et al. 2011; Laganière et al. 2011). Soil microaggregates (i.e., organo–mineral complexes) form through the interaction of clay colliding with organic compounds and inorganic cementing materials (Blanco-Canqui and Lal 2004). Macroaggregates are composed of microaggregates bonded by root hair and fungal hyphae (Oades. 1984; Shukla et al. 2004). Natural zeolites are extensively used to improve soil physical environment, particularly, in sandy and clay poor soils (Abdi et al. 2006). The purpose of this study was to compare the effects of nanozeolite and zeolite on mean weight diameter of water-stable (MWDw) aggregates and OC in each fraction of MWDw in a treated soil by some plant residues.

Materials and methods

Site description

This study was conducted in Hamedan province, western part of Iran. This area is located between longitudes 47°42′ and 48°45′E and latitudes 33°28′ and 34°29′N. The climate of the region is semiarid with a mean annual precipitation of 300 mm and a mean annual temperature of 10 °C. Agriculture is an industry and constitutes principal land use in Hamadan. Major crops grown in Hamadan are winter wheat (*Triticumaestivum* L.), Alfalfa (*Medicago sativa*), potato (*Solanumtubersum*), and garlic (*Allium sativum*). The soil of this area is mostly classified as typic Haplocalcids.

Sampling, treatment, and analysis of soil

The soil samples were collected from the top layer of soil (30 cm) from an agricultural land. Wheat straw, alfalfa straw, nanozeolite pH, and electrical conductivity (EC) were measured in a 1:10 soil, plant residues, and nanozeolite separately: water extract after shaking for 2 h. Their OCs were measured by wet oxidation (Walkley and Black 1934). The total phosphorous (P) was measured in an acidic (HCl) solution of ashes of plant residues spectrophotometrically as blue molybdate–phosphate complexes under partial reduction with ascorbic acid (Peperzak et al. 1959). The total nitrogen (N) content was measured by the Kjeldahl method (Hinds and Lowe 1980). Cation exchange capacity (CEC) of the soils were determined by the 1 M NaOAc, pH 8.2 method (Pratt 1965). Subsequently, the soil was then air dried and mixed with different amounts of ($d < 2$ mm) wheat straw and alfalfa straw (0 and 5 % w/w) separately; afterward, the samples were mixed with different amounts of nanozeolite (0, 10, and 30 % w/w) and then were kept in field capacity moisture content and lab condition.

Untreated soils were also incubated as controls. The treated soils, wheat straw, alfalfa straw, and nanozeolite were uniformly wetted with a spray to a water content near field capacity. The approximate field capacity of each mixture of soil, wheat straw, alfalfa straw, and nanozeolite was measured gravimetrically, and the volume of water required to bring each of the soils to the approximated field capacity was determined by weighing daily. The treated and moistened soils were incubated in lab condition

(20–25 °C) for 90 days. After 1, 5, 10, 20, 30, 45, 60, 75, and 90 days of incubation, a portion of each soil was taken for the study of MWDw and its OC fractions.

Aggregate fractionation

The classical procedure described by Kemper and Rosenau (1986) was used to separate water-stable aggregates. In brief, 20 g of <4 mm air–dried soil samples was put on the topmost sieve of a nest of three sieves of 2, 1, 0.25, and 0.053-mm mesh sizes and pre-soaked in distilled water for 30 min. Then, the nest of sieves was oscillated vertically in water 20 times, using a 4-cm amplitude at the rate of one oscillation per second. Care was taken to ensure that soil particles on the topmost sieve were always below the water surface during each oscillation. After wet sieving, the water-stable soil materials left in each sieve were quantitatively transferred into beakers, dried in the oven at 50 °C for 48 h, weighed, and stored for later analysis. OC content in each fraction was measured by Walkley and Black (1934) method. MWDw was calculated by the following equation:

$$\mathrm{MWDw} = \sum_{i=1}^{n} (X_i.W_i)$$

where X_i is the mean diameter of the ith sieve, and W_i is the amount of total aggregates in ith fraction.

Statistical data analysis

The experiment was a complete randomized factorial design with three replicates. The factors applied were alfalfa straw (0 and 5 % w/w), wheat straw (0 and 5 % w/w), Nanozeolite (0, 10, and 30 % w/w), and incubation times (1, 5, 10, 20, 30, 45, 60, 75, and 90 days). Experimental data of MWD and OC fractions were subjected to analysis of variance and the means compared with the Duncan's new multiple range test using SAS Ver.9.2 (SAS Institute 2008).

Results and discussion

Table 1 shows the sand, clay, and silt contents were 69, 12, and 19 % in the studied soil, respectively. The soil texture was loamy sand. The soil was not saline (EC 1.1 dS m^{-1}); equivalent calcium carbonate and pH values were 1.79 % and 7.2, respectively, with low CEC (4.80 Cmol Kg^{-1}) and Total OC (3.41 g Kg^{-1}).

Table 2 presents some properties of the applied plant residues. Alfalfa and wheat straw had a neutral pH (6 and 7.97), high OC (511 and 532 g Kg^{-1}) values and C/N (23.30 and 90.75) and C/P (85.20 and 123.50) ratios, respectively. Some properties of applied nanozeolite and zeolite are shown (Table 3). The pH values in both the nanozeolite and zeolite are neutral, and nanozeolite and zeolite are not saline. CEC in nanozeolite is much higher than that in zeolite.

Table 4 shows the analysis of variance of the effects of nanozeolite, plant residues application, incubation time, and their interaction on MWDw and OC fractions in soil. Although the effects of applying nanozeolite, alfalfa, and wheat straws, their interaction, and incubation time on MWDw and OC fractions in soil were significant ($p < 0.01$), the interactions between nanozeolite and incubation time; plant residues and incubation time, and between nanozeolite, plant residues, and incubation time did not have significant effects on MWDw and OC fractions in the soil.

Table 5 shows the analysis of variance of the effects of zeolite and plant residues' application, incubation times, and their interactions on MWDw and OC fractions in soil. The effects of zeolite, alfalfa, and wheat straws

Table 1 Some chemical and physical properties of applied soil

EC (dS m^{-1})	PH	CEC (Cmol + Kg^{-1}_{soil})	Total organic C (g Kg^{-1})	CCE[a] %	Sand	Clay	Silt
1.1	7.2	4.80	3.41	1.79	69	12	19

[a] Carbonate calcium equivalent

Table 2 Some physical and chemical properties of used plant residues in this study

	PH	EC (dS m^{-1})	Total organic C (g Kg^{-1})	Total nitrogen (g Kg^{-1})	Total phosphorous (g Kg^{-1})	C/N	C/P
Alfalfa straw	6	9.5	511	22	5.98	23.30	85.20
Wheat straw	7.97	4.3	532	7	4.31	90.75	123.50

applications, their interactions, and incubation times on MWDw and OC fractions in soil were significant ($p < 0.01$). Also, the interactions between the application of zeolite and incubation time, plant residues and incubation time, and between zeolite and plant residues and incubation time did not have significant effects on MWDw and OC fractions in the soil.

Figure 1 shows Duncan's tests of means of MWDw and OC fractions in the soil as affected by nanozeolite, zeolite, alfalfa, and wheat straws percentages applied in the soil.

Table 3 Some properties of applied nanozeolite and zeolite

Parameter	Nanozeolite	Zeolite
pH	7.17	7.20
EC(dS m^{-1})	0.98	1.20
CEC (Cmol + Kg_{soil}^{-1})	400.39	169.3
Total organic C (g Kg^{-1})	1.03	0.93

MWDw increased by the addition of higher percentages of nanozeolite and zeolite (10 and 30 %) and 5 % plant residues into the soil. The amounts of MWDw on the treated soil with 30 % zeolitic materials and 5 % plant residues increased by 0.735 and 0.685 mm in 30 % nanozeolite plus 5 % alfalfa straw (compound A), and 30 % zeolite plus 5 % alfalfa straw (compound B) compared with the control, respectively.

The highest amounts of MWDw were observed as 1.24 and 1.19 mm in compound A and compound B, respectively. The comparison of treatments obtained, in mixed samples that applied nanozeolite, zeolite, and different plant residues, compound A was more effective than compound B (Fig. 1), because the amount of MWDw was higher in compound A into the compound B.

Although MWDw is known to be related to soil aggregate stability, the results of recent researches indicated that the SOC is closely related to the formation and stability of soil aggregates (Tisdall and Oades 1982). Thus, the

Table 4 Analysis of variance (mean square) of the effects of nanozeolite and plant residues application, incubation time, and their interactions on MWDw and organic carbon fractions in soil

Source	DF	MWDw (mm)	Organic carbon (g Kg_{soil}^{-1})			
			Agg. size >2 mm	Agg. size (1–2 mm)	Agg. size (0.25–1 mm)	Agg. size (0.053–0.25 mm)
Nanozeolite	2	9.7**	52.28**	46.97**	40.03**	27.79**
Plant residues	3	0.45**	2.47**	1.65**	1.4**	0.97**
Time	8	1.16**	17.54**	7.19**	6.12**	4.25**
Nanozeolite × time	16	1.10 ns	9.76 ns	5.82 ns	4.96 ns	3.44 ns
Plant residues × time	24	0.30 ns	4.22 ns	1.78 ns	1.52 ns	1.05 ns
Nanozeolite × plant residues	6	1.88**	10.21**	9.01**	7.68**	5.33**
Nanozeolite × plant residues × time	48	0.28 ns	2.44 ns	1.46 ns	1.24 ns	0.86 ns
Error	216	0.003	0.001	0.002	0.0015	0.0011

ns Mean square of the treatment is not significant

** Mean square of the treatment is significant at the 0.01 level

Table 5 Analysis of variance (mean square) of the effects of zeolite, plant residues' application, incubation time, and their interactions on MWDw and organic carbon fractions in soil

Source	DF	MWDw (mm)	Organic carbon (g/Kg soil)			
			Agg. size >2 mm	Agg. size (1–2 mm)	Agg. size (0.25–1 mm)	Agg. size (0.053–0.25 mm)
Zeolite	2	7.03**	27.85**	20.91**	17.82**	13.37**
Plant residues	3	0.68**	2.20**	1.62**	1.31**	1.08**
Time	8	1.49**	10.55**	7.18**	6.13**	4.75**
Zeolite × time	16	1.002 ns	5.40 ns	3.82 ns	3.25 ns	2.26 ns
Plant residues × time	24	0.40 ns	2.60 ns	1.74 ns	1.49 ns	1.01 ns
Zeolite × plant residues	6	1.47**	5.7**	4.27**	3.64**	2.52**
Zeolite × plant residues × time	48	0.26 ns	1.37 ns	0.97 ns	0.83 ns	0.67 ns
Error	216	0.0004	0.01	0.007	0.008	0.005

ns Mean square of the treatment is not significant

** Mean square of the treatment is significant at the 0.01 level

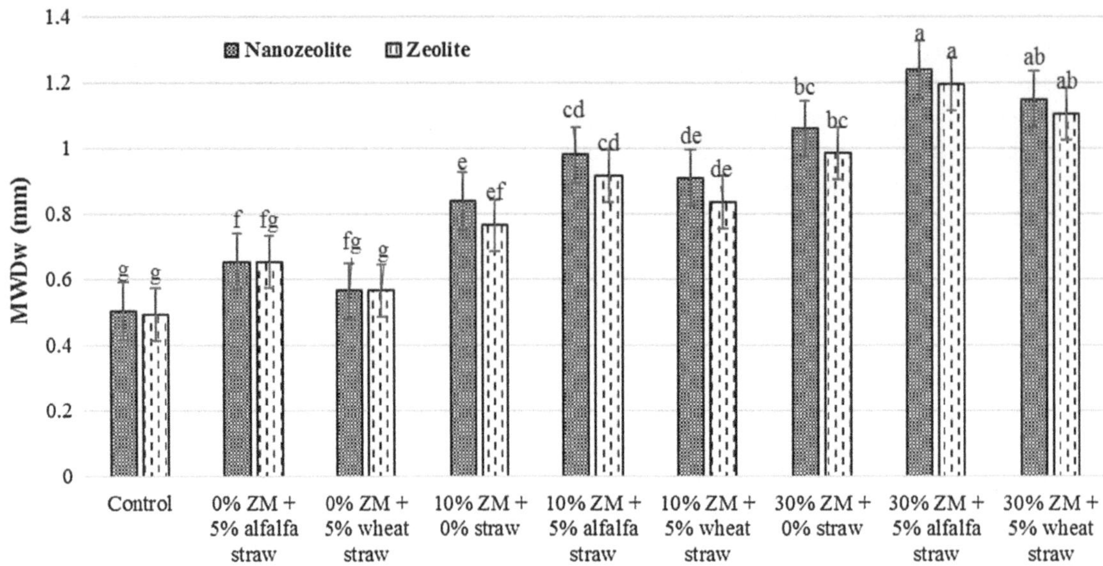

Fig. 1 Comparison of MWDw (mm) in treated soil by zeolitic materials (ZM) and plant residues

amendment of organic residues can improve soil structure and increase aggregate stability (Hati et al. 2008). Soil management systems that leave more plant residues on the soil surface generally allow for improvements in soil aggregation and aggregate stability (Carpenedo and Mielniczuk 1990).

Aggregate stability and SOC have also been shown to be positively correlated with polyphenol concentration (Griffiths and Burns 1972; Martens 2000, 2002; Martens and Loeffelmann 2002), which has been thought to be an important precursor of humic substances that stabilize aggregates (Dell'Agnola and Ferrari 1971; Stevenson 1994).

Blair et al. (2005) observed slower decomposition and greater aggregate stability (MWD) following the addition of intermediate quality *Flemingia macrophylla* (>4 % polyphenols) versus the addition of high quality *Medicago truncatula*. The reason for the stability of the aggregate hierarchy is most probably the continuous incorporation of young organic matter (Puget et al. 2000; Six et al. 2000; Helfrich et al. 2008), like carbohydrates and glomalin-related soil protein (GRSP), which decreased less than the TOC.

Among the various components of SOM, carbohydrates have particular importance in increasing aggregate stability (Angers and Mehuys 1989; Feller and Beare 1997), and as such, may act to a certain degree in protecting soil organic carbon against mineralization.

It was suggested that, for instance, macroaggregates are stabilized mainly by carbohydrate-rich roots or plant debris occluded within aggregates (Golchin et al. 1995). Various authors have also observed the positive effects of long-term manure application on MWD (Singh et al. 2007; Su et al. 2006) and better physical condition of soil.

Nanozeolite and zeolite are clay minerals, and they have considerable calcium (Ca) content (Kazemian 2002). For this reason, greater concentration of Ca, which acts as a potential cationic bridge between the inorganic and organic colloids' surfaces (Bronick and Lal 2005; Six et al. 2004). The availability of polyvalent cations, such as Ca^{2+}, is an important factor in the chemical protection of TOC. In addition to its role as a binding agent, polyvalent cations enhance the formation of cationic bridges between organic matter and clay (Bronick and Lal 2005; Six et al. 2004); a high saturation of clay particles with these cations helps the organic–mineral complex to remain more flocculated and condensed as well as reducing the efficiency of microbial and enzymatic attacks (Baldock and Skjemstad 2000).

Calcium exerts an influence on organo–mineral complexation and its stability is observed mainly at the microaggregate level (Six et al. 2004). According to Baldock et al. (1994), the process of Ca bridge formation is the dominant factor in the positive long-term effect of the addition of Ca on the structural stability of the soil. Meanwhile, the results reported by Muneer and Oades (1989) provide support for an association between the highest content of TOC with liming, and increased availability of Ca^{2+}. Those authors also concluded that the additions of $CaCO_3$ and glucose result in the formation of macroaggregates larger than 2 mm in 80 % of the soil, and only a small proportion of aggregates that are smaller than 0.05 mm.

Briedis et al. (2012) reported that Ca^{2+} from the surface liming was positively correlated with the TOC content and acted as a cationic bridge between the clay particles' surfaces and the OC of the soil, contributing to the sequestration of carbon. Water-stability of soil aggregates has been reported to increase with revegetation in a landslide area (Burri et al. 2009).

Figure 2 shows the amounts of MWDw on the treated soil with nanozeolite, zeolite, and plant residues, increased significantly during the 90 days of incubation. The amount of MWDw on the treated soil with nanozeolite and plant residues was 0.65 mm at the start of incubation, which increased to 1.16 mm after 90 days of incubation. Also the amounts of MWDw on the treated soil with zeolite and plant residues increased from 0.58 mm on the 1st day of incubation to 1.1 mm on the 90th day of incubation. Also Fig. 2 reveals that the amounts of MWDw on treated soil with nanozeolite and plant residues were higher compared with those on treated soil with zeolite and plant residues. These changes were reasonably related to the OC variation in the soil.

Figure 3 shows that the amount of OC in all of aggregate size fractions increased significantly ($p < 0.01$) with the additions of nanozeolite, zeolite, and plant residues to the soil, especially that of alfalfa straw, into the control. The amounts of OC were significantly great with the addition of higher percentages of zeolitic materials (10 and 30 %), and plant residues (5 %), especially alfalfa straw, in all of the aggregate sizes, in both the compound A and the compound B treatments. However, the amounts of OC was higher in each aggregate size fraction in the treated soil with nanozeolite and alfalfa straw than that in the treated soil with zeolite and alfalfa straw (Fig. 3, 4, 5, 6).

The amount of OC increased from 2.16 (g Kg_{soil}^{-1}) in control to 3.93 and 3.54 (g Kg_{soil}^{-1}) in compound A and compound B, respectively, in aggregate size fraction (>2 mm). Also, in this aggregate size class, the amounts of OC increased from 10 % nanozeolite 3.27 (g Kg_{soil}^{-1}) and zeolite 2.91 (g Kg_{soil}^{-1}) in (10 % zeolitic materials plus 5 % alfalfa straw) treatment to 30 % nanozeolite 3.93 (g Kg_{soil}^{-1}) and zeolite 3.54 (g Kg_{soil}^{-1}) in (30 % zeolitic materials plus 5 % alfalfa straw) treatment. The amount of OC in aggregate size fraction (>2 mm) increased with additions of 5 % alfalfa straw 3.93 and 3.54 (g Kg_{soil}^{-1}) in (30 % zeolitic materials plus 5 % alfalfa straw) treatment toward 5 % wheat straw 3.69 and 3.31 (g Kg_{soil}^{-1}) in (30 % zeolitic materials plus 5 % wheat straw) treatment in nanozeolite and zeolite, respectively.

The achieved results from previous researches showed that long-term field studies in the North America, and China indicate that manuring increases the SOC (Liu et al. 2003; Whalen and Chang 2002). In a 25-year study, SOC progressively increased with manure applications at an average rate of 0.181 (g Kg^{-1} mg^{-1}) of manure (Hao et al. 2003). Management practices such as reduced tillage and increased C inputs through residue management and manuring improve soil structure, reduce erosion, C loss through mineralization, and CO_2 emissions (Hao et al. 2002; Lal 2003a, 2003b).

Carpenedo and Mielniczuk (1990) reported a medium-to-high correlation between the geometric mean diameter (GMD), the MWD, the amount of aggregates >2 mm, and the TOC content of Latosols from Brazil; also another research achieved similar result of OC content of Latosols from Brazil (Castro Filho et al. 2002). Kouakoua et al. (1999) and Dutartre et al. (1993) indicated a strong correlation between aggregate stability in water and the carbon content of bulk clayey Ferralsols from Africa.

Figure 4 shows that the amount of OC increased with the additions of the higher percentages of nanozeolite, zeolite, and plant residues, particularly alfalfa straw, in aggregate size 1–2 mm; in A and B compounds, OC was maximum compared with the control and other treatments. However, comparison of the OC contents in zeolitic materials showed that proportion of nanozeolite to the

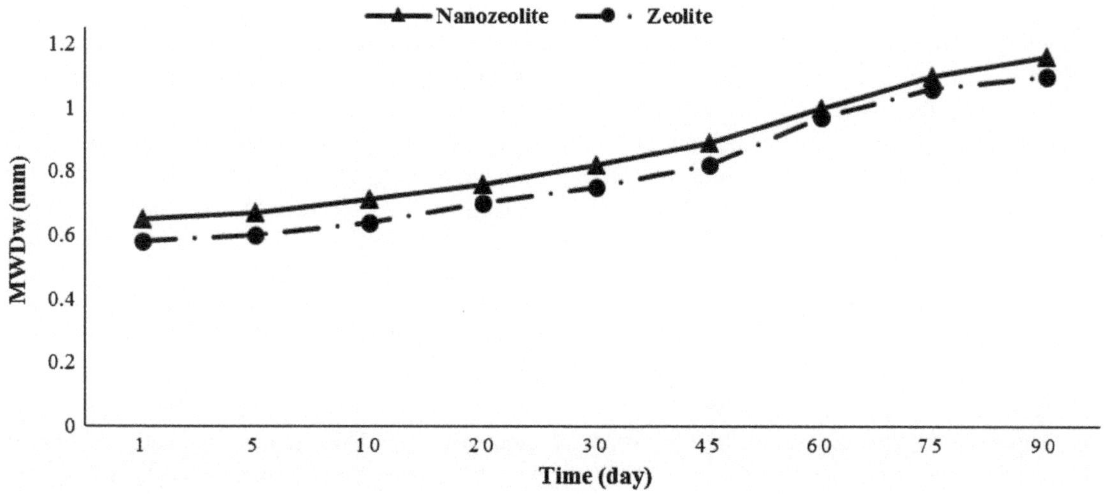

Fig. 2 The amounts of MWDw on the treated soil with nanozeolite, zeolite, and plant residues during 90 days of incubation

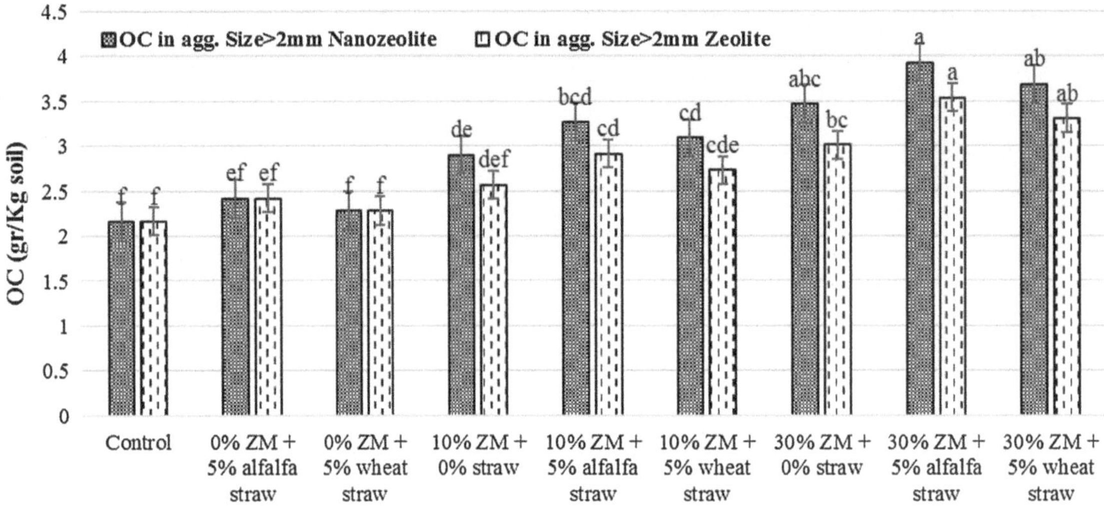

Fig. 3 Comparison of OC (g/Kg soil) in aggregate size >2 mm in treated soil by zeolitic materials (ZM) and plant residues

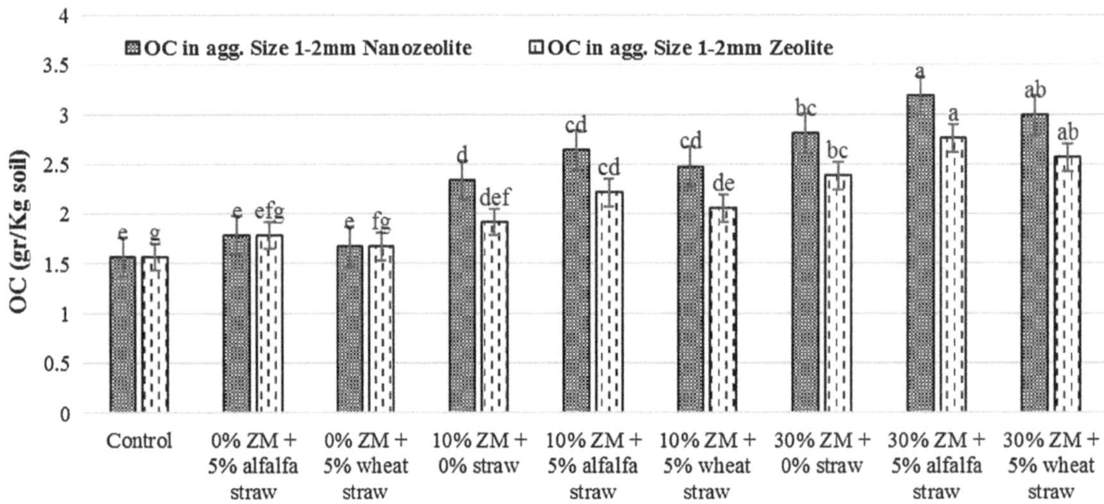

Fig. 4 Comparison of OC (g/Kg soil) in aggregate size (1–2) mm in treated soil by zeolitic materials (ZM) and plant residues

increasing amount of OC was greater than that of zeolite (Fig. 4).

The results showed the amount of OC increased in aggregate size range of 0.25–1 mm with the additions of nanozeolite and zeolite, especially nanozeolite, and also addition of plant residues, particularly alfalfa straw; the amounts of OC were 2.76 and 2.36 (g Kg_{soil}^{-1}) in compound A and compound B, respectively (Fig. 5). The results indicated that the amount of OC in aggregate size fraction (0.053–0.25 mm) increased with the additions of nanozeolite, zeolite, and plant residue. The amounts of OC were 2.2 and 1.87 (g Kg_{soil}^{-1}) in compound A and compound B, respectively. However, comparison of zeolitic materials indicated that nanozeolite was more effective than zeolite for increasing the OC (Fig. 6).

In soils organic matter of which is acting as the major binding agent, OC concentrations may increase with the increasing aggregate size fractions, because larger aggregate size fractions are composed of smaller aggregates plus organic binding agents (Elliott 1986). TOC found in microaggregates is considered to be stabler and more resistant to microbial decomposition, leading to an accumulation of SOM over a long period (Angers et al. 1997; Oades 1984).

Microaggregates have been considered a key factor in the formation of new macroaggregates (Six et al. 2000) due to the protection of intra-aggregate TOC. The results revealed that larger aggregate size fractions had greater amount of OC than smaller aggregate size fractions. This may be due to either an increased aggregates' stability and strength caused by the addition of zeolitic materials, or

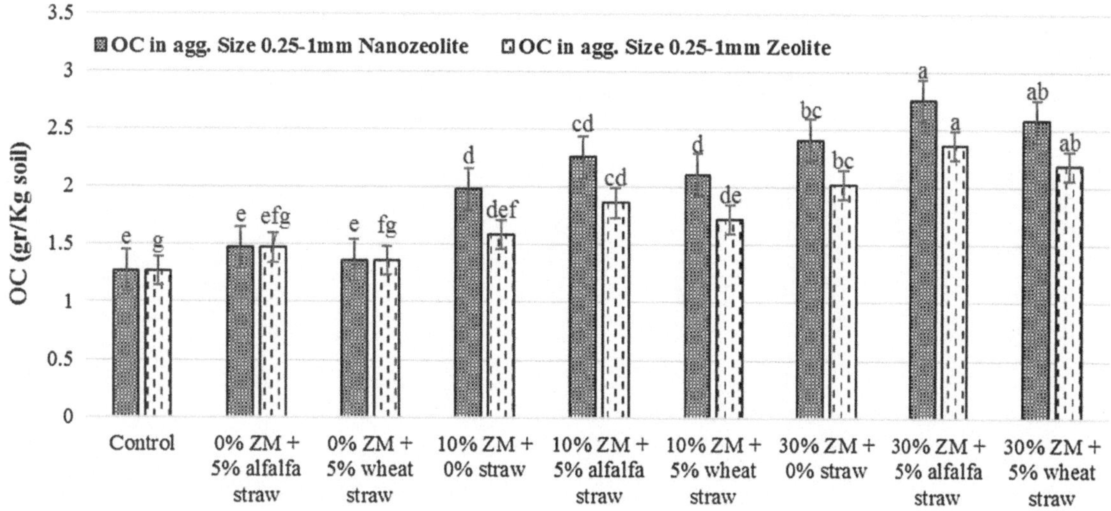

Fig. 5 Comparison of OC (g/Kg soil) in aggregate size (0.25–1) mm in treated soil by zeolitic materials (ZM) and plant residues

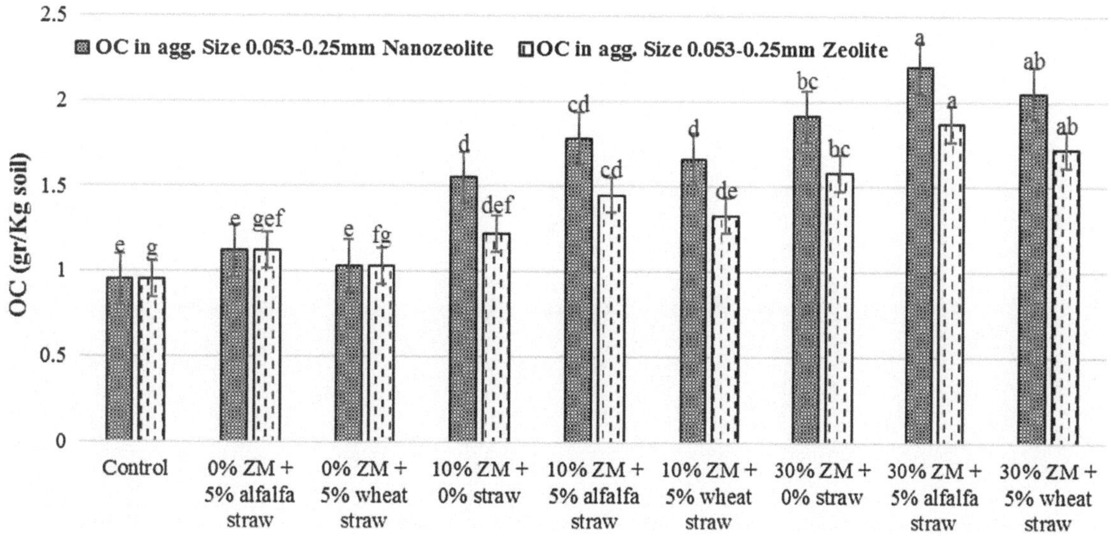

Fig. 6 Comparison of OC (g/Kg soil) in aggregate size (0.053–0.25) mm in treated soil by zeolitic materials (ZM) and plant residues

because the humification of plant residues increased the amount of OC in macroaggregates during incubation (Koushwaha et al. 2001) in the treated soil. Also Bronick and Lal (2005) concluded that smaller aggregates (0.25–0.5 mm) had smaller C aggregate compared with the larger aggregates.

Figure 7 reveals that the increased OC content of the treated soil during incubation in the lab condition. The OC content increased with the increasing time of incubation in different aggregate sizes >2 mm (Fig. 7a), 1–2 mm (Fig. 7b), 0.25–1 mm (Fig. 7c), and 0.053–0.25 mm (Fig. 7d). Overall, Fig. 7 indicates that the increasing OC content in the subsequent days of incubation in nanozeolite treatments were higher than that in zeolite treatments in all the aggregate size fractions. The amount of OC in larger aggregate size fractions was higher than that in smaller size fractions.

The manure which was applied for 2 years increased the SOC mineralization with the passage of time in different aggregate sizes (Pare et al. 2000; Aulakh et al. 2000). Aoyama and Kumakura (2001) investigated an increase in SOM with animal manure application, and consequently, the formation of macroaggregates (250–1,000 μm). They further pointed out that manure application increased the accumulation of macroaggregate-protected C and thus sequestrated more organic C in soils.

Organic and inorganic fertilizers have complex effects on soil's physical, chemical, and biological properties, resulting in variable effects on soil aggregation (Bronick and Lal 2005). Also those authors reported that application

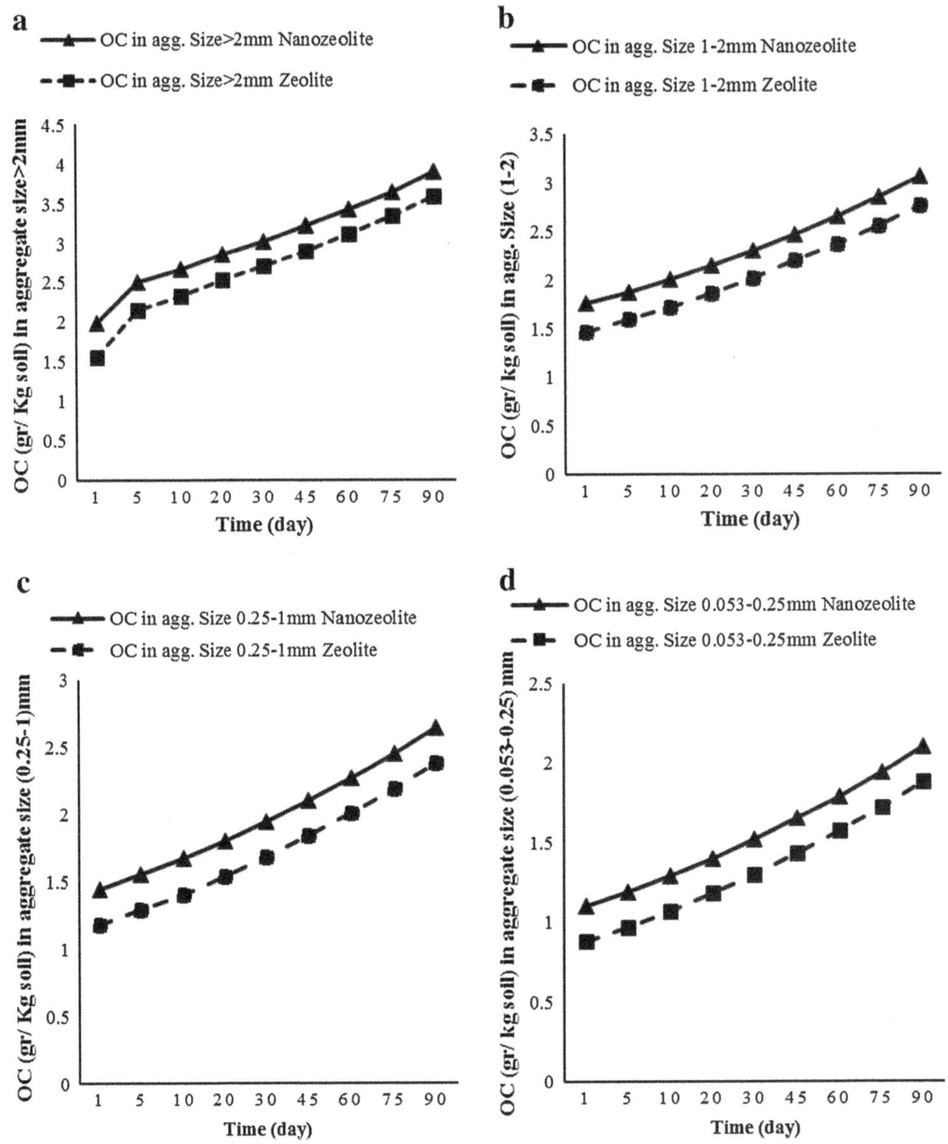

Fig. 7 The amount of OC (g/Kg soil) in **a** aggregate size >2 mm, **b** aggregate size (1–2) mm, **c** aggregate size (0.25–1) mm, and **d** in aggregate size (0.053–0.25) mm, on the treated soil with nanozeolite, zeolite, and plant residues during 90 days of incubation

of organic manure directly increases SOM, leading to improved soil aggregation and enhanced stability over time. These trends are usually indicated by increases in macroaggregate content and the MWDw, as well as by decreases in bulk density (Bronick and Lal 2005; Haynes and Naidu 1998; Sainju et al. 2003; Whalen and Chang 2002).

Huang et al. (2010) observed significantly more macroaggregates and higher SOM content in the bulk soil and the >2 mm aggregate fraction for the NPK plus OM treatment compared with the other treatments. Zhou et al. (2013) concluded that the application of NPK plus OM increased the size of sub-aggregates that comprised the macroaggregates. Also, they observed that long-term application of NPK plus OM improves soil aggregation and alters the three-dimensional microstructure of macroaggregates, while NPK alone does not. In other research studies, it was reported that the enrichment of the SOC, either from crop residues or from directly applied organic manure, has been documented to improve soil aggregation (Six et al. 1998, 2004).

Chen and Shrestha (2012) observed that the SOC increased with the passage of time (year) and caused an increase in percentage of water-stable aggregate. Higher C accumulation in macroaggregates could be due to the lower decomposable SOM associated with these aggregates and also the direct contribution of SOM to the stability of macroaggregates resulting in only C-rich macroaggregates being able to withstand slaking due to zeolitic materials in the soil. The achieved results from the present study are in accordance with the results of Tripathi et al. (2014).

Conclusion

Aggregation process in the soil is important, and it plays a considerable role in improving the soil physical characteristics such as hydraulic conductivity, infiltration, ventilation, etc. Also, the aggregation process is important in improving the carbon sequestration in soil. This report shows the effects of application of nanozeolite and zeolite on MWDw as an index of aggregate stability and strength, and OC aggregate size fractions in a soil treated with some plant residues during the incubation period. The results revealed that the amount of MWDw increased with the additions of nanozeolite, zeolite, and plant residues.

The highest amount of MWDw was measured when 30 % zeolitic materials and 5 % plant residues were added to the soil. The results showed that nanozeolite was more effective in increasing the OC contents in MWDw than zeolite. Also alfalfa straw efficiently produced more OC than wheat straw did. In other words, higher percentages of nanozeolite, zeolite, and plant residues increased the amount of MWDw in treated soils. During the incubation period, MWDw increased from the first day until the 90th day of incubation.

The highest amounts of OC were found in compound A and compound B treatments in all the aggregate size fractions. However, comparison of OC contents in zeolitic materials showed that nanozeolite plays an important role in increasing the OC contents in different aggregate size fractions compared with the zeolite. Also the amount of OC in larger aggregates size fractions was higher than that in smaller fractions. The amount of OC increased with the increasing time of incubation in different aggregates size fractions. Thus, the highest amount of OC was measured in soil treated by compound A. This compound can be more effective to improve soil carbon sequestration. However, it must be mentioned that additional and more researches need for the assessing of the effects of zeolitic materials, plant residues, and their interactions on OC fractions and other characteristics of soil under different incubation conditions.

Acknowledgments The authors sincerely thank the editor and the anonymous reviewers for reviewing the manuscript and providing critical comments to improve the paper. They also offer many thanks to the Fadak institute in Isfahan and the Afrazand factory in Tehran.

References

Abdi GH, Khui MK, Eshghi S (2006) Effects on natural zeolite on growth and flowering on straw berry. Int J Agric Resour 1:384–389

Angers DA (1992) Changes in soil aggregation and organic carbon under corn and alfalfa. Soil Sci Soc Am J 56:1244–1249

Angers DA, Mehuys GR (1989) Effects of cropping on carbohydrate content and water-stable aggregation of a clay soil. Can J Soil Sci 69:373–380

Angers DA, Recous S, Aita C (1997) Fate of carbon and nitrogen in water-stable aggregates during decomposition of ^{13}C ^{15}N-labeled wheat straw in situ. Eur J Soil Sci 48:295–300

Aoyama M, Kumakura N (2001) Quantitative and qualitative changes of organic matter in an ando soil induced by mineral fertilizer and cattle manure application for 20 years. Soil Sci Plant Nutr 47:241–252

Aulakh MS, Khera TS, Doran JW (2000) Mineralization and denitrification in upland, nearly saturated and flooded subtropical soil—II. Effect of organic manures varying in N content and C:N ratio. Biol Fertil Soils 31:168–174

Baldock JA, Skjemstad JO (2000) Role of the soil matrix and minerals in protecting natural organic materials against biological attack. Org Geochem 31:697–710

Baldock JA, Aoyanna M, Oades JM, Grant CD (1994) Structural amelioration of a South Australian red-brown earth using calcium and organic amendments. Aust J Soil Res 32:571–594

Beare MH, Cabrera ML, Hendrix PF, Coleman DC (1994) Aggregate protected and unprotected organic matter pools in conventional and no-tillage soils. Soil Sci Soc Am J 58:787–795

Blair N, Faulkner RD, Till AR, Sanchez P (2005) Decomposition of ^{13}C and ^{15}N labelled plant residue materials in two different soil types and its impact on soil carbon, nitrogen and aggregate stability, and aggregate formation. Aust J Soil Res 43:873–886

Blanco-Canqui H, Lal R (2004) Mechanisms of carbon sequestration in soil aggregates. Crit Rev Plant Sci 23:481–504

Briedis C, MoraesSá JC, Fávero Caires E, Navarro JF, Inagaki TM, Boer A, Neto CQ, Oliveira Ferreira A, Canalli LB, Burkner dos Santos J (2012) Soil organic matter pools and carbon-protection mechanisms in aggregate fractions influenced by surface liming in a no-till system. Geoderma 170:80–88

Bronick CJ, Lal R (2005) Soil structure and management: a review. Geoderma 124:3–22

Burri K, Graf F, Böll A (2009) Revegetation measures improve soil aggregate stability: a case study on a landslide area in Central Switzerland. FOSNOLA 82:45–60

Carpenedo V, Mielniczuk J (1990) Estado de agregação e qualidade de agregados de LatossolosRoxos, submetidos a diferentessistemas de manejo. R Bras Ci Solo 14:99–105

Carpenter-Boggs L, Kennedy AC, Reganold JP (2000) Organic and biodynamic management: effects on soil biology. Soil Sci Soc Am J 54:1651–1659

Carter MR, Angers DA, Gregorich EG, Bolinder MA (2003) Characterizing organic matter retention for surface soils in eastern Canada using density and particle size fractionations. Can J Soil Sci 83:11–23

Castro Filho C, Lourenco A, Guimaraes MDF, Fonseca ICB (2002) Aggregate stability under different soil management systems in a red latosol in the state of Parana. Braz Soil Tillage Res 65:45–51

Chen HYH, Shrestha BM (2012) Stand age, fire and clearcutting affect soil organic carbon and aggregation of mineral soils in boreal forests. Soil Biol Biochem 50:149–157

Chivenge P, Vanlauwe B, Gentile R, Six J (2011) Organic resource quality influences short-term aggregate dynamics and soil organic carbon and nitrogen accumulation. Soil Biol Biochem 43:657–666

Christensen BT (2001) Physical fractionation of soil and structural and functional complexity in organic matter turnover. Eur J Soil Sci 52:345–353

Dell'Agnola G, Ferrari G (1971) Molecular sizes and functional groups of humic substances extracted by 0.1 M pyrophosphate from soil aggregates of different stability. Eur J Soil Sci 22:342–349

Dutartre Ph, Bartoli F, Andreux F, Portal JM, Ange A (1993) Influence of content and nature of organic matter on the structure of some sandy soils from West Africa. In: Brussaard L, Kooistra MJ (eds) Proceedings of the international workshop on methods of research on soil structure/soil biota interrelationships vol 56. Geoderma, p 459–478

Elliott ET (1986) Aggregate structure and carbon, nitrogen and phosphorus in native and cultivated soils. Soil Sci Soc Am J 50:627–633

Feller C, Beare MH (1997) Physical control of soil organic matter dynamics in the tropics. Geoderma 79:69–116

Golchin A, Oades JM, Skjemstad JO, Clark P (1995) Structural and dynamic properties of soil organic matter as reflected by ^{13}C natural abundance, pyrolysis mass spectrometry and solid-state ^{13}C NMR spectroscopy in density fractions of an oxisol under forest and pasture. Aust J Soil Res 33:59–76

Griffiths E, Burns RG (1972) Interaction between phenolic substances and microbial polysaccharides in soil aggregation. Plant Soil 36:599–612

Gupta VVSR, Germida JJ (1988) Distribution of microbial biomass and its activity in different soil aggregate size fractions as affected by cultivation. Soil Biol Biochem 20:777–786

Hao Y, Lal R, Owens LB, Izaurralde RC, Post WM, Hothem DL (2002) Effect of cropland management and slope position on soils organic carbon pool at the Appalachian experimental watersheds. Soil Tillage Res 68:133–142

Hao XY, Chang C, Travis GR, Zhang FR (2003) Soil carbon and nitrogen response to 25 annual cattle manure applications. J Plant Nutr Soil Sci 166:239–245

Hati KM, Swarup A, Mishra B, Manna MC, Wanjari RH, Mandal KG, Misra AK (2008) Impact of long-term application of fertilizer, manure and lime under intensive cropping on physical properties and organic carbon content of an alfisol. Geoderma 148:173–179

Haynes RJ, Naidu R (1998) Influence of lime, fertilizer and manure applications on soil organic matter content and soil physical conditions: a review. Nutr Cycl Agroecosyst 51:123–137

Haynes RJ, Swift RS (1990) Stability of soil aggregates in relation to organic constituents and soil water content. J Soil Sci 41:73–83

Helfrich M, Ludwig B, Potthoff M, Flessa H (2008) Effect of litter quality and soil fungi on macroaggregate dynamics and associated partitioning of litter carbon and nitrogen. Soil Biol Biochem 40:1823–1835

Hillel D (1982) Introduction to soil physics, vol 364. Academic Press, San Diego

Hinds A, Lowe LE (1980) Ammonium-N determination soil nitrogen Berthelot reaction. Soil Sci Plant Anal 11:469–475

Huang S, Peng X, Huang Q, Zhang W (2010) Soil aggregation and organic carbon fractions affected by long-term fertilization in a red soil of subtropical China. Geoderma 154:364–369

Kazemian H (2002) Zeolite science in Iran: a brief review. International conference on the occurrence, properties and utilization of natural zeolites. Greece, p 162–164

Kemper WD, Rosenau RC (1986) Aggregate stability and size distribution. In: Klute A (ed) Methods of soil analysis, Part I, physical and mineralogical methods. American Society of Agronomy, Madison

Kouakoua E, Larré-Larrouy MC, Barthès B, Freitas PL, Neves C, Sala GH, Feller C (1999) Relations entre stabilité de l'agrégationetmatiéreorganiqetotaleet soluble à l'eauchaudedans des sols ferrallitiquesargileux (Congo, Brésil). Can J Soil Sci 79(4):561–569

Koushwaha CP, Tripathi SK, Singh KP (2001) Soil organic matter and water-stable aggregates under different tillage and residue conditions in a tropical drylandagroecosystem. Appl Soil Ecol 16:229–241

Laganière J, Angers DA, Paré D, Bergeron Y, Chen HYH (2011) Black spruce soils accumulate more uncomplexed organic matter than aspen soils. Soil Sci Soc Am J 75:1125–1132

Lal R (2003a) Soil erosion and the global carbon budget. Environ Int 29:437–450

Lal R (2003b) Offsetting global CO_2 emissions by restoration of degraded soils and intensification of world agriculture and forestry. Land Degrad Deviat 14:309–322

Liu XB, Han XZ, Song CY, Herbert SJ, Xing BS (2003) Soil organic carbon dynamics in black soils of China under different agricultural management systems. Community Soil Sci Plant Anal 34:973–984

Malik A, Scullion J (1998) Soil development on restored opencast coal sites with particular reference to organic matter and aggregate stability. Soil Usage Manage 14:234–239

Martens DA (2000) Management and crop residue influence soil aggregate stability. J Environ Qual 29:723–727

Martens DA (2002) Relationship between plant phenolic acids released during soil mineralization and aggregate stabilization. Soil Sci Soc Am J 66:1857–1867

Martens DA, Loeffelmann KL (2002) Improved accounting of carbohydrate carbon from plants and soils. Soil Biol Biochem 34:1393–1399

Muneer M, Oades JM (1989) The role of Ca–organic interactions in soil aggregate stability. III. Mechanisms and models. Aust J Soil Res 27:411–423

Oades JM (1984) Soil organic matter and structural stability mechanisms and implications for management. Plant Soil 76:319–337

Oades JM (1988) The retention of organic-matter in soils. Biogeochemistry 5:35–70

Pachepsky YA, Rawls WJ (2003) Soil structure and pedotransfer functions. Eur J Soil Sci 54:443–452

Pare T, Dinel H, Schnitzer M (2000) Carbon and nitrogen mineralization in soil amended with non-tabletized and tabletized poultry manure. Can J Soil Sci 80:271–276

Paul EA, Clark FE (1996) Soil microbiology and biochemistry. Academic Press, San Diego

Peperzak P, Caldwell AG, Hunziker R, Black CA (1959) Phosphorus fractions in manures. Soil Sci 87:293–302

Piccolo GA, Galantini JA, Rosell RA (2004) Organic carbon fractions in a yerba mate plantation on a subtropical Kandihumult of Argentina. Geoderma 123:333–341

Pikul JJL, Osborne S, Ellsbury M, Riedell W (2007) Particulate organic matter and water-stable aggregation of soil under contrasting management. Soil Sci Soc Am J 71:766–776

Pratt PF (1965) Potassium. In: Black CA (ed) Methods of soil analysis: Part 2. Chemical and microbiological properties. American Society of Agronomy, Madison, pp 1022–1030

Puget P, Chenu C, Balesdent J (2000) Dynamics of soil organic matter associated with particle-size fractions of water-stable aggregates. Eur J Soil Sci 51:595–605

Sainju U, Whitehead W, Singh B (2003) Cover crops and nitrogen fertilization effects on soil aggregation and carbon and nitrogen pools. Can J Soil Sci 83:155–165

SAS Institute (2008) SAS procedures guide. Version 9.2, SAS Institute, Cary

Shiralipour A, Mc Connell W, Smith WH (1992) Physical and chemical properties of soil as affected by municipal solid waste compost application. Biomass Bioenergy 3:195–211

Shukla MK, Lal R, Ebinger M (2004) Soil quality indicators for reclaimed mine soils in southeastern Ohio. Soil Sci 169:133–142

Singh G, Jalota SK, Singh Y (2007) Manuring and residue management effects on physical properties of a soil under the rice–wheat system in Punjab, India. Soil Tillage Res 94:229–238

Six J, Elliott ET, Paustian K, Doran JW (1998) Aggregation and soil organic matter accumulation in cultivated and native grassland soils. Soil Sci Soc Am J 62:1367–1377

Six J, Paustian K, Elliot ET, Combrink C (2000) Soil structure and organic matter: I. Distribution of aggregate–size fractions and aggregate–associated carbon. Soil Sci Soc Am J 64:681–689

Six J, Conant RT, Paul EA, Paustian K (2002) Stabilization mechanisms of soil organic matter: implication for C-saturation of soils. Plant Soil 241:155–176

Six J, Bossuyt H, Degryze S, Denef K (2004) A history of research on the link between (micro) aggregates, soil biota, and soil organic matter dynamics. Soil Tillage Res 79:7–31

Spaccini R, Piccolo A, Conte P, Haberhauer G, Gerzabek MH (2002) Increased soil organic carbon sequestration through hydrophobic protection by humic substances. Soil Biol Biochem 34:1839–1851

Stevenson FI (1994) Characterization of soil organic matter by NMR spectroscopy and analytical pyrolysis. Humus chemistry, genesis composition reaction. Willey, New York, pp 259–284

Su YZ, Wang F, Suo DR, Zhang ZH, Du MW (2006) Long-term effect of fertilizer and manure application on soil-carbon sequestration and soil fertility under the wheat–wheat–maize cropping system in northwest China. Nutr Cycl Agroecosyst 75:285–295

Tisdall JM, Oades JM (1982) Organic matter and water-stable aggregates in soils. J Soil Sci 33:141–163

Tripathi R, Nayak AK, Bhattacharyya P, Shukla AK, Shahid M, Raja R, Panda BB, Mohanty S, Kumar A, Thilagam VK (2014) Soil aggregation and distribution of carbon and nitrogen in different fractions after 41 years long-term fertilizer experiment in tropical rice–rice system. Geoderma 213:280–286

Valmis S, Dimoyannis D, Danalatos NG (2005) Assessing interrill erosion rate from soil aggregate instability index, rainfall intensity and slope angle on cultivated soils in central Greece. Soil Tillage Res 80:139–147

Von Lützow M, Koegel-Knabner I, Ekschmitt K, Matzner E, Guggenberger G, Marschner B, Flessa H (2006) Stabilization of organic matter in temperate soils: mechanisms and their relevance under different soil conditions—a review. Eur J Soil Sci 57:426–445

Walkley A, Black IA (1934) An examination of digestion method for determining soil organic matter and a proposed modification of the chromic acid titration. Soil Sci 37:29–38

Wander MM, Bidart MG (2000) Tillage practice influences on the physical protection, bioavailability and composition of particulate organic matter. Biol Fertil Soils 32:360–367

Whalen JK, Chang C (2002) Macroaggregate characteristics in cultivated soils after 25 annual manure applications. Soil Sci Soc Am J 66:1637–1647

Yamashita T, Flessa H, John B, Helfrich M, Ludwig B (2006) Organic matter in density fractions of water-stable aggregate in silty soils: effect of land use. Soil Biol Biochem 38:3222–3234

Yoo G, Wander MM (2008) Tillage effects on aggregate turnover and sequestration of particulate and humified soil organic carbon. Soil Sci Soc Am J 72:670–676

Zhou H, Peng X, Perfect E, Xiao T, Peng G (2013) Effects of organic and inorganic fertilization on soil aggregation in an ultisol as characterized by synchrotron based X-ray micro-computed tomography. Geoderma 195–196:23–30

12

Reduction of bioavailability and leachability of heavy metals during agitated pile composting of *Salvinia natans* weed of Loktak lake

W. Roshan Singh[1] · Shashi Kumar Pankaj[2] · Ajay S. Kalamdhad[2]

Abstract

Purpose Composting of harvested *Salvinia natans* weed of Loktak lake of Manipur, India can protect the precious lake from the unwanted growth of the weeds but there is a need to assess the total amount and bioavailable forms of heavy metals and minerals of the final compost before application as a soil conditioner.

Methods Studies were conducted on the transformations of the physico-chemical parameters and the total amount, bioavailable and leachable forms of heavy metals and minerals during agitated pile composting of *S. natans* with rice husk and cattle manure in five different combinations [trial 1 (5 *S. natans*: 4 cattle manure: 1 rice husk), trial 2 (6 *S. natans*: 3 cattle manure: 1 rice husk), trial 3 (7 *S. natans*: 2 cattle manure: 1 rice husk), trial 4 (8 *S. natans*: 1 cattle manure: 1 rice husk) and trial 5 (10 *S. natans*: 0 cattle manure: 0 rice husk)].

Results The highest temperature (52.2 °C) was monitored in trial 2 having 30 % cattle manure during the process. Highest reduction of moisture content and volatile solid in trial 2 corroborated the temperature profile. Total concentrations of heavy metals (Zn, Cu, Mn, Fe, Ni, Pb, Cd and Cr), total concentration of nutrients (Na, K, Ca, and Mg) and the water-soluble forms of the nutrients were enhanced significantly. The water-soluble, plant-available and leachable metals decreased favorably with maximum reduction observed mostly in trial 2. The study also revealed that the total concentrations of Zn (211–254 mg/kg), Ni (310–345 mg/kg) and Pb (805–891 mg/kg) of the final composts were higher than that of Cr (140–207 mg/kg) but the water-soluble forms of Zn (3.6–4.7 % of total Zn), Ni (NIL) and Pb (0.7–1.0 % of total Pb) were lower than that of Cr (13.3–19.3 % of total Cr) indicating that Cr in the composts had higher toxicity potential. The leachable heavy metals were within the threshold limits prescribed for agricultural application.

Conclusions Composting of *S. natans* biomass with appropriate proportion of cattle manure reduced the bioavailable and leachable forms of the heavy metals in the biomass.

Keywords *Salvinia natans* · Composting heavy metals · Bioavailability · DTPA extraction · TCLP test

Introduction

Northeast India has vast natural productive wetlands which support valuable biodiversity or heterogeneity. Loktak lake is one of the important wetlands in the region and plays an important role in providing regional ecological and economic security. The lake is situated in the southern part of the Manipur valley in Bishnupur district (24°25′–24°42′N; 93°46′–93°55′E) and is the largest natural fresh water lake of northeast India. It covers an area of about 287 km^2 which varies with the maintenance of water level of Ithai barrage across the outlet of the lake (768.5 m above mean sea level). The main characteristic feature of the lake is the presence of floating islands, locally called 'phumdis' which are heterogeneous mass of soil, vegetation and organic matter at various stages of decomposition and occurring in

✉ W. Roshan Singh
wroshan123@gmail.com

1 Manipur Pollution Control Board, Imphal West DC Office Complex, Lamphelpat, Imphal 795004, Manipur, India

2 Department of Civil Engineering, Indian Institute of Technology Guwahati (IITG), Guwahati 781039, Assam, India

different sizes. There are also many aquatic species in the clear water zone of the lake and the dominating invasive species, *Salvinia natans* has profusely outgrown and damaged the aquatic ecosystem by its rapid growth and displacement of native plants that provide food and habitat to the native animals and waterfowl.

Proliferation and change in the character of phumdi and the aquatic weeds have led to the problems of eutrophication; water quality deterioration; decline in fish production; reduction of water storage capacity; interfering with the production of agricultural crops; and blockage of channels leading to retardation of natural water flow. These all have resulted in seriously affecting the overall health of the lake (Santosh and Bidan 2002). The authors have demonstrated the potential for effective composting of phumdi biomass (Singh and Kalamdhad 2014c; Singh et al. 2014). The *S. natans* of Loktak lake may also be suitably used for bioprocessing or biomass conversion due to high organic content (Devi et al. 2002; Mande and Lata 2005). Therefore, composting followed by land application can be one of the economical options for the treatment and disposal of the hugely available *S. natans* biomass harvested to control its proliferation in the lake.

Composting of *S. natans* is not reported anywhere and there is scope for investigation in using it as organic manure and as a soil conditioner. But the phumdis and weeds act as a biological sink of the pollutants coming into the lake and there is possible presence in their biomass—heavy metals which have the potential to cause adverse effect on the environment from the application of the resulting composts. High and excessive accumulation of heavy metals in the soil, their uptake by plants, successive accumulation in human tissues, and biomagnifications through the food chain are the main concerns for health and environment (Wong and Selvam 2006; Iwegbue et al. 2007; Chiroma et al. 2012). The heavy metals exert toxic effects on soil microorganisms resulting in the change of diversity, population and overall activity of the soil microbial communities (Ashraf and Ali 2007). It is the bioavailable and leachable forms of the heavy metals rather than the total heavy metal concentrations that provide useful information about the risk of toxicity and remobilization of the heavy metals in the environment (Liu et al. 2007). Bioavailable forms of heavy metals may be: (1) water-soluble heavy metals which are the most biologically active heavy metals. They have the highest potential for contaminating the surface and ground water and entering the food chain (Iwegbue et al. 2007); (2) diethylenetriamine penta-acetic acid (DTPA)-extractable heavy metals which are the plant-available metal forms at regular or higher concentrations (Samuel et al. 2013). DTPA is a chelating agent that mimics plant uptake of heavy metals by extracting carbonate-bound and organic-bound metal fractions of calcareous soils (Walter et al. 2006).Therefore, assessment of DTPA-extractable metal fraction is carried out as a supplemental approach to check the plant availability of heavy metals (Bragato et al. 1998; Fang and Wong 1999; Fuentes et al. 2006); and (3) leachable heavy metals which are the mobile forms of the metals associated with the movement of organic and inorganic analytes present in the liquid, solid and multi-phasic forms of the wastes. The leachable heavy metals are determined through toxicity characteristic leaching procedure (TCLP) tests. If the TCLP extract of the test indicates that the heavy metals are present at concentrations above their regulatory levels after accounting for dilution from other fractions of the extract, then the waste is hazardous (US Environmental Protection Agency 1992).

Much work has been carried out on the bioavailability of heavy metals during composting of water hyacinth and phumdi biomass using different techniques (Singh and Kalamdhad 2012, a, b; Singh et al. 2014). However, no such study has been reported during composting of *S. natans* weed. Therefore, the aim of the study was to determine the variations of total heavy metals and minerals and their bioavailability during the agitated pile composting of *S. natans* amended with cattle manure and rice husk. The study also investigated the leachability of the heavy metals during the composting process.

Materials and methods

Feedstock materials

Salvinia natans was collected from Loktak lake near Thanga village, Bishnupur district, Manipur, India by the local boatmen and brought to the composting shed of Manipur Pollution Control Board, Imphal, Manipur, India. Cattle manure and rice husks were obtained from Central Agriculture University, Imphal, India. The *S. natans* was prepared for composting through cutting/shredding (maximum size restricted to 10 mm to provide better aeration and moisture control) and uniform premixing with cattle manure and rice husk. The initial characteristics of the composting materials and the proportion of the materials of the five trials are shown in Table 1.

Agitated pile composting

Composting materials were formed into trapezoidal piles (length 240 cm, base width 40 cm, top width 10 cm and height not less than 30 cm) with length to base width (L/W) ratio 6 (Singh and Kalamdhad 2012, 2014c). The five different agitated piles containing 150 kg of waste mixture were composted for 30 days and sampling was carried out

Table 1 Initial characteristic of the composting materials

Parameters	*Salvinia natans*	Cattle manure	Rice husk
Trial 1 (kg)	75	60	15
Trail 2 (kg)	90	45	15
Trial 3 (kg)	115	20	15
Trial 4 (kg)	120	15	15
Trial 5 (kg)	150	0	0
Moisture content (%)	90.5 ± 1.2	86.4 ± 0.9	9.6 ± 0.5
pH	5.6 ± 0.01	6.8 ± 0.05	6.14 ± 0.05
Electrical conductivity (EC) (dS/m)	3.5 ± 0.03	3.7 ± 0.03	1.9 ± 0.02
Volatile solids (VS) (%)	74.8 ± 0.25	72.1 ± 0.22	79.3 ± 0.23
Na (mg/kg dry matter)	6995 ± 35.5	2400 ± 5.5	1925 ± 12.3
K (mg/kg dry matter)	18,920 ± 95.0	987.5 ± 22.5	8250 ± 10.4
Ca (mg/kg dry matter)	9337 ± 90.0	8193 ± 37.2	3885 ± 12.4
Mg (mg/kg dry matter)	7277 ± 55.0	4957 ± 24.5	999 ± 7.8
Zn (mg/kg dry matter)	210 ± 0.50	182.4 ± 1.95	100 ± 2.34
Cu (mg/kg dry matter)	53.5 ± 0.50	57.8 ± 0.8	20 ± 0.45
Mn(mg/kg dry matter)	999 ± 7.0	1128 ± 1.5	550 ± 3.5
Fe (mg/kg dry matter)	6754 ± 32.5	1861 ± 2.8	3350 ± 11.4
Ni (mg/kg dry matter)	265 ± 1.75	266 ± 2.6	140 ± 2.1
Cd (mg/kg dry matter)	53.8 ± 2.2	49.9 ± 1.07	67 ± 0.20
Pb (mg/kg dry matter)	756 ± 0.75	766 ± 0.5	60 ± 0.15
Cr (mg/kg dry matter)	143.2 ± 1.22	124.4 ± 0.5	10 ± 0.08

Mean ± SD ($n = 3$)

after manual turning on 0, 3, 6, 9, 12, 15, 18, 21, 24, 27 and 30th days. Grab samples were collected from five different locations mainly from the mid and end portions of the piles to make up 1 kg and thoroughly mixed to form a homogenous sample. Collected samples were immediately air dried at 105 °C in oven, ground to pass through 0.2 mm sieve and stored for analysis of the physico-chemical parameters.

Physico-chemical analysis

A digital thermometer was used for measuring temperature during the composting process. Moisture content (MC) was determined from weight loss of the wet compost sample (105 °C for 24 h) using the gravimetric method (BIS 1982). The pH and electrical conductivity (EC) of the compost samples were measured using a pH and a conductivity meter (BIS 1982) on the filtrate (Whatman filter paper no. 42) of 10 g of the sample shaken with 100 mL distilled water in a horizontal shaker for 2 h and kept for ½ h for settling down of the solids. Volatile solids (VS) were determined by the ignition method (550 °C for 2 h in muffle furnace) (BIS 1982). The Flame Photometer (Systronic 128) was used for analysis of sodium (Na), potassium (K) and calcium (Ca) concentrations, whereas atomic absorption spectrometer (Varian Spectra 55B) was used for analysis of magnesium (Mg), zinc (Zn), copper (Cu), manganese (Mn), iron (Fe), nickel (Ni), lead (Pb), cadmium (Cd) and chromium (Cr) concentrations after digestion of 0.2 g dry sample with 10 mL of 5 H_2SO_4 and 1 $HClO_4$ mixture in block digestion system (Pelican Equipments Chennai-India) for 2 h at 300 °C (Singh and Kalamdhad 2013a, b).

Water-soluble nutrients and heavy metals were determined after extraction of 2.5 g sample with 50 mL of distilled water at room temperature for 2 h in a shaker at 100 rpm (Singh and Kalamdhad 2013b). Diethylene triamine penta-acetic acid (DTPA) extraction of the metals was carried out by mechanically shaking 4 g ground sample (screened through 0.22 mm sieve) with 40 mL of 0.005 M DTPA, 0.01 M $CaCl_2$ and 0.1 M (triethanolamine) buffered to pH 7.3 at 100 rpm (Guan et al. 2011). Toxicity characteristic leaching procedure (TCLP) test was performed according to USEPA Method 1311 (US Environmental Protection Agency 1992). Accordingly, 5 g compost sample (size less than 9.5 mm) with 100 mL of acetic acid at pH 4.93 ± 0.05 (pH adjusted with 1 N NaOH) was taken in 125 mL reagent bottle and kept at room temperature for 18 h in a shaker at 30 ± 2 rpm. The suspensions were centrifuged for 5 min at 10,000 rpm, filtered through Whatman filter paper no. 42 and the filtrate was stored in a plastic reagent bottle at 4 °C for analysis of selected heavy metals. All the results reported are the means of three replicates. Repeated measures were treated with analysis of variance (ANOVA) using SPSS software to establish statistical significance.

Results and discussion

Physico-chemical analysis

Figure 1 shows the temperature profile of the five piles during the composting. The temperatures of the piles increased rapidly within 2 days of the composting process. The highest temperature (52.2 °C) was measured on 4th day in trial 2 pile whereas the maximum temperature of trial 1 pile was 49.8 °C on the 6th day. Piles of trials 3 and 4 showed maximum temperatures of 43.2 and 42.3 °C on the 8th day of the composting, respectively. The pile of control trial 5 showed maximum temperature of only 32.2 °C on the 5th day of the process indicating that that pile composting of *S. natans* without addition of cattle manure and rice husk was not feasible. The temperature increased in all the piles due to the release of heat caused by microbial catabolism.

The variation in pH, MC, EC and VSs is shown in Fig. 2. The pH of all the trials increased significantly (ANOVA, $P < 0.05$) during the composting process from 6.2 to 7.5, 6.1 to 7.4, 6.2 to 7.4, 6.2 to 7.1 and 5.6 to 6.8 in trials 1, 2, 3, 4 and 5, respectively (Fig. 2a). The increase in pH can be attributed to ammonification or mineralization of organic nitrogen (Wong et al. 2001) and release of basic salts or degradation (decarboxylation) of organic acids (Cayuela et al. 2008) due to the microbial activity. The pH of trial 4, having highest proportion of cattle manure and least void spaces, increased only after 6 days due to the comparatively higher absorption of CO_2 and formation of organic acids preventing pH rise. In case of control trial 5 without cattle manure and saw dust, the initial pH of *S. natans* was below 6. The pH increased rapidly due to availability of sufficient oxygen from regular turning and the low temperature profile of the pile (below 40 °C) (Smars et al. 2002). The observed pH values were within the optimal range required for the development of bacteria (6.0–7.5) and fungi (5.5–8.0), and the variation in pH is similar with the results of pile composting of water hyacinth (Prasad et al. 2013; Singh and Kalamdhad 2013a, b).

US Composting Council prescribes a maximum limit of 60–80 % MC for the composting materials (TMECC 2002). The composting material should have bare minimum MC for the survival of microorganisms, whereas the rate of oxygen diffusion of the pores decreases when MC is high. Moisture is liberated as water vapor during the composting process and the loss of moisture can be viewed as an index of decomposition rate (Liao et al. 1996). The order of moisture loss in the piles was: trial 2 (30.8 %) > trial 1 (24.9 %) > trial 3 (20.8 %) > trial 4 (16.8 %) > trial 5 (13.1 %) (Fig. 2b). Therefore, trial 2 had the highest decomposition rate which was in agreement with the temperature profile of the piles. The MCs in all the piles were mostly above 60 % during the active phase. ANOVA analysis shows significant variation of MC among the trials ($P < 0.05$).

Figure 2c shows that there was an initial increase in the EC of the *S. natans* composts of all trials due to the release of mineral salts such as phosphates and ammonium ions during the decomposition of organic matter (Fang and Wong 1999). The volatilization of ammonia and precipitation of mineral salts could be the reasons for the subsequent decrease of EC in trials 1, 2, 3 and 4 (Wong et al. 2001). In the final composts, the ECs measured were 3.2, 2.7, 2.8 and 3.1 dS/m in trials 1, 2, 3 and 4, respectively. The final EC (3.7 dS/m) of trial 5 was higher than the initial EC (3.4 dS/m) after 30 days. EC of the compost reflects its degree of salinity (Huang et al. 2004). A

Fig. 1 Temperature profiles of the piles during composting

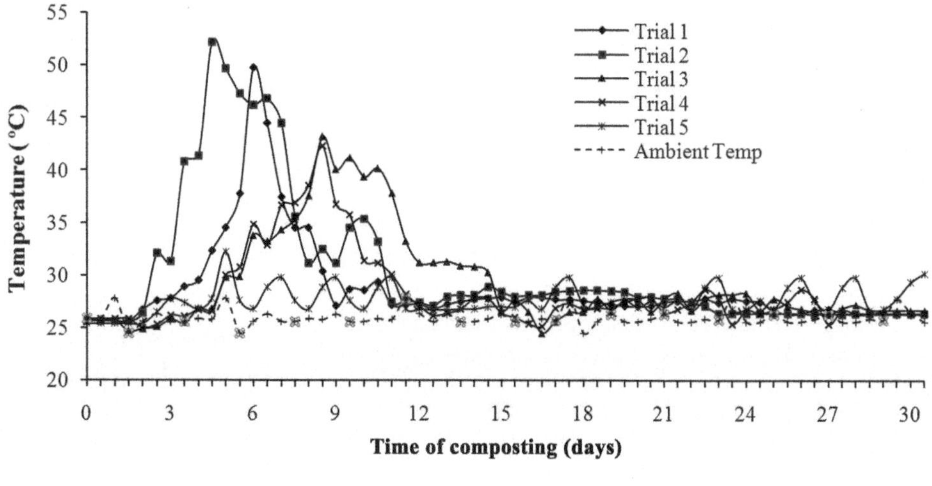

Fig. 2 Temporal variation of pH, electrical conductivity (*EC*), moisture content (*MC*) and volatile solids (*VS*)

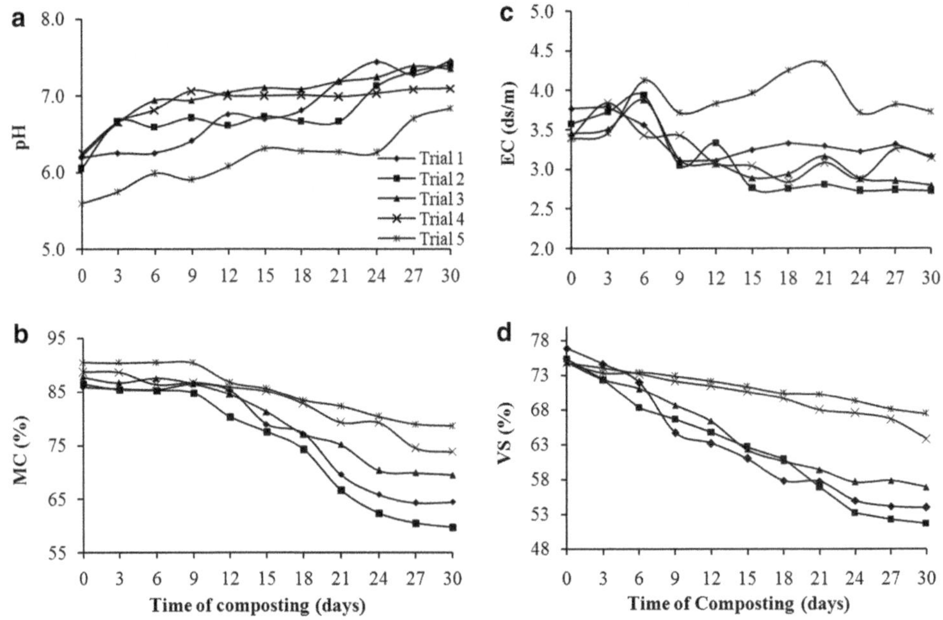

common disadvantage of compost as fertilizer is weakening of the water intake ability of plants due to osmotic effect from high salt concentrations (Koivula et al. 2004). EC about 4 dS/m or higher in composts will adversely influence plant growth, e.g., low germination rate, withering, etc. (Chen 1999; Lin 2008; Gao et al. 2010). Therefore, composts with high EC value must be mixed well with soil or other materials with low EC before application for growing crops. The variation of EC among all trials was statistically significant ($P < 0.05$).

The VS content is an indicator of organic matter present in the compost and decreases during the composting process due to degradation of the organic components by the microorganisms and loss of carbon in the form of CO_2. In the present study, the highest reduction of VS (31.4 %) was observed in trial 2 and the order of reduction was trial 2 (31.4 %) > trial 1 (29.7 %) > trial 3 (24.1 %) > trial 4 (14.7 %) > trial 5 (9.9 %) (Fig. 2d). The reductions in the concentration of the VS in all the five trials were significant ($P < 0.05$).

The changes in concentrations of nutrients (total and water-soluble forms) and heavy metals (total and water-soluble forms) during the composting process are shown in Tables 2 and 3, respectively. The total concentrations of nutrients (Na, K, Ca and Mg) and heavy metals (Zn, Cu, Mn, Fe, Ni, Pb, Cd and Cr) increased due to weight loss of the dry matter from the decomposition of the organic matter resulting in release of CO_2 and the subsequent mineralization during the composting process (Singh and Kalamdhad 2013a). The order of total nutrients and heavy metal concentrations of *S. natans* compost was as follows:

Nutrients: K > Ca > Mg > Na
Heavy metals: Fe > Mn > Pb > Ni > Zn > Cr > Cu > Cd

In case of phumdi biomass composting, the order of metal concentrations of phumdi compost was Fe > Pb > Mn > Zn > Ni > Cr > Cd > Cu (Singh et al. 2014). Increase of concentration of trace elements in soils due to prolonged application of composts may cause toxicity to plants, animals and humans. Therefore, bioavailability studies of the heavy metal contents of the *S. natans* composts are extremely important before applying the compost to the agriculture field.

Bioavailability of nutrients and heavy metals

Water-soluble nutrients and heavy metals

The water-soluble nutrients of the final composts of all trials increased in the range of about 10.9–26.4 % for Na, 14.8–55.1 % for K, 14.2–40.1 % for Ca and 17.4–66.2 % for Mg (Table 2). Lowest enhancement in water solubility of K, Ca and Mg was in trial 5 which contained only *S. natans* biomass. Similar results were also reported during the agitated pile composting of phumdi biomass (Singh et al. 2014). The increase in concentration of water-soluble nutrients was also due to weight loss of the dry matter during the decomposition and mineralization of the organic matter (Singh and Kalamdhad 2012). The variation in concentration of water-soluble nutrients among all trials was significant ($P < 0.05$).

Water-soluble fraction of heavy metals are the most easily bioavailable fraction and belongs to the most toxic constituents of composts (Singh and Kalamdhad 2013b, c, 2014a). The concentration of the water-soluble forms of heavy metals (Zn, Cu, Mn, Fe Pb and Cr) reduced significantly ($P < 0.05$) in all trials during the composting of

Table 2 Variation of total and water-soluble nutrients concentration during the composting process

Days	Nutrients concentration (mg/kg dry matter)									
	Trial 1	Trial 2	Trial 3	Trial 4	Trial 5	Trial 1	Trial 2	Trial 3	Trial 4	Trial 5
	Total Na					*Total K*				
0	4825 ± 65	4820 ± 95	4990 ± 90	5817 ± 27	6175 ± 35	15,780 ± 45	18,135 ± 90	17,190 ± 85	18,601 ± 15	18,825 ± 95
6	5525 ± 85	5165 ± 55	5785 ± 55	6160 ± 20	6375 ± 35	16,333 ± 107	19,340 ± 18	18,125 ± 15	18,935 ± 20	18,606 ± 38
12	7175 ± 35	6365 ± 55	6580 ± 20	6685 ± 75	6570 ± 40	17,523 ± 18	19,344 ± 112	19,713 ± 152	19,405 ± 10	18,900 ± 40
18	7850 ± 30	7180 ± 60	6755 ± 5	6675 ± 205	6965 ± 155	18,415 ± 30	21,168 ± 63	19,148 ± 77	19,673 ± 38	19,255 ± 25
24	7985 ± 95	8210 ± 85	7240 ± 70	7115 ± 15	7280 ± 70	19,503 ± 28	22,430 ± 0	19,943 ± 3	19,930 ± 28	19,745 ± 10
30	8350 ± 70	8440 ± 70	7460 ± 50	7505 ± 25	7510 ± 5	20,108 ± 48	23,548 ± 18	21,085 ± 20	20,725 ± 56	20,150 ± 14
	Water-soluble Na					*Water-soluble K*				
0	1632 ± 14	1739 ± 7	1725 ± 3	1535 ± 3	1545 ± 21	2860 ± 10	3044 ± 22	3336 ± 38	3392 ± 16	3536 ± 6
6	1800 ± 9	1747 ± 3	1816 ± 3	1515 ± 2	1528 ± 4	3144 ± 6	3334 ± 38	3572 ± 6	3421 ± 3	3615 ± 7
12	1814 ± 3	1809 ± 3	1852 ± 5	1607 ± 2	1634 ± 11	3154 ± 22	3572 ± 6	3580 ± 4	3697 ± 9	3508 ± 64
18	1906 ± 3	1987 ± 2	1893 ± 6	1748 ± 3	1652 ± 3	3471 ± 41	3722 ± 12	3852 ± 10	3920 ± 48	3751 ± 13
24	1948 ± 3	2160 ± 4	1947 ± 1	1753 ± 2	1704 ± 3	4364 ± 12	3937 ± 39	4370 ± 6	3903 ± 51	3900 ± 6
30	1953 ± 2	2198 ± 9	2087 ± 1	1792 ± 5	1718 ± 3	4437 ± 11	4381 ± 9	4542 ± 8	4036 ± 12	4133 ± 23
	Total Ca					*Total Mg*				
0	7385 ± 50	7718 ± 53	8130 ± 25	8400 ± 25	8920 ± 103	6445 ± 05	6848 ± 128	6520 ± 90	6958 ± 57	7215 ± 55
6	8204 ± 55	7903 ± 08	8670 ± 50	8833 ± 13	9668 ± 58	6682 ± 99	7450 ± 140	7046 ± 90	7530 ± 115	7361 ± 35
12	8340 ± 80	8638 ± 18	9848 ± 28	9063 ± 18	9693 ± 13	6983 ± 98	7720 ± 110	7250 ± 92	7631 ± 170	7459 ± 15
18	8635 ± 05	9070 ± 15	9861 ± 25	9176 ± 05	9863 ± 23	8403 ± 72	7884 ± 54	7661 ± 23	7796 ± 20	7580 ± 25
24	9763 ± 43	10,103 ± 68	9778 ± 43	9460 ± 50	10,283 ± 48	8530 ± 70	8208 ± 238	7802 ± 139	8026 ± 34	7795 ± 25
30	10,895 ± 75	11,550 ± 03	10,533 ± 88	10,710 ± 95	10,293 ± 13	8502 ± 64	9214 ± 167	7851 ± 184	8064 ± 41	7888 ± 48
	Water-soluble Ca					*Water-soluble Mg*				
0	2874 ± 08	2666 ± 24	3097 ± 05	3179 ± 05	3262 ± 22	1878 ± 36	1770 ± 20	1914 ± 60	1659 ± 29	2063 ± 19
6	3115 ± 09	2958 ± 04	2875 ± 03	3237 ± 13	3301 ± 07	1832 ± 10	1853 ± 09	2070 ± 03	17,780 ± 03	2069 ± 05
12	3255 ± 07	3080 ± 02	3362 ± 12	3448 ± 04	3277 ± 13	2127 ± 15	1827 ± 02	2433 ± 30	1851 ± 03	2249 ± 24
18	3252 ± 10	3339 ± 07	3755 ± 21	3422 ± 16	3568 ± 14	2433 ± 05	2361 ± 08	2414 ± 15	1874 ± 08	2353 ± 00
24	3529 ± 41	3632 ± 10	3943 ± 07	3676 ± 32	3617 ± 05	2550 ± 18	2751 ± 22	2553 ± 20	2102 ± 35	2226 ± 23
30	3679 ± 29	3736 ± 14	3968 ± 00	3677 ± 07	3635 ± 07	2713 ± 25	2942 ± 28	2741 ± 11	2248 ± 00	2476 ± 32

Mean ± SD (n = 3) (ANOVA: $P < 0.05$)

Table 3 Variation of total and water-soluble heavy metal concentrations during the composting process

Days	Heavy metal concentrations (mg/kg dry matter)									
	Trial 1	Trial 2	Trial 3	Trial 4	Trial 5	Trial 1	Trial 2	Trial 3	Trial 4	Trial 5
	Total Zn					*Total Cu*				
0	192.0 ± 1.0	159.5 ± 1.0	196.8 ± 2.3	203.0 ± 1.5	210.0 ± 0.5	51.0 ± 1.0	63.5 ± 7.0	52.8 ± 1.3	47.3 ± 1.8	53.0 ± 0.5
6	195.5 ± 3.0	167.3 ± 1.8	194.0 ± 1.5	206.0 ± 2.0	211.8 ± 5.8	58.0 ± 1.0	76.5 ± 0.5	54.3 ± 1.3	55.5 ± 1.0	54.3 ± 1.3
12	208.5 ± 2.0	190.3 ± 2.8	206.0 ± 1.5	211.8 ± 2.3	232.5 ± 1.5	62.0 ± 0.5	79.5 ± 1.0	70.5 ± 1.5	63.3 ± 0.8	63.3 ± 0.8
18	209.8 ± 1.3	195.1 ± 1.9	218.3 ± 2.3	216.8 ± 2.3	247.5 ± 1.0	67.8 ± 1.3	87.8 ± 1.8	72.5 ± 1.5	77.5 ± 1.5	77.3 ± 1.3
24	218.5 ± 2.5	207.5 ± 2.0	234.8 ± 0.8	219.8 ± 0.8	243.3 ± 2.3	85.0 ± 1.0	87.3 ± 3.8	83.0 ± 6.0	76.5 ± 1.0	79.5 ± 1.0
30	218.3 ± 1.3	211.3 ± 0.8	234.3 ± 2.3	225.5 ± 1.3	254.3 ± 1.3	88.5 ± 1.0	92.3 ± 1.8	86.8 ± 0.8	85.5 ± 1.5	82.8 ± 0.8
	Water-soluble Zn					*Water-soluble Cu*				
0	10.5 ± 0.08	10.3 ± 0.11	13.5 ± 0.05	13.5 ± 0.1	14.4 ± 0.07	7.6 ± 0.1	6.4 ± 0.0	7.6 ± 0.1	7.7 ± 0.1	7.9 ± 0.1
6	10.3 ± 0.08	9.9 ± 0.01	12.3 ± 0.08	13.1 ± 0.1	13.7 ± 0.02	6.3 ± 0.1	6.3 ± 0.1	7.0 ± 0.0	7.8 ± 0.1	7.4 ± 0.1
12	10.4 ± 0.06	9.2 ± 0.14	12.4 ± 0.04	12.2 ± 0.0	13.3 ± 0.09	5.6 ± 0.1	5.2 ± 0.2	6.4 ± 0.1	7.3 ± 0.1	7.0 ± 0.3
18	9.6 ± 0.05	9.2 ± 0.14	11.6 ± 0.07	11.8 ± 0.1	12.1 ± 0.08	5.5 ± 0.1	4.4 ± 0.0	5.7 ± 0.1	6.8 ± 0.1	6.5 ± 0.0
24	8.7 ± 0.11	8.3 ± 0.08	11.1 ± 0.08	11.1 ± 0.1	11.7 ± 0.06	4.8 ± 0.0	4.4 ± 0.0	5.1 ± 0.0	5.7 ± 0.1	6.3 ± 0.0
30	7.8 ± 0.06	7.8 ± 0.04	11.0 ± 0.05	10.3 ± 0.1	11.4 ± 0.05	4.8 ± 0.1	4.3 ± 0.0	4.7 ± 0.1	5.6 ± 0.1	6.3 ± 0.0
	Total Mn					*Total Fe*				
0	804 ± 2.0	1076 ± 20.8	1167 ± 11.5	1165 ± 7.8	1000 ± 7.0	6786 ± 25	5736 ± 0.3	6652 ± 26.8	7559 ± 51.5	6755 ± 32.3
6	839 ± 6.3	1322 ± 12.3	1319 ± 1.8	1308 ± 7.3	1038 ± 5.5	6679 ± 13.3	5929 ± 18.5	6894 ± 12.3	7236 ± 29.8	7110 ± 2.8
12	932 ± 4.3	1433 ± 5.0	1318 ± 12.3	1234 ± 10.5	1112 ± 8.8	7410 ± 33	6078 ± 1.3	7335 ± 09.0	7962 ± 15.0	7455 ± 23.3
18	868 ± 10.0	1498 ± 4.5	1383 ± 2.8	1425 ± 15.5	1191 ± 3.0	7745 ± 5.8	6635 ± 7.5	7441 ± 00.3	8449 ± 06.8	7545 ± 27.8
24	959 ± 1.3	1553 ± 7.8	1375 ± 16.8	1465 ± 8.5	1255 ± 0.3	8416 ± 9.3	7280 ± 17.5	7712 ± 01.5	8532 ± 19.0	7625 ± 12.5
30	1028 ± 12	1609 ± 1.8	1584 ± 9.8	1466 ± 15.5	1222 ± 4.3	8608 ± 2.8	7905 ± 3.0	8120 ± 09.3	9294 ± 32.0	7678 ± 4.3
	Water-soluble Mn					*Water-soluble Fe*				
0	51.9 ± 1.5	56.4 ± 0.1	56.1 ± 0.3	61.0 ± 0.8	55.4 ± 1.7	249.6 ± 0.10	257.9 ± 0.4	266.9 ± 2.1	231.1 ± 6.2	267.4 ± 2.9
6	51.1 ± 0.1	44.2 ± 1.1	55.6 ± 0.3	58.9 ± 0.8	54.8 ± 0.4	243.8 ± 18.7	229.1 ± 4.0	249.8 ± 3.0	207.1 ± 0.2	248.0 ± 1.1
12	48.4 ± 0.2	40.3 ± 0.1	53.0 ± 0.1	58.1 ± 0.1	49.1 ± 0.2	176.3 ± 0.95	189.0 ± 3.9	260.6 ± 0.3	200.7 ± 5.9	249.3 ± 2.7
18	43.2 ± 0.5	36.7 ± 0.1	51.5 ± 0.3	53.0 ± 0.6	53.5 ± 0.2	183.8 ± 1.30	146.1 ± 1.0	186.6 ± 0.3	185.3 ± 1.0	235.3 ± 1.8
24	43.7 ± 0.1	36.2 ± 0.2	51.2 ± 0.2	51.0 ± 0.7	45.1 ± 3.1	152.3 ± 0.90	142.2 ± 0.1	182.5 ± 0.2	165.5 ± 2.7	222.3 ± 0.5
30	37.4 ± 0.5	34.2 ± 0.3	47.0 ± 1.1	52.3 ± 0.7	44.8 ± 0.1	137.5 ± 0.20	126.1 ± 1.0	172.0 ± 1.1	143.9 ± 1.0	205.4 ± 0.9
	Total Ni					*Total Pb*				
0	265.5 ± 9.5	261.3 ± 0.3	280.8 ± 9.8	289.3 ± 1.3	265.8 ± 1.8	766.5 ± 2.0	746.0 ± 2.5	717.3 ± 2.3	739.0 ± 2.0	756.3 ± 0.8
6	268.3 ± 1.3	265.8 ± 1.3	275.0 ± 3.0	303.5 ± 1.5	274.3 ± 13.3	765.3 ± 3.3	763.0 ± 2.5	767.5 ± 4.5	730.5 ± 3.0	754.3 ± 1.3
12	314.1 ± 8.4	288.8 ± 1.8	275.5 ± 7.5	312.3 ± 1.8	289.3 ± 1.3	792.8 ± 4.3	707.8 ± 9.8	779.5 ± 4.0	757.0 ± 1.5	801.3 ± 2.8
18	321.5 ± 4.5	308.3 ± 1.3	307.3 ± 1.3	336.3 ± 7.3	375.8 ± 7.3	789.3 ± 4.8	799.5 ± 1.0	791.8 ± 2.8	776.0 ± 2.0	807.3 ± 1.8
24	344.0 ± 1.5	326.8 ± 1.3	302.3 ± 1.8	346.3 ± 2.8	302.5 ± 6.0	837.0 ± 3.5	839.0 ± 2.0	802.8 ± 1.8	804.0 ± 3.5	834.0 ± 2.0

Table 3 continued

Days	Heavy metal concentrations (mg/kg dry matter)									
	Trial 1	Trial 2	Trial 3	Trial 4	Trial 5	Trial 1	Trial 2	Trial 3	Trial 4	Trial 5
30	344.5 ± 4.0	333.0 ± 2.5	316.8 ± 0.3	345.3 ± 0.8	309.8 ± 3.8	890.5 ± 3.5	844.0 ± 3.5	805.3 ± 1.3	805.0 ± 0.5	848.3 ± 3.8
	Water-soluble Ni					*Water-soluble Pb*				
0	–	–	–	–	–	8.9 ± 0.36	10.8 ± 0.40	8.6 ± 0.3	10.5 ± 0.1	9.7 ± 0.1
6	–	–	–	–	–	8.5 ± 0.09	10.3 ± 0.10	8.5 ± 0.0	9.9 ± 0.2	8.8 ± 0.1
12	–	–	–	–	–	7.7 ± 0.05	8.6 ± 0.10	7.8 ± 0.0	9.1 ± 0.6	9.2 ± 0.0
18	–	–	–	–	–	7.5 ± 0.16	8.9 ± 0.00	7.7 ± 0.0	9.3 ± 0.2	8.6 ± 0.1
24	–	–	–	–	–	7.7 ± 0.04	7.4 ± 0.20	6.0 ± 0.0	8.3 ± 0.1	9.0 ± 0.1
30	–	–	–	–	–	6.4 ± 0.10	6.5 ± 0.30	5.5 ± 0.2	8.3 ± 0.1	8.3 ± 0.0
	Total Cd					*Total Cr*				
0	47.8 ± 1.3	47.5 ± 1.8	42.8 ± 1.8	49.5 ± 1.0	53.8 ± 2.3	96.9 ± 0.9	89 ± 1.5	82.5 ± 1.0	109.3 ± 2.2	143.3 ± 1.2
6	52.0 ± 0.5	43.5 ± 1.5	47.5 ± 1.5	52.0 ± 0.5	52.3 ± 1.8	109.0 ± 2.0	96.8 ± 0.7	89.7 ± 1.5	122.3 ± 1.2	104.5 ± 2.5
12	56.0 ± 1.0	55.8 ± 0.2	51.0 ± 0.5	58.5 ± 0.5	58.5 ± 1.0	131.9 ± 1.2	108.8 ± 1.8	94.5 ± 2.0	137.8 ± 0.3	164.4 ± 2.4
18	57.8 ± 1.3	57.0 ± 0.5	54.0 ± 2.8	59.0 ± 0.5	59.5 ± 1.0	126.6 ± 4.5	122.5 ± 0.5	124.3 ± 3.2	164.5 ± 7.5	132.3 ± 4.8
24	61.5 ± 0.5	59.0 ± 0.5	58.5 ± 1.0	62.3 ± 1.3	58.8 ± 0.3	159.5 ± 2.5	132.8 ± 4.3	133.6 ± 0.6	185.8 ± 1.8	194.0 ± 1.5
30	63.8 ± 0.3	61.5 ± 0.5	58.0 ± 1.0	62.0 ± 1.5	63.0 ± 1.0	167.0 ± 8.9	142.0 ± 1.5	139.8 ± 2.3	189.5 ± 2.5	206.8 ± 0.8
	Water-soluble Cd					*Water-soluble Cr*				
0	–	–	–	–	–	29.4 ± 0.40	34.5 ± 0.17	34.4 ± 0.07	32.4 ± 1.04	33.7 ± 0.04
6	–	–	–	–	–	28.0 ± 0.12	30.3 ± 0.16	33.0 ± 0.59	30.7 ± 0.11	31.6 ± 0.33
12	–	–	–	–	–	25.7 ± 0.69	28.4 ± 0.06	28.9 ± 0.31	30.2 ± 0.05	31.9 ± 1.00
18	–	–	–	–	–	24.5 ± 0.17	26.8 ± 0.33	30.4 ± 0.25	28.8 ± 0.40	30.5 ± 0.17
24	–	–	–	–	–	25.6 ± 0.13	25.4 ± 0.11	27.8 ± 0.06	27.3 ± 0.10	29.5 ± 0.14
30	–	–	–	–	–	22.2 ± 0.07	24.7 ± 0.10	27.1 ± 0.25	27.2 ± 0.19	28.9 ± 0.30

–Not detected (mean ± SD, $n = 3$, $P < 0.05$)

Table 4 Variation of DTPA-extractable metals during the composting

Days	DTPA-extractable metal concentrations (mg/kg dry matter)				
	Trial 1	Trial 2	Trial 3	Trial 4	Trial 5
	Zn				
0	83.4 ± 0.3	79.4 ± 0.3	88.8 ± 0.1	88.8 ± 0.1	92.5 ± 0.4
6	75.6 ± 0.3	80.5 ± 0.4	86.9 ± 0.3	88.3 ± 0.1	91.5 ± 0.4
12	74.7 ± 0.1	72.6 ± 0.3	81.6 ± 0.2	83.9 ± 1.5	88.8 ± 0.2
18	68.7 ± 0.1	68.9 ± 0.1	79.6 ± 0.3	81.1 ± 1.6	84.6 ± 0.1
24	64.6 ± 0.3	64.3 ± 0.4	78.2 ± 0.5	77.8 ± 0.1	82.2 ± 0.5
30	63.4 ± 0.1	59.4 ± 0.5	72.4 ± 0.2	72.4 ± 0.2	82.0 ± 0.2
	Mn				
0	181.7 ± 0.4	179.7 ± 0.3	185.6 ± 1.1	191.0 ± 0.2	192.4 ± 0.1
6	171.9 ± 0.5	174.4 ± 1.2	171.3 ± 0.9	177.1 ± 1.3	187.6 ± 2.1
12	164.4 ± 1.1	166.9 ± 0.5	172.6 ± 0.5	175.3 ± 0.9	178.7 ± 0.1
18	151.8 ± 0.5	157.9 ± 0.4	169.5 ± 0.4	171.9 ± 0.8	175.3 ± 1.0
24	144.6 ± 4.2	148.5 ± 0.5	163.8 ± 0.4	170.8 ± 0.3	175.3 ± 0.9
30	144.1 ± 0.3	139.3 ± 0.2	161.8 ± 0.4	170.4 ± 0.6	171.9 ± 0.8
	Ni				
0	18.7 ± 0.1	17.9 ± 0.0	17.6 ± 0.2	17.1 ± 0.0	17.4 ± 0.1
6	17.7 ± 0.2	17.6 ± 0.2	17.3 ± 0.1	16.9 ± 0.0	17.1 ± 0.0
12	16.9 ± 0.1	15.7 ± 0.0	16.8 ± 0.1	16.3 ± 0.0	16.6 ± 0.0
18	16.8 ± 0.0	14.8 ± 0.0	16.3 ± 0.0	15.8 ± 0.0	16.5 ± 0.1
24	15.9 ± 0.0	14.7 ± 0.1	15.8 ± 0.0	15.2 ± 0.1	16.5 ± 0.0
30	15.4 ± 0.2	13.8 ± 0.1	15.1 ± 0.2	14.5 ± 0.0	16.3 ± 0.1
	Cd				
0	–	–	–	–	–
6	–	–	–	–	–
12	–	–	–	–	–
18	–	–	–	–	–
24	–	–	–	–	–
30	–	–	–	–	–
Days	DTPA-extractable metal concentrations (mg/kg dry matter)				
	Trial 1	Trial 2	Trial 3	Trial 4	Trial 5
	Cu				
0	7.8 ± 0.0	8.8 ± 0.1	9.6 ± 0.1	9.5 ± 0.1	9.7 ± 0.1
6	7.5 ± 0.1	8.5 ± 0.0	9.4 ± 0.1	9.6 ± 0.0	9.0 ± 0.0
12	7.2 ± 0.1	7.7 ± 0.0	8.8 ± 0.0	9.3 ± 0.1	8.7 ± 0.0
18	6.2 ± 0.1	6.2 ± 0.0	8.6 ± 0.3	8.7 ± 0.1	8.5 ± 0.0
24	6.1 ± 0.1	6.2 ± 0.0	7.7 ± 0.1	8.4 ± 0.0	8.4 ± 0.0
30	5.6 ± 0.2	5.9 ± 0.0	7.7 ± 0.2	8.0 ± 0.0	8.4 ± 0.0
	Fe				
0	281.2 ± 2.3	290.9 ± 0.8	293.3 ± 1.0	300.8 ± 0.4	314.1 ± 2.6
6	263.2 ± 6.9	289.7 ± 0.3	263.6 ± 1.5	294.4 ± 1.1	301.8 ± 0.7
12	223.0 ± 0.5	264.5 ± 1.0	253.2 ± 0.1	262.8 ± 0.5	286.0 ± 2.6
18	236.0 ± 4.5	212.9 ± 0.6	231.3 ± 0.2	223.3 ± 0.2	262.9 ± 0.4
24	200.7 ± 0.5	185.0 ± 1.5	210.6 ± 0.3	211.6 ± 0.9	251.8 ± 0.6
30	188.8 ± 4.3	175.9 ± 1.5	202.1 ± 0.3	205.6 ± 0.3	241.0 ± 2.3
	Pb				
0	29.5 ± 1.4	27.2 ± 0.0	28.9 ± 0.2	28.7 ± 0.2	31.4 ± 0.1

Table 4 continued

Days	DTPA-extractable metal concentrations (mg/kg dry matter)				
	Trial 1	Trial 2	Trial 3	Trial 4	Trial 5
6	26.7 ± 0.4	25.2 ± 0.4	27.3 ± 0.4	28.1 ± 0.1	31.2 ± 2.0
12	25.4 ± 0.5	24.7 ± 0.1	25.4 ± 1.2	24.3 ± 0.2	25.6 ± 1.7
18	24.1 ± 0.5	20.7 ± 0.5	22.8 ± 0.3	21.8 ± 0.6	25.6 ± 2.4
24	22.4 ± 0.8	18.4 ± 0.2	21.8 ± 0.3	22.5 ± 0.2	26.7 ± 0.5
30	22.0 ± 0.2	17.4 ± 0.2	21.8 ± 0.2	21.9 ± 0.7	25.9 ± 0.1
	Cr				
0	45.9 ± 0.7	52.6 ± 0.5	56.0 ± 0.5	47.8 ± 1.4	50.4 ± 0.2
6	42.3 ± 0.4	47.4 ± 0.7	49.4 ± 0.8	44.5 ± 0.8	50.5 ± 0.5
12	38.8 ± 0.1	37.6 ± 1.1	36.3 ± 0.1	46.1 ± 0.7	46.7 ± 0.0
18	26.5 ± 0.1	32.3 ± 0.8	31.9 ± 0.3	38.9 ± 0.0	49.0 ± 0.2
24	24.4 ± 0.0	23.4 ± 0.3	31.9 ± 0.2	36.6 ± 0.1	46.8 ± 0.1
30	22.1 ± 0.5	23.2 ± 0.0	29.4 ± 0.5	32.73 ± 0.58	41.7 ± 0.1

–Not detected (mean ± SD, $n = 3$, $P < 0.05$)

S. natans (Table 3). The order of reduction of water-soluble heavy metals in the trials was as below:

Zn: T1 (25.4 %) > T2 (24.7 %) > T4 (23.7 %) > T5 (20.8 %) > T3 (18.4 %)
Cu: T3 (38.4 %) > T1 (36.4 %) > T2 (32.6 %) > T4 (27.7 %) > T5 (19.8 %)
Mn: T2 (39.5 %) > T1 (27.9 %) > T5 (19.1 %) > T3 (16.0 %) > T4 (14.3 %)
Fe: T2 (51.1 %) > T1 (44.9 %) > T4 (37.7 %) > T3 (35.6 %) > T5 (23.3 %)
Pb: T2 (39.7 %) > T3 (35.7 %) > T1 (28.0 %) > T4 (20.8 %) > T5 (14.1 %)
Cr: T2 (28.3 %) > T1 (24.4 %) > T3 (21.3 %) > T4 (16.0 %) > T5 (14.2 %)

Water-soluble Ni and Cd were not detected in the composts similar to the results of phumdi biomass compost (Singh et al. 2014). Reduction of water-soluble forms of Zn, Cu, Mn, Fe, Pb and Cr during the process may be attributed to the binding of the metals with the –OH and –COOH groups enriched by cattle manure. These groups and the newly formed humus increased the binding sites and combined with metals (released during mineralization of organic biomass) to form insoluble and immobile complexes (Guan et al. 2011; Singh and Kalamdhad 2013a, 2014a). The increase in concentration of water-soluble Fe was reported during phumdi biomass composting without cattle manure (Singh et al. 2014) but in the present study the concentration of water-soluble Fe was reduced in all the trials. The reduction of water-soluble fraction of heavy metals was also reported by other researchers during solid waste composting (Castaldi et al. 2005; Singh and Kalamdhad 2013b, c).

The solubility of the nutrients and heavy metals depends on the solubility equilibria for which the initial concentration of the ions, pH, complex formations, etc, are important (Larson et al. 1973; Benefield et al. 1982; Wang et al. 2009). The pH of the *S. natans* composts was in the range where the alkali or alkaline earth metals (i.e., Na^+, K^+, Ca^{2+}, Mg^{2+}, etc) were soluble. The possible reason for the decrease in water-soluble trace metals was the formation of complexes with the heavy metal ions. Composting results in humification of the organic matter with the formation of humus substances rich in humic acid-like organic carbons having increased aromatic characteristics; oxygen and nitrogen concentrations; functional groups; etc., depending on the nature and composition of the initial organic matter. The humic substances are low in fulvic acid-like organic carbon and water-extractable organic carbon (Roletto et al. 1985; Senesi 1989). Structurally, humic substances are 20–30 % aliphatic and 20–40 % aromatic. 20 % of hydrogen is bound to oxygen as carboxyl (–COOH) and phenolic hydroxyl (–OH) groups and the rest bound directly to carbon (Young 2010). The interaction of macronutrients (i.e., alkali or alkaline earth metals) and heavy metals with the humic substances is a complex phenomenon with differing views. The alkali or alkaline earth metals are held weakly by exchangeable hydrated ions, or by electrostatic forces by charged humus carboxyl groups. In case of heavy metals, a metal chelate complex is formed and two or more coordinate positions of the metal ion are occupied by donor groups of a single ligand to form an internal ring structure (Stevenson 1994; Tan 2010). The divalent and trivalent metals formed highly insoluble complexes, i.e., multidentate or multinuclear chelate sites with the functional groups of humus and are dominated by

Table 5 Variation of leachable concentrations of heavy metals during the composting process

Days	Leachable heavy metal concentrations (mg/kg dry matter)									
	Trial 1	Trial 2	Trial 3	Trial 4	Trial 5	Trial 1	Trial 2	Trial 3	Trial 4	Trial 5
	Zn					*Cu*				
0	124.6 ± 0.1	123.4 ± 0.1	141.0 ± 0.7	137.6 ± 0.1	129.1 ± 0.6	20.5 ± 0.2	24.2 ± 0.1	24.4 ± 0.2	24.7 ± 0.3	23.1 ± 0.6
6	121.0 ± 0.7	119.0 ± 0.1	136.5 ± 0.2	133.8 ± 0.1	128.0 ± 2.9	20.1 ± 0.0	22.3 ± 0.1	22.1 ± 0.1	24.2 ± 0.1	22.2 ± 0.1
12	114.6 ± 0.6	118.3 ± 0.8	133.4 ± 0.5	130.6 ± 0.3	122.8 ± 0.3	19.2 ± 0.1	19.5 ± 0.0	19.6 ± 0.1	23.3 ± 0.1	22.1 ± 0.1
18	103.5 ± 0.4	104.5 ± 0.2	134.9 ± 0.4	124.2 ± 0.1	124.2 ± 0.0	19.0 ± 0.1	18.3 ± 0.1	18.5 ± 0.1	20.2 ± 0.1	19.1 ± 0.2
24	100.5 ± 0.2	97.7 ± 0.1	132.3 ± 0.1	122.3 ± 0.2	122.0 ± 0.3	17.2 ± 0.1	17.1 ± 0.1	18.2 ± 0.2	19.2 ± 0.1	19.0 ± 0.1
30	100.5 ± 0.2	97.2 ± 0.2	126.7 ± 0.1	118.4 ± 0.0	121.3 ± 0.3	15.8 ± 0.2	15.8 ± 0.1	17.7 ± 0.1	18.6 ± 0.1	18.5 ± 0.1
	Mn					*Fe*				
0	305.1 ± 3.8	298.6 ± 5.5	289.2 ± 1.6	290.9 ± 9.0	291.0 ± 2.3	467.6 ± 1.4	494.1 ± 2.9	407.3 ± 0.2	478.9 ± 1.0	473.8 ± 9.5
6	282.7 ± 0.2	278.0 ± 1.6	286.0 ± 4.6	298.3 ± 9.7	284.4 ± 1.4	415.2 ± 0.2	464.4 ± 0.1	396.4 ± 0.6	462.6 ± 2.1	477.1 ± 2.1
12	276.8 ± 2.0	271.7 ± 0.9	271.8 ± 0.5	288.3 ± 9.3	274.6 ± 2.7	397.5 ± 0.2	407.9 ± 1.1	363.8 ± 1.3	442.7 ± 0.0	451.3 ± 0.5
18	266.1 ± 3.7	261.9 ± 0.2	261.6 ± 4.5	273.4 ± 4.7	247.0 ± 7.3	378.2 ± 0.9	353.4 ± 0.3	351.8 ± 0.5	393.2 ± 0.2	429.3 ± 0.1
24	259.0 ± 0.3	247.8 ± 3.0	247.8 ± 9.1	266.9 ± 5.7	265.4 ± 1.7	326.8 ± 0.3	347.2 ± 0.1	305.3 ± 0.2	391.7 ± 0.6	411.1 ± 2.0
30	248.4 ± 0.3	242.2 ± 4.1	235.0 ± 9.5	248.0 ± 8.5	262.6 ± 1.2	305.4 ± 1.1	337.3 ± 0.4	284.8 ± 0.3	364.0 ± 0.7	385.5 ± 1.1
	Ni					*Pb*				
0	175.0 ± 0.3	197.0 ± 0.5	209.1 ± 1.1	201.2 ± 1.0	205.6 ± 0.7	56.6 ± 0.4	63.0 ± 0.1	56.5 ± 1.0	65.2 ± 0.1	64.7 ± 0.2
6	165.0 ± 0.7	193.2 ± 0.5	193.3 ± 0.3	192.9 ± 0.5	202.3 ± 0.0	53.3 ± 0.5	59.6 ± 0.5	55.5 ± 0.2	66.2 ± 0.0	62.7 ± 0.4
12	153.0 ± 0.3	172.9 ± 0.3	183.1 ± 0.1	184.5 ± 0.3	201.7 ± 0.5	49.1 ± 0.7	54.4 ± 0.2	53.4 ± 0.1	56.9 ± 0.7	63.0 ± 0.1
18	144.4 ± 0.1	163.5 ± 0.4	163.2 ± 0.5	182.1 ± 0.0	192.7 ± 0.4	48.6 ± 2.3	51.1 ± 0.3	45.2 ± 0.3	51.4 ± 5.1	62.2 ± 0.1
24	136.9 ± 0.4	145.8 ± 1.1	155.3 ± 0.4	168.9 ± 0.1	182.1 ± 0.1	43.1 ± 0.1	48.6 ± 1.1	46.4 ± 0.1	54.5 ± 0.0	59.6 ± 0.1
30	130.3 ± 0.4	137.6 ± 0.1	152.5 ± 0.0	169.3 ± 0.4	177.5 ± 0.1	37.3 ± 0.0	41.2 ± 0.9	41.7 ± 0.1	50.5 ± 0.2	57.0 ± 0.7
	Cd					*Cr*				
0	1.51 ± 0.05	2.24 ± 0.02	2.51 ± 0.05	2.35 ± 0.03	2.40 ± 0.02	68.6 ± 4.3	77.9 ± 1.0	58.9 ± 0.4	61.2 ± 0.4	66.6 ± 0.3
6	1.25 ± 0.03	1.71 ± 0.05	2.43 ± 0.05	2.29 ± 0.03	2.42 ± 0.02	62.8 ± 1.5	54.3 ± 0.7	54.0 ± 1.1	58.1 ± 0.8	58.8 ± 1.4
12	1.11 ± 0.05	1.38 ± 0.04	2.29 ± 0.03	2.18 ± 0.04	2.33 ± 0.01	54.6 ± 0.7	37.9 ± 0.8	50.0 ± 2.8	53.3 ± 0.1	57.8 ± 1.0
18	0.89 ± 0.03	1.17 ± 0.01	2.10 ± 0.02	2.14 ± 0.04	2.34 ± 0.02	46.8 ± 0.4	42.0 ± 0.7	49.0 ± 0.2	53.2 ± 0.3	52.7 ± 1.2
24	0.75 ± 0.03	1.31 ± 0.07	1.78 ± 0.04	1.90 ± 0.04	2.16 ± 0.02	45.6 ± 0.7	31.3 ± 1.0	54.2 ± 1.3	44.4 ± 0.2	52.8 ± 0.6
30	0.77 ± 0.01	1.11 ± 0.03	1.60 ± 0.04	1.77 ± 0.01	1.92 ± 0.02	45.6 ± 0.7	31.2 ± 0.2	37.1 ± 0.6	42.6 ± 0.3	51.3 ± 1.0

(Mean ± SD, $n = 3$, $P < 0.05$)

double or triple bonds to the mix of carboxyl and phenolic hydroxyl groups (Young 2010). Cations with two positive charges such as Cu^{2+} can only be replaced by another transitional metal ion that has two positive charges. Chelation of toxic heavy metals such as mercury (Hg), lead (Pb) and cadmium (Cd) also forms organo-metal complexes which are less available for plant uptake. The stability of a metal chelate complex is determined by such factors as the number of atoms that form a bond with the metal ion, the number of rings that are formed, the nature and concentrations of the metal ions, and pH. The order of decreasing ability of metal ions to form chelating complexes with humic acids is $Fe^{3+} > Cu^{2+} > Ni^{2+} > Co^{2+} > Zn^{2+} > Fe^{2+} > Mn^{2+}$ (Stevenson 1994). Whereas, Tipping and Hurley (1992) evaluated the binding strength of the metals with humus in the increasing order $VO^{2+} > Cu^{2+} > Pb^{2+} > Zn^{2+} = Ni^{2+} > Co^{2+} > Cd^{2+} > Mn^{2+} > Ca^{2+} > Mg^{2+}$.

Diethylene triamine penta-acetic acid (DTPA) extraction of heavy metals

The variation of DTPA-extractable heavy metals (Zn, Cu, Mn, Fe, Ni, Pb, Cd and Cr) during the composting process is shown in Table 4. It has been considered that DTPA-extractable metals are potentially bioavailable for plant uptake (Chiang et al. 2007; Singh and Kalamdhad 2013b, 2014b). DTPA-extractable heavy metals were reduced significantly ($P < 0.05$) in all trials during the composting process. The order of DTPA reduction was as below:

Zn: T2 (25.3 %) > T1 (24.1 %) > T4 (18.5 %) > T3 (18.4 %) > T5 (11.3 %)

Cu: T2 (32.8 %) > T1 (28.8 %) > T3 (20.8 %) > T4 (17.2 %) > T5 (13.7 %)

Mn: T2 (22.5 %) > T1 (20.7 %) > T3 (12.9 %) > T4 (10.8 %) > T5 (10.7 %)

Fe: T2 (39.5 %) > T1 (32.8 %) > T4 (31.6 %) > T3 (31.1 %) > T5 (23.3 %)

Ni: T2 (22.9 %) > T1 (17.7 %) > T4 (15.2 %) > T3 (14.5 %) > T5 (6.2 %)

Pb: T2 (36.1 %) > T1 (25.5 %) > T3 (24.9 %) > T4 (23.8 %) > T5 (17.5 %)

Cr: T2 (56.0 %) > T1 (51.9 %) > T3 (47.4 %) > T4 (31.5 %) > T5 (17.2 %)

DTPA-extractable Cd was not detected in the *S. natans* composts of all trials. Highest reduction of DTPA-extractable heavy metals in trial 2 is consistent with the results of phumdi biomass and water hyacinth composting (Singh et al. 2014; Singh and Kalamdhad 2013a, 2014b) and is due to higher degradation and higher humus formation. The reduction of DTPA-extractable heavy metals in all trials can be attributed to the degraded organic matter forming complex compounds with heavy metals (Fang and Wong 1999; Singh and Kalamdhad 2013a, b). The bioavailable heavy metals renovated into a more stable form during composting process (Castaldi et al. 2005; Singh and Kalamdhad 2013a) due to the increase in pH, metal biosorption by the microbial biomass or metal complexation with the newly formed humic substances (Castaldi et al. 2006; Cai et al. 2007; Singh and Kalamdhad 2014a). The pH is considered as one master factor controlling ion exchange, reduction/oxidation, adsorption and complexation reactions (Walter et al. 2006). Cations are adsorbed on organic matter at high pH. The effect of organic matter amendments on heavy metal solubility also depends greatly upon the degree of humification of the organic matter and subsequent effect upon soil pH (Gupta and Sinha 2007). The variations of DTPA-extractable metals were all statistically significant for the trials ($P < 0.05$). A noticeable result of the study was that the total concentrations of Zn (211.3–254.3 mg/kg); Ni (309.8–345.3 mg/kg); and Pb (805.0–890.5 mg/kg) of the final composts were higher than that of Cr (139.8–206.8 mg/kg) but the water-soluble forms of Zn [7.8–11.4 mg/kg (3.6–4.7 % of total Zn)]; Ni (NIL) and Pb [5.5–8.3 mg/kg (0.7–1.0 % of total Pb)] were lower than Cr [22.2–28.9 mg/kg (13.3–19.3 % of total Cr)] indicating that Cr was more bioavailable and had higher toxicity potential.

Toxicity characteristic leaching procedure (TCLP) test for heavy metals

The changes in leachability of Zn, Cu, Mn, Fe, Ni, Cd, Pb and Cr during the composting process are given in Table 5. Leachability of heavy metals was reduced in all trials during the composting of *S. natans* in the range 6.1–21.2 % for Zn; 19.7–35.0 % for Cu; 10.2–18.9 % for Mn; 11.1–34.7 % for Fe; 13.7–30.1 % for Ni; 11.9–34.5 % for Pb; 20.0–50.5 % for Cd; and 22.9–60.0 % for Cr. The reduction of leachable fraction of heavy metals was also reported during phumdi biomass composting (Singh et al. 2014) and the results were also consistent with the findings of other researchers (Chiang et al. 2007; Singh and Kalamdhad 2014a, b). The threshold limit for heavy metals' contamination is: Cd—20 mg/kg, Cr—100 mg/kg and Pb—100 mg/kg (US EPA method 1311, US Environmental Protection Agency 1992). Therefore, the results of TCLP test confirmed that the heavy metal concentrations in all trials were under the threshold limits for compost use for agriculture purposes. The reduction of leachability of all selected heavy metals during the composting process may also be attributed to the metals forming complexes with humic substances (Singh and Kalamdhad 2013b, 2014a). The pH had also an important role in the reduction of the leachable concentration of the metals. The pH controls the solubility equilibria or complexation by soluble and surface legends and the increasing pH within the optimum range resulted in reducing the solubility and bioavailability of heavy metals (Cambier and Charlatchka 1999). ANOVA analysis showed significant differences in the leachable concentration of heavy metals (Zn, Cu, Fe, Mn, Fe, Ni, Pb, Cd and Cr) of all trials ($P < 0.05$).

Conclusion

Composting of *S. natans* biomass with the addition of optimum amount of cattle manure and rice husk enhanced the microbial degradation as well as the reduction of the bioavailability of heavy metals of the composts. Highest and lowest reductions of MC (30.8 and 13.1 %) and VS (31.4 and 9.9 %) were observed in trial 2 and trial 5, respectively. The highest and lowest reduction of water-soluble, plant-available and leachable metals was also mostly observed in trial 2 and trial 5, respectively. The pH of the trials was enhanced significantly and reached neutral. The total concentrations of Zn, Ni and Pb of the composts were higher than that of total Cr; but the concentrations of the water-soluble Cr of the composts were comparatively higher than that of water-soluble Zn, Ni and Pb, indicating higher toxicity of Cr. The leachable concentrations of heavy metals of the composts were within the prescribed limits. Transformation of *S. natans* biomass into compost may help in the protection of Loktak lake and at the same time decrease the use of chemical fertilizers.

Acknowledgments The authors gratefully acknowledge the financial support of the Council of Scientific and Industrial Research, Government of India, New Delhi for the study [Grant no. 24(0319)/12/EMR-II dt. 07/05/2012].

References

Ashraf R, Ali TA (2007) Effect of heavy metals on soil microbial community and mung seed germination. Pak J Bot 39:629–636

Benefield LD, Judkins JF, Weand BL (1982) Process chemistry for water and waste water treatment. Prentice-Hall Inc., Englewood Cliffs

BIS (1982) Methods for analysis of solid wastes (excluding industrial solid wastes). Indian Standards Institution, New Delhi

Bragato G, Leita L, Figliolia A, Nobili M (1998) Effects of sewage sludge pre-treatment on microbial biomass and bioavailability of heavy metals. Soil Till Res 46:129–134

Cai QY, Mo C, Wu QT, Zeng QY, Katsoyiannis A (2007) Concentration and speciation of heavy metals in six different sewage sludge-composts. J Hazard Mater 147:1063–1072

Cambier P, Charlatchka R (1999) Influence of reducing conditions on the mobility of divalent trace metals in soils. In: Selim HM, Iskandar IK (eds) Fate and transport of heavy metals in the vadose zone. Lewis Publishers, Boca Raton

Castaldi P, Santona L, Melis P (2005) Heavy metal immobilization by chemical amendments in a polluted soil and influence on white lupin growth. Chemosphere 60:365–371

Castaldi P, Santona L, Melis P (2006) Evolution of heavy metals mobility during municipal solid waste composting. Fresenius Environ Bull 15:1133–1140

Cayuela ML, Mondini C, Sánchez-Monedero MA, Roig A (2008) Chemical properties and hydrolytic enzyme activities for the characterisation of two phase olive mill wastes composting. Bioresour Technol 99:4255–4262

Chen JH (1999) Characteristic and applications of domestic animal wastes. Animal waste compost products quality and treatment alternatives manual. Soil Survey and Testing Center, Department of Soil and Environmental Science, National Chung Hsing University, Taiwan, pp 15–22

Chiang KY, Huang HJ, Chang CN (2007) Enhancement of heavy metal stabilization by different amendments during sewage sludge composting process. J Environ Eng Manag 17:249–256

Chiroma TM, Ebewele RO, Hymore FK (2012) Levels of heavy metals (Cu, Zn, Pb, Fe and Cr) in bushgreen and roselle irrigated with treated and untreated urban sewage water. Int Res J Environ Sci 1:50–55

Devi V, Chanakya HN, de Alwis AAP, Deepa GB, Modak J (2002) Management of non-point inorganic and organic pollution of water bodies using decentralized installation of bioreactors. In: Proceedings on Symposium on Conservation, Restoration and Management of Aquatic Eco-System, Lake 2002, Session 2 Paper 5. Centre for Ecological Sciences, Indian Institute of Sciences, Bangalore

Fang M, Wong JWC (1999) Effects of lime amendment on availability of heavy metals and maturation in sewage sludge composting. Environ Pollut 106:83–89

Fuentes A, Llorens M, Saez J, Aguilar MI, Marın ABP, Ortuno JF, Meseguer VF (2006) Ecotoxicity, phytotoxicity and extractability of heavy metals from different stabilised sewage sludges. Environ Pollut 143:355–360

Gao M, Liang F, Yub A, Li B, Yang L (2010) Evaluation of stability and maturity during forced-aeration composting of chicken manure and sawdust at different C/N ratios. Chemosphere 78:614–619

Guan TX, He HB, Zhang XD, Bai Z (2011) Cu fractions, mobility and bioavailability in soil–wheat system after Cu-enriched livestock manure applications. Chemosphere 82:215–222

Gupta AK, Sinha S (2007) Phytoextraction capacity of the plants growing on tannery sludge dumping sites. Bioresour Technol 98:1788–1794

Huang GF, Wong JWC, We QT, Nagar BB (2004) Effect of C/N on composting of pig manure with sawdust. Waste Manag 24:805–813

Iwegbue CMA, Emuh FN, Isirimah NO, Egun AC (2007) Fractionation, characterization and speciation of heavy metals in composts and compost-amended soils. Afr J Biotechnol 6:067–078

Koivula N, Raikkonen T, Urpilainen S, Ranta J, Hanninen K (2004) Ash in composting of source-separated catering waste. Bioresour Technol 93:291–299

Larson ET, Sollo FW, Mcgurk FF (1973) Complexes affecting the solubility of calcium carbonate in water. Research Report No. 68, Illinois State Water Survey, University of Illinois, Urbana

Liao PH, Jones L, Lau AK, Walkemeyer S, Egan B, Holbek N (1996) Composting of fish wastes in a full-scale in-vessel system. Bioresour Technol 59:163–168

Lin C (2008) A negative-pressure aeration system for composting food wastes. Bioresour Technol 99:7651–7656

Liu Y, Ma L, Li Y, Zheng L (2007) Evolution of heavy metal speciation during the aerobic composting process of sewage sludge. Chemosphere 67:1025–1032

Mande S, Lata K (2005) Potential of Converting Phumdi Waste of Loktak Lake into Briquettes for Fodder and Fuel Use. In: Mathur SM, Mathur AN, Trivedi RK, Bhat YC, Mohonot P (eds) Aquatic weeds: problems, control and management. Himanshu, New Delhi, pp 100–108

Prasad R, Singh J, Kalamdhad AS (2013) Assessment of nutrients and stability parameters during composting of water hyacinth mixed with cattle manure and sawdust. Res J Chem Sci 3:1–4

Roletto E, Barberis R, Consiglio M, Jodice R (1985) Chemical parameters for evaluating compost maturity. Biocycle 26:46–48

Samuel P, Ingmar P, Boubie G, Daniel L (2013) Trivalent chromium removal from aqueous solution using raw natural mixed clay from Burkina Faso. Int Res J Environ Sci 2:30–37

Santosh S, Bidan C (2002) Distribution of aquatic vegetation in Loktak lake. In: Proceedings of Management of Phumdis in Loktak lake, Manipur, India

Senesi N (1989) Composted materials as organic fertilisers. Sci Total Environ 81(82):521–542

Singh J, Kalamdhad AS (2012) Concentration and speciation of heavy metals during water hyacinth composting. Bioresour Technol 124:169–179

Singh J, Kalamdhad AS (2013a) Bioavailability and leachability of heavy metals during water hyacinth composting. Chem Speciat Bioavailab 25:1–14

Singh J, Kalamdhad AS (2013b) Assessment of bioavailability and leachability of heavy metals during rotary drum composting of green waste (water hyacinth). Ecol Eng 52:59–69

Singh J, Kalamdhad AS (2013c) Reduction of bioavailability and leachability of heavy metals during vermicomposting of water hyacinth (*Eichhornia crassipes*). Environ Sci Pollut Res 20:8974–8985

Singh J, Kalamdhad AS (2014a) Effect of lime on speciation of heavy metals during agitated pile composting of water hyacinth. Front Environ Sci Eng 10:10–22

Singh J, Kalamdhad AS (2014b) Effects of natural zeolite on speciation of heavy metals during agitated pile composting of water hyacinth. Int J Recycl Org Waste Agricult 3:1–17

Singh WR, Kalamdhad AS (2014c) Potential for composting of green phumdi biomass of Loktak lake. Ecol Eng 67:119–126

Singh WR, Pankaj S, Singh J, Kalamdhad AS (2014) Evaluation of bioavailability of heavy metals and nutrients during agitated pile composting of green Phumdi. Res J Chem Environ 18:1–8

Smars S, Gustafsson L, Beck-Friis B, Jönsson H (2002) Improvement of the composting time for household waste during an initial low pH phase by mesophilic temperature control. Bioresour Technol 84:237–241

Stevenson FJ (1994) Humus chemistry. John Wiley Sons, New York

Tan KH (2010) Principles of soil chemistry, 4th edn. CRC Press, Boca Raton

Tipping E, Hurley MA (1992) A unifying model of cation binding by humic substances. Geochim Chosmochim Acta 56:3627–3641

TMECC (2002) Test methods for the examination of composting and compost. US Composting Council, Bethesda

US Environmental Protection Agency (1992) Method 1311—toxicity characteristic leaching procedure (TCLP) In: SW-846 Test Methods for Evaluating Solid Waste, Physical/Chemical Methods. US Environment Protection Agency. http://www.epa.gov/solidwaste/hazard/testmethods/sw846/index.htm. Accessed 21 Jan 2014

Walter I, Martinez F, Cala V (2006) Heavy metal speciation and phytotoxic effects of three representative sewage sludge for agricultural uses. Environ Pollut 139:507–514

Wang SG, Sun XF, Gong WX, Ma Y (2009) Biosorption of metals on to granular sludge. In: Wang LK, Chen JP, Huag YT, Shammas NK (eds) Heavy metals in the environment. Taylor and Francis Group/CRC Press, Boca Raton, pp 201–223

Wong JWC, Selvam A (2006) Speciation of heavy metals during co-composting of sewage sludge with lime. Chemosphere 63:980–986

Wong JWC, Mak KF, Chan NW, Lam A, Fang M, Zhou LX, Wu QT, Liao X (2001) Co-composting of soya bean residues and leaves in Hongkong. Bioresour Technol 76:99–106

Young SD (2010) Chemistry of Heavy Metals and Metalloids in Soils. In: Alloway BJ (ed) Heavy metals in soils: trace metals and metalloids in soils and their bioavailability. Springer, Dordrecht, pp 51–97

13

Simultaneous solidification of potassium and phosphorus using rice straw charcoal and saturated phosphorus adsorbent

Akagi Koutarou · Uemoto Seigo · Hiroyuki Harada · Seichiro Oura · Biplob Kumar Biswas · Takanori Watari · Yoshiharu Mitoma

Abstract Phosphorus (P) ore is an expensive and limited resource that will be depleted in a few decades if the current global consumption rate continues. Japan, one of the most developed countries in the world, relies completely on imported phosphate rock for P. Potassium (K) ore, which is equally important for continuous development, is also becoming increasingly expensive. The recovery of P and K is therefore important for continuous and sustainable development. In this study, concentrated P (obtained from eluent) and K (obtained through alkali leaching of rice straw charcoal) were recovered as potassium magnesium phosphate (PMP) through the controlled addition of magnesium (Mg). A PMP crystal was produced when an equimolar (with respect to P) Mg solution was added to the leaching solution (rich in P and K) at a pH range of 11–12. Thus, the production of PMP in a crystalline form demonstrates the huge scope for the recycling of limited resources.

Keywords Potassium magnesium phosphate · Desorption · Extraction · Solidification

A. Koutarou · U. Seigo · H. Harada (✉) · Y. Mitoma
Department of Life and Environmental Science,
Prefectural University of Hiroshima, 562 Nanatuka-cho,
Shoubara, Hiroshima 727-0023, Japan
e-mail: ho-harada@pu-hiroshima.ac.jp

S. Oura · T. Watari
Department of Applied Chemistry, Saga University,
1 Honjo Saga, Saga 840-8502, Japan

B. K. Biswas
Department of Chemical Engineering, Jessore University
of Science and Technology, Jessore 7408, Bangladesh

Introduction

The global human population is expanding at a rapid rate, requiring more food and resulting in increased demand for chemical fertilizers. The raw materials used to produce chemical fertilizers are not in danger of being depleted in the short term; however, because of population growth the raw materials required for fertilizer production will be consumed at a rapid rate. The three main elements of fertilizer are nitrogen (N), phosphorus (P) and potassium (K). K and P are obtained from underground resources and regional reserves that are unevenly distributed. Japan relies solely on imported raw materials for chemical fertilizer production, which results in high prices for chemical fertilizers in this country.

Potash ore, the raw material for potassium production, is unevenly distributed around the globe, with 53 % of total global potash reserves located in Canada (US Geological Survey 2012). The majority of potash produced in 2008 was from Canada (31 %), Russia (19 %) and Belarus (14 %). Based on current usage trends, potash reserves will last for nearly a century (Avnimelech et al. 1992). The majority of phosphate produced in 2008 was from unevenly distributed resources in Morocco (70 %), Iraq (8 %), China (5.2 %) and Algeria (3.0 %).

Depletion of phosphate rock is estimated to occur sooner than depletion of potash ore, based on the assumption that the amount of fertilizer used will increase by 3 % annually (Kuroda et al. 2005). With the geometric increase in populations throughout the world and the increased usage of these two resources by developing countries, current supplies of P and K will likely be exhausted in 2060 and 2120, respectively (Kuroda et al. 2005). It is imperative to develop a sustainable method for recycling these resources.

Since neither K nor P is produced in Japan, the development of such recycling technology would be very useful in the field of fertilizer production and overall social and industrial development in this country. Although the need for recycling of K is high, there have been few studies on potential methods of K recovery (Zhou et al. 2007). Zhou et al. (2007) recovered 44.3 % K from a dried stem–leaf sample of water hyacinth during the formation of $KC_4H_5O_6$ by adding tartaric acid. Utilization of agricultural waste or plant biomass is necessary for a sustainable society. Biomass resources have been used as bio-fuels or as industrial raw materials in recent years. In the present study, we have developed an adsorbent using agricultural waste biomass. We have made a P adsorbent from orange juice residue. In this study, we examined the solidification of P as a potassium salt by adjusting the solubility. Rice straw charcoal was used to supply K. The potassium–magnesium complex salt of P was produced by the addition of magnesium and by adjusting the pH. This compound was easily solidified. The use of P and K together raises the prospect of producing potassium magnesium phosphate (PMP) if magnesium is also added (Eq. 1). Precipitated or crystallized PMP can be used as fertilizer and can positively benefit the agricultural sector (Xu et al. 2011).

$$K^+ + PO_4^{3-} + Mg^{2+} + 6H_2O \Rightarrow KMgPO_4 \cdot 6H_2O \qquad (1)$$

The typical K content of rice straw is 1.7 % (Summer and Williams 2001). This amount of K can be used if it is recovered in an efficient and cost-effective way. K can be leached from rice straw using an alkaline solution, especially when the rice straw is in carbonized form. In our previous study, we found that a bio-adsorbent prepared from orange juice residue exhibited very high capacity to adsorb P. The residue-adsorbed P was successfully desorbed using NaOH solution (Biswas et al. 2008). Therefore, the alkaline solution (rich in P after the desorption process) could be easily used for leaching K from rice straw charcoal. From the above viewpoint we set our objective for this study, which was to prepare PMP by adding magnesium to P- and K-rich solution obtained from extraction of K from rice straw charcoal by the P-rich alkaline desorption liquid.

Methods

Analytical methods

The concentrations of phosphate, and ammonium were determined by the molybdenum blue ascorbic acid method and indophenol method using ultraviolet–visible spectrophotometry (Jasco, V-630BIO). The concentrations of potassium, magnesium and sodium ions were measured by a polarized Zeeman atomic absorption spectrophotometer (HITACHI Z-2000). Collected white precipitates were dissolved in 3.0 M HCl to calculate percentages of phosphorous, potassium, magnesium, and sodium. While scanning electron microscope is used to observe the surface morphology of the crystals. Characterization of the white precipitate was carried out by X-ray diffraction analysis equipment. The morphology and chemical composition of the dried crystals were determined using scanning electron microscope (SEM: JEOL, JCM-5100) and energy dispersive X-ray spectroscopy (EDX: HITACHI S 3000) apparatus at an emission voltage of 15 kV. All experiments were conducted twice and the samples were analyzed to ensure reproducibility of experimental data.

Materials

Sodium di-hydrogen phosphate (Cat. No. 32379-00), Magnesium chloride (Cat. No. 103) were obtained from Kanto Chemical Co., Inc. while, sodium hydroxide (No. 198-104 13765) was obtained from Wako Pure Chemical Industries, Ltd. All other chemicals and reagents were of analytical grade and used without any purification. Rice straw was obtained from a local farm and cut into pieces 2–3 cm long. Pieces were then air dried for 20 days and placed in a small crucible which was then placed in a large crucible. The space between the small and large crucible was filled with powdered graphite. The crucible was then heated at an initial temperature of 540 °C and the temperature was gradually increased over an 1 h period to 800 °C in an electric furnace. After cooling, the carbonized rice straw was crushed into particles of less than 150 μm in size. Figure 1 and Table 1 show the results of the XRD and EDX analyses. The adsorbent was made from immobilized

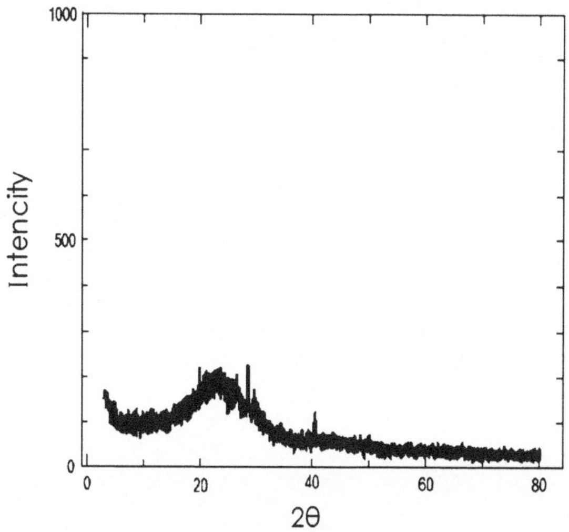

Fig. 1 Component analysis of rice straw charcoal analysis (XRD)

Table 1 Component analysis of rice straw charcoal analysis (XRD)

Elements	C	O	Si	K
Wt (%)	49.9	29	19.5	1.6

Zr from unused bio-material using the method described by Biswas et al. (2008). This adsorbent was saturated in 1000 mg/L sodium di-hydrogen phosphate solution. The capacity of the saturated adsorbent was approximately 0.8–1.1 mol-P/kg.

Methods

Desorption of phosphorus

Batch-wise desorption tests were carried out using the adsorbent saturated with phosphorus. The pH of the desorption liquid was adjusted to between 11 and 13 using 1 M NaOH and 1 M HCl solution at a solid-to-liquid (S/L) ratio of 33. The solutions were then stirred for 24 h at 40 rpm. Beforehand the effect of solvent pH and that of S/L ratio (from 15 to 33) were examined. The concentration relationships between phosphorus, potassium and magnesium were adjusted to be greater than the solubility product of magnesium phosphate potassium. Simultaneous recovery of both phosphorus and the potassium was resulted to obtain a solid product. However, it was necessary to clarify the elute concentrations under various conditions for obtaining a precipitate.

Therefore, by changing the amount of adsorbent in the solvent, it was possible to determine the concentration. In order to achieve a leaching at a lower pH, desorption operation was conducted again under the same condition using the same adsorbent.

Behavior of the potassium extraction

To determine the behavior of potassium extraction from rice straw charcoal, batch-wise extraction tests were carried out at varying pH and stirring time at a specified S/L ratio. The pH of the extracted potassium solution was adjusted to between 11 and 13 using 1 M NaOH and 1 M HCl. The solvent and rice straw charcoal solution which had S/L ratio of 80 was stirred for 1 h at 40 rpm. The elution test was carried out at a specific S/L ratio and at varying time (from 5 to 90 min) to determine the effect of elution time.

Solidification

During potassium extraction from the rice straw charcoal using phosphorous desorption liquid after separation of the desorption liquid, potassium was extracted by mixing this solution at an 25 S/L. The whole solution was stirred at 40 rpm for 24 h. The suspensions were filtered through filter paper type 5C (ADVANTEC Co.) and the equilibrium concentration was measured. After that different amount of 100 mM magnesium chloride solution was added to fixed amount of obtained solution (rich in P and K). This was performed to determine optimum molar ratio of magnesium added. However, the subtle change in phosphorus concentration due to the increase in volume can be ignore. The PMP precipitates along with rice straw charcoal was washed with deionized water, dried in oven at 60 °C for 24 h and finally analyzed by EDX. The concentration of P was measured using ion-chromatography (Mettler Toledo, Japan) and the concentrations of magnesium, potassium and sodium were determined using an atomic absorption spectrophotometer.

Results and discussion

Desorption of P with NaOH

In our previous study (Biswas et al. 2008), high P desorption rates from zirconium-immobilized orange waste residue were observed when a 0.2 M NaOH solution was used as the desorption eluent. The adsorption and desorption reactions resulted from the exchange between the hydroxyl groups and different phosphate species on the coordinated site of the Zr. If desorption of P was conducted using a high concentration NaOH solution, evaporation of the solvent would be required to solidify tri-sodium phosphate. Moreover, using a relatively low concentration of NaOH solution, KMP precipitate can be easily obtained because of its low solubility product (Taylor et al. 1963). In this study, 0.05–0.1 M NaOH solutions were used to avoid the effect of sodium in the formation of the KMP precipitate/crystal. The pH of the desorption liquid was adjusted in the range of 12.5–13 when mixed with the absorbent.

The changes in the desorption rate at varying S/L ratios (from 15 to 33) are shown in Fig. 2. Desorption increased because of the repetitive operation at every S/L ratio tested. Repetitive operations applied of the new alkali solution were conducted after separation of adsorbent. However, it was found that total desorption increased slightly with the decrease in the S/L ratio. This desorption rate corresponded to the range of 250–300 mg/L of P. Table 2 shows the saturation adsorption capacity (for P) of various adsorbents. From this table it is evident that the saturated adsorption capacity of the adsorbent of this study is higher than many other adsorbents used in other studies.

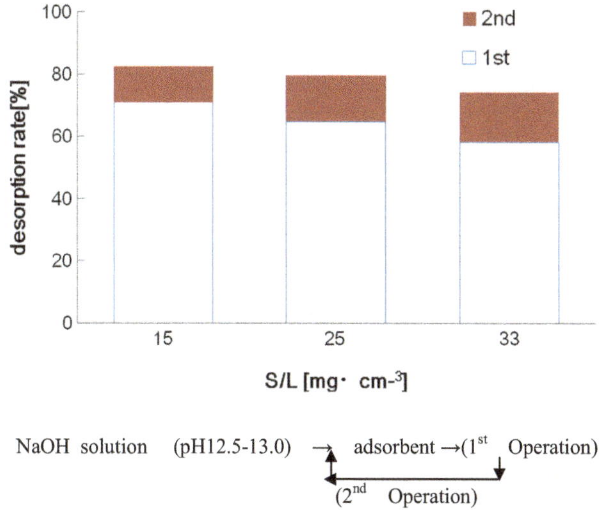

Fig. 2 Relationship between S/L and desorption rate

Table 2 Maximum phosphorous adsorption capacity of various adsorbent in bibliography

Type of adsorbent	Maximum adsorption capacity (mg/g)	References
Fe–Al–Mn trimetal oxide adsorbent	48.3	Lŭ et al. 2013
Peat	8.91	Xiong and Mahmood 2010
Lanthanum doped vesuvianite	1.32	Li et al. 2009
Synthetic hydrotalcite	47.3	Kuzawa et al. 2006
This research (zirconium doped on orange juice residual)	37.2	Biswas et al. 2008, 2009

K extraction from rice straw charcoal

Figure 3 shows the relationship between the K concentrations extracted from different rice straw charcoal prepared at different carbonization temperatures. The concentration of extracted K was lower at relatively low carbonization temperature. The highest concentration of K was obtained from the rice straw charcoal prepared at a carbonization temperature of 540 °C. However, at carbonization temperatures higher than 500 °C, KCl was found to occur in rice straw charcoal. Figure 4 depicts changes in the extraction of K from rice straw charcoal over time. This figure also shows that about 20 min is required to reach equilibrium. After this time no significant increase in extraction efficiency is observed.

The change in K extraction from rice straw charcoal with changing pH has been illustrated in Fig. 5. The amount of extracted K from rice straw charcoal was found

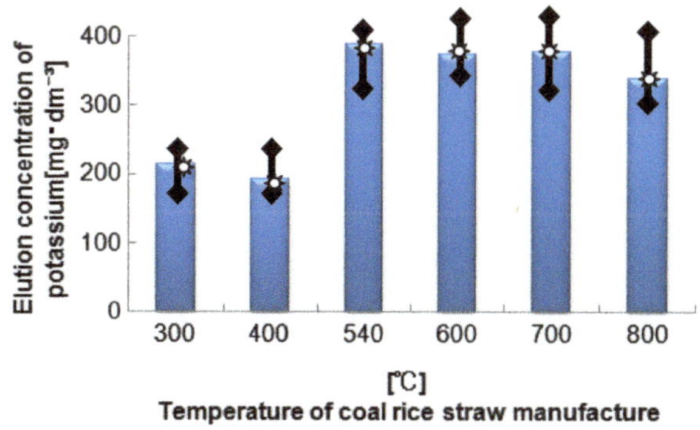

Fig. 3 The relationship between the concentration of potassium elution and temperature on rice straw carbonization. *Range of *up-down arrow* indicates the highest and the lowest, *circle* indicates four data's average

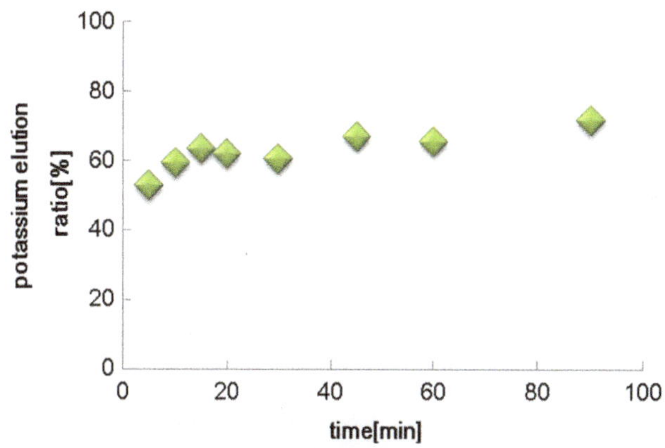

Fig. 4 Changes in the extraction rate of potassium with elapsed time

to be higher at lower pH (pH <3) whereas the extraction was almost constant at 60–70 % at pH between 3 and 13. The pH was found to increase after equilibrium, which might be attributed to basic substances eluted from the rice straw charcoal. In addition, a KMP precipitate was produced when the pH was 10.5–12. The results indicated that K extraction from rice straw charcoal is possible using the alkaline desorption liquid containing concentrated P. Figure 6 shows the effects of the S/L ratio on the K extraction rate. The K extraction rate was maintained at 60–70 % at different L/S ratios and at initial pH values between 10 and 13. This low K extraction efficiency can be attributed to the fact that it is difficult to dissolve KCl, which was formed during high temperature carbonization. The K concentrations were 2000 and 1200 mg/L for S/L ratios of 25 and 12.5, respectively.

Solidification of P and K

Analyses of the precipitate were conducted by energy dispersive X-ray spectrometry and by measurement of P, K, Na

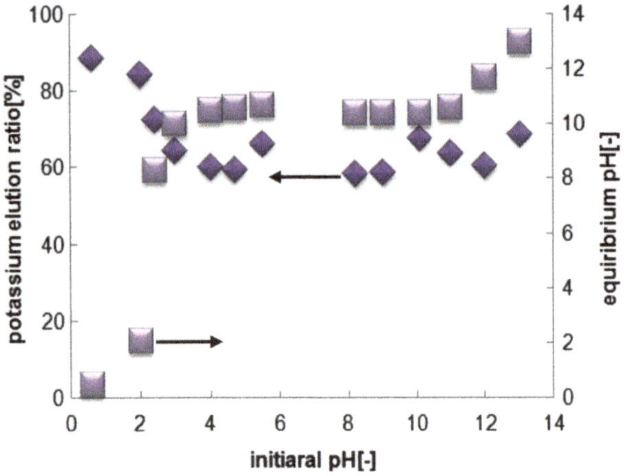

Fig. 5 The potassium extracted from the rice straw charcoal by adding the solvent

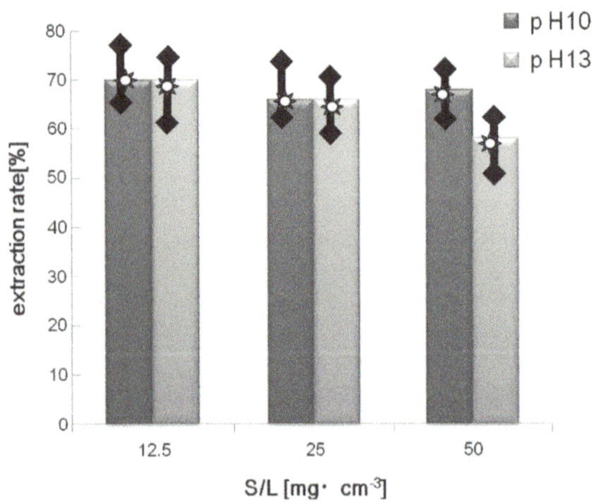

Fig. 6 Relationship between S/L ratio and extraction ratio. *Range of *up-down arrow* indicates the highest and the lowest, *circle* indicates four data's average

Table 3 Changes of the components in the solution

	Run	Initial condition	Equilibrium condition
pH	Run-1	11.12	10.71
	Run-2	11.63	11.23
PO_4-P (mM)	Run-1	9.1	0.74
	Run-2	7.6	0.94
K^+ (mM)	Run-1	29.5	23.2
	Run-2	29.2	23.4
Na^+ (mM)	Run-1	22.2	28
	Run-2	27	31.4
Mg^{2+} (mM)	Run-1	9.2	1.1
	Run-2	7.6	0.88

Table 4 Ratio of the number of atoms in the solid

	Run	(%)
P (%)	Run-1	35.02
	Run-2	35.31
K (%)	Run-1	35.27
	Run-2	35.42
Na (%)	Run-1	1.94
	Run-2	1.75
Mg (%)	Run-1	27.77
	Run-2	27.52

and Mg in the liquid. Table 3 shows the concentrations of P, K, Na and Mg in the solution while Table 4 shows the relative amounts of each of the elements. The rate of P extraction was greater than 90 % (Table 3). This high rate of P extraction indicated that the P concentration in the desorption liquid was appropriate for use in the solidification process. In addition, 90 % of the added Mg was removed. The K concentration in the treatment liquid remained high, which might require other methods of recycling. The ratio of K–Mg–P in the PMP salt was almost equimolar.

Conclusion

In this study, the K component of rice straw charcoal in agricultural waste was used for the solidification of P in an alkaline desorption solution. The P adsorbent was made from orange juice production residues, a previously unused waste biomass. Therefore, the present study offers the potential for recovery of fertilizer resources from biomass-based agricultural waste. This study recommended the formation of a P complex solid salt containing K and Mg. Using agricultural waste the fertilizer components could be recovered, which contributes to the development of recycling processes.

Potassium magnesium phosphate (PMP) was obtained from recovered P and K along with Mg in bench-scale experiments. In this study, a solution of 0.1 M NaOH (pH 13) was used for P desorption from the orange waste residue adsorbent. This ensured that: (1) concentrated P was obtained and (2) the adsorbent could be used repeatedly for adsorption of P from solution. This study demonstrated that using this method, P desorption rates of 70–80 % were obtained, which was sufficient to achieve the desired concentration of P. The study also revealed that a higher concentration of K was extracted from rice straw charcoal at lower pH values, while the extraction efficiency ranged between 60 and 70 % at pH values ranging from 3 to 13. The result indicated that PMP precipitation is an efficient method for yielding a multinutrient product through simultaneous removal of K and P.

References

Avnimelech Y, Bazelet M, Weber B (1992) Replacement of sodium in home and industrial uses by potassium: importance and possibilities. Potassium Ecosyst Biochem Flux Cation Agro For Syst: 377–382

Biswas BK, Inoue K, Ghimire KN, Harada H, Ohto K, Kawakita H (2008) Removal and recovery of phosphorus from water by means of adsorption onto orange waste gel loaded with zirconium. Bioresour Technol 99:8685–8690

Biswas BK, Inoue K, Harada H, Ohto K, Kawakita H (2009) Leaching of phosphorus from incinerated sewage sludge ash by means of acid extraction followed by adsorption on orange waste gel. Environ Sci 21:1753–1760

Kuroda A, Noboru T, Kato J, Ohtake H (2005) Development of technologies to save phosphorus resources in response to phosphate crisis. J Environ Biotechnol 4:87–94

Kuzawa K, Jung YJ, Kiso Y, Yamada T, Nagai M, Lee TG (2006) Phosphate removal and recovery with a synthetic hydrotalcite as an adsorbent. Chemosphere 62:45–52

Li H, Ru J, Yin W, Liu X, Wang J, Zhang W (2009) Removal of phosphate from polluted water by lanthanum doped vesuvianite. J Hazard Mater 168:326–330

Loblich KR (1985) US Patent No. US4496526A

Lŭ J, Liu H, Zhao X, Sun L, Jiuhui Q (2013) Adsorptive removal of phosphate by a nanostructured Fe–Al–Mn trimetal oxide adsorbent. Powder Technol 233:146–154

Summer MD, Williams J (2001) Proceeding: rice straw management update. UCCE, Yuba

Taylor AW, Gurney EL, Frazier AW (1963) Solubility products of magnesium ammonium and magnesium potassium phosphates. T Faraday Soc 59:1580–1584

US Geological Survey (2012) Mineral Commodity Summaries: 118–122

Xiong JB, Mahmood Q (2010) Adsorptive removal of phosphate from aqueous media by peat. Desalination 259(1–3):59–64

Xu K, Wang C, Wang X, Qian Y (2011) Simultaneous removal of phosphorus and potassium from synthetic urine through the recipitation of magnesium potassium phosphate hexahydrate. Chemosphere 84:207–212

Xu K, Wang C, Wang X, Qian Y (2012) Laboratory experiments on simultaneous removal of K and P from synthetic and real urine for nutrient recycle by crystallization of magnesium–potassium–phosphate–hexahydrate in a draft tube and baffle reactor. Chemosphere 88(2):219–223

Zhou W, Zhu Duanwei, Tan Liangfeng (2007) Extraction and retrieval of potassium from water hyacinth, (Eichhornia crassipes). Bioresour Technol 98:226–231

Growth performance study of Koi fish, *Anabas testudineus* (Bloch) by utilization of poultry viscera, as a potential fish feed ingredient, replacing fishmeal

Pratyush Bhaskar · Saroj Kumar Pyne · Arun Kumar Ray

Abstract

Background In India, fishmeal is habitually considered as the major protein source for fish culture. Rising cost, deteriorating quality and unavailability of fishmeal caused huge difficulty in modern aquaculture, particularly for fish nutrition. Nearly all *Anabas testudineus* cultures faced loss, due to high feed costs, inappropriate feed formulation and management. So it was necessary to find a profitable replacement of fishmeal, which could provide better growth performance to *A. testudineus.*

Results In this study, different types of poultry wastes were tested, and poultry viscera confirmed satisfactory results, having 60.67 % crude protein (% Dry matter basis). Further, a feeding trial was conducted for 60 days in 90-L circular fibre tanks with proper aeration, to evaluate poultry viscera, in the formulated diet for *A. testudineus*. Triplicate groups of fingerlings each were fed four isonitrogenous diets, at 5 % of wet body weight basis. Feed readjusted biweekly. Comparatively, fish accumulated highest and significant ($P < 0.05$) increase in body weight and fat deposition, when dietary fishmeal completely replaced by poultry viscera.

Electronic supplementary material The online version of this article (doi:10.1007/s40093-014-0082-y) contains supplementary material, which is available to authorized users.

P. Bhaskar (✉) · S. K. Pyne
Department of ASEPAN P.S.B, Visva-Bharati, Sriniketan, Birbhum, West Bengal, India
e-mail: pratyush.bhaskar@rediffmail.com

S. K. Pyne
e-mail: sarojpyne@yahoo.com

A. K. Ray
Department of Zoology, Visva-Bharati, Santiniketan, Birbhum, West Bengal, India

Discussions Thus, this study revealed that better growth performance in Koi, *A. testudineus* (Bloch) could be achieved through utilization of poultry viscera in the formulated fish feed, compared to fishmeal. Hence, poultry waste recycling could be stimulated also.

Keywords Crude protein · Fishmeal · Koi · Poultry viscera · Proximate composition

Abbreviations

90-L	90 Litre capacity, of the circular fibre fish rearing tank
cm	Centimetre (Length)
CP	Crude protein
DO	Dissolved oxygen in water
E longitude	Eastern longitude
FCR	Feed conversion ratio
FM	Fishmeal
gm	Gram (weight)
NFE	Nitrogen-free extract
N latitude	Northern latitude
PVisc	Poultry viscera
SGR	Specific growth rate
T	Treatment
Temp	Temperature
WB	West Bengal

Introduction

Fish as the prime animal protein source in the nutritional budgeting of West Bengal (Eastern India) deliver high protein, low cholesterol and tasty-soft flesh, etc. About

85–90 % fish protein is digestible and all the dietary essential amino acid is in fish flesh (Cappell et al. 2007).

Climbing Perch (*Anabas testudineus*) is locally known as 'Koi' in West Bengal. It is one of the most popular fishes in West Bengal, due to its traditional role as a medically prescribed diet for sick and convalescents. It contains high amount of available iron and copper essentially needed for haemoglobin synthesis. In addition, it also contains all the essential amino acids (Saha 1971).

Anabas testudineus encompasses a wide range of geographical distribution. This species is also found in India, Pakistan, Burma, Sri Lanka, Thailand, China, Hong Kong, Philippines and Malaysia (Jayaram 1981). It is found in fresh and brackish waters mostly in ponds, swamps and lakes. Climbing perch appears to be visual feeders, feeding primarily during the day (Patra 1993). It is a ferocious predator, omnivorous in feeding habit and tolerate wide range of environmental conditions.

In adverse atmospheric circumstances, this fish proves to be a very good challenger of fish culture. But, often its production fails to fulfil its high market demand, due to lack of proper culture. However, the results of the survey in 2002, more than 50 % climbing perch culture farmers were lost due to high feed costs (Tihn 2003).

There are very little information available based on the nutritional requirement of *A. testudineus*. It is important to know the minimum protein requirement for best growth in formulating a balanced diet as protein is the key nutrient offering growth and other metabolic activities.

Usage of combination of various animal or plant protein ingredients in the place of fish meal in fish diet has been demonstrated successfully (Millamena 2002). Viscera are the large organs inside the body of animals, such as the heart, lungs and stomach, etc. Research findings have revealed that Recycling of wastes from poultry slaughter houses is of economical, biological and environmental importance (Cai et al. 1994; Steffens 1994). However, poultry wastes recycling could be stimulated by the utilization of poultry viscera.

Considering the above, the present study was conducted with a view to set up the best dietary protein requirement of *A. testudineus* and to assess the result of replacing dietary fishmeal protein by poultry viscera, on the growth performance of *A. testudineus*.

Materials and methods

Experimental sites and study period

During the study period, from April, 2012 to May, 2013 fish feed ingredients were collected and then transported to the Zoology Laboratory, Visva-Bharati, Birbhum, W.B, India (23°41′30″N Latitude and 87°41″20E Longitude).

Collection and storage of samples

Samples of feed ingredients such as Poultry viscera, fish meal, rice bran, wheat bran, Mustard oil cakes, wheat flour, etc., were collected, and sun dried properly, then packed in polyethylene bags to prevent initial spoilage and brought to the ASEPAN laboratory Visva-Bharati, then stored in refrigerator.

Proximate composition analysis

Proximate analysis is usually the first step in the chemical evaluation of a feed ingredient, where the material is subjected to a series of relatively simple chemical tests so as to determine the content of moisture, crude protein, lipid, crude fibre, ash, etc.

Estimation of moisture (%)

Moisture is commonly determined by drying a sample at some elevated temperature (100 ± 5 °C) for 30 min and further at 60 °C, until a constant weight was obtained, following [Association of Official Analytical Chemicals (AOAC 1995)].

Determination of ash (%)

Ash is readily determined by ignition from dried sample at about 550 ± 50 °C for 6–8 h in muffle furnace. The residue is weighed and reported as ash, following AOAC (1995).

Determination of crude protein (%)

The crude protein determined by Micro-Kjeldahl method using a Kjeltec system (Tecator, Sweden) through digestion and distillation steps. Kjeltab (containing potassium sulphate and catalysts), H_2SO_4, NaOH, Phenolphthalein, etc., used in this method, following Pearson (1999).

Determination of crude lipid (%)

The lipid content was determined by Soxhlet apparatus. 2gm of sample wrapped in whatman filter paper (No-1) and placed in a thimble connected with Soxhlet apparatus. Initial weight of soxhlet flask recorded and filled with 200 ml petroleum ether, which boiled for 8 h at 60–80 °C through the thimble, by siphoning process. Flask was taken out and allowed to evaporate. The difference in the two weights of the round joint flask gave the weight of the lipid, following AOAC (1995).

Analysis of fibre (%)

Done by following the method of Pearson (1976).

Nitrogen-free extract (%)

{100 − (% moisture + % crude protein + % crude lipid + % crude fibre + % ash)} following AOAC (1995).

Carbohydrate (%)

[Nitrogen-free extract % + Crude Fibre %] following Hastings (1976).

Caloric value/gross energy (kcal/100 g)

[(Carbohydrate × 4.1/100) + (Protein × 5.65/100) + (Lipid × 9.45/100)]. After (Cho et al. 1982).

Growth parameters

The following parameters were used to evaluate the growth, by using the following formulas according to Castell and Tiews (1980):

Survival rate (%) = 100 × (final number of fish/initial number of fish).

Specific growth rate (SGR) (%/day) = 100 x (Ln Wf − Ln Wi)/Δt.

Feed conversion ratio (FCR) = feed consumed/(Wf − Wi). Where: Wf = final weight and Wi = initial weight. Feed consumed = feed given − feed not eaten. Δt = duration (days).

Experimental design

90-L circular fibre tanks were used in this experiment, with proper aeration (Fig. 1). Triplicate groups of fingerlings each were fed four isonitrogenous diets (above 34 % C.P) namely T-1, T-2, T-3 and T-4. Acclimatization period of 15 days was applied to all fishes in laboratory condition, with the feed of 1: 1 mixture of rice bran and mustard oil cake. Stocking rate was 25 fish/tank. Feed ingredients were cleaned under running tap-water, then sun dried for weeks and then stored in air tight plastic container in refrigerator. Diet particle size was used as 0.4 mm to 1.25 mm. Feeds were applied twice a day (08:00 a.m. and 03:00 p.m.) at the rate of 5 % wet body weight basis of the fry. Feed readjusted biweekly. Experimental trial period was of 60 days.

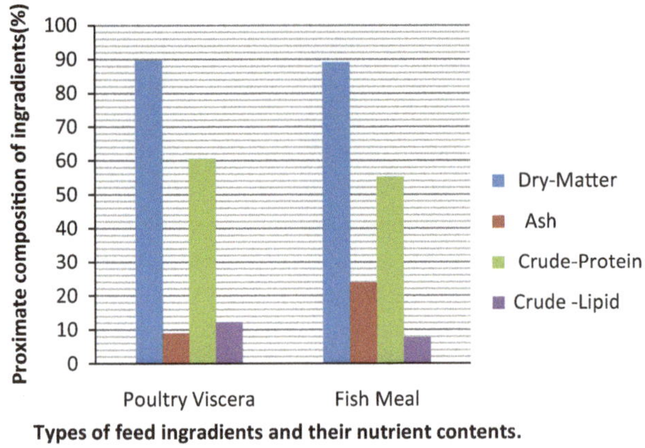

Fig. 1 Nutritional comparison between PVisc and FM (% dry matter basis): *PVisc* poultry viscera, *FM* fishmeal

Sampling and data analysis

25 fishes were sampled (Randomly) every 14th day to determine fish growth rate from each treatment, by the help of glass nylon hapa. 1 h after feeding, faeces, waste particles of food and dead bodies of fish were siphoned out for measurement. The survival rate of fish was calculated at 60th day. Feed ingredients, formulated diets, fish body carcass, etc., were analysed for proximate composition following the AOAC (1995) procedures. Water quality parameters were determined by APHA (1992) procedures. All the structured designs and data were analysed using MS-Excel one-way ANOVA. This was followed by Duncan's new multiple range test (Duncan 1955) to identify the level of significance of variance ($P < 0.05$) among the treatment means. Standard deviations (±SD) of treatment means were also calculated.

Results

Table 1 shows highest protein content in poultry viscera (60.67 %) that is much more than in Fishmeal (55.19 %). Highest lipid contents observed in poultry skin (12.89 %). Highest ash contents observed in poultry littre (15.98 %). In this study, the estimated protein, lipid and ash contents of poultry viscera are 60.67, 12.25 and 08.93 %, respectively those are slightly higher than the findings of Cai et al. (1994). The estimated Lipid contents of Poultry Intestine are 10.41 % which is more or less similar with the findings (07.64 %) of Tabinda and Butt (2012).

Feed formulation and proximate chemical composition of the control and experimental diet (%)

Replacement of fish meal with poultry viscera in compound diet was performed on the culture of *A. testudineus*

Table 1 Proximate composition of different poultry wastes and FM (% dry matter basis)

Ingredients	Dry matter	Ash	Crude protein	Crude lipid
Poultry viscera	$89.70^{b} \pm 0.15$	$08.93^{a} \pm 0.40$	$60.67^{a} \pm 0.41$	$12.25^{a} \pm 0.22$
Poultry intestine	$88.4^{b} \pm 0.73$	$06.25^{a} \pm 0.07$	$53.77^{a} \pm 0.53$	$10.41^{a} \pm 0.65$
Poultry skin	$89.59^{ab} \pm 0.52$	$07.97^{a} \pm 0.19$	$31.09^{a} \pm 0.35$	$12.89^{ab} \pm 0.28$
Poultry littre	$87.03^{ab} \pm 0.64$	$15.98^{a} \pm 0.56$	$19.01^{a} \pm 0.74$	$02.13^{b} \pm 0.33$
Fish meal	$89.10^{a} \pm 0.23$	$23.95^{a} \pm 0.38$	$55.19^{b} \pm 0.53$	$07.77^{b} \pm 0.42$

Values are ±SD of three replications. Figures in the same row having different superscript are significantly different ($P < 0.05$)
FM fish meal

for 60 days. In the feeding treatments fish meal was replaced by 33.3, 66.6, and 100 % of poultry viscera, respectively, as the experimental diets. The diet containing 100 % fish meal (0 % poultry viscera) was set as the control diet to compare performances with those fed with other experimental diets (Tables 2, 3, 4).

Notes: different levels of replacement of fish meal with poultry viscera.

63.44 %, more or less similar to the findings of Fisheries Research Institute, FRI (1989). During the present study the crude lipid content was recorded in the range between 09.22 and 16.49 % in poultry viscera, more or less similar to the findings of Hasan and Amin (1997).

Triplicate groups of fingerlings each were fed four iso-nitrogenous (34–35 % C.P) diets namely T-1, T-2, T-3 and T-4 which is, however, similar with the treatments of

Feed Treatments	Levels of Replacement of F.M with P.Visc	Diet-Type
T-1	100 % Fish meal, 0 % Poultry viscera	**Control Diet.**
T-2	66.6 % Fish meal, 33.3 % Poultry viscera	**Experimental Diet.**
T-3	33.3 % Fish meal, 66.6 % Poultry viscera	
T-4	0 % Fish meal, 100 % Poultry viscera	

Discussion

In this study, the common fish feed ingredients like fish meal, mustard oil cake, rice bran, wheat bran and wheat flour were used for proximate analysis, which are usually available throughout the year and all over India. Low cost and good quality supplementary feeds proved essential for superior practices, and it was the key demand of farmers to reduce production cost. For formulation of fish feed, information about feed ingredient's price, nutritive value, application strategies, and its seasonal variation on quality and availability were crucial. The present study attempted to collect this information.

Due to rising cost, uncertainty and unavailability of fishmeal, it was an immense problem in modern aquaculture, especially in relation to fish nutrition, to find a desirable replacement for fish meal. Researches on poultry viscera have revealed interesting results and more works were being done by scientists from all over the world.

The analysed crude protein content of the poultry viscera was estimated in the range between 57.90 and

Adhikary et al. (2009) and this level is higher than that of 30 % reported for *A. testudineus* by Chareontesprasit and Jiwyam (1996).

The results showed that the growth of *A. testudineus* fry varied significantly ($P < 0.05$) with different diets. At the end of feeding trial, highest S.G.R (1.39) and survival rate (85 %) were found where fish were fed on T-4 (treatment where fishmeal 100 % replaced by poultry viscera). The poorest growth rate was shown by T-1 (control treatment, where fishmeal was not replaced by poultry viscera). These results are more or less similar with the findings of Doolgindachabaporn (1994).

FCR was highest (3.82) in T-1 and gradually decreased (2.10) in T-4 with the increasing dietary protein levels. Similar trend observed in Tilapia (Jauncey 1982) and *Puntius gonionotus* (Wee and Ngamsnae 1987). Doolgindachabaporn (1994) found that the FCR value of *A. testudineus* ranges from 1.8 to 3.0. Similarly, FCR values in the present study were comparable with FCR values (01.29–01.62) in rainbow trout, by the nutritional diet of co-dried fish silage (Hardy et al. 1984). Potongkam (1972)

Table 2 Formulation and proximate chemical composition of the control and experimental diets (%)

Ingredients	Control	Experimental (replacement of FM by PVISC)		
		33.3 %	66.6 %	100 %
	T-1	T-2	T-3	T-4
Feed formulation of the control and experimental diets (%)				
Fish meal (FM)	30.00	20.00	10.00	00
Poultry viscera (PVisc)	00	10.00	20.00	30.00
Mustard oil cake	15.50	15.50	15.50	15.50
Rice bran	29.90	29.10	29.60	29.50
Wheat bran	19.00	19.80	19.30	19.40
Vitamin premix	00.50	00.50	00.50	00.50
Mineral premix	00.10	00.10	00.10	00.10
Sunflower oil	05.00	05.00	05.00	05.00
Total	100	100	100	100
Proximate chemical composition of the control and experimental diets (% dry matter basis)				
Dry matter	90.97	91.29	90.92	91.13
Crude protein	34.33	34.37	34.90	35.88
Crude fat	07.62	08.8	09.20	09.68
Crude ash	06.58	06.92	07.66	06.36
Crude fibre	02.51	02.30	02.30	02.13
Nitrogen-free extract	39.93	38.90	36.86	37.08
Gross energy (kcal/100 g)	439.97	446.27	444.68	454.95

Per kg vit premix: 4,000,000 IU vitamin A, 480,000 IU vitamin D3, 40,000 mg vitamin E, 2,400 mg vitamin K3, 4,000 mg vitamin B1, 6,000 mg vitamin B2, 40,000 mg niacin, 10,000 mg Ca-panthothenate, 4,000 mg vitamin B6, 10 mg vitamin B12, 100 mg D-biotin, 1,200 mg folic acid, 40,000 mg vitamin C and 60,000 mg inositol

Per kg min premix: 23,750 mg Mn, 75,000 mg Zn, 5,000 mg Zn, 2,000 mg Co, 2,750 mg I, 100 mg Se, 200,000 mg Mg

Table 3 Growth parameters and survival of *A. testudineus*

Parameters	Treatments			
	T-1	T-2	T-3	T-4
Initial weight (g)	$0.91^{a} \pm 0.04$	$0.91^{a} \pm 0.04$	$0.92^{a} \pm 0.07$	$0.92^{a} \pm 0.07$
Final weight (g)	$4.61^{b} \pm 0.07$	$5.79^{a,b} \pm 0.05$	$5.87^{a} \pm 0.14$	$6.31^{a,b} \pm 0.07$
FCR	$3.82^{a} \pm 0.03$	$2.70^{a,b} \pm 0.02$	$2.63^{b,c} \pm 0.08$	$2.10^{c} \pm 0.04$
Survival rate (%)	$83^{a} \pm 0.08$	$84^{a} \pm 0.10$	$83^{a} \pm 0.05$	$85^{a} \pm 0.07$
(S.G.R)	$1.17^{b} \pm 0.04$	$1.33^{a,b} \pm 0.06$	$1.34^{a,b} \pm 0.03$	$1.39^{b} \pm 0.06$

Values are ±SD of three replications. Figures in the same row having different superscript are significantly different ($P < 0.05$)

Table 4 Water quality parameters for different treatments

Parameters	Treatments			
	T-1	T-2	T-3	T-4
Average Temp. (°C)	$29.8^{a} \pm 0.03$	$29.7^{a} \pm 0.22$	$29.8^{a} \pm 0.43$	$29.8^{a} \pm 0.12$
Average pH	$7.9^{b} \pm 0.04$	$7.6^{a,b} \pm 0.03$	$7.7^{a} \pm 0.05$	$7.8^{a,b} \pm 0.23$
Average D.O (mg/l)	$6.1^{a} \pm 0.06$	$6.3^{a,b} \pm 0.04$	$6.0^{b,c} \pm 0.04$	$6.1^{c} \pm 0.04$

Water quality parameters were determined by APHA (1995) procedures. Values are ±SD of three replications. Figures in the same row having different superscript are significantly different ($P < 0.05$)

reported that FCR of climbing perch fed on trash fish and pellet was 2.07 and 1.89, respectively.

In the present study, the water quality parameters were found to be within the acceptable level for fish culture. Recommendation for water quality usually specifies DO greater than 3 mg/l for growth of channel catfish (Weeks and Ogburn 1973). The pH values obtained in this study (7.6–7.9) fell within the suitable range according to Swingle (1967) and Boyd (2000).

My current research focused on the evaluation of poultry viscera as a prospective replacement for fish meal in the culture of Koi, *A. testudineus*. It is hoped that the results will reveal more clues that will justify poultry viscera as a potential replacement for fish meal. The experimental results suggested that T-4 (treatment where fishmeal 100 % replaced by poultry viscera) can be recommended for the intensive culture of *A. testudineus*.

Conclusion

Poultry viscera can 100 % replace fish meal potentially, in the compound diet for Koi, *A. testudineus*, only by proper formulation and sun-dry processing. Applying these feed fish farming becomes more profitable to the poor fish farmers by lowering the feed costs to a certain degree.

Author contribution All authors, have made reasonable effort on all parts of the work necessary for the development of this manuscript in accordance with their expertise. All authors read and approved the final manuscript.

Acknowledgments The authors are thankful to the Head, Department of ASEPAN, Palli Siksha Bhavana, Visva-Bharati, Sriniketan and Department of Zoology, Siksha Bhavana, Visva-Bharati Santiniketan, for providing necessary facilities for this research.

Conflict of interest The authors declare that there is no conflict of interest.

References

Adewoye SO, Ayoola PB (2010) Nutrient composition of some freshwater fishes from Ikosi dam, Ogbomoso, Nigeria. Electr J Environ Agric Food Chem 10(9):1579–4377

AOAC (Association of Official Analytical Chemicals), (1995) Official Method of Analysis, 12th edn. Association of Official Analytical Chemists, Washington 832

APHA (1992) Standard methods for the examination of water and waste-water. American Public Health Association, 1015 Eighteenth street, Washington pp 1134

Belal IEH, Al-Owaifeir A, Al-Dosari M (1995) Replacing fish meal with chicken offal silage in commercial *Oreochromis niloticus* (L.) feed. Aquac Res 26:855–858

Bhaskar Pratyush, Pyne Saroj Kumar, Ray Arun Kumar (2014) Evaluation of poultry viscera as potential fish feed ingredient, compared to fishmeal. Int J Curr Res 6(02):5241–5243

Boyd CE (2000) Water quality: an introduction. Kluwer, Boston

Cai T, Pancorbo OC, Barnhart HM, Sander JE, Merka WC (1994) Chemical and microbiological characteristics of poultry processing by-products, waste, and poultry carcasses during lactic acid fermentation. J Appl Poult Res 3:49–60

Cappell R, Wright S, Nimmo F (2007) Sustainable production and consumption of fish and shellfish. Environmental impact analysis. Final report of Defra LCA project. Haskoning UK Ltd, UK

Castell JD, Tiews K (1980) Report of the EIFAC, IUNS and ICES working group on the standardization of methodology in fish nutrition research. EIFAC technical paper 36. Hamburg, Federal Republic of Germany, 21–23 March, 1979, 24 pp

Chareontesprasit N, Jiwyam W (1996) Dietary protein requirement on growth for climbing perch (*A. testudineus*). Khon Kaen J Agri 24(3):116–120

Cho CY, Slinger SJ, Bayley HS (1982) Bioenergetics of Salmonid fishes: energy intake, expenditure and productivity. Comp Biochem Physiol B 73:25–41

Davies SJ, Williamson J, Robinson M Robert IB (1989) Practical inclusion levels of common animal by-products in complete diets for tilapia (*Oreochromis mossambicus*, Peters). Proc Third Int Symp on Feed Nutr Fish, Toba, Japan pp 326–332

Davis DA, Arnold CR (2000) Replacement of fishmeal in practical diets for the Pacific white shrimp, *Litopenaeus vannamei*. Aquaculture 185:291–298

Dong FM, Hardy RW, Haard NF, Borrows FI, Rasco BA, Fairgrieve WT, Forster IP (1993) Chemical composition and protein digestibility of poultry by-product meals for salmonid diets. Aquaculture 116:149–158

Doolgindachabaporn S (1994) Development of optimal rearing and culturing system for Climbing perch, *Anabas testudineus* (Bloch). Doctoral Thesis, University of Manitoba, Canada

Duncan DB (1955) Multiple range and multiple *F* tests. Biometrics 11:1–42

El-Sayed A-FM (1998) Total replacement of fish meal with animal protein sources in Nile tilapia, *Oreochromis niloticus* (L.) feeds. Aquac Res 29:275–280

Fowler LG (1991) Poultry by product meal as a dietary protein source in fall chinook salmon diets. Aquaculture 99:309–321

FRI (1989) Survey of potential fish feed ingredients of Bangladesh on the basis of their availability and biochemical composition. Research Project Report No. 1: Fisheries Research Institute, Mymensingh, Bangladesh pp. 70

Hardy RW (1996) Alternate protein sources for salmon and trout diets. Animal Feed Sci Chronol 59:71–80

Hardy RW, Shearer KD, Spinelli J (1984) The nutritional properties of co-dried fish silage in rainbow trout diets. Aquaculture 38:35–44

Hasan MR, Amin MR (1997) Effect of processing techniques on the nutritional quality of poultry offal meal. Bangladesh J Fish 20:139–144

Hastings WH (1976) Fish nutrition and fish feed manufacture. In: FAO technical conference on aquaculture, p. 13, Kyoto, Japan, FIR : Aq. Cof. 76. R 23

Henken AM, Lucas H, Tijseen PAT, Michiels MAM (1986) A comparison between methods used to determine the energy content of feed, fish and faecal samples. Aquaculture 58:195–201

Jauncey K (1982) The effect of varying dietary protein level on the growth, food conversion, protein utilization and body composition of juvenile tilapia (*Tilapia mossambicus*). Aquaculture 27:43–54. doi:10.1016/0044-8486(82)90108-9

Jayaram KC (1981) The freshwater fisheries of India, Pakistan, Bangladesh, Burma and Sri Lanka—a handbook. Zoological Survey of India, Calcutta

Kaushik SJ (1995) Nutrient requirements, supply and utilization in the context of carp culture. Aquaculture 129:225–241

Millamena OM (2002) Replacement of fish meal by animal by-product meals in a practical diet for grow-out culture of grouper, *Epinephelus coioides*. Aquaculture 204:75–84

Nengas I, Alexis MN, Davies SJ (1999) High inclusion levels of poultry meals and related by products in diets for gilthead sea bream, *Sparusaurata* L. Aquaculture 179:13–23

Patra BC (1993) Satiation time, appetite and daily pattern of feed intake and faces release by an air-breathing fish (*A. testudineus* Bloch). J Aquac Tropic 8(1):41–46

Pearson D (1976) The chemical analysis of food. Churchil Livingstone, Edinburgh, London and New York, pp 375

Pearson D (1999) Pearsons composition and analysis of foods. University of Reading, Reading

Phumee P, Hashim R, Paiko MA, Shu-Chien AC (2009) Effects of dietary protein and lipid content on growth performance and biological indices of iridescent Shark (Pangasius hypophthalmus, Sauvage 1878) fry. Aquac Res 40:456–463

Potongkam K (1972) Experiment on feeding climbing perch, *Anabas testudenius* (Bloch) with ground trash fish and pellets. Department of Fisheries Annual Report, Bangkok, Thailand

Saha KC (1971) Fisheries of West Bengal. West Bengal Government Press, Alipore

Steffens W (1994) Replacing fish meal with poultry by product meal in diets for rainbow trout, *Oncorhynchus mykiss*. Aquaculture 124:27–34

Swingle HS (1967) Standardization of chemical analysis for waters and pond muds. FAO Fish Res 4(4):397–421

Tabinda AB, Butt A (2012) Replacement of fish meal with poultry by-product meal (chicken intestine) as a protein source in grass carp fry diet. Pakistan J Zool 44(5):1373–1381

Tihn LV (2003) Culture of climbing perch in ponds with different crude protein diets, Vietnam. MS Thesis, Can Tho University

Webster CD, Thompson KR, Morgan AM, Grisby EJ, Gannam AL (2000) Use of hempseed meal, poultry by-product meal and canola meal in practical diets without fish meal for sunshine bass (*Moronechrysop* × *M. saxatilis*). Aquaculture 188:299–309

Wee KL, Ngamsnae P (1987) Dietary protein requirement of fingerlings of the herbivorous *Puntius gonionotus* (Bleeker). Aquac Fish Manag 18(2):121–129

Weeks JP, Ogburn CB (1973) Catfish production guide. Circ. E-18, Alabama Cooperative Extension Service, Auburn University, Auburn, USA

15

Recycling of sewage sludge as production medium for cellulase by a *Bacillus megaterium* strain

Adel Ali Saeed Al-Gheethi[1]

Abstract

Background Cellulase is one of the enzymes commonly used in several agricultural, industrial and sewage sludge treatment processes. The present study aimed to investigate the potential use sludge generated from sewage treatment plants as a production medium for cellulase by *B. megaterium* strain that was isolated from a sewage treatment plant. The production of cellulase in the sludge medium was compared to different cellulosic materials: cotton, filter paper, bagasse and sawdust as well as to galactose, fructose, lactose, maltose, mannitol, mannose, ribose, sucrose and xylose. The production of cellulase was conducted at optimum conditions (0.4 mL of the bacterial inoculum, 45 °C, 72 h, pH 6.5 and citrate phosphate buffer) that were determined in this study.

Results The sludge medium has induced the cellulase production by *B. megaterium* strain compared to cotton, filter paper, bagasse and sawdust. However, *B. megaterium* produced high cellulase in the presence of carbohydrate compounds as carbon source. More cellulase was produced in the sludge medium containing low concentrations of Ni^{2+}, Zn^{2+} and Cu^{2+} ions.

Discussion The ability of *B. megaterium* strain to produce cellulase in the sewage sludge medium was due to that the strain has acclimatized to resist heavy metals and produce the enzyme genetically. Moreover, *B. megaterium* has an important environmental role for reuse of sewage sludge as production medium for cellulase that could be used in many of applications, including production of animal feed, formulation of detergents, juice clarification, paper industry and wine production.

Keywords Cellulase · *Bacillus megaterium* · Sludge · Heavy metals · Recycle

Introduction

Cellulose is the most abundant renewable natural biological resource produced in the biosphere (about 100 billion dry tons per years) (Zhang et al. 2006). Municipal solid wastes contain 40–50 % cellulose, 12 % hemi-cellulose and 10–15 % lignin by dry weight (Wang et al. 1994). Several anaerobic bacteria have the ability to degrade cellulose in anaerobic digestion (Chynoweth and Pratap 1996). The degradation of cellulose by cellulase(s) enzymes produced by numerous microorganisms is very important in several agricultural and waste treatment processes (Hamer 2003; Angenent et al. 2004; Schloss et al. 2005). The aerobic microorganisms usually secrete copious amounts of free cellulase, which acts synergistically to degrade cellulose (Blouzard et al. 2007; Mingardon et al. 2007).

The widely accepted mechanism for enzymatic cellulose hydrolysis involves synergistic actions by endo-glucanase or CMCase (EC 3.2.1.4), exoglucanase or cellobiohyrolase (EC 3.1.1.91) and β-glucosidase (EC 3.2.1.21) (Lynd et al. 2002; Zhang and Lynd 2004; Sadhu et al. 2013). Endoglucanase hydrolyzes accessible intra-molecular β-1-4 glucosidic bonds of cellulose chain ends. Exoglucanases processively cleave cellulose soluble cellobiose or glucose and β-glucosidase hydrolyzes cellobiose to glucose (Krishna 1999; Zhang et al. 2006).

✉ Adel Ali Saeed Al-Gheethi
adelalghithi@gmail.com; aasa1982@yahoo.com

1 Faculty of Medical Science, Yemeni Jordanian University, Sana'a, Yemen

Applications of cellulase include production of animal feed, formulation of detergents, juice clarification, paper industry and wine production. Cellulase contributes to 8 % of the worldwide industrial enzyme demands and the demand is expected to increase by 100 % in future (Costa et al. 2008). However, production of cellulase(s) by different microorganisms was found to be affected by many factors, e.g. inocula size, incubation temperature, incubation period, pH value, buffers, carbon and nitrogen sources (Mawadza and Zvauya 1996; Camassola et al. 2004; Silva et al. 2005; Immanuel et al. 2006).

The type and concentration of carbohydrate are critical for maximal cellulase production in the production medium. Krishna (1999) has suggested that the glucose has enhanced cellulase synthesis to a significant level. However, addition of cellulose, lactose or glucose at concentration above 1 % level led to a significant reduction in enzyme synthesis. Alam et al. (2004) have studied the production of extracellular cellulase by *S. omiyaensis* under different carbon source availability. Four carbon sources have been investigated: carboxyl methyl cellulose (CMC), avicel, rice bran and sawdust at the rate of 1.2 %. They have revealed that the highest CMCase enzyme production was recorded when CMC was used as a carbon source and the lowest CMCase production when sawdust and rice bran were used as carbon sources.

Emtiazi et al. (2007) have revealed that *Paenibacillus* strain produced high CMCase when CMC was used as only carbon sources. Karim et al. (2014) have revealed that B. *licheniformis* produced a significant amount of cellulase when wheat bran and orange peel were used as a sole carbon source. However, the production of CMCase in the sludge medium has not been studied before. Therefore, the current work aimed to investigate the recycling of sludge as production medium of CMCase by *Bacillus megaterium* strain in comparison to cellulosic materials such as cotton, filter paper, bagasse and sawdust as well as to different carbon sources (galactose, fructose, lactose, maltose, mannitol, mannose, ribose, sucrose and xylose). The effect of different concentrations of nickel ions was also tested to investigate the potential of bacterial strain to produce the enzyme in different types of sludge contaminated with heavy metals.

Materials and methods

Collection of sewage sludge samples

Twelve sewage sludge samples were collected weekly from four STPs (referred to as ISTP, TSTP, ASTP and SSTP) in Yemen. TSTP was located in Taiz and treat Industrial waste generated from Industrial and Commercial Company. ISTP, ASTP and SSTP were located in Ibb, Aden and Sana'a, respectively. The sewage flows to these plants comprise residential, commercial and industrial. TSTP and ASTP are oxidation ponds. The treatment of sewage in the ISTP and SSTP is based on primary and secondary processes. The sludge samples were collected from each STP in sterile paper bags. The samples were transported to the laboratory in a cooler box and the microbiological analysis was carried out within 2 h of collection.

Determination of heavy metals (Cu^{2+}, Ni^{2+} and Zn^{2+}) concentrations

Heavy metals were extracted from dewatered sludge samples by nitric acid digestion method (APHA 1998). The heavy metal concentrations (Zn^{2+}, Cu^{2+} and Ni^{2+}) in the digested samples were analysed and determined by atomic absorption. An atomic absorption spectrophotometer (AAS) was used for this purpose.

Screening of bacterial isolates for resistance to nickel ions and production of cellulase

Hundred and twenty-seven (127) bacterial isolates were isolated from sludge samples collected from four sewage treatment plants in Yemen. These bacterial isolates were purified according to APHA, 9225B (1999). The screening for the bacterial isolates resistant to nickel ions was conducted according to Hernández et al. (1998) and Abdel-Monem et al. (2010). The bacterial isolates were sub-cultured in BHI agar medium containing 15 and 10 mM Ni^{2+} for 24–48 h at 35 °C. The plates were exposed to sulphhydryl gas (resulted from the reaction of 2 g of sodium sulphide with 10 mL of concentrated HCl in a sealed container, the reagent was prepared before each experiment) in a sealed container for the formation of the metal sulphide. Plates were carefully screened to detect any change in the region surrounding or inside the bacterial colonies.

The bacterial isolates were screened for the growth on CMC–Yeast Extract (CYE) agar medium containing (g L^{-1}): $NaNO_3$, 2; KCl, 0.5; $MgSO_4$, 0.5; K_2HPO_4 (buffer), 1.0, yeast extract (growth factor), 1 and CMC as carbon source, 10; pH 7.0 ± 0.2. After incubation at 37 °C for 48 h, the isolates that exhibited good growth were recorded. To detect production of cellulase, the surface of the plates was floated with iodine $ZnCl_2$ solution (3 % $ZnCl_2$ was added to gram's iodine solution). Diameters of blue zones were measured. The bacterial isolates that exhibited positive results for the production of cellulase were grown on CMC–Yeast Extract (CYE) broth medium for 5 days at 37 °C. Cellulase production was determined by CMC cup-plate clearing zone (CCZ) assay.

In this assay, 2 % (w/v) CMC was dissolved in citrate buffers (pH 7) and supplemented with 1.5 % agar for solidification. After sterilization, equal amounts of assay medium (20 mL) were poured in sterilized petri dishes (12 cm in diameter). Cups (10 mm in diameter) were made in each plate using a sterile cork borer. Equal amounts of the enzyme solution (cultural supernatant) were put into each cup. Plates with cups containing enzyme solutions were incubated at 37 °C for 24 h, then the surface of the plates was floated with iodine $ZnCl_2$ solution. Diameters of blue zones were measured. The mean values of three readings were calculated.

Identification of the most potent bacterial isolates

The bacterial isolates that exhibited the ability in the accumulation of Ni^{2+} ions were identified based on morphological, culture and biochemical tests according to Noel (1984), Sneath et al. (1986), Garrity et al. (2002) and Brenner et al. (2004).

Cellulase production under catabolite repression

The purpose of this experiment was to investigate the ability of bacterial strains to produce cellulase enzyme as inducible or genetically. The medium used was as mentioned previously with glucose (1 % w/v) equivalent to CMC as the only carbon source. After 48 h incubation at 37 °C, the cellulase enzyme assay was done by C.C.Z assay as mentioned previously.

Factors affecting CMCase production by bacterial isolate No. 1295S

The bacterial isolate No. 1295S, which secretes the highest yield of cellulase under catabolic repression, was selected for studying different factors affecting CMCase production: inocula size, temperatures, incubation periods, pH values, buffers and nitrogen sources.

Inoculum preparation

The Bacterial isolate No. 1295S was maintained as stock culture on Brain Heart Infusion agar (BHIA). The bacterial strain was grown at 37 °C for 24 h and then stored at 4 °C for regular sub-culturing. The bacterium inoculum was prepared using BHI broth in 250 mL conical flask. The inoculum was kept in shaker (200 rpm) at 37 °C for 14 h before it was used for the CMCase production.

Production medium

Carboxyl methyl cellulose yeast extract (CYE) liquid medium was used for studying factors affecting the CMCase production. The constituents of 50 mL of this medium were dispensed in flasks of 250 mL capacity. The pH was adjusted at 6.5 and autoclaved at 121 °C for 15 min.

Biomass yield

The bacterial growth was determined by dry weight method. After collection of supernatant, the biomass residue was dried at 80 °C for 24 h and the yield was expressed as mg g^{-1} of substrate.

β-1,4 Endoglucanase (CMCase activity) determination

β-1,4 endoglucanase (CMCase activity) (EC 3.2.1.4) was determined by measuring the amount of reducing sugars released in the reaction mixtures containing 1.7 mL of 0.1 M acetate buffer (pH 5.5), 0.8 mL of 2 % (w/v) CMC (sigma) solution and 0.5 mL of the culture supernatant. The mixture was maintained at 50 °C for 50 min. The reducing sugars in the supernatant were assayed by 3,5-dinitrosalicylic acid (DNSA) methods (Miller 1959). Glucose was used as standard. Measurements were made in a spectrophotometer (Win. Aspect T 20, 031-2004, Germany) at 540 nm wavelength in the presence of the blank. The reaction mixtures containing heat-inactivated post-culture liquids (boiled for 5 min) were used as blanks. The cellulolytic enzyme activities were expressed in units defined as the quantity of enzyme required to produce 1 μmol/h of glucose, under the conditions of the assay.

Effect of different inocula sizes

The effect of different inocula sizes on the production of cellulase enzyme was investigated at 0.1, 0.2, 0.3, 0.4, 0.6, 0.8, 1, 1.5, 2 and 2.5 mL of bacterial inoculum ($\sim 2.7 \times 10^9$ CFU mL^{-1}). After the inoculation, the production medium was incubated at 37 °C and pH 6.5 for 4 days. Detection of CMCase production was performed by the determination of reducing sugar using colorimetric technique as previously mentioned.

Effect of temperature

For this purpose, the production medium was dispensed into flasks of 250 mL capacity; each flask containing 50 mL of liquid medium was adjusted at pH 6.5. The flasks were autoclaved at 121 °C for 15 min. The sterile medium was inoculated with 0.4 mL ($\sim 2.7 \times 10^9$ CFU mL^{-1}) of a standardized bacterial inoculum, and incubated at the following temperatures 20, 30, 37, 45 and 60 °C. At the end of the incubation period of 4 days, the CMCase produced was performed as described previously.

Effect of different incubation periods

The production medium was prepared as mentioned above where bacterial isolate no. 1295S was incubated for 1, 2, 3, 4, 5, 6 and 7 days at 45 °C and pH 6.5. At the end of each incubation period, the production of CMCase was determined using the colorimetric technique.

Effect of pH values

The pH values of the production medium were adjusted at 4, 4.5, 5, 5.5, 6, 6.5, 7, 7.5 and 8 by the careful addition of drop of (0.1 N) HCl and (0.2 N) NaOH using pH meter (IA 31-1114 WTW Germany). The media were dispensed in flasks of 250 mL capacity each containing 50 mL and then autoclaved at 121 °C for 15 min, then inoculated with 0.4 mL of a standardized bacterial inoculation. The inoculated media were incubated at 45 °C for 72 h.

Effect of different buffers applied at various pH ranges

CYE liquid medium was used for detecting the suitable buffer for CMCase production. The following buffers with their pH ranges were used for such a purpose, citrate buffer with pH range from 5.6 to 6.2, citrate phosphate buffer with pH range from 6 to 7 and phosphate buffer with pH range from 6 to 8. All these buffers with their different pH were prepared according to Collee et al. (1989). The constituents of the production medium were dissolved in the buffer solution and poured into the flask contained 50 mL of the buffered medium at the particular pH. The media were inoculated with 0.4 mL of a standardized bacterial inoculation. Then, the medium was incubated at 45 °C for 72 h. Extraction of the crude enzyme was carried out as mentioned above at the end of the incubation period and determined using colorimetric technique.

Effect of different cellulosic materials supplied at different concentration

This experiment was designed to test the effect of different cellulosic materials on the CMCase production. DCY liquid production medium was used without CMC. Substrates were singly supplied to production medium. These substrates were CMC, cellulose powder, cotton, filter paper, sawdust, bagasse and tobacco leaves. All substrates were supplied at concentrations 2, 4, 6, 8 and 10 mg mL^{-1} (w/v), respectively. pH value was adjusted to 6.5. The production medium was inoculated with bacterial inoculum and then incubated at 45 °C for 72 h.

Effect of different carbon sources applied at different concentrations

For this purpose, galactose, fructose, lactose, maltose, mannitol, mannose, ribose, sucrose and xylose were used at different concentrations (2, 4, 6, 8 and 10 mg mL^{-1}) (w/v). They were supplied singly to the sterilized carbon-free production medium. The carbon sources used were sterilized by membrane filter. The production medium with CMC was used as control; other steps were carried out as mentioned previously.

Effect of different sludge media applied at different concentrations of sludge

The ability of *B. megaterium* strain to produce enzyme in the presence of sludge as only carbon and nitrogen source at different concentrations (2, 4, 6, 8 and 10 mg mL^{-1}) (w/v) was studied. The production medium was inoculated with 0.4 mL of a standardized bacterial inoculum. At the end of the incubation period 72 h, the CMCase produced was detected using colorimetric technique.

Effect of Ni^{2+} ion concentrations

To study the effect of Ni^{2+} ion concentrations on CMCase production, the (CYE) liquid medium was prepared with concentrations 117.2, 234.4, 351.6, 468.8, 586, 703.2 and 820.4 μg Ni^{2+} mL^{-1}. The medium was inoculated with bacterial strain and then incubated at 45 °C for 72 h. At the end of the incubation period, CMCase concentration was determined as described previously.

Data analysis

All analyses were conducted in triplicate and values were reported as means with standard deviations. Data were subjected to one-way analysis of variance (ANOVA) in the general linear model using the SPSS 11.5 statistical package. The statistical package (EASE, M-STAT) was used to perform the analyses of least significance difference (LSD). ANOVA was used to determine the significance ($p < 0.05$) of the differences between results.

Results and discussion

Selection of the most potent bacterial strains

One-hundred and twenty-seven bacterial isolates were obtained from sewage sludge samples collected from ISTP, TSTP, ASTP and SSTP at republic of Yemen. These isolates were purified and screened for nickel resistance at concentrations of 15 and 10 mM of nickel ions, where the highest value of nickel ions was 15 mM in sludge sample at AWTP (Table 1).

Among the 127 bacterial isolates, three bacterial isolates (586S, 1295S and 222W) exhibited good growth in the presence of nickel concentrations of 15 mM (876 μg Ni^{2+} mL^{-1}) and two bacterial isolates (117S and 120S) showed high growth at 10 mM (584 μg Ni^{2+} mL^{-1}).

Bacterial isolates forming a dark colour around or inside the colony (possible reduction and precipitation of the metal) were considered to be possible nickel bio-sorbents. Five bacterial isolates producing dark colour colonies grown in the presence of nickel were selected for further studies. Neither clear halos nor dark colours were observed around the colony when the clinical strains (*E. coli*, control) were grown in the presence of nickel ion concentrations. The bacterial isolate Nos. 586S, 1295S and 222W were characterized by a higher nickel tolerance than the bacterial isolate Nos. 117S and 120S.

Survey for nickel resistance among the obtained bacterial isolates was chosen because nickel is among the most toxic heavy metals and could be found in different industries (Kaewchai and Prasertsan 2002). Ni^{2+} is one of the most frequently encountered heavy metals in sewage streams (Padmavathy 2008). Ni^{2+} is a trace element and plays a role as cofactors for some of the bacterial enzymes (Nies 1999; Dosanjh and Michel 2006).

Only few bacterial strains were described to grow in the presence of higher concentration than 10 mM of Ni^{2+}. Hernández et al. (1998) found that from 52 bacterial strains isolated from contaminated soils of an oil refinery, only two strains of bacteria *Escherichia hermannii* and *Enterobacter cloacae* were capable of accumulating either nickel, vanadium or both metals at 10 mM. Leung et al. (2000) isolated nineteen metal resistant and non-resistant bacteria of Ni^{2+}, Pb^{2+}, Cu^{2+} and Zn^{2+} from the activated sludge. In the present work, among the 127 bacterial isolates obtained from sludge, five bacterial isolates exhibited resistance to 10 mM of Ni^{2+} ions. Al-Gheethi et al. (2014) reported that *Bacillus* sp., *Pseudomonas* sp., *Chryseomonas* sp., and *Burkholderia* sp. isolated from sewage effluent have the ability to tolerate 6 mM of Ni^{2+} ions.

Table 1 Heavy metal contents (mg kg^{-1} dry wt.) in sewage sludge at four sewage treatment plants (STPs) in Yemen ($N = 3$)

STPs	Heavy metals concentrations (mg kg^{-1})		
	Cu^{2+}	Ni^{2+}	Zn^{2+}
ISTP	575.6	369.3	3920.0
TSTP	733.3	156.0	9213.3
ASTP	366.3	391.0	2425.6
SSTP	10.6	26.0	77.2

ISTP Ibb sewage treatment plant, *TSTP* Taiz sewage treatment plant, *ASTP* Aden sewage treatment plant, *SSTP* Sana'a sewage treatment plant, *S* sludge

The results of screening for cellulase production showed that 69 (almost 54.33 %) from 127 isolates exhibited high growth, 59 from 69 bacterial isolates produced clearing zones at varying degree and considered as positive for cellulase(s) production. It has been reported that the clear zone methodology was developed to isolate polymer-degrading strains from mixtures of microorganisms typically found in environments such as compost or soil. Microorganisms capable of degrading a polymeric material are determined by looking for zones of clearing around microbial colonies on agar media that are opaque due to the presence of powdered polymer (Pettigrew and Johanson 1996; Ten et al. 2004). This screening process was performed in this work to choose the bacterial isolates, which have the potential to grow in the CYM medium containing CMC as carbon source.

The second screening was carried out by growing the bacterial isolates, which have positive results in the previous screening in the liquid medium for confirmation of the cellulase production using C.C.Z method. In this screening, the diameters of the clearing zones surrounding the wells on the plate screening medium ranged from 11 ± 0.4 to 77.5 ± 7.4 mm. This screening step was found to give reliable indication of exhibited cellulolytic activities. However, the enhancement of cellulase production was not linked with the amount of bacterial growth. The size of clearing zone diameter of each isolate is depicted in Fig. 1. Results showed that among the 59 bacterial isolates which exhibited clear zone around the colony in the last screening, 42 isolates (71.2 %) exhibited potential to produce cellulase enzyme as determined by C.C.Z technique. Among the forty-two cellulolytic bacterial isolates, five bacterial isolates (586S, 1295S, 117S, 120S and 222W) showed nickel tolerance and selected for further studies as will be described below.

Identification of the most potent bacterial isolates

The bacterial isolates Nos. 586S, 1295S, 222W, 117S and 120S which showed high nickel tolerance were identified as *Sporosarcina pasteurii* 586S, *Bacillus megaterium* 1295S, *Staphylococcus xylosus* 222W, *Bacillus subtilis* 117S and *Pseudomonas cepacia* 120S (Table 2).

Production of cellulase under catabolic repression

The considerations in the enzymatic treatment of sludge are the presence of substrates such as glucose, which may inhibit the production of enzymes by the added bacterial strains. The ability of bacterial strains to produce

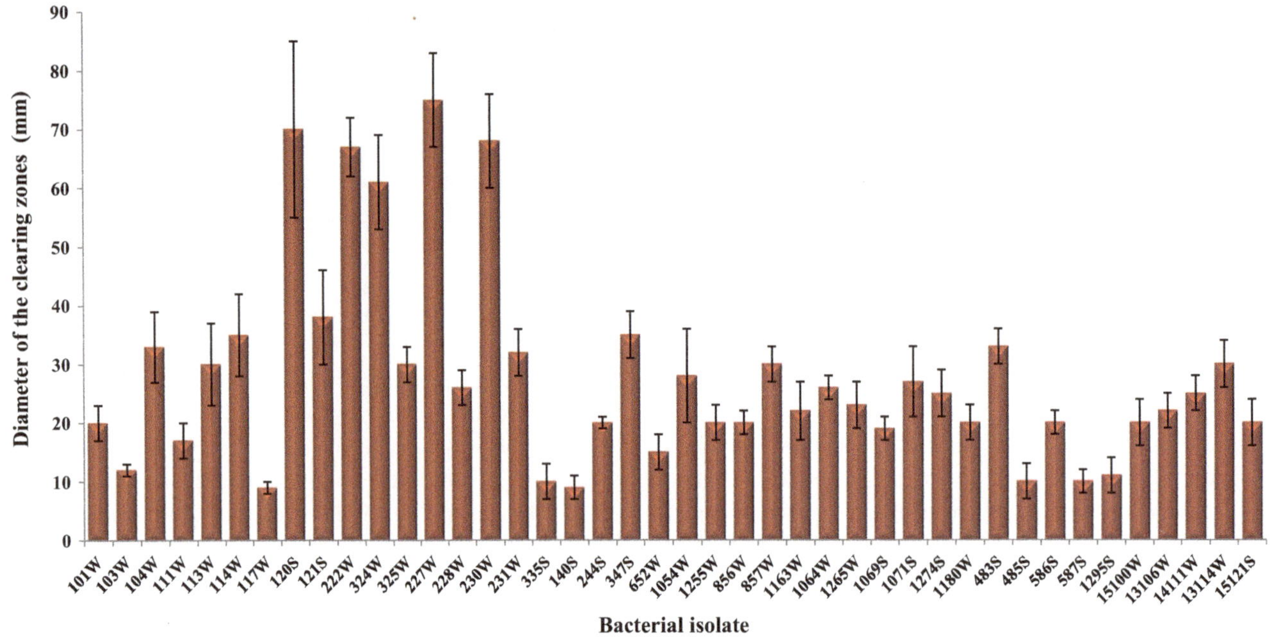

Fig. 1 Cellulase(s) production by 42 bacterial isolates obtained from four sewage treatment plants in Yemen in the presence of 1 % CMC as a sole carbon source. *Each point* is the average of three determinations

detectable amounts of CMCase under catabolic repression (genetically) was screened in the presence of glucose (1 % w/v) equivalent to CMC as sole carbon source. As shown in Fig. 2, it could be observed that *B. megaterium* and *B. cepacia* could still synthesize cellulase with varying degrees. These results were in agreement with Allcock and Woods (1981) who stated that the CMCase in *Clostridium acetobutylicum* induced by molasses and it was not repressed by glucose. *S. pasteurii*, *B. subtilis* and *S. xylosus* appeared to lose the ability to produce the enzyme in the presence of glucose. Bakare et al. (2005) demonstrated that cellulase activity by *P. fluorescens* increased when cellulose material was added to the culture medium than when glucose was used as sole carbon source. In this work, *B. megaterium* 1295S produced CMCase in CYE2 medium (CMC–yeast extract agar medium containing glucose as carbon source) more than that in CYE1 (CMC–yeast extract agar medium containing CMC as carbon source) medium whereas *B. cepacia* 120S produced highest amounts of enzyme on CYE1 medium. *B. megaterium* 1295S has produced cellulase on CYE2 twofold than on CYE1 and indicating the ability to produce cellulase genetically. Conversely, *B. cepacia* 120S and *S. xylosus* 222W have described as inducible enzyme production.

Al-Gheethi and Norli (2014) studied the production of β-lactamase by bacteria isolated from sewage effluents in the presence or absence of cephalexin antibiotic as inducible substrate. They found that *B. stearothermophilus* and *Burkholderia cepacia* could still synthesize β-lactamase at varying conditions. *Chryseomonas luteola* has lost the ability to produce the enzyme in the absence of cephalexin (as inducibly). *B. subtilis* has produced β-lactamase in antibiotic-free and induced medium and considered as producing β-lactamase genetically. In the current study, the bacterial species, which produced highest cellulase in the presence of glucose, were regarded as catabolite repression resistant and have constitutively CMCase production.

Factors affecting production of CMCase by *B. megaterium* strain

B. megaterium strain was selected for studying the factors affecting CMCase production. The selection of *B. megaterium* strain was based on that the strain has produced cellulase under catabolite repression and has the ability to grow at 15 mM Ni^{2+} ions. The factors investigated were inocula size, incubation temperatures, incubation periods, pH values, different buffers applied at various pH ranges. In these experiments, the amount of CMCase was determined by measuring reducing sugars using the 3,5-dinitrosalicylic acid (DNSA) methods as described in "Materials and methods" (Miller 1959). The DNSA methods were used because the determination of differences in enzyme activity levels less than twofold is difficult using C.C.Z techniques (Zhang et al. 2006). The DNSA method is one of the most common assays for measuring reducing sugars for cellulase activity assays because of their relatively high sugar detection range (Coward-Kelly et al. 2003; Kongruang, et al. 2004; Zhang and Lynd 2005).

Table 2 Morphological, physiological and biochemical tests of five bacterial strains isolated from four sewage treatment plants at Yemen

Test	Bacterial isolates				
	Sporosarcina pasteurii 586S	*Bacillus megaterium* 1295S	*B. subtilis* 117S	*Pseudomonas cepacia* 120S	*Staphylococcus xylosus* 222 W
Gram stain	+(ve)	+(ve)	+(ve)	−(ve)	+(ve)
Cell shape	Bacilli	Bacilli	Bacilli	Rod	Cocci
Spore formation	+	+	+	−	−
Motility	+	+	+	+	−
Anaerobic growth	+	+	+	+	+
Growth at					
30 °C	+	+	+	+	+
37 °C	+	+	+	+	+
45 °C	+	+	+	+	+
60 °C	+	+	−	−	−
Catalase production	+	+	+	−	+
Indole production	−	−	+	−	−
Citrate utilization	−	+	+	+	+
MR	+	−	−	−	−
VP	−	+	+	−	−
TSIA	R/Y	R/Y	R/Y	Y/Y	Y/Y
Urease production	+	+	−	−	+
Hydrolysis of					
Gelatin	+	+	+	+	+
Starch	+	+	+	+	−
Casein	+	+	+	+	−
Acid from					
Glucose	+	+	+	+	+
Lactose	−	+	+	−	−
Mannose	−	+	+	+	−
B-galactose	−	+	+	−	+
Sucrose	−	−	−	−	−
Fructose	−	+	+	+	−
Xylose	−	−	+	+	+
Maltose	−	+	+	−	+
Ribose	−	+	+	+	+
Mannitol	−	+	+	+	−
Gas from glucose	−	−	−	−	−
H_2S production	−	−	−	−	−
Esculin hydrolysis	−	−	−	−	−
L-alanine	−	−	−	−	−
L-cysteine	+	+	+	+	+
L-methionine	−	−	−	−	−
L-arginine	−	−	−	−	−

+ve Gram-positive stain, *−ve* Gram-negative stain, + positive, − negative, *R* red, *Y* yellow, *MR* methyl red, *VP* Voges-Proskaure, *TSIA* triple sugars iron agar

The maximum amount of CMCase production (16.8 U mL^{-1}) was noted with 0.4 mL of the bacterial suspension (Table 3). The optimum inocula size for heavy growth was 0.1 mL. Krishna (1999) studied the effect of inoculum size (5–40 %) on production of exoglucanase and endoglucanase by *B. subtilis* and revealed that 15 % (v/w) of inoculum size was optimal for the production of both enzymes. Immanuel et al. (2006) reported that one millilitre inoculum of *Bacillus* spp., *Cellulomonas* spp. and *Micrococcus* spp. was optimal to produce cellulase enzyme.

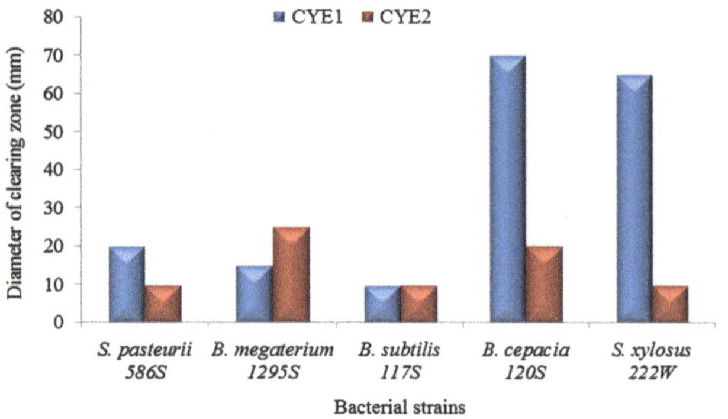

Fig. 2 Production of cellulase(s) enzymes under catabolite repression by five bacterial strains. *Each point* is the average of three determinations. *CYE1* CMC–Yeast Extract agar medium containing carboxyl methyl cellulose (CMC) as carbon source. *CYE2* CMC–Yeast Extract agar medium containing glucose as carbon source

The high production of CMCase (35.1 U mL^{-1}) and biomass yield (710 mg g^{-1}) by *B. megaterium* strain was recorded at temperature 45 °C (Table 4). Alam et al. (2004) observed similar results; the maximum production of CMCase by *Streptomyces omiyaensis* occurred at 45 °C, while the maximum growth was recorded at the range from 35 to 40 °C. The range between 35 and 60 °C was reported as optimal temperature of growth and production of cellulase by *Bacillus* sp. (Chan and Au 1987; Krishna and Varma 1990; Mawadza and Zvauya 1996; Ito 1997; Krishna 1999; Immanuel et al. 2006; Sadhu et al. 2013; Sethi et al. 2013).

The overproduction of CMCase (32.5 U mL^{-1}) was obtained after 72 h of incubation period (Table 5). The amount of growth increased simultaneously with time; the maximum yield of cell growth of *B. megaterium* was noted after 4 days. The enzyme production was unstable and decreased rapidly after 72 h. Reduction in CMCase yield was accompanied by an accumulation of reducing sugars in the medium and compatible with increasing amount of growth at 7 days. Krishna (1999) found similar trend in cellulase production using *B. subtilis* and indicated that the decrease in activity after 72 h might be due to denaturation of the enzyme, resulting from variations in pH during incubation. Maximal production in *Bacillus* strains has achieved after 2–3 days (Kawai et al. 1988; Ito et al. 1989). The production of CMCase by *B. megaterium* in a continuous-operating reactor may keep the bacterial growth at the highest stage of enzyme production. However, the current study focused on the batch scale of bacteria and continuous-operating reactor will be conducted in the possible future work.

The maximum CMCase production was noted at pH value of 6.5 (Table 6). Further increase in pH level to 8 resulted in considerable decrease in enzyme production. At pH 6, the *B. megaterium* showed heavy growth, but at pH 4 showed little growth and at pH 8 showed moderate growth, respectively. The optimum pH value of cellulolytic organisms varied from acidic condition such as *Trichoderma reesei* pH 2.8–pH 3.5 (Sternberg and Mandels 1979) to alkaline conditions such as *Bacillus* sp. pH 9–12 (Hakamada et al. 1997). Alam et al. (2004) revealed that *S. omiyaensis* showed heavy growth and high cellulase activity at pH 6.5. The optimal condition of cellulase production by *Bacillus* sp. was recorded at pH range from 5 to 7.5 (Dhillon et al. 1985; Araujo and Ward 1990; Mawadza and Zvauya 1996; Krishna 1999; Immanuel et al. 2006; Karim et al. 2014).

The effect of buffers on production of CMCase is illustrated in Table 7. Highest CMCase production was attained when the initial pH of the medium was adjusted to pH 6.2 and 6.4 with phosphate buffers, where 38.5 U mL^{-1} was produced. The production of CMCase in the medium with citrate phosphate buffers was 38.6 and 38.4 U mL^{-1} at pH 6.4 and 6.6. The minimum production of CMCase was noted with the citrate buffer 33.7 U mL^{-1} at pH 6.2 and non-buffered production medium 34.2 U mL^{-1} at pH

Table 3 Effect of inocula sizes on saccharification (%), biomass yield (mg g^{-1}) and the CMCase production (U mL^{-1}) by *B. megaterium*

Inoculum size (mL)*	CMCase concentrations (U mL^{-1} ± SD)	Saccharification (%)	Biomass yield (g g^{-1}) cellulose
0.1	10.9 ± 1.0	1.1	0.63
0.2	13.2 ± 1.8	2.3	0.28
0.3	12.7 ± 0.5	2.2	0.47
0.4	**16.9 ± 2.3**	2.4	0.53
0.6	16.2 ± 1.8	2.4	0.24
0.8	15.5 ± 2.4	2.4	0.36
1	14.3 ± 0.7	2.2	0.24
1.5	14.6 ± 0.7	2.2	0.31
2	14.1 ± 1.9	2.4	0.26
2.5	14.2 ± 0.1	2.3	0.43

Bold values indicate the maximum CMCase production by *B. megaterium* strain

* mL contains ~2.7 × 10^9 CFU mL^{-1} saline solution

Table 4 Effect of different incubation temperatures on saccharification (%), biomass yield (mg g^{-1}) and the CMCase production (U mL^{-1}) by *B. megaterium*

Temperature (°C)	CMCase concentrations (U mL^{-1} ± SD)	Saccharification (%)	Biomass yield (mg g^{-1}) cellulose
20	5.1 ± 1.9	0.8	640
30	22.8 ± 1.3	3.5	530
37	30.7 ± 2.5	4.6	520
45	**35.1 ± 3.1**	4.9	710
60	20.3 ± 0.5	3.1	530

Bold values indicate the maximum CMCase production by *B. megaterium* strain

Table 5 Effect of different incubation periods on saccharification (%), biomass yield (mg g^{-1}) and the CMCase production (U mL^{-1}) by *B. megaterium*

Incubation period (day)	CMCase concentration (U mL^{-1} ± SD)	Saccharification (%)	Biomass yield (mg g^{-1}) cellulose
1	22.5 ± 3.5	1.8	360
2	27.4 ± 0.8	4.1	390
3	**32.5 ± 0.0**	4.9	400
4	16.1 ± 1.5	2.5	530
5	12.2 ± 3.0	1.9	320
6	11.7 ± 3.3	1.8	300
7	7.9 ± 0.7	1.2	520

Bold values indicate the maximum CMCase production by *B. megaterium* strain

Table 6 Effect of pH value on saccharification (%), biomass yield (mg g^{-1}) and the CMCase production (U mL^{-1}) by *B. megaterium*

pH	CMCase concentration (U mL^{-1} ± SD)	Saccharification (%)	Biomass yield (mg g^{-1}) cellulose
4	18.6 ± 4.4	4.1	120
4.5	28.6 ± 1.6	4.2	150
5	29.7 ± 0.6	4.4	250
5.5	30.0 ± 2.0	4.5	180
6	32.0 ± 0.0	4.8	420
6.5	**34.2 ± 2.9**	5.2	240
7	26.0 ± 1.2	4.6	310
7.5	25.6 ± 0.7	4.5	350
8	24.7 ± 2.0	4.2	280

Bold values indicate the maximum CMCase production by *B. megaterium* strain

6.5. The optimum buffers of heavy growth were observed as citrate phosphate 1260 mg g^{-1} at pH 6.2 and phosphate buffers 1400 mg g^{-1} at pH 7.4 were used in culture medium, respectively. These results indicated that phosphate buffer is an important buffer for cellulase production by *B. megaterium*. Stutzenberger (1971) showed that maximum cellulase production by *Thermomonospora curvata* was attained when the initial pH of the medium was adjusted to pH 8.0 with phosphate buffer. However, Duff et al. (1985) stated that sodium citrate buffer improved the enzyme production in production medium by up to 40 %. Camassola et al. (2004) reported that production of cellulase enzyme by *Penicillium echinulatum* with citrate buffer was slightly higher than that in acetate buffer of the same pH.

Effect of different cellulosic material and carbon source on the production of CMCase

The effect of different cellulosic material was studied to determine the optimum concentration of each substrate supplied for the production of CMCase. Five concentrations of each substrate were used: 2, 4, 6, 8 and 10 mg mL^{-1} (w/v). The results (Table 8) show that the maximum amount of CMCase production (52.7 U mL^{-1}) was recorded at 8 mg mL^{-1} of CMC compared to 22.8 U mL^{-1} at 6 mg mL^{-1} of cellulose powder. The lowest cellulase production was noted at 2 mg mL^{-1} of cotton and 6 mg mL^{-1} of sawdust. The highest saccharification (10 %) was observed when CMC was used as carbon source followed by cellulose powder.

Table 7 Effect of different buffers on saccharification (%), biomass yield (mg g^{-1}) and the CMCase production (U mL^{-1}) by *B. megaterium*

pH	Buffers	CMCase concentration (U mL^{-1} ± SD)	Saccharification (%)	Biomass yield (mg g^{-1}) cellulose
5.6	Citrate buffer	30.8 ± 0.4	4.6	180
5.8		30.0 ± 0.9	5.0	420
6		31.6 ± 0.4	5.3	240
6.2		**33.7 ± 1.2**	5.1	310
6	Citrate phosphate buffer	29.3 ± 0.0	4.5	990
6.2		36.6 ± 0.0	5.6	1260
6.4		**38.6 ± 4.3**	5.8	710
6.6		38.4 ± 2.5	5.8	560
6.8		31.4 ± 1.5	4.7	890
7		24.1 ± 0.7	3.6	620
6	Phosphate buffer	31.8 ± 1.0	4.6	1130
6.2		38.5 ± 1.5	5.6	930
6.4		**38.5 ± 4.8**	6.0	930
6.6		38.0 ± 1.7	5.6	880
6.8		32.9 ± 1.5	5.3	770
7		33.6 ± 0.9	5.0	650
7.2		31.6 ± 1.7	4.9	1030
7.4		31.8 ± 0.9	4.7	1400
7.6		31.6 ± 1.0	4.6	570
7.8		31.0 ± 1.4	4.5	1020
8		29.0 ± 3.2	4.2	750

Bold values indicate the maximum CMCase production by *B. megaterium* strain

The effect of fructose, galactose, lactose, maltose, mannitol, mannose, ribose, sucrose and xylose on CMCase production is presented in Table 8. The results indicate that the maximum production of CMCase (51.4 U mL^{-1}) was recorded when 10 mg mL^{-1} of mannose was used as carbon source followed by 8 mg mL^{-1} of ribose (47.4 U mL^{-1}), 8 mg mL^{-1} of fructose (45.5 U mL^{-1}), 8 mg mL^{-1} of xylose (44.8 U mL^{-1}), 2 mg mL^{-1} of lactose (40.8 U mL^{-1}) and 4 mg mL^{-1} of mannitol (19.3 U mL^{-1}). The lowest production of CMCase (11.7 U mL^{-1}) was recorded with 8 mg mL^{-1} of galactose, maltose and 6 mg mL^{-1} of sucrose. The highest saccharification (%) was observed with xylose and ribose, 9.7 and 9.5 %, respectively. The results obtained in the current study are similar to those reported by Alam et al. (2004) who found that the highest CMCase production was recorded when CMC was used as carbon source and the lowest CMCase production with sawdust as a carbon source. *Paenibacillus* produced (4 U mL^{-1}) CMCase when it was grown on CMC as the only source of carbon (Emtiazi et al. 2007). Krishna (1999) suggested that addition of cellulose powder, lactose or glucose at concentration above 1 % level led to a significant reduction in enzyme synthesis by *B. subtilis*.

Recycling of sludge as production medium for CMCase

The production of CMCase by *B. megaterium* inoculated to sludge samples was investigated in the present study. The results are presented in Fig. 3. The maximum amount of CMCase production (14.3 U mL^{-1}) was recorded in 6 mg mL^{-1} (w/v) of sludge from SSTP, followed by sludge collected from TWTP at concentration 4 mg mL^{-1}, where 13.1 U mL^{-1} was produced. The lowest amount of CMCase production (1.4 U mL^{-1}) was observed in sludge collected from ASTP and ISTP (1.6 U mL^{-1}). The highest saccharification (7.5 %) was observed when 10 mg mL^{-1} (w/v) of sludge from SSTP was used as production medium, while the lowest saccharification (0.3 %) was noted with 8 mg mL^{-1} (w/v) of sludge from ASTP. The production of CMCase in sludge medium has not been reported before. However, Barros et al. (2013) used cassava wastewater as production medium for amylase, protease and lipase by *B. subtilis* strains and revealed that the bacteria produce detectable amounts of these enzymes in comparison to the synthetic liquid medium. Al-Gheethi and Norli (2014) investigated the production of β-lactamase in sewage effluents medium by *B. subtilis*, *C. luteola* and

Table 8 Effect of different cellulosic material and carbon source applied at different concentrations on saccharification and CMCase production by *B. megaterium*

Carbon source	Conc. (mg mL^{-1})	CMCase activity (U mL^{-1} ± SD)	Saccharification (%)
CMC	2	11 ± 0.2	10
	4	25 ± 0.5	10.1
	6	40 ± 1.3	9.8
	8	**52.7 ± 2.5**	**10**
	10	15 ± 0.9	2
Cellulose powder	2	10 ± 0.3	**8.9**
	4	13 ± 1.2	5.6
	6	**22.8 ± 0.5**	5.8
	8	15 ± 1.3	5
	10	12 ± 0.7	2
Cotton	**2**	**4 ± 1.3**	**6**
	4	1.5 ± 1.2	0.7
	6	1.5 ± 0.6	0.4
	8	1.4 ± 0.3	0.3
	10	1.3 ± 0.1	0.2
Filter paper	**2**	**8.5 ± 1.4**	**5.5**
	4	2.8 ± 0.1	1
	6	1.8 ± 0.5	0.2
	8	1.5 ± 0.2	1
	10	1.5 ± 0.0	0.1
Tobacco leaves	2	5 ± 0.1	**4.1**
	4	**23 ± 1.4**	3.5
	6	13.5 ± 0.4	3.2
	8	8.3 ± 0.1	1.2
	10	4.2 ± 0.3	0.4
Sawdust	2	2.8 ± 0.2	**2**
	4	2.7 ± 0.2	1.3
	6	**5.6 ± 0.6**	1
	8	4.2 ± 0.1	0.8
	10	3 ± 0.1	8.5
Bagasse	2	1.8 ± 0.2	1
	4	5.2 ± 0.5	**2**
	6	6.7 ± 0.2	1.8
	8	**9.8 ± 0.4**	1
	10	5 ± 0.3	1
Fructose	2	8 ± 0.2	8.5
	4	20 ± 0.3	8
	6	33.2 ± 0.5	8.5
	8	**45.5 ± 1.1**	**8.9**
	10	44.2 ± 0.4	6.8
Galactose	2	1.6 ± 0.1	1.3
	4	2.3 ± 0.5	1
	6	8.1 ± 1.2	2
	8	**11.7 ± 0.4**	**2.1**
	10	1.9 ± 0.2	0.1
Lactose	2	<0.1	0
	4	10.3 ± 2.4	3.7
	6	15.2 ± 0.5	4.1
	8	**40.8 ± 3.2**	**7.5**
	10	34.5 ± 0.3	4.9

Table 8 continued

Carbon source	Conc. (mg mL^{-1})	CMCase activity (U mL^{-1} ± SD)	Saccharification (%)
Maltose	**2**	**11.6 ± 0.6**	**8.3**
	4	8.1 ± 0.3	3.1
	6	>0.1	<0.1
	8	>0.1	<0.1
	10	>0.1	<0.1
Mannitol	2	4.8 ± 0.2	7.1
	4	**19.3 ± 1.2**	**7.5**
	6	10 ± 1.5	2
	8	2.3 ± 0.1	0.2
	10	>0.1	<0.1
Mannose	2	1.2 ± 0.0	1.1
	4	2.4 ± 0.1	2.7
	6	32 ± 0.8	**9.4**
	8	35 ± 1.5	7
	10	**51.4 ± 1.4**	8
Ribose	2	11.2 ± 0.3	**9.5**
	4	24.5 ± 0.4	9.2
	6	29.7 ± 0.6	7.5
	8	**47.4 ± 0.8**	9
	10	30 ± 0.4	4.8
Sucrose	2	5.2 ± 1.3	**4.3**
	4	6.2 ± 1.5	2.3
	6	**11.7 ± 1.5**	3
	8	10 ± 0.8	1.7
	10	7.3 ± 0.3	1
Xylose	2	13.7 ± 0.4	**9.7**
	4	18.8 ± 1.2	7
	6	30.5 ± 1.1	8
	8	**44.8 ± 1.3**	8.5
	10	40.2 ± 0.6	5.9

Bold values indicate the maximum CMCase production by *B. megaterium* strain

B. cepacia. They revealed that the bacterial strains produced β-lactamase at levels above 0.333 U mL^{-1}.

In comparison to the concentrations of heavy metals in sludge collected from the STPs, it can be noted that the concentration of Cu^{2+} and Zn^{2+} ions in the sludge from TSTP was more than that in the sludge from ASTP and ISTP. However, the production of CMCase and percentage of saccharification in sludge from TSTP were more than those in the sludge from ASTP and ISTP. These results could be explained based on the toxicity of Ni^{2+} ions in comparison to Cu^{2+} and Zn^{2+} at the high concentration. It has reported that Ni^{2+}, Cu^{2+} and Zn^{2+} improve the enzymatic reaction at low concentration (Nies 1999). However, at high concentrations, Ni^{2+} becomes more toxic than Cu^{2+} and Zn^{2+}. This is because the transport of Zn^{2+} and Cu^{2+} through bacterial cells membrane is by specific transport system (Fath and Kolter 1993; Fagan and Saier 1994), while Ni^{2+} is accumulating by the fast and unspecific system (Smith and Maguire 1995; Tao et al. 1995). Besides, the ability of bacteria to resist Cu^{2+} was reported to be more than Ni^{2+} at the same concentrations (Al-Gheethi et al. 2014). Lankinen et al. (2011) reported that the production of β-glucosidase and β-cellobiosidase by decomposing fungi in heavy nickel-contaminated soil (more than 20 mg kg^{-1}) was less than that in the non-contaminated soil. In this study, however, the ability of *B. megaterium* to produce detectable amount of CMCase in sludge contaminated with Ni^{2+} ions is because *B. megaterium* was tolerant to 15 mM of nickel. Therefore, this bacterium would be used to produce cellulase from the sludge.

Effect of different nickel ion concentrations

The effect of different nickel ion concentrations was investigated to know the minimal Ni^{2+} ion concentrations at which *B. megaterium* strain produces maximum amount of CMCase. Ni^{2+} ion concentrations of 117.2–468.8 μg Ni^{2+} mL^{-1} have provided a broad range for the growth of

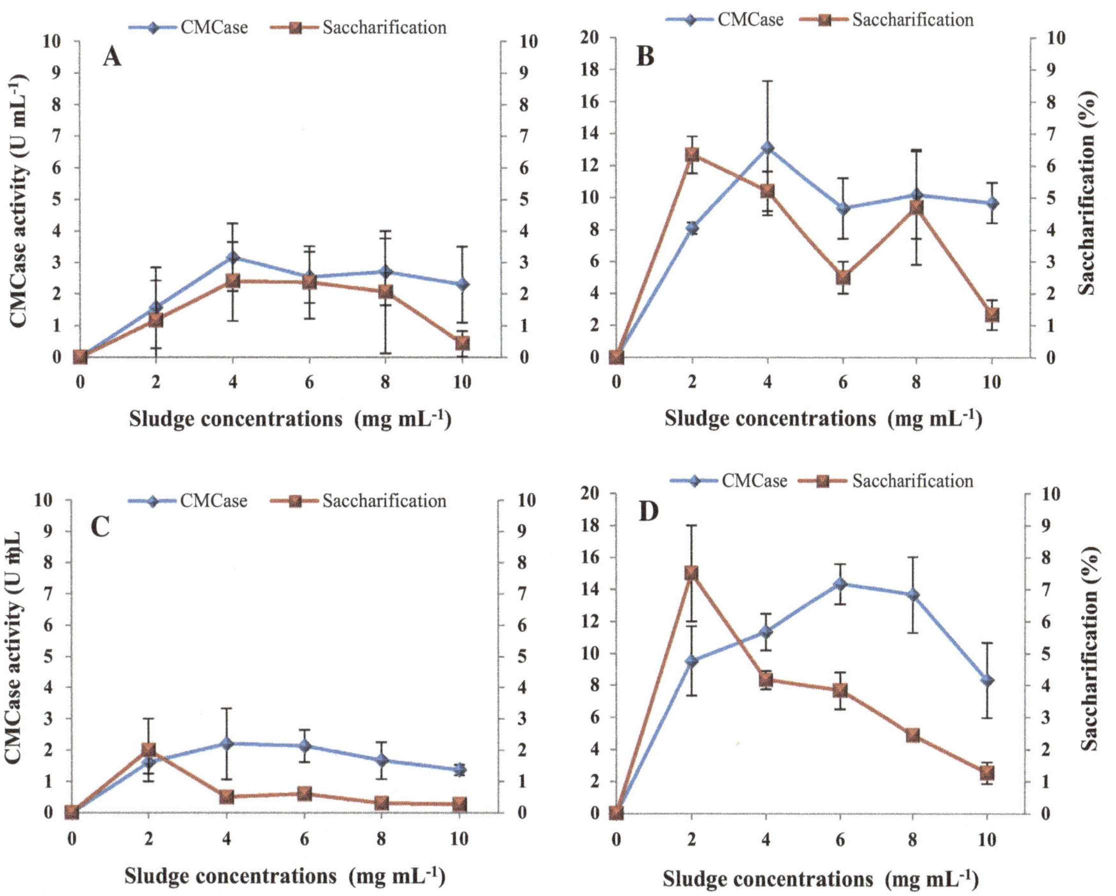

Fig. 3 Effect of different sludge media applied at different concentrations of sludge on saccharification and CMCase production by *B. megaterium*. **a** Sludge from Ibb sewage treatment plant, **b** sludge from Taiz sewage treatment plant, **c** sludge from Aden sewage treatment plant, **d** sludge from Sana'a sewage treatment plant

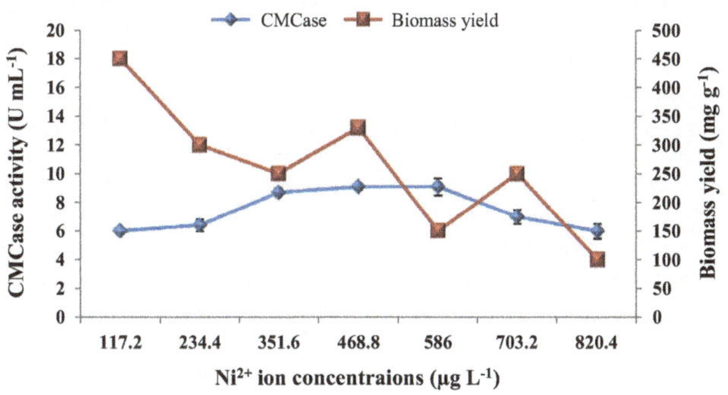

Fig. 4 Effect of different Ni^{2+} ion concentrations in carboxyl methyl cellulose yeast extract (CYE) broth medium on biomass yield and CMCase production by *B. megaterium*. *Each point* is the average of three determinations

bacteria and stimulated CMCase production with optimal production at 468.8 μg Ni^{2+} mL^{-1}. There was little detectable growth above 468.8 μg Ni^{2+} mL^{-1} or cellulase production below 351.6 μg Ni^{2+} mL^{-1} or above 586 μg Ni^{2+} mL^{-1}. The maximum biomass yield and CMCase production (450 mg g^{-1} and 9.1 U mL^{-1} respectively) was obtained at 117.2 and 468.8 μg Ni^{2+} mL^{-1} of the nickel ions, respectively (Fig. 4).

Nickel has been demonstrated for many bacteria and plant species, where eight nickel-containing enzymes present in one or more of these species have been identified (Ragsdale 1998; Watt and Ludden 1999). Nickel is identified as micronutrients at trace concentrations. On the other hand, it has been noted that the addition of Ni^{2+} ions at low concentrations enhanced biomass yield (Sujarittanonta and Sherrard 1981). Husain et al. (2013) revealed that the biomass of *P. fluorescens* increased with increasing Ni^{2+} ions in the broth medium from 250 to 1000 mg L^{-1}; the maximum growth was recorded at 1000 mg L^{-1}. However, the ability of Ni^{2+} ions to induce the CMCase production by *B. megaterium* strain has not been reported before.

In comparison to the production of CMCase in the sludge medium, the concentrations of nickel ions in the sludge from TSTP and SSTP were 156 and 26 mg kg^{-1}, while the maximum concentrations of nickel added to the production medium were 820.4 μg L^{-1}. However, the amount of CMCase produced in the sludge medium was more than that in the CYE medium. The increasing CMCase production in the sludge may be related to the sludge which is rich in nutrients and trace elements that induced the enzyme production more than CYE medium that is a synthetic medium. These findings are in consistent with those reported by Barros et al. (2013). They revealed that the production of amylase, protease and lipase by *B. subtilis* strains in cassava wastewater was more than that in the synthetic liquid medium. Another explanation of high production of CMCase in the sludge medium may be due to the nitrogen source. The sludge is rich with the amino acid and organic compounds, which represent as nitrogen source of bacteria, while in CYE the nitrogen source used was inorganic (sodium nitrate). Al-Gheethi (2008) revealed that the production of cellulase in the presence of peptone as nitrogen source was more when sodium nitrate was used as nitrogen source.

Conclusions

It can be concluded that *B. megaterium* strain isolated from the sludge possesses an important potential to produce CMCase in sewage sludge medium. The potential of bacterial strain to use sewage sludge as production medium would lead to biodegradation of cellulose in the sludge and, thus, can be used as a source of biofuel.

Acknowledgments This work was conducted at Taiz University (TU)-Yemen. I gratefully thank TU for their technical assistance and academic staff. Many thanks for Mr. Jamil Ali Saeed Al-Gheethi for his support during the study.

References

Abdel-Monem MO, Al-Zubeiry AH, Al-Gheethi AAS (2010) Biosorption of nickel by *Pseudomonas cepacia* 120S and *Bacillus subtilis* 117S. Water Sci Technol 61:2994–3007

Alam MZ, Manchur MA, Anwar MN (2004) Isolation, purification and characterization of cellulolytic enzyme produced by the isolate *Streptomyces omiyaensis* A_2. Pak J Biotechnol Sci 7(10):1647–1653

Al-Gheethi AAS (2008) Bacteriological studies on sludge from some municipals wastewater treatment plants in Yemen. MSc Thesis, Department of Applied Microbiology, Faculty of Applied Science, Taiz University, Taiz, Yemen

Al-Gheethi AAS, Norli I (2014) Biodegradation of pharmaceutical wastes in treated sewage effluents by *Bacillus subtilis* 1556WTNC. Environ Process 1:459–489

Al-Gheethi AAS, Norli I, Lalung J, Megat-Azlan A, Nur-Farehah ZA, Ab. Kadir MO (2014) Biosorption of heavy metals and cephalexin from secondary effluents by tolerant bacteria. Clean Technol Environ Policy 16:137–148

Allcock ER, Woods DR (1981) Carboxymethyl cellulase and cellobiase production by *Clostridium acetobutylicum* in an industrial fermentation medium. Appl Environ Microbiol 41(2):539–541

Angenent LT, Karim K, Al-Dahhan MH, Wrenn BA, Domíguez-Espinosa R (2004) Production of bioenergy and biochemicals from industrial and agricultural wastewater. Trends Biotechnol 22:477–485

APHA (1998) Standard methods for the examination of water and wastewater, 20th edn. American Public Health Association, Washington, DC

APHA (1999) Standard methods for the examination of water and wastewater, 9225B, 21st edn. American Public Health Association, Washington, DC

Araujo A, Ward OP (1990) Hemicellulases of *Bacillus* sp. preliminary comparative studies of production and properties of mammamases and galactanases. J Appl Bacteriol 68:1253–1261

Bakare MK, Adewale IO, Ajayi AO, Shonukan OO (2005) Purification and characterization of cellulase from the wild-type and two improved mutants of *P. fluorescens*. Afr J Biotechnol 4(9):898–904

Barros FFC, Simiqueli APR, de Andrade JC, Pastore GM (2013) Production of enzymes from agroindustrial wastes by biosurfactant-producing strains of *Bacillus subtilis*. Biotechnol Res Int 2013:1–9

Blouzard JC, Bourgeois C, Philip P, Valette O, Bélaïch A, Tardif C, Bélaïch JP, Pagès S (2007) Enzyme diversity of the cellulolytic system produced by *Clostridium cellulolyticum* explored by two-dimensional analysis: identification of seven genes encoding new dockerin-containing proteins. J Bacteriol 189(6):2300–2309

Brenner DJ, Krieg NT, Staley JT (2004) Bergeys manual of systematic bacteriology, The Proteobacteria, 2nd edn, vol 2, part A, B and C. Williams and Wilkins Awaverly Co, USA

Camassola M, De Bittencourt LR, Shenem NT, Andreaus J, Dillon AJP (2004) Characterization of cellulase complex of *Penicillium echinulatum*. Biocat Biotrans 22(5–6):391–396

Chan KY, Au KS (1987) Studies on cellulase production by *Bacillus subtilis*. Antonie Van Leeuwenhoek 53(2):125–136

Chynoweth DP, Pratap P (1996) Anaerobic digestion of municipal solid wastes. In: Palmisano AC, Barlaz MA (eds) Microbiology of solid waste. CRC Press Inc, Florida, pp 71–104

Collee JG, Duguid JP, Fraser AG, Marmion BP (1989) Practical medical microbiology, 13th edn. Longman-FE. Ltd, London

Costa RB, Silva MVA, Freitas FC, Leitao VSF, Lacerda PSB, Ferrara MA, Bon EPS (2008) Mercado e Perspectivas de Uso de Enzimas Industriais e Especiais no Brasil. In: Bon EPS, Ferrara MA, Corvo ML, Vermelho AB, Paiva CLA, Alencastro RB, Coelho RRR (eds) Enzimas em Biotecnologia, Producao, Aplicac oes e Mercados, 1st edn. Rio de Janeiro, Intercie^ncia, p 463–488

Coward-Kelly G, Alello-Mazzari C, Kim S, Granda C, Holtzapple M (2003) Suggested improvement to the standard filter paper assay used to measure cellulase activity. Biotechnol Bioeng 82:745–749

Dhillon N, Chhibber S, Saxena M, Pajni S, Vadehra DV (1985) A constitutive endoglucanase (CMCase) from *Bacillus licheniformis*. Biotechnol Lett 7(9):695–697

Dosanjh NS, Michel SL (2006) Microbial nickel metal-loregulation NikRs for nickel ions. Curr Opinion Chem Biol 10(2):123–130

Duff SJB, Cooper DG, Fuller OM (1985) Effect of colloidal materials on cellulase production by *Trichoderma reesei* Rut-C30. Appl Environ Microbiol 49(4):934–938

Emtiazi G, Pooyan M, Shamalnasab M (2007) Cellulase activities in nitrogen fixing *Paenibacillus* sp. isolated from soil in N-free media. World J Agri Sci 3(5):602–608

Fagan MJ, Saier MHJ (1994) P-type ATPases of eukaryotes and bacteria: sequence comparisons and construction of phylogenetic trees. J Mol Evol 38(1):57–99

Fath MJ, Kolter R (1993) ABC-transporters: bacterial exporters. Microbiol Rev 57(4):995–1017

Garrity GM, Johnson KL, Bell J, Searles D (2002) Bergey's manual of systematic bacteriology: taxonomic outline of the prokaryotes, 2nd edn. Eds. Release 3.0 July 2002. Springer-Verlag, New York

Hakamada Y, Koike K, Yoshimatsu T, Mori H, Kobayashi T, Ito S (1997) Thermostable alkaline cellulase from an alkaliphilic isolate *Bacillus* sp. KSM-S237. Extremophiles 1(3):151–156

Hamer G (2003) Solid waste treatment and disposal: effect on public health and environmental safety. Biotechnol Adv 22:71–79

Hernández A, Mellado RP, Martínez JL (1998) Metal accumulation and vanadium-induced multidrug resistance by environmental isolates of *Escherichia hermannii* and *Enterobacter cloacae*. Appl Environ Microbiol 64(11):4317–4320

Husain RSA, Thatheyus AJ, Ramya D (2013) Bioremoval of nickel using *Pseudomonas fluorescens*. Am J Microbiol Res 1(3):48–52

Immanuel G, Dhanusha R, Prema P, Palavesam A (2006) Effect of different growth parameters on endoglucanase enzyme activity by bacteria isolated from coir retting effluents of estuarine environment. Int J Environ Sci Technol 3(1):25–34

Ito S (1997) Alkaline cellulases from alkaliphilic *Bacillus*: enzymatic properties, genetics and application to detergents. Extremophiles 1(2):61–66

Ito S, Shikata S, Ozaki K, Kawai S, Okamoto K, Inoue S, Takei A, Ohta YI, Satoh T (1989) Alkaline cellulase for laundry detergents production by *Bacillus* sp. KSM 635 and enzymatic properties. Agri Biol Chem 53:1275–1281

Kaewchai S, Prasertsan P (2002) Biosorption of heavy metal by thermotolerant polymer-producing bacterial cells and the bioflocculant. Songklanakarin J Sci Technol 24(3):421–430

Karim A, Nawaz MA, Aman A, Qader SAU (2014) Hyper production of cellulose degrading endo (1,4) β-D-glucanase from *Bacillus licheniformis* KIBGE-IB2. J Rad Res Appl Sci. doi:10.1016/j.jrras.2014.06.004

Kawai S, Okoshi H, Ozaki K, Shikata S, Ara K, Ito S (1988) Neutrophilic *Bacillus* strain KSM.522 that products an alkaline carboxy methyl cellulase. Agri Biotechnol Chem 52:1425–1431

Kongruang S, Han MJ, Breton CIG, Penner MH (2004) Quantitative analysis of cellulase-reducing ends. Appl Biochem Biotechnol 113–116:213–231

Krishna C (1999) Production of bacterial cellulases by solid state bioprocessing of banana wastes. Biores Technol J 69:231–239

Krishna KB, Varma A (1990) Endoglucanase production by entrapped *Bacillus thermoalkalophilus*. World J Microbiol Biotechnol 6(3):267–272

Lankinen P, Kähkönen MA, Rajasärkkä J, Virta M, Hatakka A (2011) The effect of nickel concentration on growth of litter-decomposing fungi, extracellular enzyme activities and toxicity in soil. Boreal Environ Res 16(3):229–239

Leung WC, Wong MF, Chua H, Lo W, Yu PH, Leung CK (2000) Removal and recovery of heavy metals by bacteria isolated from activated sludge treating industrial effluents and municipal wastewater. Water Sci Technol 41(12):233–240

Lynd LR, Weimer PJ, Van Zyl WH, Pretorius IS (2002) Microbial cellulose utilization: fundamentals and biotechnology microbiology. Microbiol Mol Biol Rev 66:506–577

Mawadza C, Zvauya R (1996) Some factors affecting endo-β-1,4-glucanase production by two *Bacillus* strains isolated from Zimbabwean hot springs. J Basic Microbiol 36:177–185

Miller GL (1959) Use of dinitrosalicylic acid reagent for determination of reducing sugar. Anal Chem 31:426–428

Mingardon F, Chanal A, López-Contreras AM, Dray C, Bayer EA, Fierobe HP (2007) Incorporation of fungal cellulases in bacterial minicellulosomes yields viable, synergistically acting cellulolytic complexes. Appl Environ Microbiol 73(12):3822–3832

Nies DH (1999) Microbial heavy metals resistance. Appl Microbiol Biotechnol 51(6):730–750

Noel RK (1984) Bergey's manual of systematic bacteriology, vol 1. Williams and Wilkins Awaverly Co, USA, p 409–602

Padmavathy V (2008) Biosorption of nickel (II) ions by baker's yeast: kinetic, thermodynamic and desorption studies. Bioresour Technol 99(8):3100–3109

Pettigrew CA, Johanson BN (1996) Testing the biodegradability of synthetic polymeric materials in solid waste. In: Palmisano AC, Barlaz MA (eds) Microbiology of solid waste. CRC Press Inc, Florida, pp 71–104

Ragsdale SW (1998) Nickel biochemistry. Curr Opin Chem Biol 2:208–215

Sadhu S, Saha P, Sen SK, Mayilraj S, Maiti TK (2013) Production, purification and characterization of a novel thermotolerant endoglucanase (CMCase) from strain isolated from cow dung. Springer plus 2:1–10

Schloss PD, Hay AG, Wilson DB, Gossett JM, Walker LP (2005) Quantifying bacterial population dynamics in compost using 16S rRNA gene probes. Appl Microbiol Biotechnol 66:457–463

Sethi S, Datta A, Gupta BL, Gupta S (2013) Optimization of cellulase production from bacteria isolated from soil. ISRN Biotechnol 2013:1–7

Silva R, Lago ES, Merheb CW, Macchione MM, Park YK, Gomes E (2005) Production of xylanase and CMCase on solid state fermentation in different residues by *Thermoascus aurantiacus miehe*. Brazilian J Microbiol 36:235–241

Smith RL, Maguire ME (1995) Distribution of the CorA Mg^{2+} transport system in gram-negative bacteria. J Bacteriol 177(6):1638–1640

Sneath P, Mair NS, Sharpe EM (1986) Bergey's manual of systematic bacteriology, vol 2. Williams and Wilkins Awaverly Co, USA

Sternberg D, Mandels GR (1979) Induction of cellulolytic enzymes in *T. reesei* by sophorose. J Bacteriol 139:761–769

Stutzenberger FJ (1971) Cellulase production by *Thermomonospora curvata* isolated from municipal solid waste compost. Appl Microbiol 22:147–152

Sujarittanonta S, Sherrard JH (1981) Activated sludge nickel toxicity studies. J Water Poll Control Fed 53(8):1314–1322

Tao T, Snavely MD, Farr SG, Maguire ME (1995) Magnesium transport in *Salmonella typhimurium*: mgtA encodes a P-type ATPase and is regulated by Mg^{2+} in a manner similar of the mgt B P-type ATPase. J Bacteriol 177(10):2654–2662

Ten LN, Im WT, Kim MK, Kang MS, Lee ST (2004) Development of a plate technique for screening of polysaccharide-degrading microorganism by using a mixture of insoluble chromogenic substances. J Microbiol Meth 56:375–382

Wang YS, Byrd CS, Barlaz MA (1994) Anaerobic biodegradability of cellulose and hemicellulose in excavated refuse samples. J Indus Microbiol 13:147–153

Wattt RK, Ludden PW (1999) Nickel binding proteins. Cell Mol Life Sci 56:604–625

Zhang YH, Lynd LR (2004) Toward an aggregated understanding of enzymatic hydrolysis of cellulose. Non complexed cellulose systems. Biotechnol Bioeng 88:797–824

Zhang YH, Lynd LR (2005) Determination of the number average degree of polymerization cellodextrans and cellulose. Biomacromolecules 6:1510–1515

Zhang YHP, Himmel ME, Mielenz JR (2006) Outlook for cellulase improvement: screening and selection strategies. Biotechnol Adv 24:452–481

16

Dynamics of chemical changes through production of various composts/vermicompost such as farm manure and sugar industry wastes

Rahmat Ullah Shah · Muhammad Abid · Muhammad Farooq Qayyum · Rehmat Ullah

Abstract

Purpose Owing to aridity in our agro-ecosystem, mineralization of organic substrate is quite rapid and thereafter volatilized due to lower matrix affinity. In these consequences, the study has been chalked out with the hypothesis to alter the best approaches for mineralization of available organic resources as soil supplement to reduce the economic burden on the farming community. Our laboratory study showed the sequential temporal variations in physic-chemical properties of available organic substrates such as farm manure and sugar industry waste during composting/vermicomposting.

Methods The organic material obtained from the farm manure of live-stock farm and another sources of industrial organic waste {sugarcane baggase (SCB), pressmud (SPM), mixture of SCB, SPM and sugarcane effluent} were used for this mineralization perspectives. However, all organic substrates properties remained static except moisture up to a period of 21 days. Thereafter, these produced and processed matrix was subsequently composted and vermicomposted during 45 days under normal shade. No earthworms were spiked in composting while *Lumbricus rubellus* collected near vicinity of research area were inoculated at the rate of 50 g/kg of waste in vermicomposting. Contrarily, their applications was mandated as and when required policy subsequently. Moisture and temperature status of substrates were monitored regularly. However, compost/vermicompost substrates were assessed @ 15 days interval to evaluate temporal changes in physico-chemical characteristics.

Results Vermicomposting of farm manure and sugar industry wastes produced best quality manure with enriched nutritional status comprising more OC (4 %), N (3 %), P (2 %), K (7 %), Ca (3.5 %), Na (2.5 %), SO_4^{-2} (3.1 %) and B (twofold) as compared to composting.

Conclusion Inoculation of local specie for vermicomposting is a viable option to be recommended to the farming community.

Keywords Composting · Vermicomposting · Raw organic sources · Local earth worms · Chemical change

R. U. Shah (✉) · M. Abid · M. F. Qayyum
Department of Soil Science, Faculty of agricultural sciences and Technology, Bahaud din Zakariya University, Multan, Pakistan
e-mail: rahmatshah78@yahoo.com

R. Ullah
Soil and Water Testing Laboratory, Rajanpur, Punjab, Pakistan

Background

The earth being a natural planet is practically involved in the service of mankind through many attributes (Rockström et al. 2009). It has been playing pivotal role for the sustainability of this universe (Schellnhuber 2004). In this study, major role has been exploited by the mankind need. This could be emphasized by only providing the humanity as food, feed and shelter. The global population is increasing and because of this, the world may experience great fresh water scarcity. Our water resources are limited and, hence, water treatment and recycling methods are the only alternatives for getting fresh water in the forthcoming decades. Therefore, there is a great need for the development of a suitable, inexpensive and rapid wastewater treatment techniques and reuse or conservation methods in the present century. The different types of water treatment and recycling techniques have been discussed in terms of their basic principles, applications, costs, maintenance and suitability. Additionally, a systematic approach to water treatment and recycling involving their understanding, evaluation and selection parameters has identified for best humanity interest (Gupta et al. 2012a, b). Among all sub-continents, our country has been mainly known as agricultural country throughout the world (Anderson et al. 2010). Our economy central linchpin is directly linked with sustainability of production from the agricultural scope of system (Zaman et al. 2012). Pakistan is basically comprised of total area of 79.61 M ha, cultivated area of 22.51 M ha, uncultivated area of 23.25 M ha and fallow lands of 7.05 M ha (Anonymous 2011). Our country has been mainly dependent on *DoFasla* crop system (Ullah et al. 2012).

About 500 tons of industrial wastes (liquid and solid) being discharged daily from sugar factory during crushing season and presently dumped in vicinity of the sugar factory in our country. The quantity of wastes, however, depends on the crushing capacity of sugar mills (Arainet al. 2004). More than 12 million tons of sugarcane press mud (Sardar et al. 2012) and sugarcane baggase is being produced annually by running 80 sugar mills all over the country. Different types of organic wastes are available, but still cost effective technology has not yet been established to use them effectively. Yet, quality of such produced irrigation water could be improved through addition of carbon based material like compost for the elimination of pollutants. In this regard, various strategies such as the activated carbon reactor and advanced oxidation process (Gupta et al. 2012a, b), implication of matrix such as coloring techniques, Congo red and triarylmethane dye Light Green SF (Mittal et al. 2009a, b, 2010a, b), formation and classification of alumina-coated carbon nano-tubes (Gupta et al. 2011), retrieval of Chrysoidine Y (Mittal et al. 2010a, b), novel organic waste mixture with Fe_2O_3 nanoparticles (Gupta and Nayak 2012), photo-remediation of multi walled organic adsorbent (Saleh and Gupta 2012a, b), multi-walled carbon nanotubes-ionic liquid-carbon paste electrode (Khani et al. 2010), C-source material (Karthikeyan et al. 2012) and various organic sources (Jain et al. 2003) for the removal of lethal pollutants from agriculture waste water.

Moreover, compost is available at farmer's field level, cheap and best available source of substrate for compost and vermicompost production (Said et al. 2010).Average amount of nutrients depleted from soil for 100 MT sugarcane production are N, P, K, Ca, Mg, S, Fe, Mn, Zn and Cu i.e. 148.00, 123.00, 238.00, 42.00, 39.00, 38.00, 7.50, 4.12, 0.50, and 0.10 kg, respectively (Calcino 1995). Major part of these crop nutrients' is being removed from agricultural soil and discharge and dumped in the vicinity of mills in the form of sugarcane industrial wastes. These industrial wastes have sufficient amount of macro and micronutrients (Subba 2012). All these nutrients can be effectively recycled for the improvement of soil fertility index and crop yield (Hussein and Anjum 1999).

Most of literature has overlooked the importance of composting and vermicomposting technology in existing cropping schemes. However, many scientists revealed that the use of composting and vermicomposting approaches might be encouraged on large scale spectrum. Now, it is a dire need of today of our developing country that inclusion of micro-organism based technology should be mandated and suggested to our farming community strongly by our policy makers. Keeping in view of these facts, our planned study has been initiated with the objectives to evaluate the production and characterization of compost and vermicompost through their changes in our existing environmental conditions of Punjab-Pakistan. The produce and outcome of compost and vermicompost organic substrate would be efficiently used for maintaining the soil fertility as a supplementary source of fertilization in our country.

Methods

Hajra Rehman Sugar Mills, district Muzaffer Garh, Punjab-Pakistan was selected for the collection of available organic substrates. However, these waste organic resources have been used for the composting and vermicomposting procedure in the Research area of Bahaud din Zakariya University, Multan. Therefore, two hundred kilogram of sugarcane baggase (SCB), sugarcane press-mud (SCPM) and 50 L effluents from sugarcane industry (SCE) was collected in plastic bags and cans, respectively. These materials were labeled, packed in their respective material; then brought to laboratory and stored in main godown (store) of the research area of Soil Science Department. In addition to this, the livestock farm situated near the university main gate was visited. The material was physically checked properly through feel method. Then, this material was packed in 50 kg plastic bags and labeled properly with permanent marker on both sides of each bag. Only two bags were loaded and brought to laboratory and stored in the main godown (store) of research area of Soil Science Department.

Next day, the 54 (fifty-four) soil pits having dimension of 2 × 2 × 2 feet (L × W × D) were developed using shovel under the green shade area of our Research Farm. Besides this, the stored materials of sugar cane industry and dairy farm waste were air dried on polythene sheet under the normal shade. The base of each soil pit was leveled with its original soil. Each side of soil pits with plastic sheets was covered. Each of the six soil pits was filled with 15 kg of sugar cane baggase, pressmud, sugarcane baggase + pressmud + effluent (1:1:1) and farm manure.

These soil pits were covered by plastic sheets under the shade for 21 days. The moisture content was initially determined from each soil pits to maintain it up to 60 % by thermohygrometer. It would be maintained through water sprinkling up to desired moisture content level.

The covering of plastic sheet having raw material low density polyethylene (LDPE) and ethylene-vinylacetate (EVA) containing 80–170 μm thickness of each soil pits have raised the internal temperature that leads towards the sanitation of substrate matrix. Each soil pits remained covered up till 21 days. Then after this period, the upper covered plastic sheets were removed and these soil pits permanently remained opened under the shade. After 21 days, available local earthworm species (Lumbricus rubellus) from the banks of unpaved water channel and moist/green leaf debris, present near research area of Bahaud din Zakariya University Multan with acclimatized soil/material were collected in well aerated, open plastic buckets and brought to experimental site. Each vermicomposting soil pits was spiked with 50 g of local specie of earth worms and did not inoculated in all composting soil pits. It is important to note that thin layer of acclimatized material of local earth worm species were spread on substrate of each soil pits of vermicomposting. Temporal changes were stipulated by this meter by applying moisture to maintain its temperature up to 25 to 35 °C and moisture contents up to already mentioned desired level.

The samples of each organic substrate were collected at 0, 15, 30 and 45 days after inoculation of local earth worms. All the substrate samples were collected in plastic ribbon bags, marked with permanent marker and brought to laboratory for analyses of chemical changes. Under controlled environmental conditions, following parameters were determined as per established methodologies for characterization of compost and vermicompost.

The moisture content (%) of all collected organic samples was determined by gravimetric method (Hess and Angelaki 2003). The pH of the filtrate of each organic substarte was estimated by pH meter (Alves et al. 2006). Total nitrogen (ppm) was estimated by Kjeldahldistillation method (Bremner 1996). The organic carbon content (ppm) was recorded by chromic acid oxidation and titrimetric method (Walky and Black 1934). Prior to this, calcium contents (ppm) of all organic substrate samples after digestion process were estimated by versenate titration method (Hess and Angelaki 2003). Phosphorous, potassium, sodium, sulpher, boron contents (ppm) of each sample was determined by standard method of Jackson, using flame photometer (Rashid 1986). At the end, the concentrations of boron (ppm) were determined through HCL boiling procedure by using colorimetric method (Bingham 1982).

Results

Moisture retention capacity (%)

For moisture retention capacity, usage of various raw organic substrates through processes of composting and vermicomposting has been tabulated in Table 1. It was found that raw organic source from sugarcane baggase has hold moisture contents sequentially and significantly ($P < 0.01$) throughout the process of both composting practices (Table 1). However, it was observed that it retained less (48.29 and 48.33 %) during 45 days after inoculation of earth worms (DAI) in both practices. Raw organic source of pressmud also hold significantly ($P < 0.01$) moisture profile initially increased and then decreased sequentially throughout the entire process of both practices. Similarly, this source has retained more moisture content (48.41 and 48.69 %) especially at 45 DAI in both composting mechanisms. Prior to this, mixture of Sugarcane {baggase, pressmud and effluent} has also maintained moisture status significantly ($P < 0.01$) throughout the both experimentation. Likewise, their mixture has subsequently sustained more moisture profile (61.28 and 61.12 %) at 0 DAI during all process of composting. Besides this, raw organic source from farm manure has retained less moisture contents significantly ($P < 0.01$) with heterogenic trend in both practices. Less moisture retention (46.12 and 47.19 %) was noted markedly at 45 DAI during composting and vermicomposting. Subsequently, it was noticed generally that moisture contents increased initially at 0 DAI and then decreased sequentially at 15, 30 and 45 DAI in all treatments of raw organic sources during composting and vermicomposting strategies. However, our values expressed moisture behavior in this sense that net retention capacity was found significantly ($P < 0.01$) in sugarcane baggase, pressmud, mixture of sugarcane {baggase, pressmud and effluent} and farm manure under both processes. Therefore, it was noted that less net retention capacity of various available organic raw sources was observed for moisture contents in farm manure (−2.62 %) as compared to sugarcane baggase (−2.41 %), pressmud (−1.51 %) and their mixture (−1.19 %) respectively, under composting practice only. Therefore, less net retention capacity of various available organic raw sources was observed for moisture contents in farm manure (−3.19 %) as compared to pressmud (−2.48 %), sugarcane baggase (−2.32 %), and their mixture (−2.16 %) respectively, under vermicomposting practice. However, it was observed that less net retention capacity of all raw organic material for moisture contents was noticed at 45 DAI under both processes. Hence, farm manure has retained less net moisture retention capacity than to all other organic sources, respectively, under both practices.

Table 1 Moisture restoration stability in various organic strategies

Sources of organic substrates	Moisture contents (%)									
	Compost					Vermicompost				
	0 (DAI)	15 (DAI)	30 (DAI)	45 (DAI)	Net retention capacity (%)	0 (DAI)	15 (DAI)	30 (DAI)	45 (DAI)	Net retention capacity (%)
Sugarcane Baggase	49.48 ± 0.22	50.11 ± 0.86	49.14 ± 0.07	48.29 ± 0.66	−2.41	49.48 ± 0.22	50.11 ± 0.92	49.16 ± 0.13	48.33 ± 0.36	−2.32
Pressmud	49.93 ± 0.44	52.16 ± 0.17	51.42 ± 0.51	48.41 ± 0.97	−1.51	49.93 ± 0.44	52.16 ± 0.39	51.41 ± 0.54	48.69 ± 0.29	−2.48
Sugarcane Baggase+ Pressmud+ Effluent	62.47 ± 0.66	74.81 ± 0.34	64.89 ± 0.59	61.28 ± 0.38	−1.19	62.47 ± 0.66	74.81 ± 0.81	66.52 ± 0.52	61.12 ± 0.94	−2.16
Farm Manure	48.74 ± 0.64	55.48 ± 0.61	51.17 ± 0.97	46.12 ± 0.18	−2.62	48.74 ± 0.64	55.48 ± 0.54	53.17 ± 0.05	47.19 ± 0.76	−3.19

DAI (days after inoculation of earth worms)

pH trend

The pH of composting and vermicomposting materials tend to decrease with incubation time in all organic waste materials tested (Table 2). In this context, our results annotated that it varied significantly during 45 days after inoculation of earth worms (DAI) in both practices. Raw organic source of pressmud also changed significantly ($P < 0.01$) pH trend throughout the entire process of both practices. Similarly, this source has also changed pH status especially at 45 DAI in both composting mechanisms. Besides this, mixture of sugarcane {baggase, pressmud and effluent} has also lowered down the pH status significantly ($P < 0.01$) throughout the both experimentation. Analogously, their mixture has subsequently piled up more acidic pH status at 45 DAI during all process of composting. Moreover, raw organic source from farm manure has also shown gradient increment of pH level significantly ($P < 0.01$) with passage of time up to end of both practices. Lowest pH value was observed in farm manure compost and vermicmpost from pressmud. Subsequently, it was noticed generally that pH trend did not increase initially at 0 DAI and then decreased sequentially at 15, 30 and 45 DAI in all treatments of raw organic sources during composting and vermicomposting strategies. pH reaction varied in this way that net variation rate was observed significantly ($P < 0.01$) in sugarcane baggase, pressmud, mixture of sugarcane {baggase, pressmud and effluent} and farm manure under both processes. Therefore, it was observed that minimum net variation rate of various available organic raw sources was observed for pH status in farm manure (−10.39 and −12.98 %) as compared to their mixture (−6.57 and −5.26 %), Pressmud (−4.0 and −5.33 %) and sugarcane baggase (−5.19 %) respectively, under composting and vermicomposting practice only. It was observed that more net variation rate of all raw organic material for pH trend was noticed at 45 DAI under composting and vermicomposting processes. However, farm manure yielded less net variation rate amid other organic amendments under composting and vermicomposting technologies.

Nutrients dynamics in compost and vermicompost material

The data in Tables 3, 4, 5, 6, 7, 8, 9, 10 regarding to nutrient related element such as OC, N, P, K, Ca, Na^+, S and B depicted that raw organic source from sugarcane baggase, sugarcane pressmud, sugarcane effluent and farm manure mineralized sequentially and significantly ($P < 0.01$) throughout the process of mineralization under

Table 2 pH trends of various organic strategies

Sources of organic substrates	(pH trends)									
	Compost				Net retention capacity (%)	Vermicompost				Net retention capacity (%)
	0 (DAI)	15 (DAI)	30 (DAI)	45 (DAI)		0 (DAI)	15 (DAI)	30 (DAI)	45 (DAI)	
Sugarcane Baggase	7.7 ± 0.10	7.6 ± 0.08	7.5 ± 0.09	7.3 ± 0.10	−5.19	7.7 ± 0.10	7.6 ± 0.08	7.4 ± 0.08	7.3 ± 0.10	−5.19
Pressmud	7.5 ± 0.09	7.5 ± 0.10	7.4 ± 0.10	7.2 ± 0.09	−4	7.5 ± 0.09	7.4 ± 0.10	7.2 ± 0.10	7.1 ± 0.09	−5.33
Sugarcane Baggase+ Pressmud+ Effluent	7.6 ± 0.08	7.5 ± 0.09	7.3 ± 0.08	7.2 ± 0.08	−6.57	7.6 ± 0.08	7.5 ± 0.10	7.4 ± 0.08	7.1 ± 0.08	−5.26
Farm Manure	7.7 ± 0.09	7.5 ± 0.09	7.2 ± 0.09	7.1 ± 0.09	−10.39	7.7 ± 0.09	7.6 ± 0.09	7.1 ± 0.09	7.1 ± 0.09	−12.98

DAI (days after inoculation of earth worms)

both composting practices. Therefore it is revealed from our data that, OC remained up to 17.41 and 18.34 g kg^{-1}, N remained up to 1.33 and 1.34 g kg^{-1}, P remained up to 0.69 and 0.77 g kg^{-1}, K remained up to 1.93 and 1.96 g kg^{-1}, Ca remained up to 1.85 and 1.89 g kg^{-1}, Na remained up to 0.48 and 0.49 %, S remained up to 1.39 and 1.41 mg kg^{-1} and B remained up to 13.46 and 15.82 g kg^{-1} during 45 days after inoculation of earth worms (DAI) in both practices. Likewise, raw organic source of pressmud also produced significantly ($P < 0.01$) median contents of OC remained up to 17.41 and 18.34 g kg^{-1}, N remained up to 1.58 and 1.60 g kg^{-1}, P remained up to 2.73 and 2.79 g kg^{-1}, K remained up to 2.24 and 2.27 g kg^{-1}, Ca remained up to 2.61 and 2.71 g kg^{-1}, Na remained up to 0.56 and 0.58 %, S remained up to 2.12 and 2.23 mg kg^{-1} and B remained up to 16.16 and 17.62 g kg^{-1} throughout the entire process of both practices. Moreover, mixture of sugarcane {baggase, pressmud and effluent} has also enhanced the status of OC which remained (17.41 and 18.34 g kg^{-1}), N remained up to (1.42 and 1.44 g kg^{-1}), P remained up to (2.01 and 2.07 g kg^{-1}), K remained up to (3.67 and 4.27 g kg^{-1}), Ca remained up to (4.68 and 4.79 g kg^{-1}), Na remained up to (1.57 and 1.62 %), S remained up to (2.31 and 2.37 mg kg^{-1}) and B remained up to (12.34 and 15.34 g kg^{-1}) significantly ($P < 0.01$) throughout both experimentation. Besides this, raw organic source from farm manure has also shown gradient increment of OC which remained (17.41 and 18.34 g kg^{-1}), N remained up to (2.31 and 2.34 g kg^{-1}), P remained up to (5.32 and 5.61 g kg^{-1}), K remained up to (4.99 and 5.86 g kg^{-1}), Ca remained up to (6.88 and 6.97 g kg^{-1}), Na remained up to (1.18 and 1.21 %), S remained up to (2.10 and 2.13 mg kg^{-1}) and B remained up to (17.10 and 20.10 g kg^{-1}) strength significantly ($P < 0.01$) with passage of time remained up to the end of both practices.

More contents of OC, N, P, K, Ca, Na, S and B were noted markedly at 45 DAI during composting and vermicomposting. Factually, it was noticed generally that contents of all nutrients did not increase initially at 0 DAI and then increased sequentially at 15, 30 and 45 DAI in all treatments of raw organic sources during composting and vermicomposting strategies. However, our values depicted picture in this way that net mineralization rate was found significantly ($P < 0.01$) in sugarcane baggase, pressmud, mixture of sugarcane {baggase, pressmud and effluent} and farm manure under both processes.

For OC, maximum net mineralization rate of various available organic raw sources was observed in farm manure (29.64 and 32.54 %) as compared to their mixture (14.72 and 19.61 %), Pressmud (36.28 and 39.83 %) and sugarcane baggase (12.83 and 18.59 %) respectively, under both composting practices. Likewise, maximum net mineralization rate of N was noted in farm manure (7.94 and 9.34 %) as compared to Pressmud (3.26 and 4.57 %), sugarcane baggase (3.1 and 3.87 %) and mixture of sugarcane {baggase, pressmud and effluent} (2.16 and 3.59 %) respectively, under both composting practices. Analogously, maximum net mineralization rate of total phosphorous was markedly observed in farm manure (10.83 and 16.88 %) as compared to sugarcane baggase (6.25 and 7.81 %), pressmud (4.60 and 6.90 %) and their mixture (2.03 and 5.08 %) respectively, under both composting practices. Equidistantly, maximum net mineralization rate of potassium was found in farm manure (13.44 and 19.35 %) than to the sugarcane mixture (11.58 and 18.01 %), pressmud (9.74 and 16.41 %) and sugarcane baggase (4.89 and 6.52 %) respectively, under both

Table 3 Organic Carbon (%) contents by processing of various organic strategies

Sources of organic substrates	Organic carbon contents (%)									
	Compost				Net retention capacity (%)	Vermicompost				Net retention capacity (%)
	0 (DAI)	15 (DAI)	30 (DAI)	45 (DAI)		0 (DAI)	15 (DAI)	30 (DAI)	45 (DAI)	
Sugarcane Baggase	15.43 ± 0.89	15.77 ± 1.70	16.05 ± 1.65	17.41 ± 0.91	12.83	15.43 ± 0.89	15.94 ± 1.10	16.4 ± 2.76	18.34 ± 1.93	18.59
Pressmud	12.1 ± 1.00	12.94 ± 2.04	13.26 ± 0.84	16.49 ± 1.32	36.28	12.1 ± 1.00	13.25 ± 2.54	13.36 ± 2.32	16.92 ± 2.06	39.83
Sugarcane Baggase+ Pressmud+ Effluent	13.72 ± 1.53	14.01 ± 1.72	14.25 ± 1.07	15.74 ± 0.58	14.72	13.72 ± 1.53	14.14 ± 1.74	14.7 ± 1.01	16.41 ± 1.89	19.61
Farm manure	22.13 ± 1.85	23.49 ± 1.79	24.17 ± 2.05	28.69 ± 1.50	29.64	22.13 ± 1.85	23.64 ± 1.38	24.47 ± 2.58	29.16 ± 1.06	32.54

DAI (days after inoculation of earth worms)

Table 4 Total Nitrogen (g kg^{-1}) status of various organic strategies

Sources of organic substrates	Total nitrogen (g kg^{-1})									
	Compost				Net retention capacity (%)	Vermicompost				Net retention capacity (%)
	0 (DAI)	15 (DAI)	30 (DAI)	45 (DAI)		0 (DAI)	15 (DAI)	30 (DAI)	45 (DAI)	
Sugarcane Baggase	1.29 ± 0.032	1.30 ± 0.026	1.31 ± 0.020	1.33 ± 0.036	3.1	1.29 ± o.032	1.31 ± 0.026	1.32 ± 0.030	1.34 ± 0.032	3.87
Pressmud	1.53 ± 0.012	1.54 ± 0.015	1.56 ± 0.020	1.58 ± 0.021	3.26	1.53 ± 0.012	1.55 ± 0.010	1.58 ± 0.015	1.6 ± 0.252	4.57
Sugarcane Baggase+ Pressmud+ Effluent	1.39 ± 0.015	1.4 ± 0.026	1.41 ± 0.015	1.42 ± 0.020	2.16	1.39 ± 0.015	1.41 ± 0.019	1.42 ± 0.030	1.44 ± 0.044	3.59
Farm manure	2.14 ± 0.021	2.16 ± 0.025	2.19 ± 0.026	2.31 ± 0.044	7.94	2.14 ± 0.021	2.19 ± 0.026	2.26 ± 0.026	2.34 ± 0.032	9.34

DAI (days after inoculation of earth worms)

Table 5 Total Phosphorous (g kg^{-1}) status by processing of various organic strategies

Sources of organic substrates	Compost				Net retention capacity (%)	Vermicompost				Net retention capacity (%)
	0 (DAI)	15 (DAI)	30 (DAI)	45 (DAI)		0 (DAI)	15 (DAI)	30 (DAI)	45 (DAI)	
Sugarcane Baggase	0.64 ± 0.05	0.65 ± 0.05	0.67 ± 0.16	0.69 ± 0.11	6.25	0.64 ± 0.05	0.74 ± 0.15	0.75 ± 0.13	0.77 ± 0.20	7.81
Pressmud	2.61 ± 0.67	2.64 ± 0.14	2.68 ± 0.45	2.73 ± 0.57	4.60	2.61 ± 0.67	2.65 ± 0.47	2.71 ± 0.47	2.79 ± 0.34	6.90
Sugarcane Baggase+ Pressmud+ Effluent	1.97 ± 0.33	1.98 ± 0.10	1.99 ± 0.09	2.01 ± 0.17	2.03	1.97 ± 0.33	1.99 ± 0.22	2.03 ± 0.14	2.07 ± 0.05	5.08
Farm manure	4.3 ± 0.36	4.92 ± 0.54	5.09 ± 0.42	5.32 ± 0.46	10.83	4.3 ± 0.36	5.03 ± 0.23	5.3 ± 0.85	5.61 ± 0.60	16.88

DAI (days after inoculation of earth worms)

Table 6 K^{+} (g kg^{-1}) status by processing of various organic strategies

Sources of organic substrates	Potassium contents (g kg^{-1})									
	Compost				Net retention capacity (%)	Vermicompost				Net retention capacity (%)
	0 (DAI)	15 (DAI)	30 (DAI)	45 (DAI)		0 (DAI)	15 (DAI)	30 (DAI)	45 (DAI)	
Sugarcane Baggase	1.84 ± 0.10	1.86 ± 0.25	1.89 ± 0.12	1.93 ± 0.17	4.89	1.84 ± 0.10	1.87 ± 0.12	1.91 ± 0.20	1.96 ± 0.13	6.52
Pressmud	1.95 ± 0.20	1.98 ± 0.26	2.05 ± 0.17	2.14 ± 0.24	9.74	1.95 ± 0.20	2.01 ± 0.17	2.09 ± 0.18	2.27 ± 0.30	16.41
Sugarcane Baggase+ Pressmud+ Effluent	3.11 ± 0.19	3.15 ± 0.16	3.3 ± 0.21	3.67 ± 0.24	11.58	3.11 ± 0.19	3.19 ± 0.26	3.28 ± 0.20	4.27 ± 0.22	18.01
Farm manure	4.91 ± 0.23	4.94 ± 0.22	4.96 ± 0.13	4.99 ± o.13	13.44	4.91 ± 0.23	5.1 ± 0.16	5.42 ± 0.46	5.86 ± 0.25	19.35

DAI (days after inoculation of earth worms)

Table 7 Ca^{2+} (g kg^{-1}) status by processing of various organic strategies

Sources of organic substrates	Calcium contents (g Kg^{-1})									
	Compost				Net retention capacity (%)	Vermicompost				Net retention capacity (%)
	0 (DAI)	15 (DAI)	30 (DAI)	45 (DAI)		0 (DAI)	15 (DAI)	30 (DAI)	45 (DAI)	
Sugarcane Baggase	1.77 ± 0.10	1.79 ± 0.22	1.82 ± 0.41	1.85 ± 0.28	4.52	1.77 ± 0.10	1.79 ± 0.17	1.83 ± 0.15	1.89 ± 0.16	6.78
Pressmud	2.19 ± 0.05	2.28 ± 0.22	2.42 ± 0.33	2.61 ± 0.33	19.18	2.19 ± 0.05	2.33 ± 0.18	2.52 ± 0.32	2.71 ± 0.22	23.74
Sugarcane Baggase+ Pressmud+ Effluent	4.14 ± 0.17	4.26 ± 0.18	4.42 ± 0.26	4.68 ± 0.15	13.04	4.14 ± 0.17	4.31 ± 0.27	4.51 ± 0.33	4.79 ± 0.12	15.70
Farm manure	6.5 ± 0.45	6.57 ± 0.33	6.72 ± 0.27	6.88 ± 0.10	5.85	6.5 ± 0.45	6.57 ± 0.34	6.76 ± 0.21	6.97 ± 0.05	7.85

DAI (days after inoculation of earth worms)

Table 8 Sodium contents by processes of composting

Sources of organic substrates	Sodium contents (%)									
	Compost				Net retention capacity (%)	Vermicompost				Net retention capacity (%)
	0 (DAI)	15 (DAI)	30 (DAI)	45 (DAI)		0 (DAI)	15 (DAI)	30 (DAI)	45 (DAI)	
Sugarcane Baggase	0.43 ± 0.11	0.44 ± 0.10	0.45 ± 0.09	0.48 ± 0.08	11.63	0.43 ± 0.11	0.44 ± 0.10	0.46 ± 0.09	0.49 ± 0.08	13.95
Pressmud	0.50 ± 0.07	0.51 ± 0.05	0.53 ± 0.07	0.56 ± 0.07	12.00	0.50 ± 0.07	0.51 ± 0.05	0.54 ± 0.06	0.58 ± 0.08	16.00
Sugarcane Baggase+ Pressmud+ Effluent	1.39 ± 0.12	1.44 ± 0.11	1.50 ± 0.08	1.57 ± 0.11	12.95	1.39 ± 0.12	1.45 ± 0.10	1.53 ± 0.07	1.62 ± 0.04	16.55
Farm manure	0.98 ± 0.12	1.02 ± 0.06	1.09 ± 0.15	1.18 ± 0.11	20.41	0.98 ± 0.12	1.04 ± 0.03	1.12 ± 0.10	1.21 ± 0.13	23.47

DAI (days after inoculation of earth worms)

Table 9 Sulphate contents by processes of composting's

Sources of organic substrates	Sulphate contents (mg kg^{-1})									
	Compost				Net retention capacity (%)	Vermicompost				Net retention capacity (%)
	0 (DAI)	15 (DAI)	30 (DAI)	45 (DAI)		0 (DAI)	15 (DAI)	30 (DAI)	45 (DAI)	
Sugarcane Baggase	1.27 ± 0.14	1.28 ± 0.15	1.29 ± 0.15	1.39 ± 0.19	9.45	1.27 ± 0.14	1.30 ± 0.14	1.39 ± 0.03	1.41 ± 0.02	11.02
Pressmud	1.92 ± 0.17	1.94 ± 0.16	1.97 ± 0.16	2.12 ± 0.07	10.42	1.92 ± 0.17	1.98 ± 0.11	2.11 ± 0.10	2.23 ± 0.09	16.15
Sugarcane Baggase+ Pressmud+ Effluent	1.99 ± 0.12	2.01 ± 0.12	2.12 ± 0.04	2.31 ± 0.06	16.08	1.99 ± 0.12	2.02 ± 0.12	2.23 ± 0.08	2.37 ± 0.06	19.10
Farm manure	1.89 ± 0.09	1.84 ± 0.11	1.87 ± 0.12	2.10 ± 0.18	11.11	1.89 ± 0.09	1.95 ± 0.09	1.97 ± 0.05	2.13 ± 0.04	12.69

DAI (days after inoculation of earth worms)

Table 10 Boron concentrations through processes of composting

Sources of organic substrates	Boron contents (g kg^{-1})									
	Compost				Net retention capacity (%)	Vermicompost				Net retention capacity (%)
	0 (DAI)	15 (DAI)	30 (DAI)	45 (DAI)		0 (DAI)	15 (DAI)	30 (DAI)	45 (DAI)	
Sugarcane Baggase	11.13 ± 0.12	11.61 ± 0.12	11.91 ± 0.12	13.46 ± 0.12	8.54	11.13 ± 0.12	14.08 ± 1.00	14.73 ± 0.09	15.82 ± 0.13	20.49
Pressmud	13.47 ± 0.03	13.71 ± 0.10	15.02 ± 0.09	16.16 ± 0.10	10.13	13.47 ± 0.03	14.34 ± 0.11	16.66 ± 0.12	17.62 ± 0.12	30.81
Sugarcane Baggase+ Pressmud+ Effluent	11.26 ± 0.11	11.43 ± 0.10	11.81 ± 0.10	12.34 ± 0.12	9.59	11.26 ± 0.11	12.16 ± 0.09	14.38 ± 0.11	15.34 ± 0.11	36.23
Farm manure	14.6 ± 0.14	15.74 ± 0.11	16.04 ± 0.10	17.1 ± 0.15	17.12	14.6 ± 0.14	17.92 ± 0.10	19.14 ± 0.11	20.1 ± 0.33	37.67

DAI (days after inoculation of earth worms)

composting practices. For Ca contents, maximum net mineralization rate was noted in pressmud (19.18 and 23.74 %) as compared to their mixture (13.04 and 15.70 %), farm manure (5.85 and 7.85 %) and sugarcane baggase (4.52 and 6.78 %) respectively, under both composting practices. Prior to this, maximum net mineralization rate of Na was present in farm manure (20.41 and 23.47 %) than to the mixture of sugarcane {baggase, pressmud and effluent} (12.95 and 16.55 %), pressmud (12.0 and 16.0 %) and sugarcane baggase (11.63 and 13.95 %) respectively, under both composting practices. In addition to this, maximum net mineralization rate of S was annotated in mixture of sugarcane {baggase, pressmud and effluent} (16.08 %) than tofarm manure (11.11 %), pressmud (10.42 %) and sugarcane baggase (9.45 %) respectively, under composting process only. Therefore, net mineralization rate under vermicomposting process for sulphate strength was present more in mixture of sugarcane {baggase, pressmud and effluent}(19.10 %), pressmud (16.15 %), farm manure (12.69 %) and sugarcane baggase (11.02 %) respectively. For B contents, maximum net mineralization rate was expressed with its median contents infarm manure (17.12 and 37.67 %) as compared to pressmud (10.13 and 30.81 %), their mixture (9.59 and 36.23 %) and sugarcane baggase (8.54 and 20.49 %) respectively, under both composting process. It was suggested that more net mineralization rate of all raw organic material for nutrient dynamics was noticed at 45 DAI under composting and vermicomposting processes. Hence, it was pointed out finally that farm manure has given maximum nutrient dynamics in both composting processes as compared to all other treatments and suggesting its large scale implication in the field to provide the optimum balance nutrition for the maintenance of crop productivity in their existing cropping patterns.

Discussion

Vermicomposting being a friendly environment technology had been induced as a precursor for the sustainability of soil mineral liberation to the rhizosphere of many exotic species of *Lycopersicum esculentum* L. (Reséndez et al. 2012). However, composting in this consortium also made a significant linchpin redox pertaining to sustaining the soil productivity in relation to yield of many crop plants (Hemmat et al. 2010). Prior to this, adoption of these technologies had been evidenced from the recent advancement of the scientific pool of work. Such hallmark had been consolidated on the maintenance of soil fertility indices (Ullah et al. 2013). Our data in above table in this consequence has also depicted soil fertility improved significantly through the implementation of vermicomposting strategies. In this regard, Theunissen et al. (2010) have also primarily focused on the improvement of soil fertility due to inclusion of vermicomposting dilemma. Moreover, many scientific investigations have given clear picture that encoring vermicomposting approach in the arid to hyper arid environmental conditions which proved best tool for the soil fertility improvement (Bhosale et al. 2012). Our investigation favors in this context that vermicomposting valuably improved the soil nitrogen to potassium build up in the soil profile.

Initially, the soil water contents, aeration, availability of organic matter, soil temporal fluctuation and soil mesic properties have been actively involved in the mineralization of available organic source through the efficient microbial activity (Tejada et al. 2009). Our data in Tables 1 and 2 clearly reflected pertaining to availability of optimum soil water content for the active existence of soil niche. Synergistically, it was evident from the suggestion endorsed by Ullah et al. (2013) that active participation of microbial population has vitally mineralized the available organic substrate in the optimum soil water contents.

Soil reaction (pH) has also been annotated as a nerve impulse for the availability of liberated nutrients on the available soil solumn (Lauber et al. 2009). It has to play a precursor for the recommendation of amendment for the amelioration of all nature of virgin to true soils. However, our results advocated the invariability of soil reaction to efflux liberated nutrition from soil matrix. The mica based dominant parent material has naturally acquired the soil reaction in alkaline form. This fact is directly linked with the process of mineralization to yield the soil nutrition perceptiveness (Brady and Weil 2010). Our results expressed in Table 2 are in similar fashion.

Soil nitrogen has been noted markedly from the mineralization process of available organic matter substrate to the existing soil microbial habitat (Raviv et al. 2004). Nitrogen being structural and morphological constituent nutrient has been easily available in vermicomposting to composting technologies as compared to other approaches involved for the organic resources. In this regard, our results in Table 4 clearly support the feasibility of available nitrogen contents for the growth and development of the tomato crop (Fernandez et al. 2010). Another structural strengthening element such as calcium in Table 7 dominating mineral has also pointed out significant release of this element from the soil matrix by activating the redox potential through intrusion of soil biota (Mansfeldt 2004). Its oozing from the soil matrix would have reduced the solubility of lethal element like sodium which in turn hampers the growth and development mechanisms involved in tomato with stand fields. Our suggestions have enabled us from the data (Table 8) that minimum sodium contents might have negative impact on the recovery efficiency for the respiration of microbial activity as compared to other nature of organic processes. Such evidence has been amazingly validated by (Wong et al. 2008) that microbial activity remained un-effective on lower salinity stress.

Micronutrients have also been supplementary used for the maintenance of growth process of all nature of crop. Its availability from any medium has been involved for the many physiological mechanisms like photosynthesis, fertilization, xylem solute transportation, cell divisions, stomata functionalities, fruit setting, enzyme activation, carbohydrate mobility, transpiration pull, cell turgidity and immunity against disease infections (Marschner 2012). Our data in this consequence, especially, for the availability of boron and sulphate contents depicted their status (Tables 9 and 10) from the mineralized produce through active participation of earth worms.

Regulation of metabolites in the cell venation to webbing of leaf area would be entangled by the insertion of potassium contents. However, the production of potassium contents through involvement of earth worms and by many other ways might be significantly involved in the aforementioned functions. Moreover, it has primarily participated in the regulation of stomata opening and closing through balancing the Na^+:K^+ ratio (Ren et al. 2005). Our findings pertaining to potassium contents in Table 6 reflected that optimum potassium median contents have been recovered efficiently through vermicomposting amid to all other strategies for the process of mineralization of organic substrates. Many approaches have also suggested the same suggestions like to our results (Fornes et al. 2012).

Phosphorous plays a pivotal role for root establishment, energy currency, fruit quality, fruit setting, stem strengthens and early maturity of various crops. Its availability is directly linked with the existence of optimum pH. The phosphorous median content has been proliferated sufficiently from the produce of earth worm heaps as compared to all other strategies opted for the decomposition of available raw material in our study. However, this point has been raised by Jouquet et al. (2010) pertaining to availability of phosphorous contents by the implementation of many living to non-living approaches on available organic source pool. Therefore, Rigane et al. (2011) suggestions also coincide to our observation in Table 5 for the availability of phosphorous contents.

Organic matter contents has been processed by vermicomposting and/or by composting from the organic flora raw material, fauna including partially decomposed dead living entities like earthworm and micro to macrobial niche (Jack et al. 2011). This mineralization mechanism of organic matter produces partially to fully decompose organic carbon contents. The organic carbon improves the soil structure, friability, water holding capacity, porosity, tillage, soil fertility which in turn contributes in boosting up the yield of all crops. Our study suggested the suitable quantum of mineralized organic carbon contents through the implementation of vermicomposting technologies when compared with non-vermicomposting technologies (Table 3). This paradox has been significantly verified by the recommendations of research advancement. Many researchers (Prakash et al. 2010; Ngo et al. 2011; Subramanian et al. 2010) have also recommended that vermicomposting technology yielded huge pile of organic carbon contents.

Conclusion

The study conducted on composting and vermicomposting of sugarcane baggase, pressmud, their mixture along with effluent and farm manure suggested many suited approaches. It was noted that composting strategy of sugarcane base raw organic material and farm manure have produced good quality of nutrient enriched material. The vermicomposting also produced huge piles of material but its quality remained best as compared to composting methodologies. Hence, it has been concluded from the study that use of available local earth worm species should be encouraged by incorporating farm manure initially in the field to improve the soil nutrition capability by activating their mineralization activities to reduce the economic burden on the farming community as a policy preview.

Acknowledgments Authors would like to acknowledge from the core of their hearts to Higher Education Commission Islamabad, for providing financial assistance for the completion of this study. I would also thank all the staff of soil and water testing laboratory and Multan for providing technical assistance for the completion of this manuscript.

Conflict of interest The authors declared that they have no conflict of interest.

References

Alves J, Lombardi M, Lada C (2006) The mass function of dense molecular cores and the origin of the IMF. *arXiv preprint astro-ph/0612126*

Anderson K, Cockburn J, Martin W (2010) Agricultural price distortions, inequality and poverty. World Bank Publications, 1818 H street NW Washington DC, pp 209–246

Anonymous (1999) Agriculture Monthly Economic Letter. NBP 26(8):3–4

Anonymous (2011) Pakistan statistical year book 2011 by Federal Bureau of Statistics, Government of Pakistan

Arain MA, Fatima Q (2004) Recycling of sugarcane industrial waste as a biofertilizer through composting. Pak J SciInd R 47(1):34–41

Bastiaanssen WG, Ali S (2003) A new crop yield forecasting model based on satellite measurements applied across the Indus Basin, Pakistan. Agr Ecosyst Environ 94(3):321–340

Berry T (2013) The sacred universe: Earth, spirituality, and religion in the twenty-first century. Columbia University Press

Bhosale PR, Chonde GS, Nakade DB, Raut PD (2012) Studies on physico-chemical characteristics of waxed and dewaxed pressmud and its effect on water holding capacity of soil. J Biosci 1(1):35–41

Bingham FT (1982) Boron. In: Page AL (ed) Methods of soil analyses, part 2: Chemical and mineralogical properties. Am Soc Agron Madison, WI, pp 431–448

Brady NC, Weil RR (2010) Elements of the nature and properties of soils. Pearson Prentice Hall

Bremner JM, Sparks DL, Page AL, Helmke PA, Loeppert RH, Soltanpour PN, Sumner ME (1996) Nitrogen-total. In: Methods of soil analysis. Part 3-chemical methods. pp 1085–1121

Calcino DV (1995) Australian sugarcane nutrition manual. Published by sugarcane research and development corporation (SRDC). Catena 77:238–247

Conant RT, Ryan MG, Ågren GI, Birge HE, Davidson EA, Eliasson PE, Bradford MA (2011) Temperature and soil organic matter decomposition rates–synthesis of current knowledge and a way forward. Glob Change Biol 17(11):3392–3404

Fernández-Gómez MJ, Nogales R, Insam H, Romero E, Goberna M (2010) Continuous-feeding vermicomposting as a recycling management method to revalue tomato-fruit wastes from greenhouse crops. Waste Manag 30(12):2461–2468

Fornes F, Mendoza-Hernández D, García-de-la-Fuente R, Abad M, Belda RM (2012) Composting versus vermicomposting: a comparative study of organic matter evolution through straight and combined processes. Biores Technol 118:296–305

Gupta VK, Nayak A (2012) Cadmium removal and recovery from aqueous solutions by novel adsorbents prepared from orange peel and Fe_2O_3 nanoparticles. Ch Eng J 180:81–90

Gupta VK, Agarwal S, Saleh TA (2011) Synthesis and characterization of alumina-coated carbon nanotubes and their application for lead removal. J Hazard Mat 185(1):17–23

Gupta VK, Ali I, Saleh TA, Nayak A, Agarwal S (2012a) Chemical treatment technologies for waste-water recycling-an overview. RSC Adv 2(16):6380–6388

Gupta VK, Jain R, Mittal A, Tawfik Saleh A, Nayak A, Agarwal S, Sikarwar S (2012b) Photo- catalytic degradation of toxic dye amaranth on TiO_2/UV in aqueous suspensions. Mat Sci Eng 32(1):12–17

Hemmat A, Aghilinategh N, Rezainejad Y, Sadeghi M (2010) Long-term impacts of municipal solid waste compost, sewage sludge andfarmyard manure application on organic carbon, bulk density and consistencylimits of a calcareous soil in central Iran. Soil Till Res 108(1):43–50

Hess BJ, Angelaki DE (2003) Gravity modulates Listing's plane orientation during both pursuit and saccades. J Neurophysiol 90(2):1340–1345

Hussain T, Anjum S (1999) Role of Biofertilizer in Boosting Agriculture. "Daily Dawn" Economic and Business Review, 4–10 Oct, P.No.3

Jack AL, Rangarajan A, Culman SW, Sooksa-Nguan T, Thies JE (2011) Choice of organic amendments in tomato transplants has lasting effects on bacterial rhizosphere communities and crop performance in the field. Appl Soil Ecol 48(1):94–101

Jackson ML (1967) Soil, chemical analysis. Prentice Hall of India Private Limited, New Delhi

Jain R, Gupta VK, Nayak A, Agarwal S, Shrivastava M (2011) Removal of the hazardous dye- Tartrazine by photodegradation on titanium dioxide surface. Mat Sci Eng 31(5):1062–1067

Jain AK, Gupta VK, Bhatnagar A, Suhas (2003) A comparative study of adsorbents prepared from industrial wastes for removal of dyes. Sep Sci Technol 38(2): 463–481

Jouquet P, Plumere T, Thu TD, Rumpel C, Duc TT, Orange D (2010) The rehabilitation of tropical soils using compost and vermicompost is affected by the presence of endogeic earthworms. Appl Soil Ecol 46(1):125–133

Karthikeyan S, Gupta VK, Boopathy R, Titus A, Sekaran G (2012) A new approach for the degradation of high concentration of aromatic amine by heterocatalytic Fenton oxidation: kinetic and spectroscopic studies. J Mol Liq 173:153–163

Khani H, Rofouei MK, Arab P, Gupta VK, Vafaei Z (2010) Multi-walled carbon nanotubes-ionic liquid-carbon paste electrode as a super selectivity sensor: Application to potentiometric monitoring of mercury ion (II). J Hazard Mat 183(1–3):402–409

Lauber CL, Hamady M, Knight R, Fiere N (2009) Pyrosequencing-based assessment of soil pH as a predictor of soil bacterial community structure at the continental scale. Appl Environ Biol 75(15):5111–5120

Mansfeldt T (2004) Redox potential of bulk soil and soil solution concentration of nitrate, manganese, iron, and sulfate in two gleysols. J Plant Nutr Soil Sc 167(1):7–16

Marschner H (ed) (2012) Marschner's mineral nutrition of higher plants, vol 89. Academic press, Marschner

Mittal A, Kaur D, Malviya A, Mittal J, Gupta VK (2009a) Adsorption studies on the removal of coloring agent phenol red from wastewater using waste materials as adsorbents. J Colloid Interface Sci 337(2):345–354

Mittal A, Mittal J, Malviya A, Gupta VK (2009b) Adsorptive removal of hazardous anionic dye "Congo red" from wastewater using waste materials and recovery by desorption. J Colloid Interface Sci 340(1):16–26

Mittal A, Mittal J, Malviya A, Gupta VK (2010a) Removal and recovery of Chrysoidine Y from aqueous solutions by waste materials. J Colloid Interface Sci 344(2):497–507

Mittal A, Mittal J, Malviya A, Kaur D, Gupta VK (2010b) Decoloration treatment of a hazardous triarylmethane dye, Light Green SF (Yellowish) by waste material adsorbents. J Colloid Interface Sci 342(2):518–527

Ngo PT, Rumpel C, Dignac MF, Billou D, Duc TT, Jouquet P (2011) Transformation of buffalo manure by composting or vermicomposting to rehabilitate degraded tropical soils. Ecol Eng 37(2):269–276

Prakash M, Karmegam N (2010) Vermistabilization of pressmud using Perionyx ceylanensis Mich. Biores Technol 101(21): 8464–8468

Rashid A (1986) Mapping zinc fertility of soils using indicator plants and soil analyses. PhD Dissertation, University of Hawaii, HI

Raviv M, Medina S, Krasnovsky A, Ziadna H (2004) Organic matter and nitrogen conservation in manure compost for organic agriculture. Compost Sci Util 12(1):6–10

Ren ZH, Gao JP, Li LG, Cai XL, Huang W, Chao DY, Lin HX (2005) A rice quantitative trait locus for salt tolerance encodes a sodium transporter. Nature Genet 37(10):1141–1146

Reséndez AM, Aguilar FJL, Viramontes UF, Dimas NR, Arroyo JV, Carrillo JLR, Ríos PC, Valdés MHR (2012) Tomato production in sand: vermicompost mixtures compared with sand and nutritive solution. Basic Res J Agri Sc Rev 1(1):19–26

Rigane MK, Medhioub K (2011) Assessment of properties of Tunisian agricultural waste composts: application as components in reconstituted anthropic soils and their effects on tomato yield and quality. Resour Conserv Recy 55(8):785–792

Rockström J, Will S, Kevin N, Åsa P, Chapin FS, Eric FL, Timothy ML, Foley JA (2009) A safe operating space for humanity. Nature 461(7263):472–475

Said G, Khan MJ, Usman K, Rehman H (2010) Impact of press mud as organic amendment on physico-chemical characteristics of calcareous soil. Sarhad J Agric 26(4):565–570

Saleh TA, Gupta VK (2012a) Photo-catalyzed degradation of hazardous dye methyl orange by use of a composite catalyst consisting of multi-walled carbon nanotubes and titanium dioxide. J Colloid Interface Sci 371(1):101–106

Saleh TA, Gupta VK (2012b) Column with CNT/magnesium oxide composite for lead (II) removal from water. Env Sci Pollut Res 19(4):1224–1228

Sardar S, Ilyas SU, Malik SR, Javaid K (2012) Compost fertilizer production from sugar press mud (SPM). Int J Chem Environ Eng 3(1):39–43

Schellnhuber HJ (ed) (2004) Earth system analysis for sustainability. MIT Press

Srivastava SK, Gupta VK, Mohan D, Sharma S (1997) Design parameters for fixed bed reactors of activated carbon developed from fertilizer waste for the removal of some heavy metal ions. Waste Manag 17(8):517–522

Subba RB (2012) Sugar factory and allied industrial wastes. Basic Res J Agri Sci Rev. 1(1):19–26

Subramanian S, Sivarajan M, Saravanapriya S (2010) Chemical changes during vermicomposting of sago industry solid wastes. J Hazard Mat 179(1):318–322

Tejada M, García-Martínez AM, Parrado J (2009) Effects of a vermicompost composted with beet vinasse on soil properties, soil losses and soil restoration. Catena 77(3):238–247

Theunissen J, Ndakidemi PA, Laubscher CP (2010) Review: potential of vermicompost produced from plant wasteon the growth and nutrient status in vegetable-production. Int J Phy Sci 5(13):1964–1973

Ullah R, Lone MI, Mian SM, Ali S, Ullah KS, Sheikh AA, Ali I (2012) Impact of seasonal variations and cropping systems on soil microbial biomass and enzymatic activities in slope gradient moisture stressed soils of Punjab-Pakistan. Soil Environ 31(1):21–29

Ullah R, Lone MI, Khan KSK, Mehdi SM, Qazi MA (2013) Effect of cropping systems and seasonal variations on soil microbial biomass and enzymatic activities in arid soils. J Ani Plant Sci 23(2):493–499

Walkley A, Black CA (1934) Anexamination of the Degtjaraff method for determining soil organic matter and a proposed modification of the chromic acid titration method. Soil Sci 37:29–38

Wong VN, Dalal RC, Greene RS (2008) Salinity and sodicity effects on respiration and microbial biomass of soil. Biol Fert Soils 44(7):943–953

Zaman K, Khan MM, Ahmad M, Rustam R (2012) The relationship between agricultural technology and energy demand in Pakistan. Energ Policy 44:268–279

Rahmat Ullah Shah is a PhD Indigenous Scholar under the financial assistance of wining Higher Education Scholarship from HEC-Pakistan and working as a PhD student at the Department of Soil Science, Faculty of agricultural sciences and Technology, Bahaud din Zakariya University, Multan-Pakistan.

17

Evaluation of in-vessel co-composting of yard waste and development of kinetic models for co-composting

K. M. Sangamithirai[1] · J. Jayapriya[1] · J. Hema[2] · Ravi Manoj[1]

Abstract
Background Organic solid waste management is a major challenge in high population density areas like apartments, educational institutions, hospitals, etc., as their disposal with other wastes could not only lead to issues like vector menace, odour generation but also would lead to loss of resources that could be reused such as nutrients which will benefit soil. In places like educational institutions, the major portion of waste generated is organic waste, especially food waste and yard waste from the campus, apart from paper waste. Composting is an organic waste treatment method that is cost effective and leads to resource recovery. In this study, the characteristics of composting of yard waste and co-composting of yard waste with wastes generated in an institution (university campuses of Anna University, Chennai, India) are examined. The composts were characterized in terms of total solids, volatile solids, pH, electrical conductivity, carbon: nitrogen (C:N) ratio and total Kjeldahl nitrogen. The stability index and germination index of the compost were assessed.
Results The C:N ratio of all the different composts (except the one generated from yard with canteen waste consisting of vegetable waste) produced in this study has a value less than 30:1, which satisfies the basic requirements for the compost. It was also observed that the mature compost from these feedstocks did not have heavy metals at toxic levels.
Conclusion This study demonstrated that the co-composting of yard wastes with paper and canteen wastes (fruit, coffee grounds, tea leaf waste) could produce value-added products. In addition to that, kinetic models were developed to predict the decomposition rate constant of the process.

✉ J. Jayapriya
jayapriyachem@gmail.com

1 Department of Chemical Engineering, A. C. Tech, Anna University, Chennai 600 025, India

2 Department of Biotechnology, PSG College of Technology, Coimbatore 641604, India

Keywords Composting · Yard waste · Paper waste · Canteen waste · Compost · Maturity

Introduction

Municipal solid waste (MSW), commonly known as trash or garbage, is defined as any waste generated by household, commercial and/or institutional activities. These municipal wastes are generally considered to be non-toxic and non-hazardous. There has been a significant increase in the generation of MSWs in India, as is the case around the world, over the last few decades as a result of rapid urbanization and rapid population growth (Diaz et al. 1993). It is known that high population density areas such as educational institutions, shopping malls and hospitals generate high volumes of MSW, which include a mixture of vegetable wastes in cooked and uncooked stages, leftovers, packaging materials, papers, plastics, rags, glass and other fabrics, dust, ash and a variety of combustible and non-combustible matter (Elwell et al. 1996). The current study was conducted in an educational institution with student strength of over 6000 in Anna University, Chennai, India. Canteens and hostels in the university generate vegetable and food wastes of approximately 350 kg/day. Apart from this, paper wastes and hardboards of 8000 kg/month are also generated. These wastes are disposed along with other MSW, leading to not only pollution, odour generation, vector menace but also loss of valuable resources that could be recovered and reused. The efficient recovery of resources depends on source separation of organic, biodegradable waste and non-degradable waste. Nowadays, educational institutions, as is the case with many other places, focus on minimizing the amount of waste generated while maximizing the extraction of value-added products from these wastes. The waste management hierarchy suggests resource recovery before waste disposal and an environmentally friendly one would be to reduce waste generation followed by reuse and composting rather than incineration or landfill (Giusti 2009).

Incineration of waste is a thermal process, which converts the waste materials into heat, gas, steam, and ash. However, due to the low calorific value of institutional generated waste, this process will not be efficient and cannot be exploited to recover energy completely from these wastes (Rand et al. 2000). Moreover, the incineration plants are not without their concerns, as they also lead to air pollution if not managed properly to reduce the polluting gases in the effluent. This could escalate the cost of the process and therefore may not be economically viable in developing countries. Composting involves the accelerated degradation of organic matter by microorganisms under controlled conditions (aeration, temperature and humidity), in which the organic material undergoes a characteristic thermophilic stage that allows sanitization of the waste by the elimination of pathogenic microorganisms (Lung et al. 2001).

There are various methods of composting from simple windrows to complex in-vessel composting (Misra and Roy 2003). In-vessel composting requires less space for operation than traditional outdoor windrow/pile composting methods. Further, In-vessel composting is not disrupted by weather changes. The final product compost could be used as soil conditioner and has the ability to increase soil porosity, structural stability, available water content and also reduces erosion (Sartori et al. 1985; Guidi et al. 1988; He et al. 1992; Hargreaves et al.2008; Diacono and Montemurro 2010). The process of composting and the quality of the final product is affected by several factors, including temperature, aeration, moisture content, etc., as well as substrate nature, especially C/N ratio, particle size and nutrient content (Diaz et al. 2002; Jiang et al. 2011).

As the nature of waste generated from this place, mainly the food waste, paper waste and the yard waste is not expected to change with time and can be considered as representative sampling and its composting process analysis could be extended to similar situations elsewhere. The need for on-site rapid composting to reduce the space requirements demands that a controlled composting process is applied. In-vessel composting makes use of controlled conditions to augment the aerobic degradation process (American Planning Association, 2006; Renkow and Rubin 1998). In-vessel composters of various levels of sophistication are available based on proprietary technology. The current study aimed at the ascertaining the feasibility of subjecting yard waste generated in the campus to in-vessel composting and evaluate the effect of additives (food wastes and paper waste) in co-composting. The physiochemical and microbiological characterisation of the compost was done to evaluate its quality. The effect of changing various process parameters such as organic matter, temperature, moisture, pH, electrical conductivity during the aerobic composting of institutional wastes, were used to develop kinetic models that could explain the composting rate.

Materials and methods

Compost preparation

Feedstock was collected from the different campuses of Anna University such as College of Engineering, Guindy, Chennai, India; Alagappa College of Technology, Chennai, India, and School of Architecture and Planning, Chennai, India. Sample collection included (1) yard wastes (dried fall leaves, grass clippings and wooden debris), (2) canteen wastes (fruit wastes, spent coffee grounds, spent tea leaves, vegetable waste) and (3) paper wastes. Three kg of feedstock in various combinations as described below was added into the composters (perforated cylindrical composter of 25 cm diameter and 80 cm height, made of plastic) daily for a period of 5 days. The process was carried out in three replicates in parallel. The experiments were carried out during the months of December 2011 to March 2012. The feedstock combinations were:

- Yard wastes (dried fall leaves: grass clippings: wooden debris, 1:1:1), (CY).
- Yard wastes + vegetable waste (2:1), (CYV).
- Yard wastes + fruit wastes (2:1), (CYF).
- Yard wastes + spent coffee grounds (2:1), (CYCo).
- Yard wastes + spent tea leaves (2:1), (CYT).
- Yard wastes + paper waste (2:1), (CYP).

The composts were manually turned over twice a week throughout the composting period of 3 months to control the temperature rise and provide proper aeration. The compost temperature was monitored at different compost depths using glass thermometers. Adequate moisture level was maintained by sprinkling water over the compost. The moisture level was maintained using the squeeze test (Solvita 2009).

Physicochemical analysis

The chemical compositions of various raw materials used for the composting are given in Table 1. The pH and electrical conductivity were determined in compost-water extracts (1:10 w/v) (Rhoades 1982). Moisture content (MC) was determined as weight loss upon drying at 105 °C in an oven for 24 h. The Volatile solids (VS) were obtained by igniting the dried samples at 500 °C for 20 min [American Public Health Association (APHA) Standard Methods, 1992]. Total organic carbon (TOC) of compost was determined by Walkley–Black method (Nelson and Sommers 1982). Total-N was determined by the Kjeldahl method following digestion in salicylic acid with metal catalysts (Mulvaney 1996).

Phosphorus was analyzed by ammonium molybdate–ascorbic acid after acidification of the extracts with 4.5 M H_2SO_4 as described by Murphy and Riley (1962) and Joergensen and Mueller (1996). The heavy metal content in the composts was determined after being digested in a solution of HNO_3/HCl, 1:3 (v/v) by Atomic Absorption Spectroscopy (AAS).

Compost analysis

The maturity and stability of the compost generated was analysed by germination index and CO_2 evolution index, respectively. The germination index (GI) was calculated using seeds of *Cucumber Himangi* (Zucconi et al. 1981). The filter paper placed on glass Petri dishes with six seeds of cucumber was wetted with 10 ml of a compost-water extract (1:10 v/v).The Petri dishes were sealed to minimize the water loss and kept in the dark at room temperature. At the end of 5 days, the percentage of seed germination in the compost extract was compared with that of water-control.

The stability index was evaluated according to the CO_2 evolution rate (Q), in mg CO_2 g^{-1} compost day^{-1}. The compost is considered as being very stable ($Q < 2$); stable ($Q < 4$); or unstable ($Q > 4$) (United States Department of Agriculture and the United States Composting Council-Test Methods for Examination of Composting and Compost 2001). The moisture of the samples was adjusted to 70–85 % and the samples were pre-incubated at room temperature for 24 h. 5 g of the pre-incubated sample were transferred into an incubation vessel containing 20 ml of 1 M NaOH. The vessel was sealed and kept at 37 °C for 4 days. The amount of CO_2 absorbed by NaOH was determined by back titration of the residual NaOH with 1 N HCl using phenolphthalein as the indicator.

Total viable count (TVC) of bacteria in the compost was determined by plating diluted (1:10 dilution) suspension on nutrient agar, incubated at 32 °C and colonies counted. The analysis was done in triplicate.

Kinetic models of composting

The degradation of organic matter as a function of time follows first-order kinetics (Haung 1993; Hamoda et al. 1998; USEPA 2000).

$$\mathrm{d}\frac{(OM)}{\mathrm{d}t} = -k_t \mathrm{d}t$$

where OM is the amount of biodegradable volatile solids at 't' during the composting process and k_t is the decomposition rate constant. In this study eight models were used to predict the decomposition rate constant. Two statistical parameters, namely reduced Chi-square value (χ^2) and Root Mean Square value (RMSE) were used to evaluate the goodness of fit of the developed models. Reduced Chi-square is the mean square of the deviations between the experimental and calculated values for the models and is used to determine the goodness of the fit. The lower the values of the χ^2, the better the goodness of the fit (Pangavhane et al. 1999). The RMSE gives the deviation between the predicted and experimental values and it tends to zero for ideal models. The modelling efficiency (EF) also shows the ability of the model to describe data and its highest value is 1 (Yaldiz et al. 2001). A high value of EF and a low value of RMSE are chosen as the criteria for goodness of fit.

Table 1 Physicochemical characteristics of compost feedstocks

Feedstock	pH	Electrical conductivity (dS/m)	Moisture content (%)	Volatile solids (%)	Total solids (%)	Total organic carbon (%)	Total nitrogen (mg/g)	Carbon:Nitrogen
CY	4.8 ± 0.1	6.5 ± 0.2	35.2 ± 0.2	73.1 ± 0.3	64.8 ± 0.2	42 ± 0.8	38.9 ± 0.1	10.8 ± 0.1
CYF	5.8 ± 0.2	6.4 ± 0.4	49.1 ± 0.1	73.6 ± 0.5	50.9 ± 0.1	32.9 ± 0.2	21.6 ± 0.5	15.2 ± 0.3
CYCo	4.2 ± 0.4	13.8 ± 0.6	35.8 ± 0.6	72.6 ± 0.1	64.2 ± 0.6	28.7 ± 0.4	12.6 ± 0.7	22.8 ± 0.5
CYT	6.3 ± 0.1	9.8 ± 0.1	38.5 ± 0.6	75.2 ± 0.4	61.5 ± 0.6	35.7 ± 0.3	23.3 ± 0.5	15.3 ± 0.3
CYV	6.1 ± 0.0	7.2 ± 0.7	48.5 ± 0.1	74.4 ± 0.4	51.5 ± 0.1	53.51 ± 0.1	24.4 ± 0.6	21.9 ± 0.4
CYP	5.6 ± 0.3	6.9 ± 0.5	31.6 ± 0.0	88.1 ± 0.2	68.4 ± 0.0	56.0 ± 0.0	11.5 ± 0.6	48.7 ± 0.5

$$\text{RMSE} = \sqrt{\frac{\sum_{i=1}^{n}\left(k_{\text{pre},i} - k_{\text{exp},i}\right)^2}{N}}$$

where k_{exp} is the experimental decomposition rate, $k_{\text{exp,mean}}$ the mean value of experimental decomposition rate, k_{pre} the predicted experimental decomposition rate

$$\chi^2 = \frac{\sum_{i=1}^{N}\left(k_{\text{exp},i} - k_{\text{pre},i}\right)^2}{N - n}$$

$$\text{EF} = \frac{\sum_{i=1}^{n}\left(k_{\text{exp},i} - k_{\text{exp,mean}}\right)^2 - \sum_{i=1}^{n}\left(k_{\text{pre},i} - k_{\text{exp},i}\right)^2}{\sum_{i=1}^{n}\left(k_{\text{exp},i} - k_{\text{exp,mean},i}\right)^2}$$

where 'N' corresponds to the number of observations and 'n' refers to the number of constants in the model.

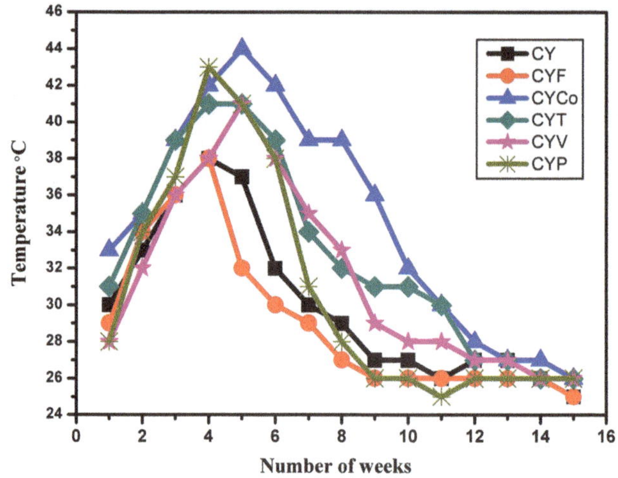

Fig. 1 Temperature changes in the in-vessel Composting process

Results and discussion

Physicochemical analysis

Composting and co-composting of yard waste with canteen and paper wastes generated in the institution were carried out under aerobic conditions and six different composts were produced. The physicochemical characteristics of feedstocks are presented in Table 1. At the end of the composting process the pH of the composted mass increased. The shift in pH towards neutral during the composting process (Table 2) favours the growth of aerobic bacteria and it drives the composting process (Agendia et al. 1997). The increase in pH values was completely attributed to the production of ammonia, presumably from protein degradation in wastes (Haug 1993). The changes in temperature during the composting process were measured at regular intervals of 24 h and varied as shown in Fig. 1. Temperature evolution reflects the microbial activity during composting and it is a good indicator of the process performance. Three temperature phases were observed: (1) an initial mesophilic phase (25–40 °C) and (2) a thermophilic phase (>40 °C) which lasted for 2–5 weeks and (3) a final mesophilic phase, that is the maturity (curing) phase was maintained for 7–10 weeks. Temperature rose from the first day of the process and became 40–44 °C on the 4th week as a consequence of the rapid decomposition of organic matter and nitrogenous compounds by the action of microorganisms. As the organic matter became more stabilized and fewer nutrients were available, the growth rate of microorganisms slowed down and the temperature in the composting process gradually decreased to ambient temperature.

The electrical conductivity reflected the degree of salinity in the compost indicating its phytotoxic/phyto-inhibitory effects on the growth of a plant, if applied to soil (Li et al. 2008). In this study, a gradual increase in EC was observed for the various composting combinations and the increase lasted from 15 to 17 days (data not shown) and it is due to the organic matter degradation and release of different mineral salts in available forms (such as phosphate, ammonium, potassium) as reported by Garg et al. (2006). However, at the later stage of composting it was observed that the EC decreased for all the different composts produced (except CYV and CYP) (Table 2), indicating that the mineral salts could have been precipitated. These results are in agreement with findings previously reported by Wong et al. (1995). In general, the acceptable level of EC in compost is a value lower than 4 dS/m since

Table 2 Physicochemical characteristics of different composts

Feedstock	pH	Electrical conductivity (dS/m)	Moisture content (%)	Volatile solids (%)	Total solids (%)	Total organic carbon (%)	Total nitrogen (mg/g)	Carbon:Nitrogen
CY	7.4 ± 0.2	5.9 ± 0.5	49.8 ± 0.2	66.5 ± 0.2	50.2 ± 0.2	18.5 ± 2.0	17.2 ± 0.1	10.7 ± 1.6
CYF	7.6 ± 0.1	4.9 ± 1.5	59.5 ± 0.4	64.3 ± 0.1	40.5 ± 0.4	22.1 ± 2.5	19.9 ± 1.1	11.10 ± 2
CYCo	6.5 ± 0.1	6.7 ± 0.8	59.9 ± 0.1	62.3 ± 0.7	40.04 ± 0.1	6.37 ± 4.0	3.76 ± 0.3	16.9 ± 3.1
CYT	6.4 ± 0.0	7.2 ± 0.2	63.14 ± 0.1	67.8 ± 0.5	36.8 ± 0.1	26.7 ± 3.4	29.4 ± 1.2	9.1 ± 2.3
CYV	7.2 ± 0.1	9 ± 0.5	60.1 ± 0.3	65.2 ± 0.4	39.9 ± 0.3	40.5 ± 0.7	20.4 ± 0.4	19.8 ± 0.4
CYP	6.1 ± 0.1	8.1 ± 1.1	58.6 ± 0.2	80.6 ± 0.6	41.4 ± 0.2	45.7 ± 3.1	9.67 ± 0.7	47.2 ± 2.8

otherwise it inhibits plant germination and growth (Rao Bhamidimarri and Pandey 1996). Since EC was high in all types of composts produced from university generated wastes, we need to dilute the produced compost before it can serve the purpose of a soil amendment. The water extracts of the compost was used for the germination assays in this study to overcome this effect.

The Total Organic Carbon in the composts was analysed at the end of the curing phase of composting and given in Table 2. Maximum TOC reduction in the composts compared to the respective feedstocks was observed for the feedstock combination CYCo (78 % reduction in TOC compared to the starting material). This might be due to the higher content of easily degradable organic matter in CYCo (C/N: 22.8), such as proteins, cellulose, and hemicellulose which are utilized by microorganisms as C and N sources. The TOC and TKN of the composts produced from different feedstock combination varies and so does their C:N ratios. After 2 weeks, it was observed that the volatile solids were reduced by 10–12 % in the matured composts compared to the respective feedstocks (Table 2).

From Table 2, a loss of TKN in compost was observed when compared to the feedstocks except for CYT. Further, these results showed that a TKN loss (70 %) was highest for CYCo when compared to CY. This could be attributed to the immobilization of NH^{4+} ions by the waste coffee grounds that possess high surface area and micropores.

Compost stability and maturity

Compost maturity and stability is analysed using C/N ratio, stability index and seed germination index. Carbon to nitrogen ratio has been used as an indicator of the potential of compost stability. A mature compost should have an C/N ratio ≤ 25 (Woods End Research Laboratory 2005). Results showed that there was no significant reduction in C:N ratio for the feedstock CY, however, the co-composting of yard with tea wastes (CYT) reduced the C:N by 41 %. In a similar trend, the reduction of C:N ratio by 25–26 % was observed with additives of fruit and coffee waste from canteen (Table 2). Previous studies reported that compost that is not well-decomposed (immature compost) has high C:N ratio (C:N > 30:1) (Saviozzi et al. 1987; Jimenez and Garcia 1989; Washington State University—Whatcom Extension Research Report 2010). If high C:N composts are used as soil amendment, carbon compounds in such composts continue to break down once incorporated into the soil and microorganisms tend to utilize the nitrogen from the soil to assist in the process (Huang et al. 2004; Tsutsuki 2009) which leaves the root zone temporarily nitrogen-poor—thereby inversely inducing *N*-immobilization process (Joern and Brichford 2006).

The C:N ratio of all the different composts (except CYP) produced in this study have C:N < 25:1, which satisfies the basic requirements for compost. Another parameter, stability index has also been used for the evaluation of compost stability. This index based on CO_2 evolution considers a compost to be stable if the index has a value lesser than 2 mg CO_2 g^{-1} compost day^{-1} (Tani 2009). In this study, the respiration rate of different composts such as CY, CYF, CYT, CYCo were lesser than 2 mg CO_2 g^{-1} compost day^{-1} (Table 3). This is in close agreement with the evaluation of composts by Wang et al. (2004), they found that composts with respiration rate less than 2 mg CO_2 g^{-1} compost day^{-1} were not phytotoxic to seedlings and would be considered stable. On the other hand, it was noted that composts, CYP and CYV have stability index higher than 2 mg CO_2 g^{-1} compost day^{-1}, indicating these composts have not yet been stabilized.

Germination index is used as an indicator of maturity and phytotoxicity. A GI of more than 80 % indicates phytotoxic-free and mature compost (Wang et al. 2004). Figure 2 shows the effect of different dilutions of the compost extracts on the germination index of cucumber seeds. The GI data suggests that direct application of compost bio-wash without dilutions showed high level of toxicity in the case of CYV and CYP and this may be attributed to the high EC of the composts. Moreover, at the dilutions of 40 and 60 % of almost all composts showed significant increase of cucumber shoot and root length. The high water holding capacity of produced composts CYT and CYP (Table 3) suggests that, it could be used as soil amendment for the plants which are less tolerant to droughts.

Synthetic fertilizers are typically high in nitrogen, phosphorous and potassium. However, the composts attained from yard waste co-composting were high in iron content (Table 4) when compared with other micronutrients. Compost is used to improve soil structure and plant nutrition (Tiquia and Tam. 1998). However, the presence of heavy metals in composts is the main cause of adverse effects on human health, transmitted through the food chain from the plants (Gigliotti et al. 1996) and it was essential to analyze the contents of heavy metal in composts. The

Table 3 Water holding capacity and stability index of the different composts

Compost	CY	CYF	CYCo	CYT	CYV	CYP
Water holding capacity (ml/100 g sample)	67.8 ± 1.2	63.5 ± 1.3	68.9 ± 1.5	72.1 ± 1.1	59.7 ± 2.0	88.6 ± 1.8
Stability index (mg CO_2 g^{-1} compost day^{-1})	1.41 ± 0.1	2.4 ± 1.2	1.06 ± 0.5	1.14 ± 0.3	8.6 ± 0.9	7.3 ± 0.4

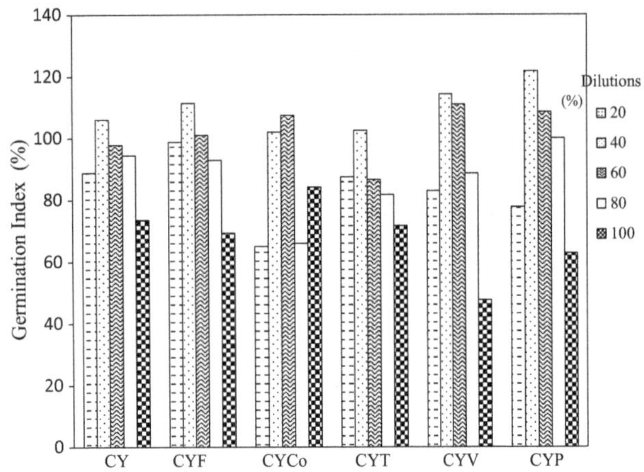

Fig. 2 Germination indexes of the different composts

heavy metal content (Cd, Cr, Ag, Cu, Ni, Pb and Zn) in the composts are shown in Table 5. It was clearly seen that heavy metal levels in the compost did not exceed the limits for application of compost as a soil amendment recommended by the USEPA (2000). Therefore, the mature compost from these feedstocks did not pose heavy metal toxicity issues.

Bacterial growth on composts

It was observed that the mesophilic bacterial density in CYF and CYCo compost (23×10^6 CFU/ml; 91×10^6 CFU/ml) was 10-fold higher when compared with CY, CYT, CYV, CYP (78×10^5, 44×10^5, 67×10^5, 48×10^5 CFU/ml) and they are equivalent to that found in fertile soils. The composts produced from institutional wastes possess good hygienic quality, since, pathogenic bacteria like, *Escherichia coli* and *Salmonella dysenteriae* were not detected in the matured compost.

Kinetic models

Proper design and operation of a composting reactor are prerequisites to compost quality and maturity. Hence, the kinetic models were developed to describe the relationship between composting rate and environmental factors over a range of practical interest. In this study eight models were used to predict the decomposition rate constant (Table 6). Models 1–3 were developed by Haung (1993) and Ekinci et al. (2001). Temperature and initial moisture content were mathematically considered in their developed models. However, different feedstocks which were tested in this study had distinct physico-chemical characteristics such as moisture, pH and EC in addition to C:N ratio. Hence, it is expected that pH and EC may also influence the rate of composting. Kinetic models have developed in this study based on these attributes and it is given in Table 6.

When the eight kinetic models were applied to the experimental results of in-vessel composting process, statistical analysis showed that the lowest RMSE value and

Table 4 Composition of micronutrients in the produced composts

Composts	Nitrogen (mg/g)	Phosphorus (mg/g)	Potassium (mg/g)	Iron (mg/g)	Magnesium (mg/g)	Calcium (mg/g)	Sodium (mg/g)
CY	17.2 ± 0.1	1.2 ± 1.2	1.05 ± 0.3	68.1 ± 0.2	1.11 ± 1.2	20.3 ± 2.7	0.63 ± 1.5
CYF	19.9 ± 1.1	1.72 ± 2.2	1.48 ± 0.1	46.4 ± 1.1	1.12 ± 2.1	33.6 ± 1.6	0.68 ± 1.8
CYCo	3.76 ± 0.3	1.78 ± 1.8	1.06 ± 0.4	68.9 ± 2.1	1.10 ± 3.3	12.2 ± 1.2	1.20 ± 2.1
CYT	29.4 ± 1.2	3.45 ± 1.4	1.08 ± 0.2	59.1 ± 1.3	1.09 ± 1.8	13.0 ± 2.2	0.84 ± 0.8
CYV	20.4 ± 0.4	1.80 ± 3.1	1.07 ± 0.0	47.9 ± 0.1	1.10 ± 2.6	11.2 ± 2.8	1.45 ± 1.3
CYP	9.67 ± 0.7	1.35 ± 0.8	1.01 ± 0.1	45.6 ± 2.2	1.09 ± 2.0	13.6 ± 1.2	0.92 ± 0.9

Table 5 Heavy metal content in the produced composts

Composts	Cadmium (mg/kg)	Chromium (mg/kg)	Silver (mg/kg)	Copper (mg/kg)	Nickel (mg/kg)	Lead (mg/kg)	Zinc (mg/kg)
CY	3.57 ± 1.1	12.30 ± 0.3	3.8 ± 0.1	140 ± 0.2	7.8 ± 0.3	27.7 ± 0.3	887 ± 0.1
CYF	1.1 ± 0.6	9.8 ± 0.1	3.1 ± 0.1	123.8 ± 0.1	4.7 ± 0.2	21.9 ± 0.4	886 ± 0.2
CYCo	2.2 ± 0.3	20.3 ± 0.2	3.5 ± 0.1	88.5 ± 0.6	8.4 ± 0.1	24.5 ± 0.2	550 ± 0.1
CYT	2.6 ± 0.4	21.6 ± 0.1	3.15 ± 0.0	67.7 ± 0.5	12.7 ± 0.1	37.5 ± 0.3	505 ± 0.2
CYV	3.5 ± 0.2	11.8 ± 0.3	2.3 ± 0.1	46.7 ± 0.2	8.20 ± 0.0	23.6 ± 0.1	453 ± 0.3
CYP	1.3 ± 0.4	50.9 ± 0.1	1.91 ± 0.0	18.3 ± 0.6	5.88 ± 0.3	34.9 ± 0.0	737 ± 0.1
Recommended heavy metal limits by USEPA (mg/kg)	35	1200	1100	1500	420	300	2800

Table 6 Various kinetic models used in this study

Sr. No.	Model	References
1.	$k_T = k_{20}a(T-20)$	Haung (1993)
2.	$k_T = k_{23}a(T-23)$	Kulcu and Yaldiz (2004)
3.	$k_T = a\exp\left(b\left[\left(\frac{m_i-c}{d}\right)+\left(\frac{T-e}{f}\right)\right]\right)$	Ekinci et al. (2001)
4.	$k_T = a\exp\left(b\left[\left(\frac{m_i-c}{d}\right)+\left(\frac{T-e}{f}\right)\right]\right) - g\frac{E_c}{T}$	This study
5.	$k_T = a\exp\left(b\left[\left(\frac{m_i-c}{d}\right)+\left(\frac{T-e}{f}\right)\right]\right)(p_H)^h$	This study
6.	$k_T = a^{(T-23)}(p_H)^b(EC)^c$	This study
7.	$k_T = \frac{a}{p_H}\exp\left[\left(bT-\left(\frac{m_i-c}{d}\right)\right)\right]$	This study
8.	$k_T = (p_H)^a\exp\left(b\left[\left(\frac{m_i-c}{d}\right)+\left(\frac{T-e}{f}\right)\right]\right) - g\frac{E_c}{T}$	This study

Table 7 Applied kinetic models and results of statistical analyses

Model No.	a	b	c	d	e	f	g	RMSE	χ^2	EF
1	1.065	$K_{20} = 0.0006$	–	–	–	–	–	0.00014	2.03×10^{-8}	0.886
2	1.065	$K_{23} = 0.0007$	–	–	–	–	–	0.00014	2.03×10^{-8}	0.889
3	0.0001	0.06	−46.2	5.47	10.7	0.99		0.00011	1.99×10^{-8}	0.990
4	0.0003	0.0025	−35.02	0.22	15.9	0.04	1.23	0.00008	1.13×10^{-8}	0.995
5	0.0005	−0.054	51.65	−9.9	36.9	15.1	0.66	0.0001	1.69×10^{-8}	0.992
6	1.054	−2.87	−0.92	–	–	–	–	0.00031	1.12×10^{-7}	0.926
7	0.0005	0.054	−13.58	−50.17	–	–	–	0.00014	2.35×10^{-8}	0.986
8	0.900	−0.043	−33690	31890	153	−0.64	0.93	0.00009	1.41×10^{-8}	0.993

the highest EF values were found in model 4 (Table 7). According to the statistical results of the models, the best model to design the in-vessel composting process of university generated wastes as given below:

$$k_T = a\exp\left(b\left[\left(\frac{m_i-c}{d}\right)+\left(\frac{T-e}{f}\right)\right]\right) - g\frac{E_c}{T}$$

In this model, 'k_T' is the decomposition rate constant (1/day), 'm_i' the initial moisture content (weight %), 'T' the process temperature (°C), 'E_c' the electrical conductivity (dS/m) and a, b, c, d, e, f and g are empirical constants.

Conclusions

This study demonstrated that the co-composting of yard wastes with paper and canteen wastes produced value added products. The respiration rates of different composts such as CY, CYF, CYT, CYCo which were lesser than 2 mg CO_2 g^{-1} compost day^{-1}, suggests that they were not phytotoxic to seedlings and would be considered stable. Whereas, co-composting with paper and vegetable waste did not achieve maturity in the same duration. The high cellulosic content in paper waste which resulted in the high initial C:N ratio and the presence of difficult to degrade (coconut husks, coir, peals, etc.) materials in the vegetable waste could be the reason for their slow composting. The dilutions of 40 and 60 % of almost all composts showed significant increase of cucumber shoot and root length. The composts produced from different feedstocks had different levels of heavy metals, but it did not exceed the limits for application of compost as a soil amendment. The nature of soil determines what kind of nutrient amendments have to be added to it and this might indicate based on compost quality, which compost will be suited for a particular soil. Finally, the developed kinetic models for the composting process revealed that the in-vessel composting rates of the institutional wastes used in this study were best represented by a model as a function of temperature, electrical conductivity and initial moisture content.

References

Agendia LP, KengneNoumsi YM, Fonkou T, Mefenya R, Sonwa JD (1997) Production of compost from biomass of Pistiastratiotes:

treatment of domestic waste water in Yaoundé (Cameroon). Cah Agric 6:15–19
American Planning Association (2006) Planning and Urban design Standards. Wiley, New York
APHA (American Public Health Association) (1992) Standard Methods for the Examination of Water and Wastewater, APHA-AWWA-WEF, 18th edn
Diacono M, Montemurro F (2010) Long-term effects of organic amendments on soil fertility. A review. Agron Sustain Dev 30:401–422
Diaz LE, Savage GM, Eggerth LL, Golueke CG (1993) Composting and recycling municipal solid waste. Lewis Publisher, CRC. Press Inc., pp 160–162
Diaz MJ, Madejon E, Lopez F, Lopez R, Cabrera F (2002) Optimization of the rate vinasse/grape marc for co-composting process. Process Biochem 37:1143–1150
Ekinci K, Keener HM, Michael FC, Elwell DL (2001) Effects of temperature and initial moisture content on the composting rate of short paper fiber and broiler litter. ASAE Annual Meeting, California
Elwell DL, Keener HM, Hansen RC (1996) Controlled, high rate composting of mixtures of food residuals, yard trimmings and chicken manure. Compost Sci Util 4:6–15
Garg P, Gupta A, Satya S (2006) Vermicomposting of different types of waste using Eiseniafoetida: a comparative study. Biores Technol 97:391–395
Gigliotti G, Businelli D, Giusquiani PL (1996) Trace metals uptake and distribution in corn plants grown on a 6-year urban waste compost amended soil. Agr Ecosyst Environ. 58:199–206
Giusti L (2009) A review of waste management practices and their impact on human health. Waste Manage 29:2227–2239
Guidi G, Pera A, Giovannetti M, Poggio G, Bertoldi M (1988) Variations of soil structure and microbial population in a compost amended soil. Plant Soil 106:113–119
Hamoda MF, Qdais HAA, Newham J (1998) Evaluation of municipal solid waste composting. Resour Conserv Recycl 23:209–223
Hargreaves JC, Adl MS, Warman PR (2008) A review of the use of composted municipal solid waste in agriculture. Agric Ecosyst Environ 123:1–14
Haug RT (1993) The Practical Handbook of Compost Engineering. Lewis Publishers, Boca Raton
Haung RT (1993) The Practical Handbook of Compost Engineering. Lewis Publishers, Boca Raton
He XT, Traina SJ, Logan TJ (1992) Chemical properties of municipal solid waste composts. J Environ Qual 21:318–319
Huang GF, Wong JW, Wu QT, Nagar BB (2004) Effect of C/N on composting of pig manure with sawdust. Waste Manage 24:805–813
Jiang T, Schuchard F, Li G, Guo R, Zhao Y (2011) Effect of C/N ratio, aeration rate and moisture content on ammonia and greenhouse gas emission during the composting. J Environ Sci 23:1754–1760
Jimenez E, Garcia VP (1989) Evaluation of city refuse compost maturity: a review. Biol Waste 27:115–142
Joergensen RG, Mueller T (1996) The fumigation extraction method to estimate soil microbial biomass: calibration of the k EN value. Soil Biol Biochem 28:33–37
Joern BC, Brichford SL (2006) Calculating manure and manure nutrient application rates. Purdue Univ Coop Ext Serv Agron Guide 8:1–6
Kulcu R, Yaldiz O (2004) Determination of aeration rate and kinetics of composting some agricultural wastes. Bioresour Technol 93(1):49–57
Li XJ, Zhang RH, Pang YZ (2008) Characteristics of dairy manure composting with rice straw. Biores Technol 99:359–367
Lung AJ, Lin CM, Kim JM, Marshall MR, Nordstedt R, Thompson NP, Wei CI (2001) Destruction of *Escherichia coli O157:H7* and *Salmonella enteritidis* in cow manure composting. J Food Protect 64:1309–1314
Misra RV, Roy RN (2003) On-farm composting methods. Food and Agricultural Organisation of United Nations, Rome
Mulvaney RL (1996) Nitrogen—inorganic forms. In: Bigham JM et al (eds) Methods of soil analysis, part 3: chemical methods. Soil Science Society of America, Madison, p 1123
Murphy J, Riley JP (1962) A modified single solution method for the determination of phosphate in natural waters. Anal Chim Acta 27:31–36
Nelson DW, Sommers LE (1982) Total carbon, organic carbon, and organic matter. In: Page AL, Miller RH, Keeney DR (eds) Methods of soil analysis, part 2, second edn, Agron. Monogr. 9. ASA-CSSA-SSSA. Madison, pp 539–577
Ohtaki A, Sato N, Nakasaki K (1998) Biodegradation of poly-caprolactone under controlled composting conditions. Polym Degrad Stabil 61:499–505
Pangavhane DR, Sawhney RL, Sarsavadia PN (1999) Effect of various dipping pretreatment on drying kinetics of Thompson seedless grapes. J Food Eng 39:211–216
Rand T, Haukohl J, Marxen U (2000) Municipal Solid Waste Incineration, A Decision Maker's Guide. The International Bank for Reconstruction and Development, World Bank, Washington, DC
RaoBhamidimarri SM, Pandey SP (1996) Aerobic thermophilic composting of piggery solid wastes. Water Sci Technol 33:89–94
Renkow M, Rubin AR (1998) Does municipal and solid waste composting make economic sense? J Environ Manage 53:339–347
Rhoades JD (1982) Soluble salts. In: Page AL, Miller RH, Keeney DR (eds) Methods of soil analysis, part 2, second edn, Agron. Monogr. 9. ASA-CSSA-SSSA. Madison, pp 167–180
Sartori G, Ferrari A, Pagliai M (1985) Changes in soil porosity and surface shrinkage in a remolded, saline clay soil treated with compost. Soil Sci 139:523–530
Saviozzi A, Riffaldi R, Levi-Minzi R (1987) Compost maturity by water extract analyses. In: De Bertoldi M, Ferranti MP, Hermite PL, Zucconi F (eds) Compost: production, quality and use. Elsevier, London
Senesi GS, Baldassarre G, Senesi N, Radina B (1999) Trace element inputs into soils by anthropogenic activities and implications for human health. Chemosphere 39:343–377
Solvita (2009) Official Solvita Guideline Compost Emissions Test-V.6.0. Woods End Laboratories, Inc
Tani M (2009) Analysis of Total Carbon and Nitrogen in Composts and Manures. Obihiro University of Agriculture and Veterinary Medicine, Obihiro
Tiquia SM, Tam NFY (1998) Elimination of phytotoxicity during co-composting of spent pig-manure sawdust litter and pig sludge. Biores Technol 65:43–45
Tsutsuki K (2009) Fundamentals in compost preparation and Utilization. Obihiro University of Agriculture and Veterinary Medicine, Obihiro
United States Department of Agriculture and the United States Composting Council (2001) Carbon dioxide evolution rate 05.08-B. In: Thompson WH (ed) Leege PB, Milner PD, Watson ME (co. eds) Test methods for examination of composting and compost. Jointly Published by the United States Department of Agriculture and the United States Composting Council. Accessed 05 August 2011
United States Environmental Protection Agency (USEPA) (2000) Biosolids technology fact sheet—in vessel composting of biosolids. EPA 832-F-00-061
Wang P, Changa CM, Watson ME, Dick WA, Chen Y, Hoitink HAJ (2004) Maturity indices for composted dairy and pig manures. Soil Biol Biochem 36:767–776

Wong JWC, Li SWY, Wong MH (1995) Coal fly ash as a composting material for sewage sludge: effects on microbial activities. Environ Technol 16:527–537

Woods End Research Laboratory (WERL) (2005) Interpreting Waste and Compost Test. J Woods End Res Lab 2:1–10

WSU (Washington State University) (2010) Fundamentals of composting: why compost, materials and methods to ensure quality compost (Whatcom Extension Research Report). Washington State University, Washington

Yaldiz O, Ertekin C, Uzun HI (2001) Mathematical modelling of thin layer solar drying of seedless sultana grapes. Energy 26:457–465

Zucconi F, Pera A, Forte M, De Bertolli M (1981) Evaluating toxicity of immature compost. Biocycle 22:44–46

18

Cultivable bacterial diversity and early plant growth promotion by the traditional organic formulations prepared using organic waste materials

Rangasamy Anandham[1,2] · Nagaiah Premalatha[1] · Hyeong Jin Jee[2] · Hang Yeon Weon[3] · Soon Wo Kwon[3] · Ramasamy Krishnamoorthy[1] · Pandiyan Indira Gandhi[4] · Yong Ki Kim[2] · Nellaiappan Olaganathan Gopal[1]

Abstract

Purpose Traditional organic formulations are widely used as a plant growth promoters; however, the knowledge on the microbial aspect of traditional organic formulations is still limited. The aim of this study was to illustrate the cultivable bacterial diversity of various traditional organic formulations and their potential for early plant growth promotion.

Methods Five different traditional organic formulations such as 100 % *panchagavya*, 33 % *panchagavya*, plant extract with native microorganisms, commercial organic fertilizer extract with two percent leaf soil extract and commercial organic fertilizer extract with 2 % yogurt were prepared and used in this study. The liquid fraction of these traditional organic formulations were used to analyze the beneficial effect on plant growth by seed treatment and foliar applications.

Results Bacterial 16SrDNA analysis revealed that the isolates fell into forty-three different genera, which can be grouped into seven different classes, such as *Alphaproteobacteria*, *Actinobacteria*, *Bacilli* (or *Firmibacteria*), *Betaproteobacteria*, *Cytophagia*, *Flavobacteriia* and *Gammaproteobacteria*. Higher bacterial diversity was observed in cow dung followed by 33 and 100 % *panchagavya*. Radish and Chinese cabbage seed germination and growth were significantly improved by traditional organic formulations compared to control.

Conclusion The results of this study showed that the bacterial diversity changes depend on the type and concentration of ingredients used in traditional organic formulations. Substantial increase in plant growth by the traditional organic formulations indicates the suitability of using these organic preparations in eco-friendly agriculture.

Keywords Eubacterial diversity · Organic farming · Organic formulations · *Panchagavya* · Plant growth-promoting bacteria

Electronic supplementary material The online version of this article (doi:10.1007/s40093-015-0107-1) contains supplementary material, which is available to authorized users.

✉ Rangasamy Anandham
anandhamranga@gmail.com

[1] Department of Agricultural Microbiology, Agricultural College and Research Institute, Tamil Nadu Agricultural University, Madurai 625 104, Tamil Nadu, India

[2] Organic Agriculture Division, National Academy of Agricultural Science, Rural Development Administration, Suwon 441-707, Republic of Korea

[3] Korean Agricultural Culture Collection, National Agro Biodiversity Center, National Academy of Agricultural Science, Rural Development Administration, Suwon 441-707, Republic of Korea

[4] Regional Research Station, Tamil Nadu Agricultural University, Vridhachalam 606 001, Tamil Nadu, India

Introduction

Intensive agriculture depends on expensive chemical inputs from off farm, which generates lot of pollutants and affects environment (Horrigan et al. 2002). Due to the negative impact of chemical inputs, organic farming gained more attention from last few decades (Badgley et al. 2007). The Food and Agricultural Organization of the United Nations (FAO) estimated that the area under organic agriculture was 37.2 million ha in 2011, which was three times higher than that of 1999 (FAO 2013). In organic farming, synthetic pesticides, mineral fertilizers, genetically modified organisms and sewage sludge applications are strictly excluded. Nonetheless, biofertilizers or traditional organic formulations prepared by using organic materials/wastes are encouraged to use as a plant nutrition (Badgley et al. 2007). Numerous studies stated that the yield and quality of plant products produced by organic farming are considerably better than that of conventional farming (Chemura 2014; Oliveira et al. 2013; Luthria et al. 2010). Organic farming not only improves the quality of the food products, but it also improves the soil fertility. In a long-term study in Switzerland, organic farming with composted animal manure improved soil quality significantly as compared to conventional farming with mineral fertilizer (Maeder et al. 2002).

Traditional organic formulations may contain numerous plant growth-promoting bacteria (PGPB), which may enhance plant growth by nitrogen fixation, growth hormone production and control phytopathogens (Amalraj et al. 2013; Naik and Sreenivasa 2009). In many Asian countries, farmers formulate their own organic formulations by combining different organic materials and treating them by fermentation or composting. For example, in India, *panchagavya* (PG) is one of the widely used traditional organic formulations, which is mostly prepared by farmers themselves. *Panchagavya* is a fermented product made from five ingredients obtained from cow, such as milk, urine, dung, curd and clarified butter (Amalraj et al. 2013). South Korean farmers use wild weed wineberry (*Rubus phoenicolasius*) extract with native microorganisms, which is prepared by mixing wild weed extract, salt, molasses and decomposed leaf soil extract. Thailand farmers use fermented plant extracts or fermented plant juices, which can promote plant growth and act as a biocontrol agent (Kantachote et al. 2009). *Bokashi* is a Japanese traditional organic formulation preparation which contains effective microorganisms (EM). Japanese first developed the EM concept during 1980s, and this preparation contains various PGPB and yeast (Yamada and Xu 2001). These traditional organic formulations have not been well characterized microbiologically.

In recent past, several culture-independent methods have been developed to illustrate higher percent of bacterial population than usually found with cultural-dependent methods (Videira et al. 2013). These culture-independent methods may give in-depth knowledge about bacterial diversity, but it is impossible to use this information to identify the physiology of particular bacterial strain, interaction between microorganisms and response of bacterial strains to different environmental conditions (Stewart 2012). However, the culture-dependent method can be more useful in studying bacterial physiology and their role in plant growth promotion (Palleroni 1997). In addition, a cultural-dependent method may be helpful in industrial applications such as developments of antibiotics or biofertilizers. Only a few reports elucidate the microbial population in different organic formulations prepared by farmers (Radha and Rao 2014; Amalraj et al. 2013; Girija et al. 2013). Recently, higher number of cultivable bacterial genera was obtained from the organic formulation prepared using fermented cow manure (Giannattasio et al. 2013). In addition, few novel and plant growth-promoting bacteria such as *Larkinella bovis* sp. nov. and *Microbacterium suwonense* sp. nov. were isolated from traditional organic formulations and tested for their plant growth promotion (Anandham et al. 2011a, b; Sreenivasa et al. 2009).

In this study, culture-dependent method was used to illustrate the bacterial diversity in various traditional organic formulations made from organic materials and evaluated the effect of organic formulations on early growth of radish and Chinese cabbage. Successfully cultivated bacterial isolates may be used to develop a microbial consortium to support sustainable agricultural practice in future.

Materials and methods

Traditional organic formulations preparations

Organic formulations such as 100 % PG, 33 % PG, plant extract with native microorganism (PNM), commercial organic fertilizer extract with 2 % leaf soil extract (OLE) and commercial organic fertilizer extract with 2 % yogurt (OY) were prepared and used in this study. Ingredients and their concentration of different traditional organic formulations are given in Table 1.

Commercially available pelleted organic fertilizer (OF) extract was used to prepare OLE and OY. Pelleted OF contains 55 % caster bean waste, 40 % rice bran and 5 % palm waste with the nutrient content of 45 % N, 1.5 % P and 1.0 % K. For preparing an organic fertilizer extract, forty gram OF was boiled in 400 ml of water (sterile distilled water) for 30 min and then the filtrate was collected using the filter paper (Whatman No. 2–90 mm). Leaf soil extract was prepared by mixing 20 g of soil [containing decomposed leaf, collected from the hilly region Gangwon Province of South Korea (37°49′22.64″N; 128°09′23.00″E)] with 120 ml of water (sterile distilled water) and shaken for 30 min, then filtered through filter paper (Whatman No. 2–90 mm). The ingredients of the each traditional organic formulation were mixed in a flask and allowed for fermentation for 5 days in shaking condition (120 rpm) at 30 °C. During fermentation, microorganisms can degrade the organic materials and enhance the production of organic formulations. All five organic

Table 1 Ingredients and their concentrations of different organic formulations

100 % PG	33 % PG	PNM	OLE	OY
Cow dung (1.5 g) Cow urine (4.5 ml) Cow milk (3.0 ml) Yogurt (3.0 g) Molasses (4.5 g) Potato (3.0 g) and make up to 1 l with water	Cow dung (0.5 g) Cow urine (1.49 ml) Cow milk (0.99 ml) Yogurt (0.99 g) Molasses (4.5 g) Potato (0.99 g) and make up to 1 l with water	0.2 % wild weed wineberry extracts, 0.1 % sodium chloride 0.5 % molasses and make up to 1 l with water	40 % OF extract, 2 % molasses 0.2 % sodium chloride 2 % leaf soil extract and make up to 1 l with water	2 % yogurt 40 % OF extract 2 % molasses and make up to 1 l with water

formulations were separately filtered through a cloth filter (performed under aseptic condition to avoid contamination), and then, the liquid fraction was used for further experiments.

Cultivable bacterial diversity analysis

Bacteria were isolated from the liquid fraction of various traditional organic formulations using trypticase soybean agar (TSA; Difco), Lysogeny agar (LA; Difco) and R_2A agar (Difco). In addition, PG ingredients (cow urine, cow dung and cow milk) were also used for bacterial isolation (by serial dilution and plating) and diversity analysis. Isolated bacterial cultures (169 isolates) were grouped into different genera based on their short-length sequencing of 16S rDNA with 895F (CRCCTGGGGAGTRCRG), 47F (GCGCCAGGGAATTCTAAAAC) and 783R (ACCMGG GTATCTAATCCKG) primers. Only 96 bacterial isolates were selected (by avoiding duplicate within the treatments), and nearly full-length sequence of 16S rDNA was sequenced for species-level identification. Genomic DNA of 96 bacterial isolates was extracted using Ultra Clean DNA isolation kit (MO BioLaboratorics Inc., Carlsbad, CA, USA) and amplified using universal 16S rDNA primers (27F and 1492R). The 16S rDNA nucleotide sequences were identified by PCR direct sequencing by fluorescent dye terminator method (ABI Prism™ Bigdye™ Terminator cycle sequencing ready reaction kit v.3.1; Applied Biosystems, Foster, CA, USA), and the products were purified by Millipore-Montage dye removal kit (Beckman Coulter Inc., Fullerton, CA, USA). Finally, the products were run into an ABI3730XL capillary DNA sequencer (50 cm capillary). 16S rDNA sequences from the automatic sequencer were aligned, and bacterial identities deduced by using the EzTaxon-e server (http://eztaxon-e.ezbiocloud.net/) to ascertain their closest relatives. The sequences obtained in this study were submitted in GenBank, with the accession numbers from GQ246652 to GQ246748. 16S rDNA sequences were aligned using the CLUSTALW program, and the phylogenetic tree was constructed using MEGA version 6.0 (Tamura et al. 2013). Neighbor-joining tree was constructed with the bootstrap values of 1000.

Pathogen detection in organic formulations preparations and their ingredients

Five traditional organic formulations and PG ingredients were tested for the presence of pathogens such as *Staphylococci* and *Enterobacteriaceae* using the STAPH PLUS (Thermo Fisher Scientific Inc., Waltham, MA, USA) and RapID One (Thermo Fisher Scientific Inc., Waltham, MA, USA), respectively, according to the manufacturers instruction.

Plant growth experiments

Radish (*Raphanus sativus* L. cultivar Hong Feng No. 1) and Chinese cabbage (*Brassica rapa* subsp. *pekinensis* cultivar Norangchusurk) seeds were surface sterilized and imbibed in 5 ml liquid fraction of five different organic formulations with 0, 50 and 100 times dilutions and incubated at 25 ± 2 °C for 72 h for seed germination. The seeds were taken as germinated if the radicle emerges from the seeds. Three-day-old germinated seeds were transplanted into 40-L pot containing 20-kg pot mixture (pH 7.0; EC—1.2; NH_4^+N—170 mg kg^{-1}; NO_3^-N—110 mg kg^{-1}; available P_2O_5—300 mg kg^{-1}; K_2O—340 mg kg^{-1}; CEC—10 cmol$_c$ kg^{-1}) and maintained for 30 days. Pot experiment was performed with sixteen treatments (five organic fertilizers with three dilutions and one control). The plants were irrigated once in 2 days until the pot mixture reached 70 % of its water-holding capacity. Additional to seed treatment, 25 ml of respective traditional organic formulations was sprayed on the plants until runoff using the handheld pneumatic sprayer, on seventh day after transplantation. In the case of control, sterile distilled water was sprayed to maintain the same conditions as other treatments. There were three replications for each treatment and arranged in a completely randomized block design in the greenhouse. Root length, shoot length and dry weight of the plants were measured 30 days after transplantation (DAT).

Statistical analysis

Data on microbial count and plant parameters were subjected to analysis of variance (ANOVA) and Duncan's multiple range test (DMRT) using the SAS package, version 9.1.3 (SAS Institute Inc., Cary, NC, USA) and tested significant differences at 0.05. Data on plant growth of individual traditional organic formulation treatment and their dilutions were compared with control plants in order to find out the effect of traditional organic formulations on radish and Chinese cabbage growth. Microbial count data were log10-transformed before statistical analysis. Heat map was constructed using Microsoft Excel conditional formatting.

Results

Bacterial count and cultivable bacterial diversity of various traditional organic formulations and PG ingredients

LA and TSA were found to support a higher number of bacterial group than that of R2A medium (Online Resource 1). No significant differences in bacterial count were observed among different organic formulations prepared and used in this study. Short-length sequencing of 169 bacterial isolates obtained from different organic formulations and PG ingredients (cow urine, cow dung and cow milk) revealed that the isolates were belonged to 43 different genera (Fig. 1). These 43 genera can be grouped into seven different classes, such as *Alphaproteobacteria*, *Actinobacteria*, *Bacilli* (or *Firmibacteria*), *Betaproteobacteria*, *Cytophagia*, *Flavobacteriia* and *Gammaproteobacteria* (Fig. 2). Fifty-four percentage of the cultivable bacterial population in 100 % PG preparation was *Bacilli* (or *Firmibacteria*). In particular, *Bacillus*, *Lactobacillus*, *Lactococcus*, *Leuconostoc* and *Lysinibacillus* genera were dominated. The higher proportion of *Alphaproteobacteria* was observed in 33 % PG followed by *Firmibacteria* and *Betaproteobacteria*. Other organic formulations such as PNM, OLE, OY and PG ingredients were found to be dominated by *Firmibacteria* and *Alphaproteobacteria* or *Actinobacteria* (Fig. 2). Higher number of cultivable bacterial species were observed in cow dung (54 species) followed by 33 % (37 species) and 100 % PG (28 species).

Out of 169 bacterial isolates, 96 were carefully selected for full-length sequencing by avoiding duplicates within the treatments. All sequenced isolates with closest neighbor and their accession numbers are given in Online Resource 2. Our sequencing results revealed that PG preparation had thirteen unique bacterial genera, which were belong to various PGPB and fermentative groups of bacteria. *Azospirillum*, *Brevundimonas*, *Lactobacillus*, *Xenophilius*, *Xylophilus*, *Comamonas* and *Acetobacter* are the representatives of PGPB genera found in PG preparation (Fig. 3).

Pathogen detection in traditional organic formulations and their ingredients

RapID STAPH PLUS system and RapID One analysis revealed the presence of *Staphylococcus aureus* and *Escherichia coli* in cow dung which was one of the ingredients in PG preparation, but those pathogens were not observed in PG organic formulations. In addition, 16S rDNA amplification confirmed the presence of foodborne pathogens (*Shigella flexneri*, *Bergyella zoohelcum* and *Enterococcus durans*) in cow dung and cow milk (Online Resource 2). However, these pathogens were not observed in 33 and 100 % PG organic formulations.

Germination assay and plant responses to different organic formulations and their dilution

Germination percentage of radish and Chinese cabbage treated with different organic formulations is presented in Fig. 4. Dilution of 33 % PG, PNM, OLE and OY fertilizers showed a significant increase in radish seed germination. On the other hand, traditional organic formulations did not show any significant increase in Chinese cabbage seed germination (Fig. 4).

Organic formulation-treated plants showed higher growth compared to control plants (except in 100 % PG treatment). Zero times dilution of 100 % PG had negative impact on plant growth, which registered the lower root length, shoot length and dry weight of radish than that of diluted treatments (Fig. 5; Online Resource 3, 4). Higher root length of radish (25.6 cm) and Chinese cabbage (33.33 cm) was observed in OY (50 times diluted) and 33 % PG (zero times diluted) treatment, respectively (Online Resource 3). Radish and Chinese cabbage shoots length was found to be higher in 33 % PG and OY (zero times diluted) treatment, respectively (Online Resource 4). Higher dry matter accumulation was found in OLE-treated (100 times diluted) radish and Chinese cabbage plants (Fig. 5). Dry weight of radish was significantly improved by 100 % PG, 33 % PG, PNM and OLE treatments than that of control plants. Traditional organic formulations have no significant impact on Chinese cabbage dry weight.

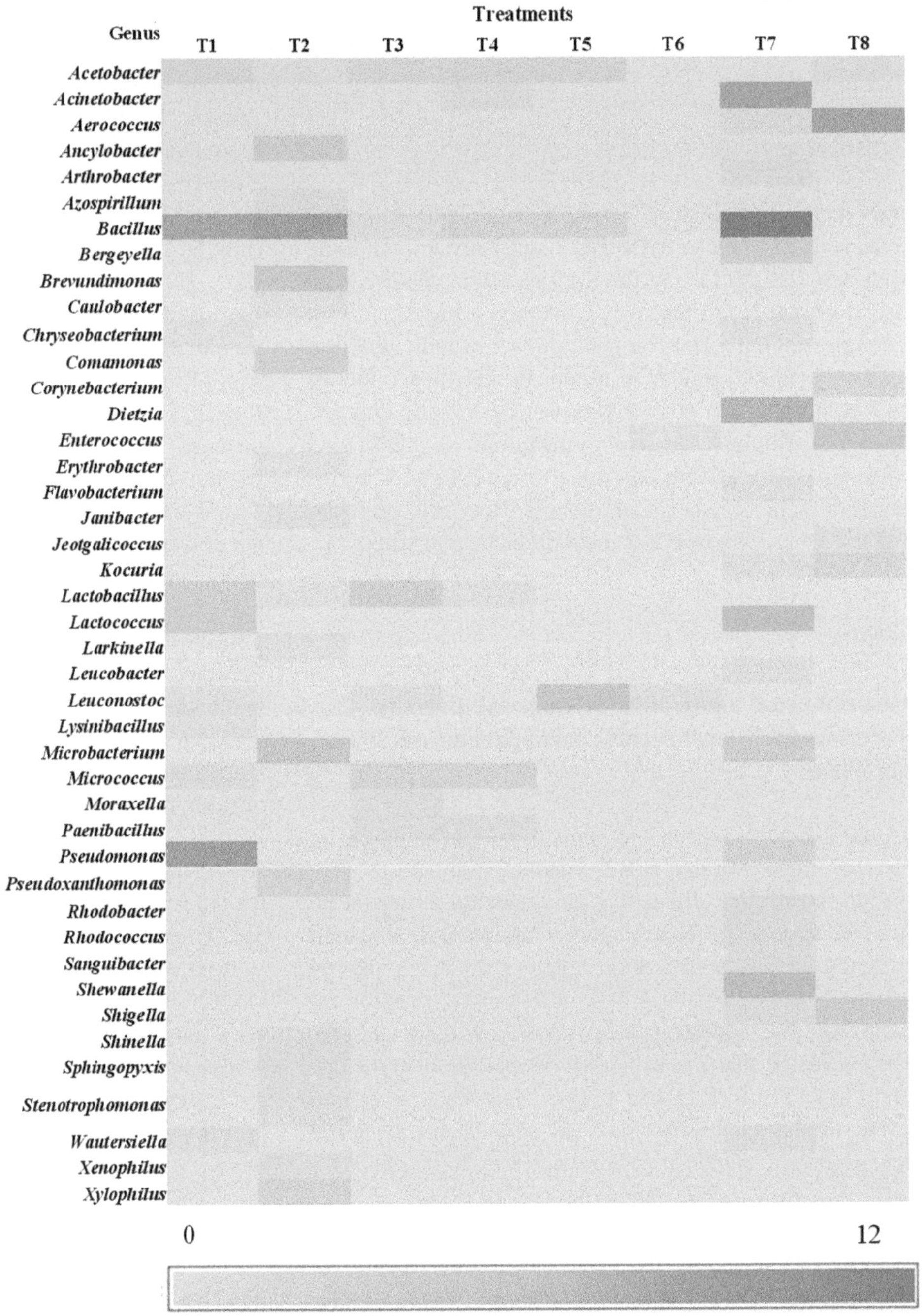

Fig. 1 Heat map shows the number of bacterial species isolated from different organic formulations and their ingredients. *T1* 100 % *panchagavya*; *T2* 33 % *panchagavya*; *T3* plant extract with native microorganisms; *T4* commercial organic fertilizer extract with 2 % leaf soil extract; *T5* commercial organic fertilizer extract with 2 % yogurt; *T6* cow urine; *T7* cow dung; *T8* cow milk

Discussion

Farmer utilizes locally available organic materials for preparing organic formulations, and there are many well-developed traditional organic formulations which are available worldwide, but most of them are unpopular. Chemical and microbiological aspect of many traditional organic formulations is not well studied. Recently, nuclear magnetic resonance analysis of organic formulations revealed the presence of various bioactive compounds, which helps in plant growth improvement (Spaccini et al. 2012). In this work, we have chosen five different traditional organic formulations such as 100 % PG, 33 % PG, PNM, OLE and OY to analyze microbiological aspects and the effect of organic formulations on early plant growth was also illustrated.

Three diverse microbiological media have been used to obtain higher number of cultivable bacterial population from different organic formulations. This cultivation-dependent approach exhibited the presence of novel bacterial isolates such as *L. bovis* sp. nov. and *M. suwonense* sp. nov. in organic formulations and PG ingredients, respectively

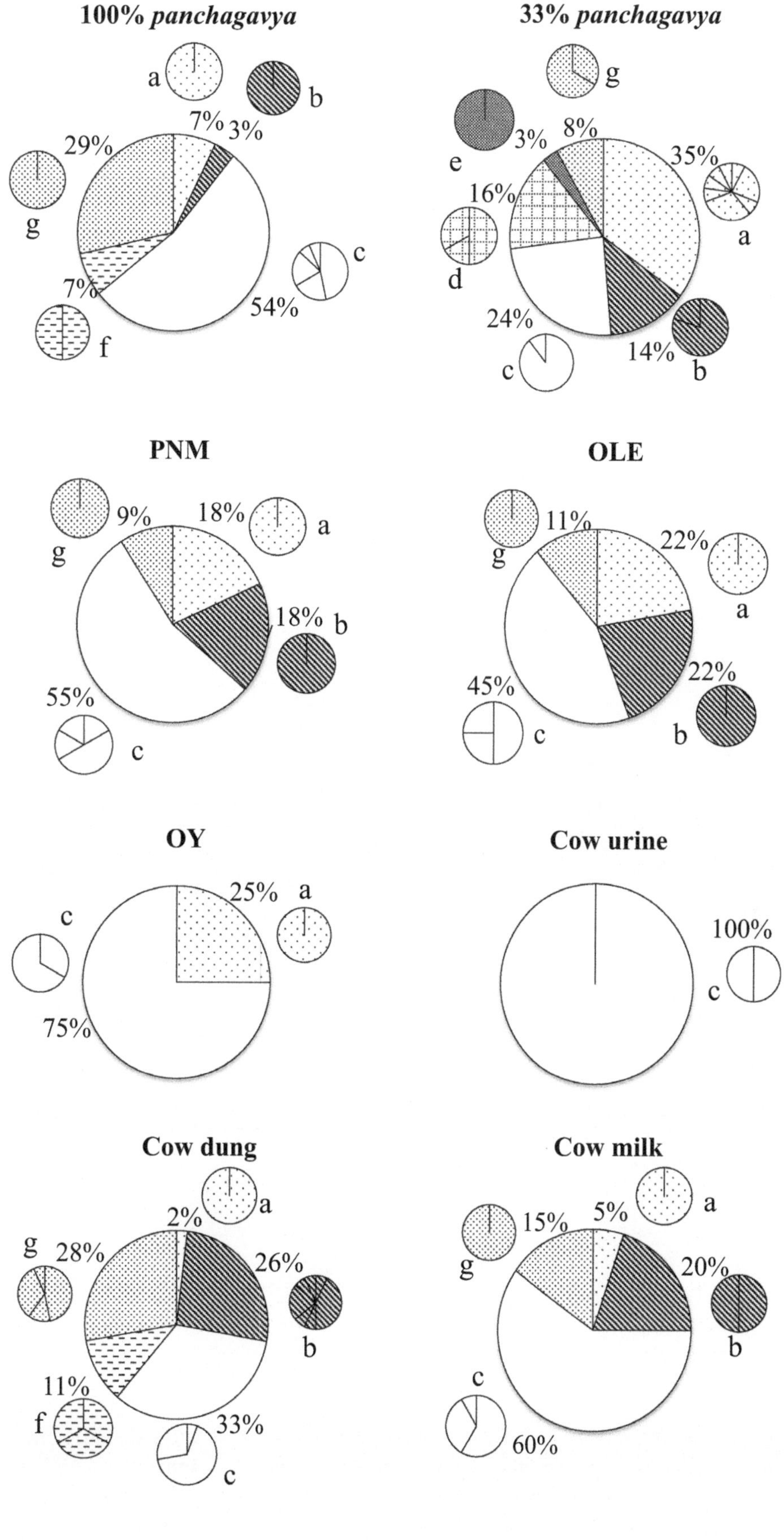

Fig. 2 Class and genus-level distribution of bacteria in different organic formulations and their ingredients. **a** *Alphaproteobacteria*; **b** *Actinobacteria*; **c** *Bacilli* (or *Firmibacteria*); **d** *Betaproteobacteria*; **e** *Cytophagia*; **f** *Flavobacteriia*; **g** *Gammaproteobacteria*. *Larger circle* represents percentages of different classes of bacteria present in organic formulations and ingredients. Partitioning of *small circle* represents number of genera present in a particular class

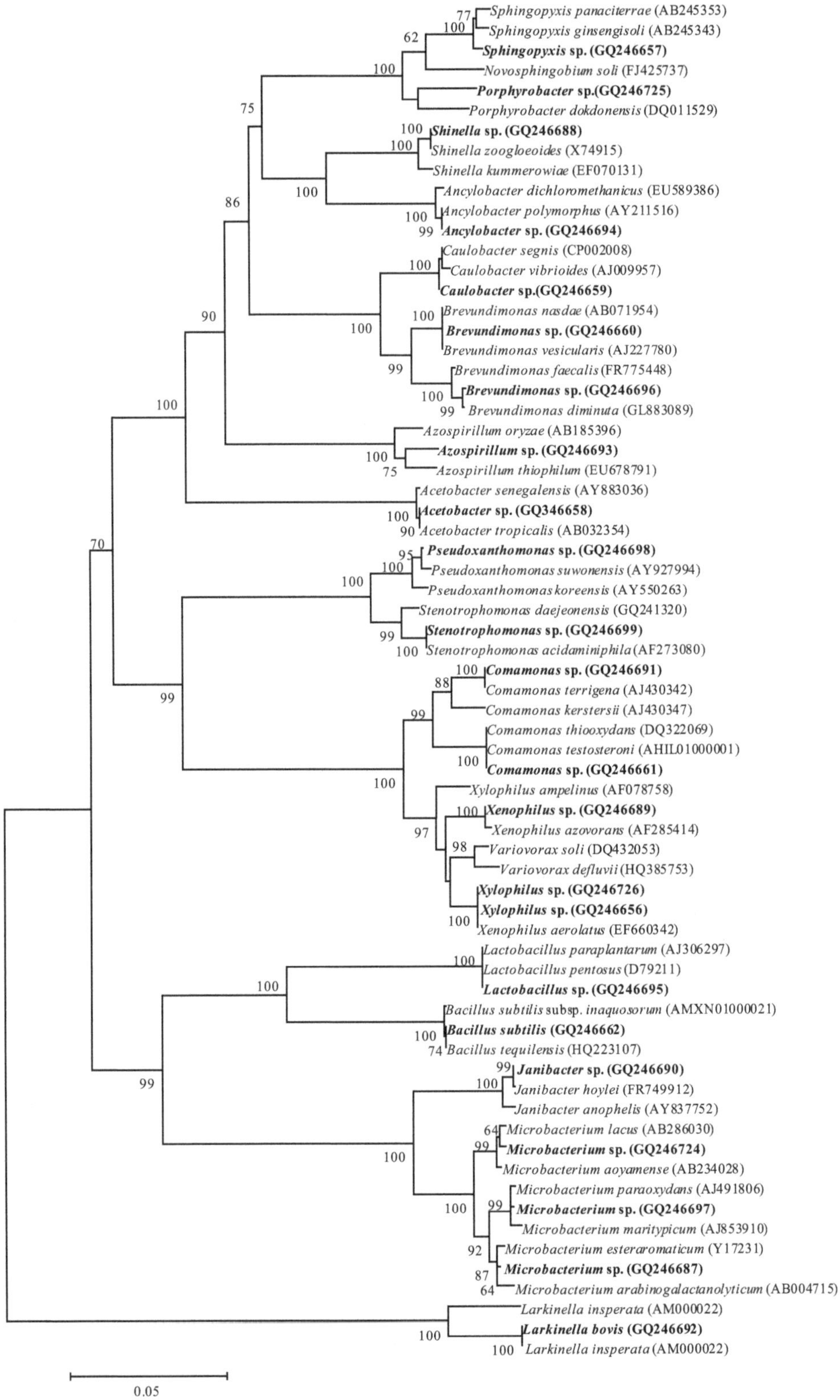

Fig. 3 Neighbor-joining phylogenetic tree shows the relationship among the bacterial species isolated from 33 % *panchagavya* preparations (highlighted in *bold*) and their closest neighbors. Tree was constructed using near-full-length sequences of bacterial 16S rDNA sequences with 1000 bootstraps values

(Anandham et al. 2011a, b). Fermentation process during the preparation of traditional organic formulations encourages different group of microorganisms to develop (Yan et al. 2013). These microorganisms may be grouped into three different groups such as plant growth promoters, fermentative microorganisms and antibiotic producers based on their functions. Fermentative group of microorganisms break down complex nutrients into simple

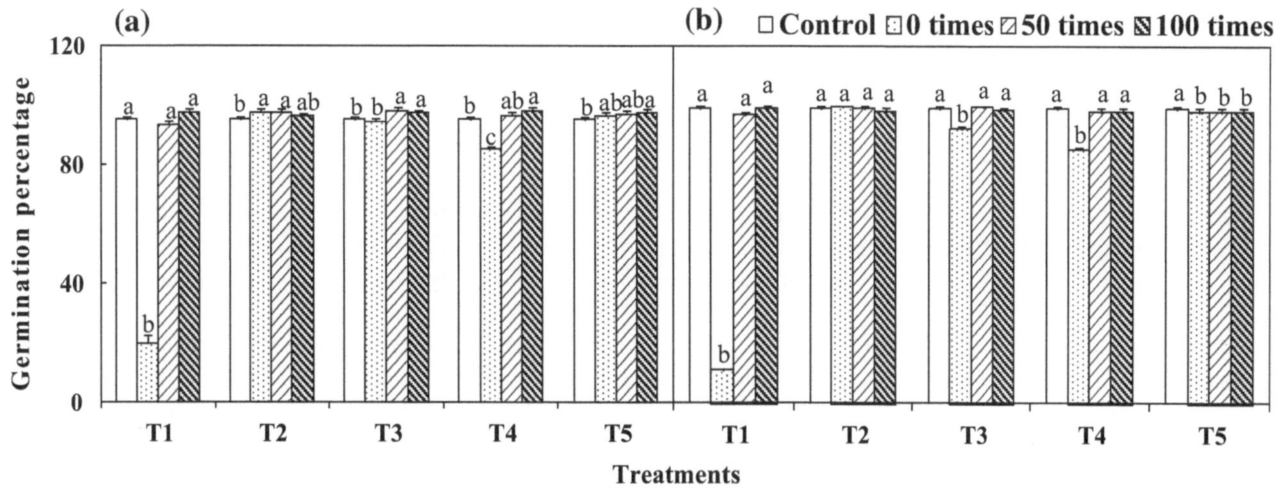

Fig. 4 Effect of seed treatment of different organic formulations on radish (**a**) and Chinese cabbage (**b**) seed germination percentage under in vitro condition. *T1* 100 % *panchagavya*; *T2* 33 % *panchagavya*; *T3* plant extract with native microorganisms; *T4* commercial organic fertilizer extract with 2 % leaf soil extract; *T5* commercial organic fertilizer extract with 2 % yogurt. Control, 0, 50 and 100 times dilution of organic formulations are represented in different shades. Values are mean ± standard error (SE) of three replications, and *same letters* in each treatment are not significantly different from each other at $P < 0.05$ according to Duncan's multiple range test (DMRT)

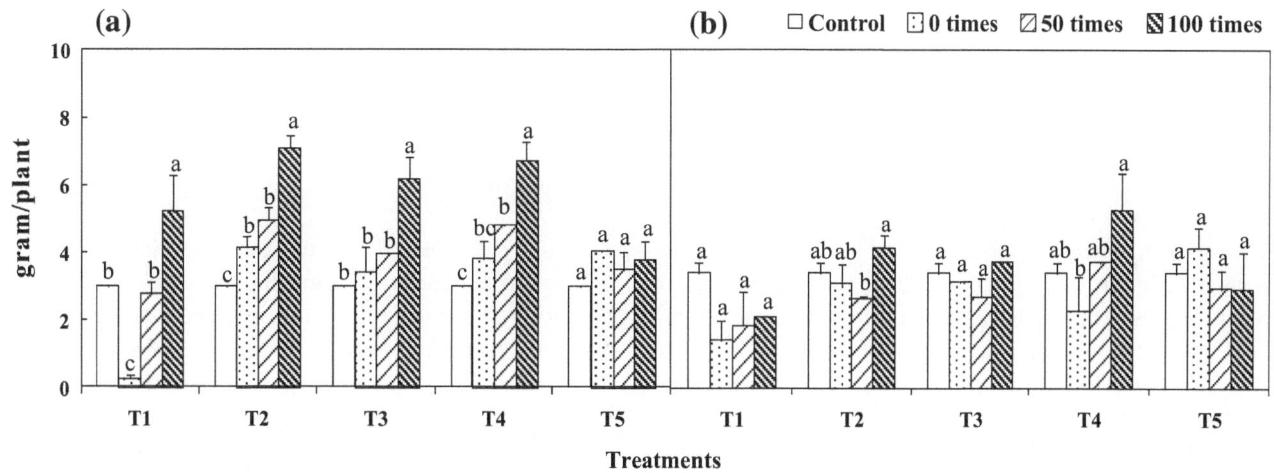

Fig. 5 Effect of different organic formulations on dry weight of radish (**a**) and Chinese cabbage (**b**) under pot culture condition. *T1* 100 % *panchagavya*; *T2* 33 % *panchagavya*; *T3* plant extract with native microorganisms; *T4* commercial organic fertilizer extract with 2 % leaf soil extract; *T5* commercial organic fertilizer extract with 2 % yogurt. Control, 0, 50 and 100 times of dilution organic formulations are represented in different shades. Values are mean ± SE of three replications, and *same letters* in each treatment are not significantly different from each other at $P < 0.05$ according to DMRT

nutrients, and antibiotic producers inhibit pathogenic bacterial growth (Yadav et al. 2013; Gea et al. 2009). On the other hand, PGPB harbored in the organic formulations improves plant growth by various mechanisms (Naik and Sreenivasa 2009). PGPB are the group of bacteria found in the rhizosphere and/or phyllosphere region of plants, which can enhance plant growth by direct and indirect mechanisms (Vacheron et al. 2013). Presence of various PGPB and slow release of nutrients are the two main reasons for plant growth promotion and improving soil health by organic formulations (Amalraj et al. 2013).

In our study, total bacterial populations in raw and processed organic formulations were not significantly changed. However, bacterial genus/species between organic formulations and their raw ingredients were changed. This clearly indicates the occurrence of bacterial succession during fermentation process. In recent past, use of pyrosequencing technique revealed the presence of 142 bacterial genera in cow dung, which includes cultivable and non-cultivable bacteria (Dowd et al. 2008). However, only 19 cultivable bacterial genera from the cow dung were able to cultivate under laboratory condition. Less number of cultivable bacterial genera obtained in this study proven the uncultivable nature of numerous bacterial genera in common laboratory media (Joseph et al. 2003). Other than 169 bacterial isolates, we were also obtained some yeast strains, which might also played an important role in the breakdown of complex sugars into simple components in the fermentation process during organic formulations preparation (Kantachote et al. 2009).

One of the PG ingredients (cow dung) was considered to be an carrier for foodborne pathogens like *S. aureus*, *E. coli*, *Salmonella* sp. and *Campylobacter* sp. (Simujide et al. 2013). In line with this, our results also showed the presence of *S. aureus* and *E. coli* pathogens in cow dung. However, these pathogens were eliminated by fermentation process during organic formulations preparation. Similar to our results, El Fels et al. (2015) showed the elimination of coliforms during the composting of organic waste material. Particularly, antibiotic-producing microorganisms in organic formulations might have played an important role in eliminating foodborne pathogens and phytopathogens (Qiu et al. 2012). Similar to this, composting of cow dung was also reported to be effective in removing foodborne pathogens such as *S. aureus* and *E. coli* (Simujide et al. 2013). Recently, Liu et al. (2013) reported that the inoculation of antagonistic bacteria with organic formulations was found to be effective in phytopathogen control.

Dominance nature of *Firmibacteria* in organic formulations used in this study might be due to the use of soil and plant extracts as one of the ingredients. Recently, Chen et al. (2013) reported the dominance of *Firmibacteria* in plant and soil ecosystem. Similarly, *Firmibacteria* was also dominated in compost prepared using agricultural waste materials (Chandna et al. 2013). PGPB such as *Acetobacter* sp. (Sevilla et al. 2001), *Azospirillum* sp. (Fages and Arsac 1991) and *Brevundimonas* sp. (Kumar and Gera 2014) found in 33 % PG were belong to *Alphaproteobacteria*. In addition, *Firmibacteria* class PGPB like *Bacillus subtilis* (Mohamed and Gomaa 2012) and *Lactobacillus* sp. (Limanska et al. 2013) were also observed (Fig. 3). Soil bacteria like *Micrococcus* sp., *Paenibacillus* sp. and *Acetobacter* sp. were dominated in PNM and OLE preparations. This may be due to the use of soil as one of the ingredient in these organic formulations preparations. Whereas *Leuconostoc mesenteroides* was found to be dominant in treatment containing OFE with 2 % yogurt, this bacterial species was mainly involved in milk fermentation process (Kihal et al. 2007).

Traditional organic formulations exhibited a significant positive impact on radish growth except 100 % PG. In line with this, many authors reported the beneficial effect of organic formulations on plant growth improvement (Naik and Sreenivasa 2009; Amalraj et al. 2013). Similarly, Pigeon pea treated with PG, vermicompost and farm yard manure improved plant growth compared to control, but the results were not significantly different among various organic formulations used (Gupta et al. 2014; Amalraj et al. 2013). Organic compost application significantly improved the dry biomass accumulation in wheat (Carpenter-Boggs et al. 2000). In agreement with this, our study also showed significantly higher radish dry biomass on organic formulations application. One hundred percent PG-treated plants showed significantly lower radish root length, shoot length, dry weight and root length of Chinese cabbage compared to that of diluted treatments and control. This might be due to the presence of plant growth inhibitors, acidic pH (5.6) and nutrients in undiluted 100 % PG. In line with this, Kadam and Pathade (2014) also showed the negative effect of higher concentration of vermicompost application on French bean growth compared to lower concentration of compost application. There were no consistent positive results for plant growth improvement on dilution of original organic formulations, except for 100 % PG. Similarly, different rate of compost application to vegetable crops showed no significant differences in plant growth between high and low concentrations (Chang et al. 2007).

Conclusion

This study illustrated the cultivable bacterial diversity of organic formulations traditionally prepared and used by farmers. The results of this study revealed that the bacterial diversity changes depend on the type and concentration of ingredients used in organic formulations. Significant increases in plant growth by organic formulations indicate the appropriateness of substituting organic formulations in the place of chemical fertilizers to enhance eco-friendly agriculture. Future study focusing on the plant growth-promoting characters of the isolates and symbiotic association among the isolates might help in developing microbial consortium for sustainable agriculture.

Acknowledgments This work was supported by the National Academy of Agricultural Science (NAAS) and the Rural Development Administration (RDA), Republic of Korea. R.A. thanks RDA for award of a postdoctoral fellowship.

References

Amalraj ELD, Praveen Kumar G, Mir Hassan Ahmed SK, Abdul R, Kishore N (2013) Microbiological analysis of panchagavya, vermicompost, and FYM and their effect on plant growth promotion of pigeon pea (*Cajanus cajan* L.) in India. Org Agric 3:23–29

Anandham R, Kwon SW, Weon HY, Kim SJ, Kim YS, Gandhi PI, Kim YK, Jee HJ (2011a) *Larkinella bovis* sp. nov., isolated from fermented bovine products, and emended descriptions of the genus *Larkinella* and of *Larkinellainsperata* Vancanneyt et al. 2006. Int J Syst Evol Microbiol 61:30–34

Anandham R, Tamura T, Hamada M, Weon HY, Kim SJ, Kim YS, Suzuki K, Kwon SW (2011b) *Microbacterium suwonense* sp. nov., isolated from cow dung. J Microbiol 49:852–856

Badgley C, Moghtader J, Quintero E, Zakem E, Chappell MJ, Vazquez KA, Samulon A, Perfecto I (2007) Organic agriculture and the global food supply. Renew Agric Food Syst 22:86–108

Carpenter-Boggs L, Reganold JP, Kennedy AC (2000) Biodynamic preparations: short-term effects on crops, soils, and weed populations. Am J Altern Agric 15:110–118

Chandna P, Mallik S, Kuhad RC (2013) Assessment of bacterial diversity in agricultural by-product compost by sequencing of cultivated isolates and amplified rDNA restriction analysis. Appl Microbiol Biotechnol 97:6991–7003

Chang EH, Chung RS, Tsai YH (2007) Effect of different application rates of organic fertilizer on soil enzyme activity and microbial population. Soil Sci Plant Nutr 53:132–140

Chemura A (2014) The growth response of coffee (*Coffea arabica* L) plants to organic manure, inorganic fertilizers and integrated soil fertility management under different irrigation water supply levels. Int J Recycl Org Waste Agric 3:59. doi:10.1007/s40093-014-0059-x

Chen F, Wang M, Zheng Y, Li S, Wang H, Han D, Guo S (2013) The effect of biocontrol bacteria on rhizosphere bacterial communities analyzed by plating and PCR-DGGE. Curr Microbiol 67:177–182

Dowd SE, Callaway TR, Wolcott RD, Sun Y, McKeehan T, Hagevoort RG, Edrington TS (2008) Evaluation of the bacterial diversity in the feces of cattle using 16S rDNA bacterial tag-encoded FLX amplicon pyrosequencing (bTEFAP). BMC Microbiol 8:125. doi:10.1186/1471-2180-8-125

El Fels L, Ouhdouch Y, Hafidi M (2015) Use of the co-composting time extract agar to evaluate the microbial community changes during the co-composting of activated sludge and date palm waste. Int J Recycl Org Waste Agric 4:95–103

Fages J, Arsac JF (1991) Sunflower inoculation with *Azospirillum* and other plant growth promoting rhizobacteria. Plant Soil 137:87–90

FAO (2013) Statistical yearbook world food and agriculture. Food and Agriculture Organization of the United Nations, Rome, pp 1–307

Gea FJ, Navarro MJ, Tello JC (2009) Potential application of compost teas of agricultural wastes in the control of the mushroom pathogen *Verticillium fungicola*. J Plant Dis Prot 116:271–273

Giannattasio M, Vendramin E, Fornasier F, Alberghini S, Zanardo M, Stellin F, Concheri G, Stevanato P, Ertani A, Nardi S, Rizzi V, Piffanelli P, Spaccini R, Mazzei P, Piccolo A, Squartini A (2013) Microbiological features and bioactivity of a fermented manure product (preparation 500) used in biodynamic agriculture. J Microbiol Biotechnol 23:644–651

Girija D, Deepa K, Xavier F, Antony I, Shidhi PR (2013) Analysis of cow dung microbiota—a metagenomic approach. Indian J Biotechnol 12:372–378

Gupta R, Yadav A, Garg VK (2014) Influence of vermicompost application in potting media on growth and flowering of marigold crop. Int J Recycl Org Waste Agric 3:47. doi:10.1007/s40093-014-0047-1

Horrigan L, Lawrence RS, Walker P (2002) How sustainable agriculture can address the environmental and human health harms of industrial agriculture. Environ Health Perspect 110:445–456

Joseph SJ, Hugenholtz P, Sangwan P, Osborne CA, Janssen PH (2003) Laboratory cultivation of widespread and previously uncultured soil bacteria. Appl Environ Microbiol 69:7210–7215

Kadam D, Pathade G (2014) Effect of tendu (*Diospyros melanoxylon* RoxB.) leaf vermicompost on growth and yield of French bean (*Phaseolus vulgaris* L.). Int J Recycl Org Waste Agric 3:44. doi:10.1007/s40093-014-0044-4

Kantachote D, Kowpong K, Charernjiratrakul W, Pengnoo A (2009) Microbial succession in a fermenting of wild forest noni (*Morinda coreia* Ham) fruit plus molasses and its role in producing a liquid fertilizer. Electron J Biotechnol 12. doi:10.2225/vol12-issue3-fulltext-12

Kihal M, Prevost H, Henni DE, Benmechernene Z, Divies C (2007) Carbon dioxide production by *Leuconostoc mesenteroides* grown in single and mixed culture with *Lactococcus lactis* in skimmed milk. World J Dairy Food Sci 2:62–68

Kumar V, Gera R (2014) Isolation of a multi-trait plant growth promoting *Brevundimonas* sp. and its effect on the growth of Bt-cotton. 3 Biotech 4:97–101

Limanska N, Ivanytsia T, Basiul O, Krylova K, Biscola V, Chobert JM, Ivanytsia V, Haertle T (2013) Effect of *Lactobacillus plantarum* on germination and growth of tomato seedlings. Acta Physiol Plant 35:1587–1595

Liu Y, Shi J, Feng Y, Yang X, Li X, Shen Q (2013) Tobacco bacterial wilt can be biologically controlled by the application of antagonistic strains in combination with organic fertilizer. Biol Fertil Soils 49:447–464

Luthria D, Singh AP, Wilson T, Vorsa N, Banuelos GS, Vinyard BT (2010) Influence of conventional and organic agricultural practices on the phenolic content in eggplant pulp: plant-to-plant variation. Food Chem 121:406–411

Maeder P, Fliebbach A, Dubois D, Gunst L, Fried P, Niggli U (2002) Soil fertility and biodiversity in organic farming. Science 296:1694–1697

Mohamed HI, Gomaa EZ (2012) Effect of plant growth promoting *Bacillus subtilis* and *Pseudomonas fluorescens* on growth and pigment composition of radish plants (*Raphanus sativus*) under NaCl stress. Photosynthetica 50:263–272

Naik N, Sreenivasa MN (2009) Influence of bacteria isolated from panchagavya on seed germination and seed vigour in wheat. Karnataka J Agric Sci 22:231–232

Oliveira AB, Moura CFH, Gomes Filho E, Marco CA, Urban L, Miranda MRA (2013) The impact of organic farming on quality of tomatoes is associated to increased oxidative stress during fruit development. PLoS One 8. doi:10.1371/journal.pone.0056354

Palleroni NJ (1997) Prokaryotic diversity and the importance of culturing. Antonie Van Leeuwenhoek J Microbiol 72:3–19

Qiu M, Zhang R, Xue C, Zhang S, Li S, Zhang N, Shen Q (2012) Application of bio-organic fertilizer can control *Fusarium* wilt of cucumber plants by regulating microbial community of rhizosphere soil. Biol Fertil Soils 48:807–816

Radha TK, Rao DLN (2014) Plant growth promoting bacteria from cow dung based biodynamic preparations. Indian J Microbiol 54:413–418

Sevilla M, Burris RH, Gunapala N, Kennedy C (2001) Comparison of benefit to sugarcane plant growth and $^{15}N_2$ incorporation following inoculation of sterile plants with *Acetobacter diazotrophicus* wild-type and Nif-mutant strains. Mol Plant Microbe Interact 14:358–366

Simujide H, Aorigele C, Wang CJ, Manda B, Lina M, Wu MY, Li Y, Ge Bai TR (2013) Reduction of food borne pathogens during cattle manure composting with addition of calcium cyanamide. J Environ Eng Landsc 21:77–84

Spaccini R, Mazzei P, Squartini A, Giannattasio M, Piccolo A (2012) Molecular properties of a fermented manure preparation used as field spray in biodynamic agriculture. Environ Sci Pollut Res 19:4214–4225

Sreenivasa MN, Naik N, Bhat SN (2009) Beejamrutha: a source for beneficial bacteria. Karnataka J Agric Sci 22:1038–1040

Stewart EJ (2012) Growing unculturable bacteria. J Bacteriol 194:4151–4160

Tamura K, Stecher G, Peterson D, Filipski A, Kumar S (2013) MEGA6: molecular evolutionary genetics analysis version 6.0. Mol Biol Evol 30:2725–2729

Vacheron J, Desbrosses G, Bouffaud ML, Touraine B, Moenne-Loccoz Y, Muller D, Legendre L, Wisniewski-Dye F, Prigent-Combaret C (2013) Plant growth-promoting rhizobacteria and root system functioning. Front Plant Sci 4:356. doi:10.3389/fpls.2013.00356

Videira S, Pereira e Silva M, Souza Galisa P, Dias A, Nissinen R, Divan V, Elsas J, Baldani J, Salles J (2013) Culture-independent molecular approaches reveal a mostly unknown high diversity of active nitrogen-fixing bacteria associated with *Pennisetum purpureum*—a bioenergy crop. Plant Soil 373:737–754

Yadav A, Gupta R, Garg VK (2013) Organic manure production from cow dung and biogas plant slurry by vermicomposting under field conditions. Int J Recycl Org Waste Agric 2:21. doi:10.1186/2251-7715-2-21

Yamada K, Xu HL (2001) Properties and applications of an organic fertilizer inoculated with effective microorganisms. J Crop Prod 3:255–268

Yan ZH, Jie L, Jing LJ, Cai LY, Fen WX, Jun CZ (2013) Microbial community dynamics during biogas slurry and cow manure compost. J Integr Agric 12:1087–1097

Effect of plant growth on some physical properties of potting culture media

A. Mohammadi-Ghehsareh[1]

Abstract

Background Physical properties of substrates considered appropriate for plant growth at planting may change over time in containers as a result of several processes. Changes include air space reduction, shrinkage of the substrate, organic matter decomposition, and physical breakdown of particles. The objective of this research was to evaluate the physical properties of date palm waste as culture media and its effect on number and yield of tomato fruit.

Results The experiment was conducted as factorial in a completely randomized block design with 9 treatments and 3 replications. Treatments included three sizes (S_1 = <0.5, S_2 = 0.5–1, and S_3 = 1–2 cm) and three composting times (C_1 = 0, C_2 = 3, and C_3 = 6 months) of date palm waste. Statistical analysis showed that the values of bulk density (BD) and water holding capacity (WHC) were significantly increased at the end of cultivation from culture media without plant in comparison to before and after planting ($p < 0.05$). Amounts of F_t (total porosity) in culture media without plant were significantly higher than those in culture media before planting and with plant ($p < 0.05$). Also during the experiment, an apparent shift in physical properties was recorded.

Conclusion The overall results of this research indicated that composting process changed the physical properties of the media before planting. Also during the experiment, an apparent shift in physical properties was recorded. The results of the study showed that composting processes continued in culture media with and without plant, in the mean time, the composting processes were higher in culture media without plant.

Keywords Soilless culture · Culture media · Date palm waste · Physical properties

Introduction

Beginning in the early 1970s, the search for organic soilless substrates for container production has been an important horticultural research topic with the introduction of hardwood and softwood barks as the primary component in nursery container substrates (Aaron 1982; Hoitink and Poole 1979). Container substrates should maintain adequate air space (AS) and container capacity [CC; which is the equivalent to the water holding capacity (WHC) of a substrate] during crop production so that growing conditions remain favorable for plant growth. Physical properties of substrates considered appropriate for plant growth at planting may change over time in containers as a result of several processes (Allaire-Leung et al. 1999; Lemaire 1995). Changes include air space reduction as a result of settling and segregation of particles of variable sizes (Bilderback and Lorscheider 1995), shrinkage of the substrate (Bruckner 1997), organic matter decomposition, and physical breakdown of particles (Bollen and Glennie 1961; Nash and Laiche 1981). Current guidelines for nursery container substrates suggest that after irrigation and drainage, substrates should have 10–30 % AS, 45–65 % CC, 25–35 % available water, 25–35 % unavailable water, 50–85 % total porosity (TP), and 0.19–0.70 g/cm^{-3} bulk density (BD) (Yeager et al. 2007). The success of a horticultural substrate is mainly

✉ A. Mohammadi-Ghehsareh
mghehsareh@yahoo.com

[1] Department of Soil Science, Faculty of Agriculture, Islamic Azad University of Isfahan, P.O. Box 81595-158, Isfahan, Iran

based on the behavior of the plants grown in them; high-quality substrates with the proper physical and chemical properties can result in high yields and excellent plant quality (Verdonck and Gabriels 1992). The objective of this research was to evaluate the physical properties of date-palm waste as culture media and its effect on number and yield of tomato fruit.

Materials and methods

This research was performed in the greenhouse research site of Isfahan Azad University (Khorasgan). The experiment was conducted as factorial in a completely randomized block design with 9 treatments and 3 replications. Palm wastes were chopped into smaller sizes by combine, and chopped wastes are separated in three sizes (<0.5, 0.5–1, and 1–2 cm) by sieve. Then, they were kept in 1.5 m^3 plastic bags for controlling the moisture and temperature. Some amounts of animal fertilizer (10 g/kg) and N and P fertilizers (4 g/kg from urea and 4 g/kg super phosphate fertilizers) were added to them as a fermentation starter, and these bags were placed in hot (25–30 °C) condition. For respiration, some air holes were made on the bags and the moisture was adjusted to 65 %. Every week, these materials were mixed together and put into the bags again (during the 3 and 6 months). Then, these date palm wastes were used as culture media for tomato cultivation. Treatments included three composting times (*C*) and three sizes (*S*) as explained in Table 1.

Physical characteristics of the culture media including bulk density and article density (Baruah and Barthakur 1998; Daifullah et al. 2003), water holding capacity (WHC) (Verdonck and Gabriels 1992), and porosity (Baruah and Barthakur 1998) were measured before planting and after plant harvested from the culture media without and with plant. Seeds of tomato (Izmir cultivar) were planted in cocopeat. The F_t (total porosity) was measured using $F_t = 1 - (\rho_b/\rho_s)$ (ρ_b and ρ_s are bulk density and article density, respectively) (Baruah and Barthakur 1998). After transplanting growth, they were transferred to 10-l pots filled with above treatments. Irrigation was done by hand and Papadopolus (1991) formula with fertigation method was used for nutrient solution. Average temperature of day and night were 30 and 18 °C, respectively, in greenhouse. During plant growth, irrigation rate, temperature, humidity, and pest control for all treatments were similar. Leaching (20 %) of culture media was performed every 15 days. Plant height (from pot top to plant tip), root dry weight (plant root separated from culture media by washing with water and dried in oven), and stem diameter (from pot top) (by caliper) of tomato plants were measured. Experimental data normality was verified, and then data were submitted to analysis of variance, using SAS (1997) software package. Means were compared using Duncan multiple test ($p < 0.05$).

Table 1 Definition of experiment treatments

Treatments	Definition
C_1S_1	Size 0–0.5 cm + 0 month composing period
C_2S_1	Size 0–0.5 cm + 3 months composing period
C_3S_1	Size 0–0.5 cm + 6 months composing period
C_1S_2	Size 0.5–1 cm + 0 month composing period
C_2S_2	Size 0.5–1 cm + 3 months composing period
C_3S_2	Size 0.5–1 cm + 6 months composing period
C_1S_3	Size 1–2 cm + 0 month composing period
C_2S_3	Size 1–2 cm + 3 months composing period
C_3S_3	Size 1–2 cm + 6 months composing period

Results and discussion

Tables 2, 3, and 4 show some physical properties of culture media before planting, without and with plant, respectively. Maximum BD values in culture media before planting (Table 2) and without (Table 3) and with plant (Table 4) were related to culture media C_1S_1 ($p < 0.05$). It was because of more entered dust in this size during the crashing and sieving. With increasing size fractions, bulk density decreased. It could be due to bigger pores and higher porosity in larger sizes. In each size, culture media with more composting time (6 months) had more bulk density than culture media with lower composting time (3 months). Carbon (C) compounds present in organic matters are used by microorganisms as an energy source. As C is lost from the compost pile, the compost becomes

Table 2 Physical properties of the culture media before planting (before experiment)

Treatment	BD (g/cm^3)	F_t (%)	WHC (%)
C_1S_1	0.25^a	83^e	89.65^c
C_2S_1	0.18^c	88^c	94.26^a
C_3S_1	0.19^b	87^d	92.62^b
C_1S_2	0.17^d	89^b	57.5^g
C_2S_2	0.18^c	88^c	74.57^d
C_3S_2	0.19^b	87^d	58.91^f
C_1S_3	0.15^f	90^a	37.75^i
C_2S_3	0.16^e	89^b	53.48^h
C_3S_3	0.17^d	88^c	59.31^e

Means followed by the same letter are not significantly different according to Duncan's multiple range test, $p < 0.05$

BD bulk density, F_t total porosity, *WHC* water holding capacity

Table 3 Physical properties of the culture media without plant (after experiment)

Treatment	BD (g/cm^3)	F_t (%)	WHC (%)
C_1S_1	0.27[a]	81[d]	93.05[c]
C_2S_1	0.19[f]	86[a]	94.03[a]
C_3S_1	0.20[e]	86[a]	94.26[a]
C_1S_2	0.20[e]	86[a]	62.13[f]
C_2S_2	0.19[f]	86[a]	78.11[d]
C_3S_2	0.19[f]	86[a]	63.03[e]
C_1S_3	0.25[b]	82[c]	40.09[i]
C_2S_3	0.24[c]	82[c]	56.11[h]
C_3S_3	0.22[d]	84[b]	61.13[g]

Means followed by the same letter are not significantly different according to Duncan's multiple range test, $p < 0.05$

BD bulk density, F_t total porosity, *WHC* water holding capacity

Table 4 Physical properties of the culture media with plant (after plant harvested)

Treatment	BD (g/cm^3)	F_t (%)	WHC (%)
C_1S_1	0.20[a]	86[e]	91.00[c]
C_2S_1	0.17[c]	89[c]	95.07[a]
C_3S_1	0.18[b]	88[d]	95.16[a]
C_1S_2	0.16[d]	90[b]	58.10[g]
C_2S_2	0.17[c]	89[c]	75.06[d]
C_3S_2	0.18[b]	88[d]	59.6[f]
C_1S_3	0.14[f]	91[a]	38.77[h]
C_2S_3	0.15[e]	90[b]	54.30[h]
C_3S_3	0.16[d]	89[c]	60.10[e]

Means followed by the same letter are not significantly different according to Duncan's multiple range test, $p < 0.05$

BD bulk density, F_t total porosity, *WHC* water holding capacity

more condensed and air spaces within the pile become smaller (Smith Hirrel and Riley 2012). Kaila (1956) reported that bulk density could be used as a basis for estimating the degree of decomposition of peats. Before and after planting, the highest amounts of porosity were observed in culture media C_1S_3 because of lower bulk density and bigger pores between them. The porosity percentage is an index for root aeration in culture media. When root media aeration is sufficient, supply of water and nutrient elements for plant is easier. In culture media before planting, the maximum amount of WHC was related to culture media C_2S_1. WHC in culture media is related to particle size and pore size. Smaller particles have a larger surface area than those with larger particles; a large surface area allows a medium to hold more moisture and to increase WHC which causes poor air–water relationship, leading to low aeration within the medium and effect on oxygen diffusion to the roots. The highest amounts of WHC in culture media without and with plant were significantly recorded in culture media C_2S_1 and C_3S_1 ($p < 0.05$). Organic matter degradation caused particles to be more chopped in compost pile and so surface area and WHC were increased.

Comparison means of physical properties of culture media before planting, at the end of cultivation from culture media without and with plant in each treatment are shown in Table 5. Values of BD were significantly increased at the end of cultivation from culture media without plant in comparison to culture media before planting and after planting ($p < 0.05$). The lowest BD was significantly observed in culture media with plant ($p < 0.05$). Plants can affect substrate structure by creating pores in substrates and decrease the bulk density of substrates through root channeling. Maximum F_t (total porosity) was significantly related to culture media with plant ($p < 0.05$) because of lower bulk density in culture media with plant. Amounts of WHC in culture media without plant were significantly higher than those in culture media with plant and before planting ($p < 0.05$). It could be due to increasing composting process in culture media without plant in comparison to culture media with plant that is related to microorganism population and activity, because in culture media without plant main section of nutritional elements were consumed by microorganisms and led to more activity in composting process in comparison to culture media with plant that is consumed by plant. Organic matter degradation caused particles to be more chopped in compost pile and so surface area and WHC were increased.

Table 6 illustrates the effect of treatments on some growing indexes of tomato plants. Culture media had a significant effect on root dry weight of tomato plants ($p < 0.05$). Highest root dry weight was related to culture media C_1S_3 because of highest porosity in this media as compared with other treatments ($p < 0.05$). Highest amount of plant height was significantly observed in treatment C_1S_3 ($p < 0.05$). Bulk density and porosity in this media had proper conditions and caused better suport of water and nutrient elements for plant leading to good growth (Olympious 1992; Kumar and Goh 1999). Culture media had no significant effect on stem diameter of tomato plants. Borji et al. (2010) showed that different substrates including date-palm peat 1 (without composting period), date-palm peat 2 (with 3 months composting period), perlite, cocopeat + date-palm peat 2 (v/v = 50 %), cocopeat + date-palm peat 1 (v/v/= 50 %), perlite + date-palm peat 2 (v/v = 50 %), perlite + date-palm peat 1 (v/v = 50 %), and cocopeat + perlite had no significant difference in stem diameter of tomato plant.

Table 5 Comparison means of physical properties of the culture media before planting (b) and at the end of cultivation from culture media without (c) and with plant (p) in each treatment

Physical properties	Treatments								
	C_1S_1	C_2S_1	C_3S_1	C_1S_2	C_2S_2	C_3S_2	C_1S_3	C_2S_3	C_3S_3
$(BD)_b$	0.25^b	0.18^b	0.19^b	0.17^b	0.18^b	0.19^a	0.15^b	0.16^b	0.17^b
$(BD)_p$	0.2^c	0.17^c	0.18^c	0.16^c	0.17^c	0.18^b	0.14^c	0.15^c	0.16^c
$(BD)_c$	0.27^a	0.19^a	0.20^a	0.20^a	0.19^a	0.19^a	0.25^a	0.24^a	0.22^a
$(F_t)_b$	83^b	88^b	87^b	89^b	88^b	87^b	90^b	89^b	88^b
$(F_t)_p$	86^a	89^a	88^a	90^a	89^a	88^a	91^a	90^a	89^a
$(F_t)_c$	81^c	86^c	86^c	86^c	86^c	86^c	82^c	82^c	84^c
$(WHC)_b$	89.65^c	94.26^c	92.62^b	57.5^c	74.57^c	58.91^c	37.75^b	53.48^c	59.31^c
$(WHC)_p$	91^b	95.16^b	95.07^b	58.1^b	75.06^b	59.6^b	38.77^b	59.3^b	60.1^b
$(WHC)_c$	93^a	94.26^a	94.03^a	62.13^a	78.11^a	58.91^c	40.09^a	56.11^a	61.13^a

Means followed by the same letter are not significantly different according to Duncan's multiple range test, $p < 0.05$

BD bulk density, F_t total porosity, *WHC* water holding capacity

Table 6 Effect of treatments on some growing indexes of tomato plants

Treatment	Plant height (cm)	Stem diameter (cm)	Root dry weight (g)
C_1S_1	105.19^e	2.19^c	3^e
C_2S_1	105.09^e	2.11^c	3.3^c
C_3S_1	105.12^e	2.41^a	3.08^{de}
C_1S_2	118.72^c	2.27^b	3.7^b
C_2S_2	118.1^d	2.34^b	3.29^c
C_3S_2	120.08^b	2.59^a	3.11^d
C_1S_3	121.16^a	2.13^c	4^a
C_2S_3	118.13^d	2.44^a	3.7^b
C_3S_3	118.08^d	2.26^b	3.22^c

Means followed by the same letter are not significantly different according to Duncan's multiple range test, $p < 0.05$

Conclusion

The overall results of this research indicated that composting process changed the physical properties of the media before planting. Also during the experiment, an apparent shift in physical properties was recorded. Physical properties generally refer to culture media ability to buffer the water held against changes in either pH or nutrition. The results showed that composting processes were continued in culture media without and with plant. It seems that the speed of composting process in culture media without plant was higher than that in culture media with plant because in culture media with plant, when media was fertigated with nutrient solution, some of the nutrient elements were used by plant and microorganisms, but in the culture media without plant these nutrient elements were used only by microorganisms and led to more activity. Mature compost increased the number and yield of tomato fruits.

References

Aaron JR (1982) Conifer bark: its properties and uses. Forest Commun. Forest Rec, p 110

Allaire-Leung SE, Caron J, Parent LE (1999) Changes in physical properties of peat substrates during plant growth. Can J Soil Sci 79:137–139

Baruah TC, Barthakur HB (1998) A textbook of soil analysis. PMT Ltd, (Vikas Publishing House), New Delhi, p 282

Bilderback TE, Lorscheider MR (1995) Physical properties of double-processed pine bark: effects on rooting. Acta Hort 401:77–83

Bollen WB, Glennie DW (1961) Sawdust, bark, and other wood wastes for soil conditioning and mulching. Forest Prod J 11:38–46

Borji H, Mohammadi-Ghehsareh A, Jafarpour M (2010) Effects of the substrate on tomato in soilless culture. Res J Agri Biol Sci 6(6):923–927

Bruckner U (1997) Physical properties of different potting media and substrate mixtures—especially air-and water capacity. Acta Hort 450:263–270

Daifullah AAM, Girgis BS, Gad HMH (2003) Utilization of agro-residues (rice husk) in small waste water treatment plans. Mater Lett 57:1723–1731

Hoitink HA, Poole HA (1979) Mass production of composted tree barks for container media. Ohio Flor Assn Bull 599:3–4

Kaila A (1956) Determination of the degree of humification of peat samples. J Sci Agri Soci Finland 28:18–35

Kumar K, Goh KM (1999) Crop residues and management practices: effects on soil quality, soil nitrogen dynamics, crop yield, and nitrogen recovery. J Adv Agron 68:197–319

Lemaire F (1995) Physical, chemical, and biological properties of growing medium. Acta Hort 396:273–284

Nash VE, Laiche AJ (1981) Changes in the characteristics of potting media with time. Commun Soil Sci Plant Anal 12:1011–1020

Olympious CM (1992) Soilless media under protected cultivation rockwool, peat, perlite and other substrates. Acta Hortic 401:443–451

Papadopolus AD (1991) Growing greenhouse tomatoes in soil and soilless. Communication Branch, Agric. Canada. Ottawa, pp 79

SAS Institute (1997) SAS user's guide statistics. SAS Institute Inc., Cary

Smith Hirrel S, Riley T (2012) Understanding the composting process. University of Arkansas, United States Department of Agriculture, and County Governments Cooperating

Verdonck O, Gabriels R (1992) Reference method for the determination of physical properties of plant substrates. II. Reference method for the determination of chemical properties of plant substrates. Acta Horticulturae 302:169–179

Yeager TH, Fare DC, Lea-Cox J, Ruter J, Bilderback TE, Gilliam CH, Niemiera AX, Warren SL, Whitwell TE, Wright RD, Tilt KM (2007) Best management practices: guide for producing container-grown plants, 2nd edn. Southern Nurserymen's Assoc, Marietta

20

Effect of vermiwash of *Eisenia foetida* produced by different methods on seed germination of green mung, *Vigna radiate*

Amita Chattopadhyay[1]

Abstract Vermiwash contains enzymes, macro and micronutrients that could stimulate the growth and yield of crops. It can be prepared by different ways. In the present study, vermiwash was produced from the earthworm, *Eisenia foetida*, under field condition and also in the laboratory under cold and heat stress methods. These washes were used in different ratios for seed germination of mung, *Vigna radiate*. The results showed that the vermiwash produced by cold stress diluted at the ratio of 1:5 showed 100 % germination of mung seed followed by 1:5 ratio vermiwash produced naturally that resulted 95 % of seed germination. Undiluted vermiwash produced naturally and by cold stress showed 90 % of germination of mung seed than the undiluted vermiwash produced by heat stress. Cold stress produced vermiwash diluted with 1:5 ratio showed a vigour index of 1650 followed by naturally produced 1:5 ratio diluted vermiwash which showed a vigour index of 1092. However, the undiluted naturally produced vermiwash showed the lowest vigour index of 450.

Keywords Vermiwash · *Eisenia foetida* · Cold stress · Heat stress · Germination · Vigour index · Seed · *Vigna radiate*

Introduction

Earthworm's role as farmer's friend is well known since time immemorial. In recent times, the commercial vermin culturists have started promoting a product, called vermiwash. This vermiwash contains enzymes, secretions of earthwarms which would stimulate the growth and yield of crops and even develop resistance in crops receiving this spray (Shield 1982; Anand et al. 1995; Suthar et al. 2005; Yadav et al. 2005). Vermiwash also has soluble plant nutrients apart from some organic acids and mucus of earthworms and microbes (Shivsubramanian and Ganeshkumar 2004). Zambare et al. (2008) showed the effectiveness of vermiwash on cowpea plant growth by laboratory-scale trial. The effect of vermiwash was observed on the growth and productivity of Marigold (Shivsubramanian and Ganeshkumar 2004). The effect of vermiwash spray significantly increased dry chilli yield (George et al. 2007). Buckerfield et al. (1999) reported that weekly applications of vermiwash increased radish yield by 7.3 %. Thangavel (2003) also showed that both growth and paddy yield increased with the application of vermiwash and vermicast extracts. The present study was carried out to evaluate the effect of vermiwash on seed germination of green mung, *Vigna radiate,* in addition to the composition of vermiwash extracted by different methods and also to find out which method was more effective in seed germination.

✉ Amita Chattopadhyay
mouami69@yahoo.com

[1] Department of Science & Technology, ICAR-Central Institute of Freshwater Aquaculture, Kausalyaganga, Bhubaneswar, Odisha 751002, India

Materials and methods

Vermiwash was extracted by three different methods which are as follows:

Method I (heat stress method)

This method was adopted from method described by Karuna et al. (1999). Well-grown adult earthworms *Eisenia foetida* were separated from casting materials by placing the worms in a plastic tub for some time. Then the earthworms were removed carefully from the casting materials and transferred to a glass beaker containing 500 ml of warm (40 °C temperature) distilled water and agitated for 5–6 min with a stirrer. Around 30 g of warms were taken. They were then removed immediately and added to another pre-sterilized plastic container filled with water at room temperature. Here the worms were rinsed thoroughly to collect the remaining excretory and secretory products adhering to the body of the worms. The earthworms were then released back to the stock culture container. The light yellow straw-coloured contents of the glass beaker and plastic beaker were mixed and the solution was stored in sterilized dark colour glass bottle at 4 °C to be used for the experimental purpose.

Method II (cold stress application)

In this case, vermiwash was produced using method described by Pramanik (2010) with slight modification. Around 25 numbers having around 30 g of total weight of well-grown matured worms (*E. foetida*) of approximately equal length were applied cold stress by keeping them in chilled conditions, in ice cubes in beaker at −5 °C for 3–4 min. The earthworms were then transferred to a 500-ml glass beaker containing cold distilled (°C) water and kept for 7–8 min with occasional stirring. The worms were transferred to a sterilized beaker containing distilled water at room temperature and rinsed for around 2–3 min and then released to the stock culture container. The light yellow-coloured exudutes from the petridish, glass beaker and plastic beaker were all mixed together and the vermiwash was stored in dark-coloured sterilized glass bottle at 4 °C for experimental purpose.

Method III (field method)

Vermiwash was collected from a vermiwash unit made up of a 15-l plastic bucket with a tap fitted near the base. The lowermost part of the bucket was filled up to 20–25 cm with broken bricks and pebbles with a layer of coarse sand (20–30 cm) above it. On the top of the sand layer around 25 cm layer of moist loamy soil was laid. Finally, the unit was filled with a mixture of partially decomposed cowdung and other organic wastes. Around 1 kg (700–800 nos.) of earthworms (*E. foetida*) was released to the unit. Water was sprinkled on this unit overnight with the help of a perforated plastic container suspended above the unit. Every morning the tap was opened to collect the vermiwash, which was brought to the laboratory and stored in sterilized dark bottle at 4 °C.

Chemical analysis of vermiwash

The organic carbon of vermiwash was determined using potassium di-chromate and concentrated sulphuric acid as extractants using hot water bath, and then followed the method of Walkley and Black (1934). Total N was determined in vermiwash samples by the Kjeldahl method using concentrated H_2SO_4, K_2SO_4 and selenium to digest sample and was estimated according to Bremner and Mulvaney (1982). Total available P was determined by the colorimetric method of Olsen et al. (1954). Total potassium, sodium and calcium were determined by flame photometer. pH was estimated by pH meter while electrical conductivity was determined by a Conductivity meter. The vermiwash was also evaluated for the plant growth hormone indole acetic acid (IAA) by modified Salkowski's method (Gordon and Paleg 1957). Chemical compositions of vermiwash produced by different methods are presented in Table 1.

Germination experiment

Each petridish was layered with Whatman no. 1 filter and 10 healthy seeds of *Vigna radiata*; green mung were placed in each petridish. Treatments were maintained in triplicates. Ten different types of treatments including control

Table 1 Chemical composition of vermiwash produced by different methods

Parameters	Vermiwash (heat stress)	Vermiwash (cold stress)	Vermiwash (field)
pH	7.98	7.40	7.52
E.C. (dS m^{-1})	0.56	0.42	1.10
Organic carbon (%)	0.042	0.036	0.009
Total nitrogen (mg L^{-1})	55.0	52.0	61.02
Total phosphorus (mg L^{-1})	19.15	17.80	18.20
Total potassium (mg L^{-1})	46.24	45.20	55.20
Sodium (mg L^{-1})	119.40	110.80	120.10
Calcium (mg L^{-1})	181.08	175.25	178.60
Magnesium (mg L^{-1})	196.34	186.90	198.00
IAA (mg L^{-1})	16.08	15.05	15.02

(with water only) were applied to the seeds. The seeds were soaked overnight in distilled water and then placed on the petridish. 5 ml of test solution was added twice daily to the petridish for 6 days. Germination percentage was calculated on the 3rd day. At the end of the experiment, root length, shoot length and vigour index were estimated. The treatments included T_1 (control, water only), T_2 (heat stress vermiwash: water in the ratio 1:1); T_3 (heat stress vermiwash: water in the ratio 1:5); T_4 (cold stress vermiwash: water in the ratio 1:1); T_5 (cold stress vermiwash: water in the ratio 1:5); T_6 (vermiwash produced normally: water in the ratio 1:1); T_7 (vermiwash produced normally: water in the ratio 1:5); T_8 (undiluted heat stress vermiwash); T_9 (undiluted cold stress vermiwash); T_{10} (undiluted normally produced vermiwash). The two dilutions 1:1 and 1:5 for all the cases were made arbitrarily. The periodic addition of vermiwash was used from the same stock solutions; only the dilutions were made freshly.

Statistical analysis

All the reported data are mean ± SD of three replicates. The one-way analysis of variance (ANOVA) was used to analyse the significant germination difference, difference in length of seedlings between vermiwash of different extraction types and dilution using statistical software SPSS (14.0 version).

Results and discussion

Germination percentage

The effect of different treatments on germination of seeds of green mung has been shown in Table 2. One-way ANOVA followed by Turkey's post hoc test showed that there was statistically significant difference ($F = 15.455$, $P < 0.05$) of germination percentage between the control (65 %) and all the other treatments, all of which showed above 80 % germination. The variance among different treatment groups was found to be homogeneous as per Levene's test of homogeneity of variance where Levene's F-statistic had a value of 0.871 which was greater than 0.05. There was 100 % germination in the seeds treated with T_5, i.e., 1:5 ratio diluted vermiwash produced from cold stress, followed by 1:5 ratio diluted vermiwash produced naturally. The vermiwash produced by cold stress was superior compared to the same produced by heat stress in seed germination. In heat stress, enzymes could be denatured. Moreover, there could be a chance of release of more amounts of micronutrients, i.e., Fe, Cu, Zn and Mn in heat stress vermiwash. Though micronutrients are helpful in seed germination, greater amounts could be toxic at the same time.

Root and shoot length

In all the cases, root and shoot length was superior over control. One-way ANOVA followed by Turkey's post hoc test revealed that there was statistical significant difference ($F = 47.122$, $P < 0.05$) of average length of seedlings between the control and all the other treatments. The variances among different treatment groups were homogeneous as per Levene's test of homogeneity of variance table where Levene's F statistic had a value of 0.456 which was greater than 0.05. The maximum root length 6.5 cm and the maximum root length 10 cm were observed at 1:5 ratio of vermiwash produced by cold stress. The minimum root length of 2–3 cm and the minimum shoot length of 3.0–3.5 cm were observed at the undiluted vermiwash

Table 2 Effect of vermiwash on seedlings, germination and vigour of green mung, *Vigna radiate*

Treatments	Root length (cm) (mean ± SD)	Shoot length (cm) (mean ± SD)	Average length of seedlings (cm) (L) (mean ± SD)	Germination (%) (G) (mean ± SD)	Vigour index (L X G)
T_1 (Control, water only)	3.0 ± 0.2	5.0 ± 0.3	8.0 ± 0.5	65 ± 5	520
T_2 (Heat stress vermiwash: water in the ratio 1:1)	3.0 ± 0.3	5.0 ± 0.3	8.0 ± 0.6	80 ± 6	640
T_3 (Heat stress vermiwash: water in the ratio 1:5)	4.0 ± 0.4	4.0 ± 0.3	8.0 ± 0.7	80 ± 5	640
T_4 (Cold stress vermiwash: water in the ratio 1:1)	4.0 ± 0.2	4.5 ± 0.3	8.5 ± 0.5	70 ± 5	595
T_5 (Cold stress vermiwash: water in the ratio 1:5)	6.5 ± 0.8	10.0 ± 0.7	16.5 ± 1.5	100	1650
T_6 (Vermiwash produced normally: water in the ratio 1:1)	4.0 ± 0.6	6.0 ± 0.6	10.0 ± 1.2	80 ± 5	800
T_7 (Vermiwash produced normally: water in the ratio 1:5)	5.5 ± 0.9	6.0 ± 0.5	11.5 ± 1.4	95 ± 5	1092.5
T_8 (Undiluted heat stress vermiwash)	2.5 ± 0.2	3.0 ± 0.3	5.5 ± 0.5	85 ± 5	467.5
T_9 (Undiluted cold stress vermiwash)	2.0 ± 0.2	3.5 ± 0.2	5.5 ± 0.4	90 ± 5	495
T_{10} (Undiluted normally produced vermiwash)	2.0 ± 0.1	3.0 ± 0.2	5.0 ± 0.3	90 ± 4	450

produced either naturally or by cold stress or by heat stress. This could be due the presence of toxicity effects of higher concentrations of nutrients in the undiluted vermiwash. The vigour index was highest at 1:5 ratio diluted vermiwash produced by cold stress followed by 1:5 ratio diluted vermiwash produced naturally. In response to vermiwash application, *V. radiate* seedlings produced a profuse growth of fine root hairs, which was not found in the seedlings that received only water. The profuse root hair growth was direct indication of involvement of IAA present in vermiwash (Gopal et al. 2010).

Provitamin D (Zrazhevsky 1957) and vitamin B (Atlavinyte et al. 1971) are reported up to a considerable level in earthworm products and probably these act as promoters for carbohydrate and protein metabolism in vermiwash-treated seedlings. Vermiwash is a collection of excretory products and excess secretions of earthworms along with micronutrients from soil organic molecules. Zambare et al. (2008) reported that boiled water extract vermiwash contained amylase and protease enzymes. They also reported that vermiwash contained nitrogen and phosphate solubilizing bacteria. Prabhu (2006) reported that vermiwash improved the germination percentage of the seeds and seedling vigour of seeds such as cowpea and paddy crops. The presence of proteases in vermiwash could help in seed germination while amylases could help for availability of simple carbon source for enhancement of plant growth and productivity (Zambare et al. 2008).

In the present study, it was evident that undiluted vermiwash was phytotoxic to *V. radiate* as it significantly reduced the percentage of germination and seedling's length of the test crop. It was also observed that in the undiluted vermiwash treatment, the roots had turned brown and had a charred appearance. The diluted vermiwash (1:1–1:5) was able to increase the seed germination percentage and the seedlings' vigour index when compared to water.

Conclusion

From the present study, it is evident that the nutrients and growth promoting substances present in the vermiwash showed its potentiality in seed germination and seedling vigour. However, the vermiwash produced by cold stress diluted at the ratio of 1:5 produced superior result followed by 1:5 ratio vermiwash produced naturally, i.e. without stress. The use and application of vermiwash could play a beneficial role in sustainable agriculture as it is environment friendly, cost effective, reliable and easily available.

Acknowledgments The author is grateful to the Department of Science and Technology, New Delhi, India for providing the Project under the Women Scientist Scheme (Science and Society) and fellowship. The author is also grateful to the Director, ICAR-CIFA, Bhubaneswar, India for providing laboratory and field facilities to carry out the work.

Compliance with ethical standards

Conflict of interest The authors declare that there is no conflict of interest.

References

Anand JA, Wilson MDP, Kale RD (1995) Effect of vermiwash on seed germination and seedling growth. J Soil Biol Ecol 15:90–95

Atlavinyte O, Daciulyte J, Luganska A (1971) Correlation between the number of earthworm, microorganisms and B_{12} in soil fertilized with straw. Liet TSR Moksulu Akad Darb Ser B 3:43–56

Bremner JM, Mulvaney CS (1982) Nitrogen ± total. In: Page AL, Miller RH, Keeney (eds) Methods of soil analysis. Part 2, Chemical and microbiological properties. American Society of Agronomy, Madison, pp 595–641

Buckerfield JC, Flavel TC, Lee KE, Webster KA, Diazcozin DJ, Jesus JB, Trigo D, Garvin MH (1999) Vermicompost in solid and liquid forms as a plant growth promoter. 6th international symposium on earthworm ecology, Vigo, Sain, 1998. Pedobiologia 43:753–759

George S, Giraddi RS, Patil RH (2007) Utility of vermiwash for the management of Thrips and Mites on chilli (*Capsicum annuum* L.) amended with soil organics. Karnataka J Agric Sci 20:657–659

Gopal M, Gupta A, Palaniswami C, Dhanapal R, Thomas GV (2010) Coconut leaf vermiwash: a bio-liquid from coconut leaf vermicompost for improving the crop production capacities of soil. Curr Sci 98(9):1202–1210

Gordon SA, Paleg LG (1957) Quantitative measurement of indole acetic acid. Physiol Plant 10:37–48

Karuna K, Patil CR, Narayanswamy P (1999) Kale stimulatory effects of earthworm body fluid (vermiwash) on Crinkle red Varity of *Vnthurium andreanum* Lind. Crop Res 17(2):253–257

Olsen SR, Cole CV, Watanabe FS, Dean LA (1954) Estimation of Available Phosphorous in Soil by Extraction with Sodium Bicarbonate. USDA Circular 939, Government Printing Office, Washington, DC

Parmanik P (2010) Quantification of hydrolic and proteolytic enzymes in the excreta of three epigeic earthworms and detection of thiocarbamic acid by GC-MS-MS. Environmentalist 30:212–215

Prabhu MJ (2006) Coconut leaf vermiwash stimulates crop yield. The Hindu Newspaper, 28th December, In: Science and Technology section

Shield EB (1982) Raising earthworms for profit. Shields Publication. PO Box 669 Eagle River Wisconsin, 128 p

Shivsubramanian K, Ganeshkumar M (2004) Influence of vermiwash on biological productivity of Marigold. Madras Agric J 91:221–225

Suthar S, Choyal RR, Singh S, Sudesh R (2005) Stimulatory effect of earthworm body fluid on seed germination and seedlings growth of two legumes. J Phytol Res 1(2):219–222

Thangavel P, Balagurunathan R, Divakaran J, Prabhakaran J (2003) Effect of vermiwash and vermicast extraction soil nutrient status, growth and yield of paddy. Adv Plant Sci 16:187–190

Walkley A, Black IA (1934) An examination of Degtjareff method for determining soil organic matter and a proposed modification of the chromic acid titration method. Soil Sci 37:29–37

Yadav AK, Kumar K, Singh S, Sharma M (2005) Vermiwash—A liquid biofertilizer. UP J Zool 25(1):97–99

Zambare VP, Padul MV, Yadav AA, Shete TB (2008) Vermiwash: biochemical and microbiological approach asecofriendly soil conditioner. ARPN J Agric Biol Sci 3(4):1–5

Zrazhevsky AI (1957) Dozdevye cervi kak factor plodorodija lesnych prov Akad.Nauk Ukr, SSR, Kiew

21

Biological fungal treatment of olive cake for better utilization in ruminants nutrition in Egypt

Mohamed Fadel[1] · Dina Helmy El-Ghonemy[1]

Abstract

Background Crop residues and agro-industrial by-products, available in appreciable quantities, can play a significant role in the nutrition of ruminants. The appropriate utilization of by-products in animal nutrition can improve the economy and the efficiency of agricultural, industrial and animal production. The present work investigates the bio-conversion of olive cake (OC), generated by the olive oil industries in Egypt, using locally isolated filamentous fungi in solid state fermentation, so as to upgrade its nutritional values and digestibility for its use as ruminants feed.

Methods Seven non-mycotoxin producing fungal strains namely *Trichoderma reesei* F-418, *T. harzianum* F-416, *T. virdie* F-520, *T. koningii* F-322, *Aspergillus oryzae* FK-923, *A. fumigatus* F-993, and *A. awamori* F-524 were cultured on OC for 7 days at 36 °C. Subsequently, the chemical composition and lignocellulolytic enzyme activities of the resultant substrate were determined.

Results The most promising result was obtained by *A. oryzae* FK-923, whereas, an increase in crude protein content ranging from 9.5 % (untreated) to 17.4 % (treated) was detected, while phenols were decreased from 3.1 to 0.92 % and fibers declined from 33 to 22.2 %. A reduction in the values of neutral detergent fiber (NDF) and acid detergent fiber (ADF) were reported. The addition of sugar cane molasses at 2 % showed an increase in crude protein to 18.9 % with a reduction in phenols and fibers to 0.69 and 21.8 %, respectively. Furthermore, the addition of active dry yeast (*Saccharomyces cerevisiae*) at 1.5 % to the fermentation medium raised the crude protein to 20.2 % (w/w), while phenols and fibers were declined to 0.55 and 19.2 %, respectively.

Conclusions Therefore, the present findings revealed *A. oryzae* FK-923 to be an efficient organism for lignocellulolytic enzymes production and simultaneous enhancement in crude protein and in vitro digestibility of OC.

Keywords Olive cake · Biological treatment · Solid state fermentation · Ruminants feed

Introduction

In many developing countries, there is a gap between available and required ruminants feed. The productivity of the animals is very low, although experimental data suggest that their production potential is higher. The greatest constraint to livestock productivity is the shortage of feeds and forages. Therefore, some countries import cereal grains in an effort to increase livestock production and meet the ever-increasing demand for animal products. The Middle East countries raise some 99 million sheep and goats, 104 million beef cattle and 202 million dairy cattle (FAO 1981). Higher livestock productivity, however, should be sought through better use of locally available feed resources. Crop residues and agro-industrial by-products, which are available in appreciable quantities from processing of fruits and vegetables, can play a significant role in the nutrition of ruminants (Economides et al. 1981). In Egypt, although there are about 30 million tons of agricultural by-products produced annually, only 4.15 million tons of crop residues are used for feeding ruminants (Ministry of

✉ Dina Helmy El-Ghonemy
delghonamy@yahoo.com

[1] Department of Microbial Chemistry, Genetic Engineering and Biotechnology Division, National Research Centre, 33 El-Bohouth St., Dokki, Giza 12622, Egypt

Agriculture 2006). The main factors limiting the utilization of crop residues in animal ration are the high crude fiber content, low crude protein, low digestibility and poor feed palatability (Abd El-Rahman et al. 2014). The gap between the availability and requirements of animal feeds in Egypt is about nine million tons of dry matter, equivalent to almost four million tons of total digestible nutrients per year (Bendray et al. 2006). Therefore, efforts should be done to efficiently use all the available by-products and wastes to decrease the ruminants feed shortage in Egypt (AboSalim and Bendary 2005).

Olive cake (OC) is the solid residue obtained after olive oil extraction. It is one of the most abundant agro-industrial by-products in the Mediterranean area constituting a source of environmental problems resulting from its accumulation and disposal (Azkar et al. 2008; Hachicha et al. 2008). In Egypt, there are about 118,382 feddan planted with olive trees, the fruitful area 96,810 feddan, produce about 314,450 ton of olive, representing 3.25 ton/feddan (Ministry of Agriculture and Land Reclamation, 2012). Basically, 100 kg of olives produced approximately 35 kg of crude OC. The utilization of OC as ruminants feed is a good way of recycling this by-product. However, the OC contains a large proportion of cell wall constituents, which renders it unpalatable and poorly digestible to the ruminants. Therefore, attempts have been made to improve the nutritive value of OC through different chemical agents, but results have been more or less satisfactory (Rowghani et al. 2008). Apart from the chemicals, the biological treatment is drawing much attention due to its potential advantages over chemical/physical treatments such as greater substrate and reaction specificity, lower energy requirements, lower pollution generation and higher yields of desired products (Misra et al. 2007; Canet et al. 2008).

Solid-state fermentation (SSF) technique using filamentous fungi has been reported to be a relatively low-cost appropriate technology for the upgrading of lignocellulosic materials in animal feeding (Ugwuanyi et al. 2009). Additionally, SSF has proven to be an efficient way to produce lignocellulolytic enzymes since it provides the filamentous fungi with environmental conditions similar to those of their natural habitat. Besides, the restricted water availability reduces the possibilities of contamination by bacteria and yeast (Neifar et al. 2009). It is well documented that the selection of a lignin degrading fungus is preferable for improving the animal feed quality without compromising much the cellulose and hemicellulose constituents. Therefore, the aim of the present study was to apply farm scaling up using biological fungal treatment of OC in SSF using *Aspergillus oryzae* FK-923 to reduce the anti-nutritional components like phenols and fibers. Besides, an increase in protein and gross energy contents in a short period of time was reported which improved the nutritive value of olive cake for its investment as animal feeds to overcome nutritional Gabe in Egypt. To the best of our knowledge, this is the first report on the nutritional evaluation of SSF olive cake with *A. oryzae* FK-923.

Materials and methods

Chemicals

All chemicals were purchased from Sigma–Aldrich or Fluka and were certified reagent grade.

Microorganism

Seven non-mycotoxin producing fungal strains namely *Trichoderma reesei* F-418 (T1), *T. harzianum* F-416 (T2), *T. virdie* F-520 (T3), *T. koningii* F-322 (T4), *Aspergillus fumigatus* F-993 (A1), *A. oryzae* FK-923 (A2), *A. awamori* F-524 (A3) and *Saccharomyces cerevisiae* F-707, obtained from the culture collection of Microbial Chemistry Department, National Research Centre (NRC), were grown for 5 days on potato dextrose agar slants (PDA) at 28 °C, and maintained at 4 °C. The slants were subcultured routinely every 3–4 weeks interval.

Preparation of spore suspension

Ten mL of sterilized water with 0.1 % Tween-80 solution was added to 5-day-old culture slants and agitated thoroughly using a shaker to suspend the spores. The number of the spore was adjusted to 1×10^6 spores/mL and was used as an inoculum throughout the study.

Screening medium

One mL aliquot (v/v) of inoculum size (1×10^6 spores/mL) was used to inoculate 500 mL Erlenmeyer flasks each containing 50 g of sugar beet pulp (SBP) moistened with distilled water to a moisture level of 65 % (v/w). The inoculated flasks were incubated for 4 days at 32 °C in a static condition. For fungal inoculum propagation, each flask from the above step was transferred to 10 L capacity Erlenmeyer flask contained 1 kg sterilized moistened SBP, and then incubated statically at 32 °C for 4 days.

Preparation of *A. oryzae* FK-923 inoculum

Three days old slants cultures of *A. oryzae* FK-923 was crushed into flasks containing 20 mL of sterilized water. The fungal spores suspensions were used at 10 % (v/w) to inoculate 500 mL capacity flasks containing 20 g of the ground waste moistened at a solid: liquid ratio of 1:2 with a basal medium composed of (g/L): sugar cane-molasses, 40; urea, 2.0; K_2HPO_4, 1.5 and $MgSO_4{\cdot}7H_2O$, 0.5. The inoculated flasks were then incubated at 36 °C for 72 h under static solid state fermentation system. After fermentation period, the above-prepared inocula of *A. oryzae* FK-923 were employed to inoculate 200 g of wastes under study moistened by the above basal medium at a solid: liquid ratio of 1:2 and packed in polyethylene bags (40 × 60 cm). The inoculated bags were incubated under static conditions for 7 days at 36 °C. At the end of incubation period, such bags were opened and oven dried at 70 °C for chemical analysis.

Scaling up using fungal treatments

The above-fermented bags were employed to inoculate polyethylene bags (150 × 225 cm) containing 10 kg of ground wastes moistened with the basal medium at a solid: liquid ratio of 1:2. The bags were incubated in a 3 × 3 m room maintained at 36 °C for 7 days. After the incubation period, the fermented wastes were air-dried to 12 % moisture then packed and stored until fed to lactating goats.

Agro-industrial by-products

Olive by-product

The olive cake (OC), resulted from a continuous cold process of olive oil production, were packed in polyethylene bags and kept in a freezer at −20 °C. Before drying, olive cake samples were thawed during 24 h under refrigeration conditions at 5 °C. Drying process was carried out at 90 °C in a convective dryer, at Microbial Chemistry Department, National Research Centre. The dehydrated samples were packed and kept in dark until analysis.

Molasses

Molasses were obtained from Egyptian Sugar and Integrated Company El-Hawmdia, Giza, Egypt.

Sugar beet pulp

Sugar beet pulp (SBP), a by-product of beet sugar industry, was obtained from Abokorkas Sugar Factory, El-Minia, Egypt.

Physico-chemical analysis of olive cake

Protein content

Crude protein determination involved the use of routine Kjeldhal nitrogen assay method with a conversion factor of 6.25 (AOAC 1990).

Lipid content

The lipid content was determined by exhaustively extracting samples in a Soxhlet apparatus, using petroleum ether (Merck, p.a.) as a solvent according to the method described in AOAC method no. 920.39. The crude fiber content was estimated by Weende method through an acid/alkaline hydrolysis of insoluble residues in accordance with AOAC (1990). The ash content was estimated by incineration in a muffle furnace (Felisa, FE-341, Jalisco, Mexico) at 550 °C. All methodologies allowed the recommendations of the official methods.

Determination of total phenolic content

Total phenolic content (TPC) was determined colorimetrically using Folin–Ciocalteu reagent (FC) as described by Chuah et al. (2008) with modifications. 40 mL of absolute ethanol was added to 3–5 g dried and finely crushed sample. The mixture was homogenized during 24 h on a magnetic stirrer then filtered through a Whatman filter paper No 1. The filtrate was washed twice with 40 mL of absolute ethanol then evaporated under reduced pressure at 40 °C on a rotary evaporator (Büchi RE 121, Switzerland). The obtained extract was then dissolved in 100 mL of absolute ethanol and kept refrigerated for further analysis. 0.5 mL aliquot of the OC extract was transferred to a glass tube containing 0.5 mL of reactive FC, after 5 min 2 mL of Na_2CO_3 solution (200 mg/mL) was added. The sample was then mixed thoroughly on a vortex mixer and the reaction proceeded for 15 min at ambient temperature. A total of 10 mL of ultra-pure water was then added and the precipitate formed was removed by centrifugation for 5 min at $4000{\times}g$. Finally, absorbance was measured at 725 nm in a spectrophotometer (Spectronic TM 20 Genesys TM131, IL, USA) and compared to a gallic acid (GA) calibration curve. Results were expressed as mg GA 100/g-dry matter. All measurements were conducted in triplicate and the data were expressed as means.

Chemical analysis

Dry matter (DM) loss was determined by calculating the difference between dry weight before and after fermentation and described as a percentage of initial weight. Acid detergent fiber (ADF) and neutral detergent fiber (NDF) were determined by the detergent system method, while acid detergent lignin (ADL) was measured according to the method described by AOAC (1990). Hemicellulose contents were estimated as the difference between NDF and ADF, while cellulose content was the difference between ADF and ADL. Ash was carried out on dried sample at 105 °C by ignition three samples each 50 g in muffle furnace at 800 °C for 5 h, and the residual ash was calculated as a percentage (%) from the dried initial weight according to AOAC (1970).

Enzyme extraction

After incubation period, the enzyme was extracted with sodium citrate buffer (0.05 M, pH 5.0) (10 mL buffer g^{-1} substrate), by shaking for 1 h at 180 rpm at room temperature. The extract was filtered through Whatman filter paper No. 1 and centrifuged for 20 min at 5000 rpm (Neifar et al. 2009).

Enzymes assay

Filter paper activity

According to the method described by Bai et al. (2012); filter paper activity (FP-ase) was determined photometrically by measuring the increase in absorbance at 540 nm, after 30 min incubation of 1.0 mL of properly diluted enzyme with 50 mg Whatman filter paper No.1 (1 × 1 cm) and 1.0 mL sodium citrate buffer (0.05 M, pH 5.0) at 50 °C. The enzyme reaction was stopped by the addition of 3 mL of 3, 5-dinitrosalicylic acid (DNS) reagent (Miller 1959). One mL of 40 % Rochelle salt solution was added so as to maintain the color. Absorbance was measured at 575 nm and compared with a standard curve using 0.10–1.0 mg of glucose/mL. One unit of enzyme activity is defined as the amount of enzyme which releases 1 μmol of glucose per min, under the standard assay conditions.

Xylanase activity

Xylanase activity was estimated according to Bailey et al. (1992) using 1 % xylan from oat as the substrate. Xylan was dissolved in 0.05 M Glycine NaOH buffer (pH 9.0). The reaction mixture, containing 0.1 mL of an appropriate dilution of the enzyme and 0.5 mL of the substrate, was incubated for 30 min at 37 °C. The reaction was terminated by placing tubes in a boiling water bath for 10 min. The supernatant was analyzed for reducing sugar with DNS reagent and absorbance was read at 575 nm. A standard curve of D-xylose was used as reference. Each unit of xylanase activity (U) was that amount of activity which released 1 μmol D-xylose min^{-1}.

Glucoamylase activity

Glucoamylase activity was determined according to the method reported by Miller (1959), by incubating 1 % (w/v) maltose with 0.9 mL sodium citrate buffer (0.05 M, pH 5.0) and 0.1 mL of diluted enzyme solution at 50 °C for 30 min. The reaction was terminated by placing tubes in a boiling water bath for 10 min. The released reducing sugars were measured with DNS reagent using glucose as a standard. Glucoamylase activity unit (U) was expressed as the amount of enzyme releasing 1 μmol of glucose per min under assay conditions.

Alpha-mylase activity

Alpha-mylase activity was determined in the culture filtrate by measuring the amount of starch hydrolyzed in the reaction mixture by the iodine method (Manning and Campbell 1961). One unit of enzyme activity has been defined as the amount of enzyme that hydrolyses 1 μmol of starch min^{-1} under assay conditions.

Protease activity

Protease activity was determined as described by Agrawal et al. (2005). 1 mL of suitably diluted enzyme was mixed with 5 mL of (2 %) casein solution (0.05 M Tris–HCl buffer, pH 8.0), and incubated at 40 °C on a gyratory shaker (200 rpm) for 30 min. An aliquot (0.5 mL) of the reaction mixture was withdrawn and the reaction was quenched by adding 1.5 mL pre-chilled trichloroacetic acid (10 %). The reaction tube was immersed in an ice bath (5 min) to completely precipitate the protein. The supernatant was recovered by centrifugation (10,000×*g*, 10 min). Tyrosine liberated during casein hydrolysis was measured in the supernatant using the method of Lowry et al. (1951). One unit of protease activity was defined as the amount of enzyme liberating 1 μmol of tyrosine per min under the standard assay conditions. The protease activity is reported per gram of dry solids used in the initial extraction.

Phytase activity

Phytase production was measured using a colorimetric method by following the release of inorganic phosphate from phytic acid according to the method presented by Dhiman et al. (2002). For this purpose, 150 μL of enzyme solution were mixed with 600 μL Tris–HCl (0.1 M, pH 7.0) supplemented with 2 mM sodium phytate and 2 mM of $CaCl_2$ and incubated at 37 °C for 30 min. The reaction was terminated by the addition of 750 μL of 5 % trichloroacetic acid, after which 1.5 μL of the color reagent were added to generate phosphomolybdate. The concentration of inorganic orthophosphate (pi) in this mixture was determined colorimetrically by measuring the absorbance of the solution at 700 nm using a Beckman Coulter DU640 Spectrophotometer (Fullerton, CA, USA). The color reagent was freshly prepared by mixing four volumes of 1.5 % (w/v) ammonium molybdate solution supplemented with 5.5 % (v/v) sulfuric acid and one volume of 2.7 % (w/v) ferrous sulfate solution. The results were compared to a standard curve prepared using K_2HPO_4 as a source of inorganic phosphate at a concentration ranging from 0.0448 to 2.8706 μM. One unit (U) of phytase activity was defined as the concentration of inorganic phosphate, in μmol, released per min under the defined reaction conditions.

Laccase assay

Laccase activity was measured by monitoring the oxidation of 2,2′-azinodi-3-ethyl-benzothiazoline-6-sulfuric acid (ABTS) at 436 nm (Bourbonnais et al. 1998). The non-phenolic dye ABTS is oxidized by laccase to the more stable and preferred state of the cation radical. The concentration of the cation radical responsible for the intense blue-green color can be correlated to the enzyme activity (Majcherczyk et al. 1998). The reaction mixture contained 0.5 mM substrate (ABTS), 2.8 mL sodium acetate buffer (0.1 M, pH 4.5), and 100 μL of suitably diluted enzyme and incubated for 5 min at 30 °C. Absorbance was read at 436 nm in a spectrophotometer against a suitable blank. One unit of enzyme activity was defined as the amount of the laccase that oxidized 1 μmol of ABTS substrate per min under assay conditions.

Polygalacturinase assay

To measure the activity of polygalacturinase, the assay mixture (1 mL) containing an equal volume of enzyme and 1 % (w/v) pectin dissolved in 0.1 M phosphate buffer (pH 7.0) was incubated at 50 °C for 10 min. The reducing sugar released was measured by the dinitrosalicylic acid method. Control samples were prepared with inactivated enzymes. One unit of enzymatic activity (U) was defined as one μmol of galacturoric acid released per min. All the experiments were conducted in triplicate and the data were expressed as a mean. The enzyme activity is expressed in terms of units per gram dry fermented substrate (U/g-ds).

Protein determination

Estimation of total protein was done by using the Bradford method and bovine serum albumin (BSA) as standard (Bradford 1976).

Amino acids analysis

Amino acid analyzer was used for amino acids analysis by hydrolyzing 0.5 g of protein with 10 mL of 6 N HCl under vacuum at 110 °C for 24 h. While, cystine and cysteine were determined after pretreatment with performic acid (Christias et al. 1975).

Results

Chemical analysis of the olive cake

Data presented in Table 1 revealed that, the olive waste has many valid part used in animal and poultry nutrition; it has 9.5 % protein, 5.7 % olive oil, which consider as a good source of energy and antioxidants, and low ash content (7 %). In spite of the previous advantages, there are other components that limit its supplementation in rations by conquerable amounts resample essentially in phenols which characterized by its low biodegradability that eliminate its nutritional value as it inhibits the microflora in the ruminant and the present lignin restricts the utilization of cell plant content by microflora. Besides, the olive cake contains a high crude fiber (33 %) which may be a good carbon source for many microorganisms. So we select different fungal strains known in biological treatments by producing no mycotoxins and applied in food fermentation and used as a probiotic in feeding field.

Effect of biological fungal treatment on the chemical composition of olive cake

OC bio-conversion into a better quality feed by biological treatment in solid state cultures could be an alternative to chemical and physical treatment for enhancing the OC quality for microbial fermentation in rumen. In the current work, it was obvious that the chemical composition of olive cake was affected according to the fungal strain used. Pointed to our aim, *A. oryzae* FK-923 (A2) gave the most promising results concerning the enhancement of the nutritive value of the OC (Table 2), as the following were achieved according to the importance for nutrition: (1) reduce the phenols to less than one. (2) Increase total protein

Table 1 Chemical composition of the olive cake

Content	Value
Dry matter (%)	88.0
Crude protein (%)	9.5
Ether extract (%)	5.7
Crude fiber (%)	33.0
ADF (%)	48.0
NDF (%)	62.0
Phenols (%)	3.1
Ash (%)	7.0
Lignin (%)	21.0
Gross energy (kcal/kg)	4200
TDN (%)	34.0

ADF acid detergent fiber, *NDF* neutral detergent fiber, *TDN* total digestible nutrient

Table 2 Effect of biological fungal treatment on the chemical composition of olive cake after 8 days of fermentation at 36 °C under solid state fermentation

Analysis	Control	*T1*	*T2*	*T3*	*T4*	*A1*	*A2*	*A3*
Dry matter (%)	88	88.0	88.0	88.0	88.0	88.0	88.0	88.0
Total protein (%)	9.5	12.8	13.9	14.8	12.5	15.2	17.4	16.6
Ether extract (%)	5.7	6.1	5.9	6.0	7.2	6.0	6.2	6.0
Crude fibers (%)	33	26.8	25.5	27.2	27.6	24.4	23.1	23.8
ADF (%)	48	38.2	37.6	39.1	38.9	35.8	34.6	36.2
NDF (%)	62	52.8	45.4	52.2	51.4	49.2	48.2	50.4
Phenols (%)	3.1	2.42	1.86	1.88	1.12	0.96	0.92	1.72
Ash (%)	7	9.8	9.8	10.6	9.6	9.8	9.4	10.2
Lignin (%)	21	18.6	20.6	17.6	18.2	16.4	14.6	11.5
Gross energy (kcal/kg)	4200	4220	4410	4260	4320	4520	4620	4420

to 17.4 %. (3) Decrease the crude fibers. (4) Decrease lignin. (5) Increasing gross energy. These results indicated that, the phenolic compounds, which are the more obstacles towards the use of OC by more amounts in rations, was decreased to a limit that did not affect the microflora. Besides, the increased amount of total protein by about 90 kg/ton can be used to substitute other expensive ingredients in feed rations. Moreover, the reduction in crude fibers content makes free to accomplish different feed rations.

Influence of biological fungal treatment on the fermented olive cake recovery

From the data illustrated in Fig. 1, it was clear that the treated OC recovery is correlated to the fungal strain applied. The percent of the recovery obtained with the application of *Aspergillus* sp. was higher than that with *Trichoderma* sp., since, *A. oryzae* FK-923 and *A. fumigatus* achieved the high recovery percent with 92.2 %.

Effect of biological fungal treatment on the total protein yield of olive cake

Results illustrated in Fig. 2 clearly indicated that, the total protein yield (kg/ton) of the treated OC was varied according to the fungus used and the percent of recovery. Whereas, the highest total protein yield (174 kg/ton) was obtained when *A. oryzae* FK-923 was used in biological treatment of olive waste, followed by *A. awamori* F-524 (161 kg/ton). On the other hand, lowest total protein yield was obtained with *T. koningii* F-322 treatment.

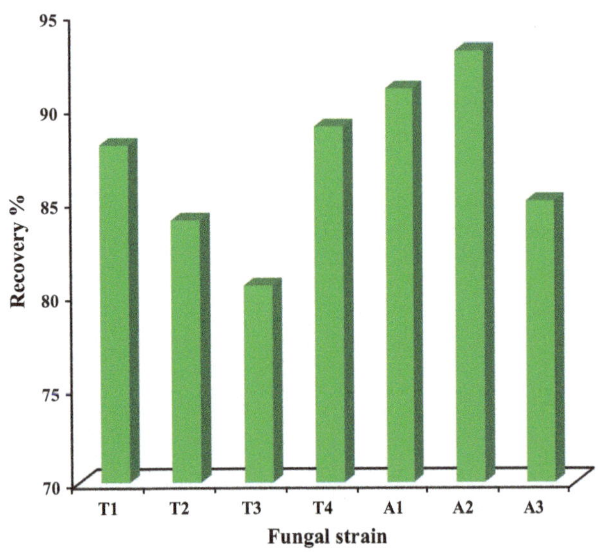

Fig. 1 Influence of biological treatment on the percent of olive waste recovery by different fungal strains after 8 days of incubation at 36 °C

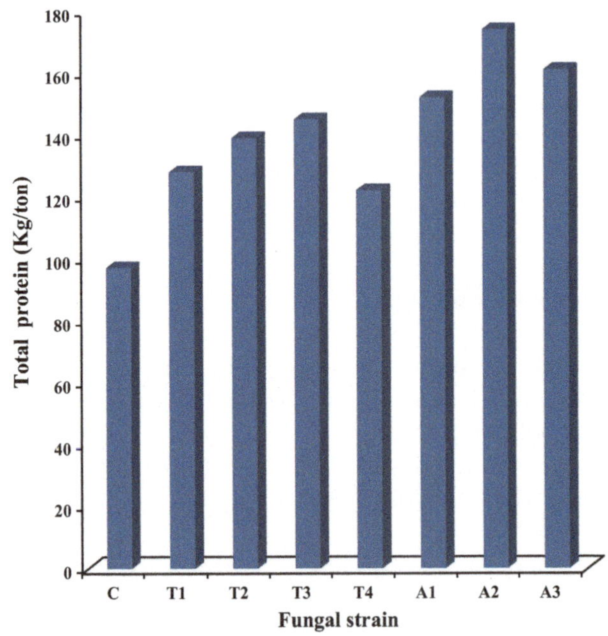

Fig. 2 Effect of biological treatment on the total protein yield in olive cake by different fungal strains

Table 3 Effect of sugar cane molasses supplementation on the chemical composition of olive cake treated by *A. oryzae* FK-923

Molasses amount (kg/ton)	Crude protein (%)	Dry matter recovery (%)	Total protein (%)	Organic matter (%)	Total fibers (%)	Phenols (%)
0	17.4	92.2	160.4	90.6	23.1	0.92
10	17.7	91.3	161.6	90.4	22.8	0.84
15	18.3	89.2	163.2	89.2	22.1	0.78
20	18.9	87.9	166.1	88.8	21.8	0.69
25	18.9	86.6	163.7	88.8	21.6	0.68

Table 4 Effect active dry yeast *S. cerevisiae* supplementation on the chemical composition of olive cake treated by *A. oryzae* FK-923

Yeast amount (kg/ton)	Crude protein (%)	Dry matter recovery (%)	Total protein (%)	Organic matter (%)	Total fibers (%)	Phenols (%)
0	18.9	87.8	166.1	88.8	21.8	0.69
0.5	19.2	87.2	167.4	88.8	20.6	0.65
1.0	20.0	86.8	175.3	88.4	19.8	0.61
1.5	20.2	85.4	170.8	88.1	19.2	0.55
2.0	19.6	85.4	167.4	88.1	19.2	0.55

Effect of sugarcane molasses supplementation on the chemical composition of treated olive cake

Form the data presented in Table 3, it was found that the addition of sugar cane molasses has an advantage effect in enhancement of feed quality of OC biologically treated by *A. oryzae* Fk-923, owing to the increase in protein content. The addition of sugar cane molasses at 2 % (w/w) was more suitable than other tested amounts, as the protein percent raised from 17.4 to 19 % in the fermented olive waste with total protein 166.1 kg/ton. On the other hand, phenols and fibers were reduced to 0.69 and 21.8 %, respectively.

Effect of active dry yeast *S. cerevisiae* supplementation on treated olive cake

Data presented in Table 4 showed that, the addition of active dry yeast has an advantage in the enhancement of feed quality of OC biologically treated by *A. oryzae* FK-923, as protein content was increased. On other hand, both phenols and fibers were decreased. Addition of dry yeast at 1.5 kg/ton olive waste was more suitable than other tested amounts, as the protein percent raised to 20.2 % in the fermented olive waste with total protein 170.8 kg/ton. On the other hand, phenols and fibers were reduced to 0.55 and 19.2 %, respectively.

Effect of biological treatment on the chemical composition of olive cake

The fermented OC showed a much lower cell wall component value than that of the untreated OC (Table 5). Whereas, organic matter, crude fiber, nitrogen free extract and total fiber contents in the biological treated olive cake were decreased than that of untreated OC (raw), while crude protein was increased from 9.5 to 20.2 % in the treated OC.

Table 5 Chemical composition, cell wall constituents and gross energy for olive cake before and after biological treatment by *A. oryzae* FK-923

Item	Olive cake	
	Raw (%)	Treated (%)
Chemical composition (%)		
Moisture	11.97	11.97
Component, % on DM basis		
OM	93.0	88.10
CP	9.5	20.2
CF	33.0	19.2
EE	5.7	6.1
NFE	76.42	42.46
Ash	7.0	11.90
Cell wall constituents, % on DM basis		
NDF	62	44.6
ADF	48	32.6
Total fibers	33	19.2
Gross energy, kcal/kg DM	4200	4320

OM organic matter, *CP* crude protein, *CF* crude fiber, *EE* ether extract, *NFE* nitrogen free extract

Enzymes involved in fermentation of olive cake by *A. oryzae* FK-923

The reduction of OC fiber content through the SSF process, using *A. oryzae* FK-923, is not surprising considering that this fungus possesses an efficient lignino-hemicellulolytic

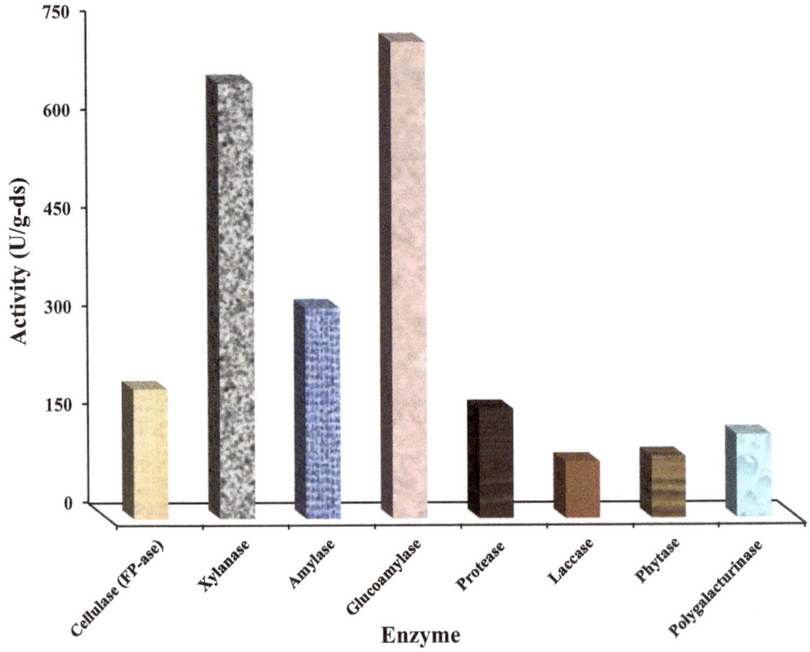

Fig. 3 Enzymes involved in fermentation of olive cake by *A. oryzae* FK-923 for 7 days at 36 °C

system as shown in Fig. 3. Since livestock and poultry cannot produce sufficient hydrolases enzymes in alimentary tracts, the presence of such enzymes add high nutritive value as well as save adding exogenous enzymes to feeds, thus increasing feed intake and production performance of livestock and poultry; increasing digestibility and absorptive of nutrients; enhancing the digestibility and utilization efficiency of feeds, and thus reducing feed costs.

The amino acid composition of microbial compared to alfalfa protein

The data presented in Table 6 showed that the amino acid profiles of *A. oryzae* FK-923 culture grown on olive cake have advantage in lysine, arginine, glutamic acid, proline, alanine, methionine and less in cysteine when compared with amino acid composition of alfalfa protein.

Table 6 Amino acid composition of microbial protein of *A. oryzae* FK-923 (g/100 g of protein) compared to Alfalfa protein

Amino acid	*A. oryzae* FK-923	Alfalfa
Lysine	9.2	6.70
Histidine	1.67	2.53
Arginine	7.12	5.54
Aspartic acid	12.8	12.54
Threonine	6.6	5.12
Serine	5.16	5.25
Glutamic acid	22.11	11.33
Proline	7.22	5.10
Glycine	6.10	5.73
Alanine	9.12	6.33
Valine	5.79	6.70
Methionine	2.05	1.36
Isoleucine	4.82	5.54
Leucine	8.86	8.43
Tyrosine	3.42	3.72
Phenylalanine	4.64	5.75
Cystine/cysteine	0.91	1.40

Discussion

In Egypt, large quantities of agriculture and agro-industrial by-products are generated and most of them are regarded as waste and non-conventional feedstuffs. The majority of researchers suggest that these lignocellulosic substrates when used as a source of protein, they are equivalent to soybean meal and other sources of protein for beef cattle diets (National Research Council (NRC) 2001; Loy and Wright 2003). The crude OC, used in this study is characterized by low protein content (9.5 %) and relatively high fat content (5.7 %). The OC is rich in fiber with NDF and ADF values of 62 and 48 %, respectively (Table 1). Since OC is rich in fiber, it has a low degradability in the rumen. OC bio-conversion into a better quality feed by *A. oryzae* FK-923 in SSF could be an alternative to chemical and physical treatment for enhancing the OC quality for microbial fermentation in rumen. The reduction in OC phenolic contents during SSF (Table 2) is an important aspect in the adaptation of *A. oryzae* FK-923, since the capacity of a substrate to resist to degradation has been partially attributed to its array of phenolic compounds (Fermor and Macauley 1991). These results were in agreement with Salmones et al. (2005) who reported a significant negative relationship between phenolic concentration in coffee pulp and laccase activity.

An increase in protein content from 9.5 to 17.4 % was observed on OC fungal treatment with *A. oryzae* FK-923 over the untreated sample (Table 2). This result is consistent with the findings of Iyayi and Aderolu (2004) that showed an increase in CP contents *Trichoderma viride* treated agro-wastes. In addition, Dhanda et al. (1994) reported that the CP content of spent straw was increased from 3.42 to 6.19 % after biological treatment. The CP increase could be the result of increased fungal biomass (Chen et al. 1995) suggesting that the treated substrate is a good protein source for livestock. The improvement of CP contents could also be due to secretion of certain extracellular enzymes into the waste during their breakdown and its subsequent metabolism (Kinfemi et al. 2009). The addition of sugar cane molasses at 2 % (w/w) showed an increase in crude protein to 18.9 % with a reduction in phenols and fibers to 0.69 and 21.8 %, respectively (Table 3). Furthermore, *S. cerevisiae* supplementation at 1.5 % (w/w) to the fermentation medium raised the crude protein to 20.2 % (w/w), while phenols and fibers were declined to 0.55 and 19.2 %, respectively (Table 4). *S. cerevisiae* possesses several properties, such as its resistance to high sugar and alcohol concentrations and a high growth rate at increasing temperatures (Dhiman et al. 2002; Jung et al. 2008).

In most bioconversion studies, compositional changes, i.e. neutral detergent fiber (NDF), acid detergent fiber (ADF) and acid detergent lignin (ADL) followed by estimation of digestibility have been carried out to estimate the nutritive value of the final fermented feed (Shrivastava et al. 2012). In the present study, the fermented OC showed a much lower cell wall component value than that of the untreated OC, due to the decrease of NDF and ADF contents by 44.6 and 32.6 %, respectively (Table 5). These results might be due to the breakdown of lignocellulosic bonds by the fungal enzymes (Bilal 2008). The reduction in the total fibers content from 33 to 19.2 % confirms and proves the ability of the selected fungal strain in producing hemicellulases, cellulases, and polygalacturinase enzymes. In this concern Baraghit et al. (2009) reported that the biological treatments of different crop residues with various fungal and bacteria strains decreased cell wall total fibers content.

The present study revealed that, the phenolic compounds which are the more obstacles towards the use of OC decreased to a limit that not affected the microflora and this confirmed that the fungus could produce laccases under the culture conditions. Laccases (EC 1.10.3.2; benzenediol: oxygen oxidoreductases) are multicopper enzymes belonging to the group of blue oxidases that catalyses oxidation of a wide variety of organic and inorganic compounds, including diphenols, polyphenols, diamines and aromatic amines. One electron at a time is removed from the substrate, and molecular oxygen is used as the electron acceptor (Gianfreda et al. 1999). Some fungal laccases degrade toxic fungal metabolites, such as aflatoxin B1 (Alberts et al. 2009) and are also useful in the field of food microbiology. Furthermore, the chemical analysis indicated that, the gross energy was increased from 4200 to 4320 kcal/kg DM.

The degradation of polysaccharide compounds was associated with the increase in reducing sugars. Enhancing the concentration of these sugars is an additional advantage besides the reduction of lignocelluloses in the improvement of biomass quality through biological treatment. Enzymes assay showed that the residual fermented substrate loaded with glucoamylase 756, alpha-amylase 422.2, xylanase 664.4, celullases (FP-ase) 198.2 and polygalagturinase 76.8 U/g-ds (Fig. 3). Since livestock and poultry can't produce sufficient xylanase and cellulases in alimentary tracts, the presence of such enzymes add high nutritive value as well as save adding exogenous xylanase and cellulases to ruminants feed, thus increasing feed intake and production performance of livestock and poultry; increasing digestibility and absorptive of nutrients; enhancing the digestibility and utilization efficiency of feeds, and thus reducing feed costs (Yang and Wyman 2004).

Cellulases and xylanases have been used in the feed of mono gastric animals to hydrolyze non starch polysaccharides such as β-glucans and arabinoxylans. Cellulases, used as feed additives alone or with proteases, can significantly improve the quality of pork meat (Kumar and Wyman 2009). Glucanases and xylanases reduce viscosity of high fiber rye- and barley-based feeds in poultry and pig. These enzymes can also cause weight gain in chickens and piglets by improving digestion and absorption of feed materials (Taniguchi et al. 2005; Shrivastava et al. 2011). The detection of phytase in the fermented olive waste may be due to the addition of *S. cerevisiae*. Phytase (myoinositol hexakisphosphate phosphohydrolase, EC 3.1.3.8) catalyzes the hydrolysis of phytate to myoinositol pentakisphosphate and orthophosphate. Improving the digestibility of proteins and increasing the availability of phosphorus and other minerals, which are usually chelated by phytic acid (Konietzny and Greiner 2002) diminishes the anti-nutritive properties of phytate and prevents environmental pollution. Phytases have been studied in different yeast strains; an extracellular acid phytase was recently purified and characterized from *S. cerevisiae* by Jin-In et al. (2009).

Therefore, it can be concluded from this study that, untreated olive cake is poor in crude protein but rich in NDF and ADF. The biodegradation of OC using *A. oryzae* FK-923 resulted in an increase of crude protein and a decrease of fiber fractions and thus an increased digestibility of the resultant substrate. This suggests that SSF–*A. oryzae* FK-923 system is suitable for conversion of solid waste from olive processing industry into a value added animal feed. However, further research work in these areas is necessary to verify the effectiveness of this degraded substrate on the performance of live animals.

Acknowledgments The authors gratefully acknowledge the National Research Centre (NRC), 33 El-Bohouth St., Giza, Egypt, for the facilities and financial support that enable the authors to accomplish this work.

Compliance with ethical standards

Conflict of interest The authors declare that they have no competing interest.

References

Abd El-Rahman HH, Abedo AA, El-Nomeary YAA, Abdel-Magid SS, Mohamed MI (2014) Effect of biological treatments of rice straw on growth performance, digestion and economical efficiency for growing calves. Glob Vet 13(1):47–54

AboSalim IA, Bendary MM (2005) Forage supplies in Egypt resources-maximizing its utilization. In: Proceedings 2nd conference of Animal Product Research Institute Sakha

Agrawal D, Partidar P, Banerjee TPS (2005) Alkaline protease production by a soil isolate of *Beauveria feline* under SSF condition: parameter optimization and application to soy protein hydrolysis. Process Biochem 40:1131–1136

Alberts JF, Gelderblom WCA, Botha A, Zyl WHV (2009) Degradation of aflatoxin B1 by fungal laccase enzymes. Int J Food Microbiol 135(1):47–52

AOAC (1970) Official methods of analysis. In: William H (ed) Association of Official Analytical Chemists, 11th edn. vol. 1, AOAC, Washington, DC

AOAC (1990) Official methods of analysis. In: Helrich K (ed) Association of Official Analytical Chemists, 15th edn. AOAC, Arlington, Virgina

Azkar T, Tosun I, Kaynak Z, Ozkara E, Yeni O, Sahin E (2008) An attractive agro-industrial by-product in environmental cleanup. J Hazard Mater 33:125–137

Bai S, Kumar MR, Kumar DJM, Balashanmugam P, Kumaran MDB, Kalaichelvan PT (2012) Cellulase production by *Bacillus subtilis* isolated from cow dung. Arch Appl Sci Res 4:269–279

Bailey MJ, Biely P, Poutanen K (1992) Inter laboratory testing of methods for assay of xylanase activity. J Biotechnol 23:257–270

Baraghit GA, Ahmed BM, El-Mahy MA (2009) Digestibility, nutritive value and rumen fermentation of rice straw and sugar cane bagasse treated with a commercial bacterial culture. Egypt J Nutr Feed 12(3):511–522

Bendray MM, Ghanem GHA, Gaafar HMA (2006) Utilization of rice straw for feeding ruminants: (productive performance of lactating buffaloes fed rice straw silage). J Agric Sci Mansoura Univ 31(8):5025–5038

Bilal MO (2008) Effect of molasses and corn as silage additives on cell wall fractions of Mott grass silage with different fermentation periods. J Anim Plant Sci 18(4):102–106

Bourbonnais R, Leech D, Paice MG (1998) Electrochemical analysis of the interactions of laccase mediators with lignin model compounds. Biochim Biophys Acta 1379(3):381–390

Bradford M (1976) A rapid and sensitive method for the quantitation of microgram quantities of protein utilizing the principle of protein dye binding. Anal Biochem 72:248–254

Canet R, Pomares B, Cabot B, Chaves C, Ferrer E, Ribó M (2008) Composting olive mill pomace and other residues from rural southeastern Spain. Waste Manag 28:2585–2592

Chen C, Fales SL, Varga GA, Royce DJ (1995) Biodegradation of cell wall components of maize stover colonized by white-rot fungi and resulting impact on in vitro digestibility. J Sci Food Agric 68:91–98

Christias C, Couvarakim C, Georgopoulos SG, Macris B, Vomvoyanai V (1975) Protein content and amino acid composition of certain fungi evaluated for microbial protein production. Appl Env Microbiol 29:250–254

Chuah AM, Lee YC, Yamaguchi T, Takamura H, Yin LJ, Matoba T (2008) Effect of cooking on the antioxidant properties of coloured peppers. Food Chem 111:20–28

Dhanda S, Kakkar VK, Garcha HS, Makkar GS (1994) Biological treatments of baddy straw and its evaluation through ruminant feeding. Indian J Anim Nutr 11(2):73–79

Dhiman TR, Zaman MS, Gimenez RR, Walters JL, Treacher R (2002) Performance of dairy cows fed forage treated with fibrolytic enzymes prior to feeding. Animal Feed Sci Technol 101:115–125

Economides S, Hadjipanayiotou M, Georghiades E (1981) The nutritive value of straw, and barley and lucenna hay, and the nitrogen supplementation on the nutritive value of straw to sheep. Technical bulletin 39, Agricultural Research Institute, Nicosia

FAO (1981) Food loss prevention in perishable crops. FAO Agricultural Service Bulletin, No. 43, FAO Statistics Division, Rome

Fermor TR, Macauley BJ (1991) Microbiology factor contributing to selectivity and nutritional quality of mushroom compost. In: Nair NGT (ed) Proceedings of AMGA/ISMS International workshop-seminar on *Agaricus* compost. AMGA Limited, Sidney, pp 26–45

Gianfreda L, Xu F, Bollag JM (1999) Laccases: a useful group of oxidoreductive enzymes. Bioremediation J 3(1):1–25

Hachicha S, Sallemi F, Medhioub K, Hachicha R, Ammar E (2008) Quality assessment of composts prepared with olive mill waste water and agricultural wastes. Waste Manag 28:2593–2603

Iyayi EA, Aderolu ZA (2004) Enhancement of the feeding value of some agro-industrial by products for laying hens after their solid state fermentation with *Trichoderma viride*. Afr J Biotechnol 3:182–185

Jin-In M, Won-Seo S, Kim DC, Soon-Oh N (2009) Purification and biochemical properties of an extracellular acid phytase produced by *Saccharomyces cerevisiae* CY strain. Process Biochem 44(1):122–126

Jung HK, Park CD, Bae DH, Hong JH (2008) Isolation of alcohol-tolerant amylolytic *Saccharomyces cerevisiae* and its application to alcohol fermentation. Food Sci Biotechnol 17(6):1160–1164

Kinfemi AA, Mohamed MI, Ayoade JA (2009) Biodegradation of cowpea shells by *Pleurotus specie* for its use as ruminant feed. World J Agric Sci 5:639–645

Konietzny U, Greiner R (2002) Molecular and catalytic properties of phytate-degrading enzymes (phytases). Int J Food Sci Technol 37(7):791–812

Kumar R, Wyman CE (2009) Effect of additives on the digestibility of corn stover solids following pretreatment by leading technologies. Biotechnol Bioeng 102(6):1544–1557

Lowry OH, Rosebrough NJ, Farr AL, Randall RJ (1951) Protein measurement with the Folin phenol reagent. J biol Chem 193:265–275

Loy DD, Wright KN (2003) Nutritional properties and feeding value of corn and its by-products. In: White PJ, Johnson LA (eds) Corn chemistry and technology. American Association of Cereal Chemists, Inc, St Paul, pp 571–603

Majcherczyk A, Johannes C, Hüttermann A (1998) Oxidation of polycyclic aromatic hydrocarbons (PAH) by laccase of *Trametes versicolor*. Enzym Microb Technol 22(5):335–341

Manning BG, Campbell LL (1961) Thermostable α-amylase of *Bacillus philus*. J Biol Chem 236:2952–2955

Miller GL (1959) Use of dinitrosalicylic acid reagent for determination of reducing sugar. Anal Biochem 3:426–429

Ministry of Agriculture (2006) Economic affairs, sector of agricultural statistics, Egypt. Ministry of Agriculture, Egypt

Ministry of Agriculture and Land Reclamation (2012) Economic affairs, sector of agricultural statistics, Egypt. Ministry of Agriculture, Egypt

Misra AK, Mishra AS, Tripathi MK, Prasad R, Vaithiyanathan S, Jakhmola RC (2007) Optimization of solid state fermentation of mustard (*Brassica campestris*) straw for production of animal feed by white rot fungi (*Ganoderma lucidum*). Asian Aust J Anim Sci 20:208–213

National Research Council (NRC) (2001) Nutrient requirements of dairy cattle, 7th rev edn. National Academyof Science, Washington, DC

Neifar M, Jaouani A, Ghorbel RE, Chaabouni SE, Penninckx MJ (2009) Effect of culturing processes and copper addition on laccase production by the white-rot fungus *Fomes fomentarius* MUCL 35117. Lett Appl Microbiol 49:73–78

Rowghani E, Zamiri MJ, Seradj AR (2008) The chemical composition, rumen degradability, in vitro gas production, energy content and digestibility of olive cake ensiled with additives. Iran J Vet Res 9:213–221

Salmones D, Mata G, Waliszewski KN (2005) Comparative culturing of *Pleurotus* sp. on coffee pulp and wheat straw: biomass production and substrate biodegradation. Bioresour Technol 96:537–544

Shrivastava B, Thakur S, Khasa YP, Gupte A, Puniya AK, Kuhad RC (2011) White-rot fungal conversion of wheat straw to energy rich cattle feed. Biodegradation 22(4):823–831

Shrivastava B, Nandal P, Sharma A, Jain KK, Khasa YP, Das TK, Mani V, Kewalramani NJ, Kundu SS, Kuhad RC (2012) Solid state bioconversion of wheat straw into digestible and nutritive ruminant feed by *Ganoderma* sp. rckk02. Bioresour Technol 107:347–351

Taniguchi M, Suzuki H, Watanabe D, Sakai K, Hoshino K, Tanaka T (2005) Evaluation of pretreatment with *Pleurotus ostreatus* for enzymatic hydrolysis of rice straw. J Biosci Bioeng 100(6):637–643

Ugwuanyi JO, McNeil B, Harvey LM (2009) Production of protein-enriched feed using agro-industrial residues as substrates. In: Pandey A, Singh-Nee Nigam P (eds) Biotechnology for agro-industrial residues utilization. Springer, The Netherlands, pp 77–103

Yang B, Wyman CE (2004) Effect of xylan and lignin removal by batch and flow through pretreatment on the enzymatic digestibility of corn stover cellulose. Biotechnol Bioeng 86(1):88–95

22

Assessment of compost quality in agitated pile composting of water hyacinth collected from different sources

Jiwan Singh[1,2] · Ajay S. Kalamdhad[1]

Abstract

Background Water hyacinth (*Eichhornia crassipes*) can cause a diversity of problems due to its fast spread and congested growth. The composting of water hyacinth can reduce the chemical fertilizer application to the agricultural field and problems related to its fast growth rate and also help in nutrient recycling. The present study was carried out on agitated pile composting of water hyacinth collected from four different areas (Bharalu River, Agriculture site, Boragaon landfill site and Industrial site). The nutrients and stability parameters were evaluated during the 30 days of agitated pile composting of water hyacinth mixed with cattle manure and sawdust. The stability of compost was evaluated using respiration techniques (CO_2 evolution and oxygen uptake rate).

Results Results showed that nutrients (Na, K, Ca, total nitrogen and phosphorus) were increased significantly in all agitated pile composting. Stability parameters such as CO_2 evolution rate and oxygen uptake rate, biochemical chemical oxygen demand and chemical oxygen demand were reduced significantly during the process. Highest reduction of moisture content (35.2 %), volatile solid (40.5 %), ammonical nitrogen (75.6 %), biochemical oxygen demand (65.3 %), CO_2 evolution rate (78.7 %), and oxygen uptake rate (74.4 %) were observed in the agitated pile of Bharalu River site as compared to other agitated piles. The higher increase of total nitrogen (83.5 %) and total phosphorus (76.0 %) was also observed in the agitated pile of Bharalu River site.

Conclusion Addition of optimum amount of cattle manure was highly efficient for making compost of water hyacinth collected from four different sites. Highest content of nutrients was found in compost of water hyacinth collected from Bharalu River site due to availability of nutrients in sewage water. In the final compost of all trials, total coliform and fecal coliform were reduced significantly, which are pathogen indicators in the compost. The best compost quality was found in water hyacinth collected from Bharalu River site.

Keywords Water hyacinth · Agitated pile composting · Nutrients · Stability · Bharalu River

Introduction

Water hyacinth (*Eichhornia crassipes*) is a free-floating macrophyte, growing in or near water, that is emergent, submerged or free floating (Singh and Kalamdhad 2012). Due to very high productivity, water hyacinth has been considered as one of the worst aquatic plants of the world. It can double its size in 5 days under optimum temperature and moisture. The dense mats of water hyacinth can interfere with navigation, recreation and irrigation as well as the blockage of canals and rivers can even cause dangerous flooding. Availability of oxygen is low beneath the mats that generate good breeding conditions for mosquito vectors of malaria, encephalitis and filariasis (Gajalakshmi et al. 2002; Malik 2007). Composting is the most promising technique for the treatment of water hyacinth; composting can solve two problems: (a) reduce chemical fertilizer load and (b) check the abnormal growth of

✉ Jiwan Singh
jiwansingh95@gmail.com

1 Department of Civil Engineering, Indian Institute of Technology Guwahati (IITG), Guwahati 781039, Assam, India

2 Department of Environmental Engineering, Kwangwoon University, Seoul 139-701, Republic of Korea

hyacinth. The formation of humic substances is the final product of any type of composting process; it holds sufficient quantity of nutrients (nitrogen, potassium, calcium and phosphorus) which are required for the plants growth Gajalakshmi et al. 2002; Prasad et al. 2013; Singh and Kalamdhad 2013a). Water hyacinth compost application in agriculture mainly results from its content of organic matter, plant nutrients, promoting plant growth and inhibiting root pathogens/soil-borne plant diseases (Balasubramanian et al. 2013).

As the composting process proceeds, the readily degradable organic mass is converted into stable humic substances (Haug 1993). The compost is considered stable if it contains mainly recalcitrant or humus-like matter, but some times unstable when it contains a high quantity of biodegradable matter that may maintain high microbial activity (Kalamdhad et al. 2009). Stability of compost is a key parameter in relation to its field application, potential of odor generation and pathogen regrowth. The stability of composts can be defined as the degree to which the organic fractions in composts have been stabilized during the process (Kalamdhad et al. 2009).

A Boragaon landfill (Guwahati, India) may cause contamination of soil and water via leaching. This landfill has covered the surrounding area of the Deepor Beel Lake. Leachate is generated by landfill possibly entering into lake and pollutes fresh water resource. Growing water hyacinth in those areas may accumulate nutrient from the polluted water. Bharalu River is highly polluted by sewage of Guwahati city, India. Water hyacinth plants are growing in this River accumulating high concentration of nutrients. Loss of nutrient from agriculture field due to runoff in low-lying areas is very common phenomenon in the agricultural field. These are the most favorable conditions for the growth of the water hyacinth which are often found in developing countries. Rising water hyacinth in these sites may contain high concentration of nutrients. Some studies have been done on agitated pile composting of different wastes (Parkinson et al. 2004; Jouraiphy et al. 2005; Singh and Kalamdhad 2012, 2013a, b), but there is no information available on composting of water hyacinth collected from different sources. Therefore, the objective of the present study was to assess the nutrients and stability parameters during agitated pile composting of water hyacinth collected from Bharalu River site (BRS), agriculture site (AS), boragaon landfill site (BLS) and industrial site (IS).

Materials and methods

Feedstock materials and design of agitated pile composting

The composting study was carried out in the summer season (from June, 2013 to July, 2013). Water hyacinth, cattle (cow) manure and sawdust were used for the preparation of different waste mixtures for the composting process. Water hyacinth was collected from four different places, i.e. Amingoan industrial site (IS) near Indian Institute of Technology Guwahati (IITG) campus, Boragaon landfill site (BLS), Bharalu River site (BRS) and low-lying area near agriculture site (AS), Guwahati, India. Water hyacinth (shoot + root) was collected from 4 to 5 different places at each sampling site to ensure the grab sampling. All collected materials were processed in the laboratory before composting process. The maximum particle size of water hyacinth in the waste mixture was restricted to 1 cm by chopping to provide better aeration. Cattle manure was obtained from dairy farm near the IITG campus. Sawdust was purchased from nearby saw mill. On the basis of our previous studies (Singh and Kalamdhad 2012, 2013a, b; Prasad et al. 2013), a mixture of 90 kg water hyacinth, 45 kg cattle manure and 15 kg sawdust (6:3:1 ratio of water hyacinth, cattle manure and sawdust) was taken for the study. This was the best combination for nutrient recycling and stability analysis during water hyacinth composting; therefore, in this study same combinations have been used with water hyacinth collected from four different sources. The initial characterizations of raw materials are given in Table 1. Four different waste combinations were composted in agitated pile (length 2100 mm, base width 350 mm, top width 100 mm and height 250 mm). A length to base width (*L*/*W*) ratio of 6 was maintained for each agitated pile. Agitated piles contained approximately 150 kg of different waste combinations and composted for 30 days. Temperature was monitored by a digital thermometer during the composting period. About 500 g of homogenized samples were collected from five different locations, mostly at the mid span and ends of the pile by compost sampler without disturbing the adjacent materials in the pile on days 0, 3, 6, 9, 12, 15, 18, 21, 24, 27, and 30 after turning. Finally all the grab samples were mixed carefully to make a homogenized sample. Triplicate samples were collected and stored at 4 °C for analysis of biochemical parameters. Sub-samples were dried at 105 °C in oven for 24 h and moisture content was calculated as given in Eq. 1, Dried samples were ground to pass to 0.2 mm sieves and stored for further analysis.

$$\text{Moisture content } (\%) = \frac{\text{Initial weight} - \text{Final weight}}{\text{Initial weight}} \times 100 \quad (1)$$

Analysis of physico-chemical and biological parameters

Each sub-sample was analyzed for the following parameters: pH and electrical conductivity (EC) (1:10 w/v

Table 1 Initial characterization of waste materials collected from different sources (Mean ± SD, $n = 3$)

Parameters	Water hyacinth BRS	Water hyacinth AS	Water hyacinth BLS	Water hyacinth IS	Sawdust	Cattle manure
pH	5.98 ± 0.16	5.86 ± 0.03	5.51 ± 0.04	5.79 ± 0.07	6.2 ± 0.005	6.7 ± 0.05
EC (dS/m)	6.04 ± 0.29	5.10 ± 0.14	4.44 ± 0.11	4.91 ± 0.01	0.4 ± 0.002	3.3 ± 0.03
Moisture content (%)	85.95 ± 0.35	88.95 ± 0.35	87.25 ± 0.65	85.94 ± 0.055	10.01 ± 0.04	80.9 ± 0.19
Volatile solids (%)	73.57 ± 1.4	74.85 ± 0.75	75.22 ± 1.66	72.64 ± 0.037	97.9 ± 0.2	72.05 ± 0.224
Ash content (%)	26.44 ± 1.4	25.15 ± 0.75	24.78 ± 1.66	27.50 ± 0.11	2.13 ± 0.23	27.9 ± 0.22
TOC (%)	40.87 ± 0.79	41.58 ± 0.75	41.79 ± 0.09	40.27 ± 0.06		
Total Nitrogen (%)	1.19 ± 0.16	1.33 ± 0.41	1.11 ± 0.12	1.15 ± 0.14	0.38 ± 0.01	1.4 ± 0.168
Ammonical nitrogen (mg/kg)	153.0 ± 0.77	87.9 ± 0.53	178 ± 10.001	152.8 ± 2.9	96.59 ± 7.5	135.6 ± 5.3
Available phosphorus (mg/kg)	3171 ± 0.5	2750 ± 150	3521 ± 20	5801 ± 200	1563 ± 11	3574 ± 119
Total phosphorus (mg/kg)	6089 ± 223	4975 ± 25	6075 ± 179	9967 ± 126	2214 ± 7	2277 ± 5.8
Sodium(mg/kg)	14405 ± 862	3427.5 ± 17	11723.5 ± 1488	8875 ± 125	1097.5 ± 2.5	2400 ± 5
Potassium(mg/kg)	26472 ± 1151	6047 ± 485	20483.5 ± 1831	18195 ± 205	697.5 ± 7.5	987.5 ± 22.5
Calcium(mg/kg)	5690.5 ± 454	2674.5 ± 221	8063.5 ± 500	6177.5 ± 7.5	2420 ± 15	9892.5 ± 17.2
COD (mg/L)	1440 ± 160	1120 ± 160	960 ± 160	1600 ± 320	255 ± 105	420 ± 0.0
BOD (mg/L)	480 ± 30	375 ± 15	465 ± 15	360 ± 0.0	950 ± 310	1920 ± 640
CO_2 evolution (mg/g VS/day)	6.07 ± 0.25	5.28 ± 0.95	5.73 ± 0.40	5.74 ± 0.1	1.95 ± 0.35	5.16 ± 0.2
OUR(mg/g VS/day)	15.33 ± 0.99	12.81 ± 0.45	14.56 ± 1.29	17.13 ± 1.5	2.38 ± 0.24	12.99 ± 0.43

BRS Bharalu river site, *AS* agricultural site, *BLS* Boragoan landfill site, *IS* industrial site, *COD* chemical oxygen demand, *BOD* biochemical oxygen demand, *TOC* total organic carbon

waste: water extract, and volatile solids (VS) (Kalamdhad et al. 2009). Volatile solid was determined by loss ignition method; 10 g sample was taken in crucible and kept in muffle furnace at 550 °C for 2 h. The total nitrogen (TN) was determined using the Kjeldahl method; 0.2 g of sample (dried) was digested with 10 ml of concentrated H_2SO_4 and catalyst mixture ($CuSO_4$:K_2SO_4, 5:1 ratio) and heated at 400 °C. After digestion, sample was diluted to 100 mL with distilled water. 10 mL of diluted sample was mixed with 20 mL of 40 % NaOH and subjected to the distillation assembly. Ammonia gas was collected in 25 mL of boric acid solution in 250 mL conical flask and then titrated with standardized H_2SO_4 (0.02 N). Boric acid solution was prepared with mixed indicator (methyl red and methylene blue). The end point is the appearance of purple colour. NH_4–N and NO_3–N were determined using KCl extraction (Tiquia and Tam 2000); a 3-g oven-dried sample was taken in a 250 mL conical flask and mixed with 30 mL 2 M KCl solution. The mixture was placed in a horizontal shaker for 2 h. The analysis for NH_4^+–N in the filtered sample was done using Phenate Method (APHA 2005). The Flame photometer (Systronic 128) was used for analysis of Na, K and Ca concentration after digestion of the 0.2-g sample with 10 mL H_2SO_4 and $HClO_4$ mixture (5:1) in block digestion system (Pelican Equipments Chennai-India) for 2 h at 300 °C. Above acid-digested sample was also taken for the determination of available and total phosphorus (acid digestion) using the stannous chloride method (APHA 2005).Calculations of VS and TN of compost samples are given in Eq. 2 and 3, as follows:

$$VS\ (\%) = \frac{\text{Initial weight} - \text{weight after burning}}{\text{Initial weight}} \times 100 \quad (2)$$

$$TN\ (\%) = \frac{14 \times (S - B) \times N}{\text{weight (g)}} \times 100, \quad (3)$$

where S mL of standard sulfuric acid used for sample, B mL of standard sulfuric acid used for blank, N normality of standard sulfuric acid

The biodegradable organic matter was measured as soluble biochemical oxygen demand (BOD) (by the dilution method; APHA 2005) and soluble chemical oxygen demand (COD) (by the dichromate method; APHA 2005). The population of total coliforms (TC) in the compost was determined in extract (compost:water extract, 1:10) by inoculation of culture tube media using the most probable number (MPN) method (APHA 2005).

The stability parameters, i.e. oxygen uptake rate (OUR) and CO_2 evolution rate were determined according to the

Kalamdhad et al. (2008).For the determination of CO_2 evolution rate, about 25 g of fresh compost sample was taken in 1 L of polyvinyl chloride airtight container. About 10 g of oven-dried (105 °C) soda lime (1.5–2.0 mm mesh size) was taken in a 100-mL beaker and placed in the above container. The initial weight of the soda lime was measured. The container with soda-lime beaker was kept in an incubator at 25 °C temperature for 24 h. After 24 h, the soda-lime was taken out and oven dried again; then the final weight was noted down.CO_2 evolution rate was calculated by the following formula Eq. (4) (Kalamdhad et al. 2008):

$$\text{Carbon dioxide evolution (mg/g VS/hr)} = \frac{(\text{W1} - \text{W2}) \times 1000}{W \times T} \times 100, \tag{4}$$

where W1 is initial weight of the soda-lime (g), W2 is final weight of the soda-lime (g), W is weight of compost sample (g) and T is time duration of incubation (h).

The OUR was measured in a liquid suspension of 5–8 g of compost sample in 500 mL of distilled water in which $CaCl_2$, $MgSO_4$, $FeCl_3$ and phosphate buffer (pH 7.2) were added. The mixture was kept in suspension by placing it on the magnetic stirrer at 30°C temperature. During this time, the dissolved O_2 in the suspension was continuously measured by the digital dissolved oxygen meter attached. The oxygen consumption rate was calculated by taking the difference of dissolved oxygen with the respect to the time intervals and calculated value was quoted as the OUR in mg O_2/g VS/h (Kalamdhad et al. 2008).

All the results reported are the means of three replicates. Repeated measures treated with analysis of variance (ANOVA) was made using SPSS software. The objective of the statistical analysis was to determine any significant difference among the parameters for different agitated piles.

Results and discussion

Physico-chemical analysis

The composting pile temperature went through three typical phases (mesophilic, thermophilic and cooling phase) and ranged from 22 to 52 °C during the entire period of composting (Fig. 1). The cattle manure affected the temperature during the agitated pile composting in all four different agitated piles. Addition of optimum amount of cattle manure to the composting materials may fasten the composting process by providing easily available carbon to the microorganisms present in the composting mass. Therefore, microbial metabolic activities have increased and temperature was increased in very short time (Singh and Kalamdhad 2012, 2014). In agitated pile composting of BRS reached the highest temperature among all four piles due to quick establishment of microbial activity during the composting process (Chen et al. 2010). Although addition of cattle manure increased the composting process in all agitated piles the results indicate that the composting microbes started utilizing the easily degradable materials at the earliest in BRS. The highest temperature was observed in BRS (52 °C) followed by BLS (51.3 °C), AS (50.8 °C) and IS (48.5 °C). After 12th day, temperature in all piles was reduced and reached near-ambient temperature. Reduction in temperature indicated all piles achieved stability within 30 days of composting period.

Initial composting mixture requires optimum moisture content for the survival of microorganisms. The reduction in moisture content during the composting process in the form of vapors by the heat generation can be viewed as an index of decomposition rate (Kalamdhad et al. 2009). Initial moisture contents were observed about 83.4, 80.9, 82.9 and 82.2 % which were reduced to 54.1, 57.7, 54.3 and 59.9 % during pile composting of water hyacinth collected from BRS, AS, BLS and IS, respectively (Fig. 2a). Highest moisture loss occurred in agitated pile of BRS (35.2 %) followed by BLS (34.5 %), AS (28.6 %) and IS (27.1 %). On analyzing the results by ANOVA, the decrease in moisture content varied significantly between the days ($P < 0.05$).

Figure 2b illustrates that the pH was increased from 6.5 to 7.5, 6.2 to 7.2, 6.4 to 7.2 and 7.1 to 7.8 in agitated pile of BRS, IS, BLS and IS, respectively. The pH values were increased during the composting process may be due to sufficient aeration provided through the turning of piles at every 3rd day. Consequently, carbon dioxide level was reduced during the composting process (Singh and Kalamdhad 2013a). Significant difference in pH was observed in all the trials ($P < 0.05$). The pH values were measured within the optimal range for the development of bacteria (6.0–7.5) and fungi (5.5–8.0) (Amir et al. 2005).

Electrical conductivity (EC) is generally measured during composting because it reflects the salinity of the compost and its appropriateness for plant growth (Singh and Kalamdhad 2013a).Fig. 2c depicts the reduction of EC from 8.1 to 6.4 dS/m, 5.7 to 3.7 dS/m and 7.7 to 5.2 dS/m in the agitated pile of BRS, IS and BLS, respectively; however, EC was increased about 2.2–3.2 dS/m in agitated pile of IS. EC was decreased during the composting process probably due to volatilization of ammonia and the precipitation of mineral salts (Kalamdhad et al. 2009). In the agitated pile of IS, EC was enhanced probably due to the net loss of weight and release of soluble salts through decomposition activity in the composting process. On analyzing the results by ANOVA, EC varied significantly between the all agitated piles ($P < 0.05$).

Figure 2d shows the trend of organic matter degradation during the 30-day composting process in four different

Fig. 1 Variation of temperature during composting process

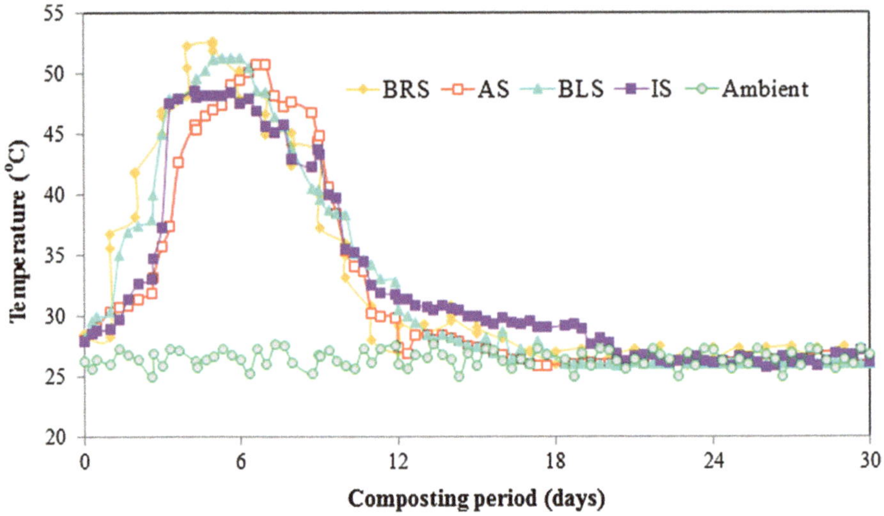

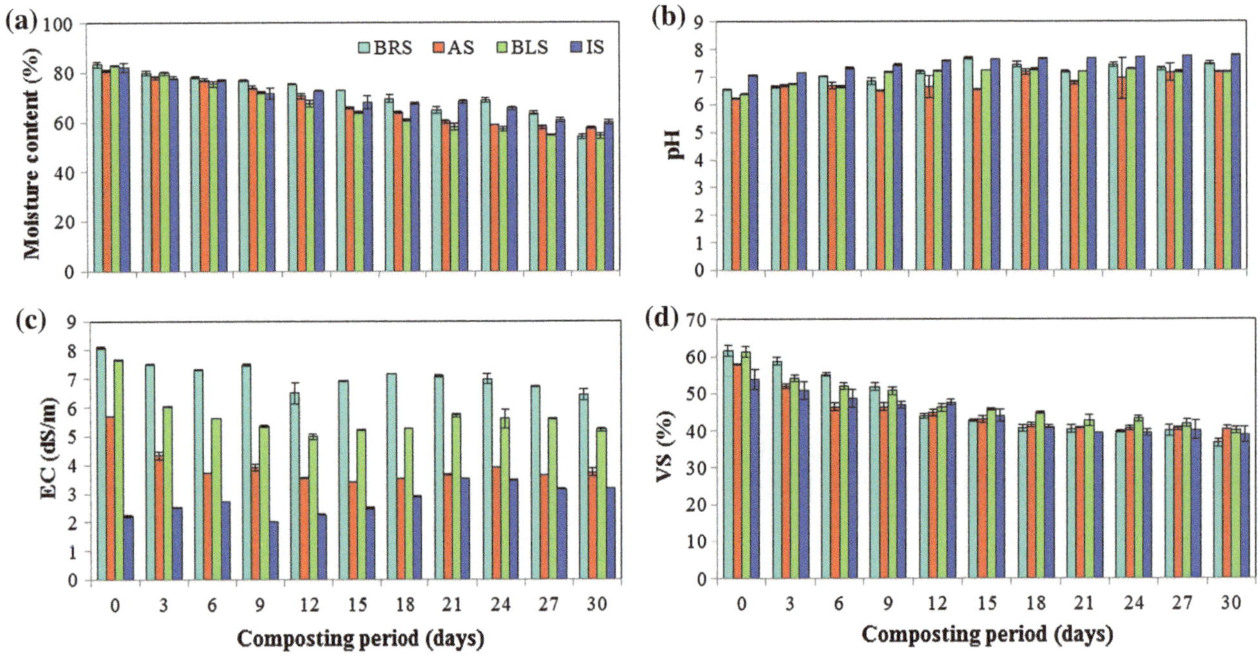

Fig. 2 Variation of moisture content, pH, electrical conductivity and organic matter during composting process

agitated piles. The content of organic matter was decreased from 61.7, 58.0, 61.4 and 53.7 % to 36.7, 40.2, 40.1 and 38.9 % in the agitated pile of BRS, AS, BLS and IS, respectively, as the decomposition progressed. Higher loss of organic matter was observed in agitated pile of BRS (40.5 %) followed by BLS (34.8 %), AS (30.6 %) and IS (27.6 %). BRS composting shows higher loss of organic matter as a result of higher temperature evolution as compared to other agitated piles. A significant variation in organic matter loss was found in all the agitated piles ($P < 0.05$). In the composting process organic matter is decomposed and transformed to stable humic substances (Prasad et al. 2013).

Table 2 shows that the total nitrogen was increased significantly ($P < 0.05$) from 0.85, 0.83, 1.05 and 0.64 % to 1.56, 1.19, 1.47 and 1.05 % in the agitated pile of BRS, AS, BLS and IS, respectively, during the composting process. The maximum increase in total nitrogen was observed in agitated pile of BRS (83.5 %) followed by IS (64.1 %), AS (43.4 %) and BLS (40.0 %). The total nitrogen was increased from 0.9 to 1.6 %, from 0.8 to 1.2 %, from 1.1 to 1.5 % and from 0.6 to 1.1 % in BRS, AS, BLS

Table 2 Changes in physico-chemical parameters during the composting process (Mean ± SD, $n = 3$)

Days	Nutrients concentration							
	BRS Total nitrogen (%)	AS	BLS	IS	BRS NH4-N (mg/kg)	AS	BLS	IS
0	0.85 ± 0.13	0.83 ± 0.07	1.05 ± 0.07	0.64 ± 0.02	253.49 ± 0.26	42.26 ± 0.53	223.31 ± 4.38	70.67 ± 0.17
3	1.05 ± 0.21	0.74 ± 0.02	1.05 ± 0.07	0.64 ± 0.03	377.9 ± 1.23	43.33 ± 0.32	206.04 ± 1.88	74.21 ± 0.03
6	1.05 ± 0.07	0.84 ± 0.14	1.05 ± 0.21	0.68 ± 0.02	156.36 ± 0.09	42.46 ± 0.0	193.03 ± 4.52	78.78 ± 0.06
9	0.98 ± 0.14	0.81 ± 0.03	1.19 ± 0.07	0.68 ± 0.02	167.69 ± 8.88	35.61 ± 0.0	182.79 ± 2.59	60.5 ± 0.13
12	1.12 ± 0.14	0.84 ± 0.0	1.26 ± 0.14	0.77 ± 0.07	105.13 ± 6.51	32.95 ± 0.50	94.08 ± 2.59	54.24 ± 0.08
15	1.19 ± 0.07	0.98 ± 0.14	1.47 ± 0.07	0.91 ± 0.07	93.32 ± 0.06	32.48 ± 0.09	92.20 ± 1.43	51.61 ± 0.34
18	1.33 ± 0.07	0.98 ± 0.0	1.33 ± 0.21	0.98 ± 0.0	91.86 ± 0.04	26.76 ± 0.57	99.18 ± 0.54	50.85 ± 0.03
21	1.26 ± 0.14	0.91 ± 0.07	1.19 ± 0.07	0.98 ± 0.0	85.3 ± 0.04	26.79 ± 4.26	94.045 ± 1.2	50.76 ± 0.10
24	1.33 ± 0.07	1.05 ± 0.07	1.267 ± 0.01	0.98 ± 0.0	69.42 ± 0.03	29.19 ± 0.07	96.49 ± 3.76	42.71 ± 0.14
27	1.47 ± 0.07	1.12 ± 0.14	1.4 ± 0.14	0.98 ± 0.0	67.76 ± 6.41	27.12 ± 0.08	103.83 ± 4.38	36.28 ± 0.35
30	1.56 ± 0.02	1.19 ± 0.1	1.47 ± 0.07	1.05 ± 0.07	61.91 ± 0.10	21.53 ± 0.01	94.30 ± 2.82	34.82 ± 0.07
Days	Total phosphorus (g/kg)				Available phosphorus (g/kg)			
0	3.05 ± 0.24	1.57 ± 0.14	3.14 ± 0.12	2.19 ± 0.0	2.8 ± 0.05	1.32 ± 0.08	1.59 ± 0.05	1.43 ± 0.14
3	3.16 ± 0.19	1.7 ± 0.01	3.41 ± 0.01	2.22 ± 0.02	3.07 ± 0.07	1.55 ± 0.16	1.51 ± 0.03	1.52 ± 0.07
6	3.16 ± 0.23	1.96 ± 0.0	3.92 ± 0.0	2.51 ± 0.0	3.1 ± 0.05	1.52 ± 0.01	1.61 ± 0.03	1.54 ± 0.08
9	3.22 ± 0.29	2.02 ± 0.01	4.04 ± 0.01	2.6 ± 0.0	3.1 ± 0.24	1.44 ± 0.04	1.64 ± 0.02	1.6 ± 0.09
12	3.38 ± 0.51	2.26 ± 0.01	4.53 ± 0.01	2.73 ± 0.01	3.01 ± 0.03	1.28 ± 0.04	1.75 ± 0.01	1.63 ± 0.03
15	3.50 ± 0.23	2.27 ± 0.03	4.53 ± 0.05	2.87 ± 0.01	3.13 ± 0.03	1.49 ± 0.17	1.81 ± 0.01	1.67 ± 0.01
18	3.46 ± 0.2	2.28 ± 0.01	4.56 ± 0.02	2.95 ± 0.0	3.45 ± 0.07	1.57 ± 0.13	1.88 ± 0.09	1.69 ± 0.0
21	3.03 ± 0.21	2.38 ± 0.0	4.76 ± 0.01	3.04 ± 0.01	3.52 ± 0.04	1.63 ± 0.14	2.37 ± 0.02	1.8 ± 0.04
24	3.53 ± 0.26	2.54 ± 0.02	5.07 ± 0.04	3.16 ± 0.0	3.64 ± 0.05	1.57 ± 0.03	2.33 ± 0.04	1.9 ± 0.11
27	4.0 ± 0.23	2.54 ± 0.0	5.07 ± 0.0	3.2 ± 0.0	3.53 ± 0.07	1.89 ± 0.11	1.95 ± 0.0	2.28 ± 0.10
30	5.37 ± 0.79	2.76 ± 0.01	5.52 ± 0.02	3.37 ± 0.0	3.57 ± 0.07	2.2 ± 0.23	1.91 ± 0.09	2.38 ± 0.10

and IS, respectively, within the composting time. It might be due to the net loss of dry mass in terms of CO_2 as well as the water loss by evaporation during oxidization of organic matter (Prasad et al. 2013). The total nitrogen generally increases during composting when organic matter loss is greater than the loss of NH_4–N (Sarika et al. 2014). The ammonical nitrogen was reduced significantly ($P < 0.05$) from 253.5, 42.3, 223.3 and 70.7 mg/kg to 61.9, 21.5, 94.3 and 34.8 mg/kg in the agitated pile of BRS, AS, BLS and IS, respectively (Table 2). The maximum reduction of ammonical nitrogen was found in agitated pile of BRS (75.6 %) followed by BLS (57.8 %), IS (50.7 %) and IS (49.1 %) during the process. At high pH, mixing and proper aeration enhance ammonia loss during maturation (Kalamdhad et al. 2009). Table 2 demonstrates that the total and available phosphorus were increased significantly ($P < 0.05$) in all agitated piles. The total phosphorus concentration was increased from 3.1, 1.7, 3.3, and 2.2 g/kg to 5.4, 2.8, 5.5 and 3.4 g/kg in BRS, AS, BLS and IS, respectively. The available phosphorus concentration was increased from 2.8 to 3.6 mg/kg, from 1.7 to 2.2 mg/kg, from 1.6 to 1.9 mg/kg and from 1.4 to 2.4 g/kg in BRS, AS, BLS and IS, respectively, in the 30 days of composting process. The concentrations of total and available phosphorus were increased during the composting process might be due to phosphorus released by micro-organisms through the mineralization of organic matter. The results of present study were consistent with those of the other study (Sarika et al. 2014). The highest percentage of total phosphorus and available were increased in the compost of BRS (76.0 %) and IS (66.3 %), respectively. Figure 3 shows that nutrient (Na, K and Ca) concentration was increased significantly ($P < 0.05$) due to degradation of organic matter during the composting process (Huang et al. 2004). The concentrations of nutrients (Na, K and Ca) were increased in the range of 1.21–6.89 g/kg for Na, 18.68–67.43 g/kg for K and 5.18–14.88 g/kg for Ca in final compost of all agitated piles. Nutrient concentrations were increased during composting process due to the net loss of dry mass (Sarika et al. 2014).

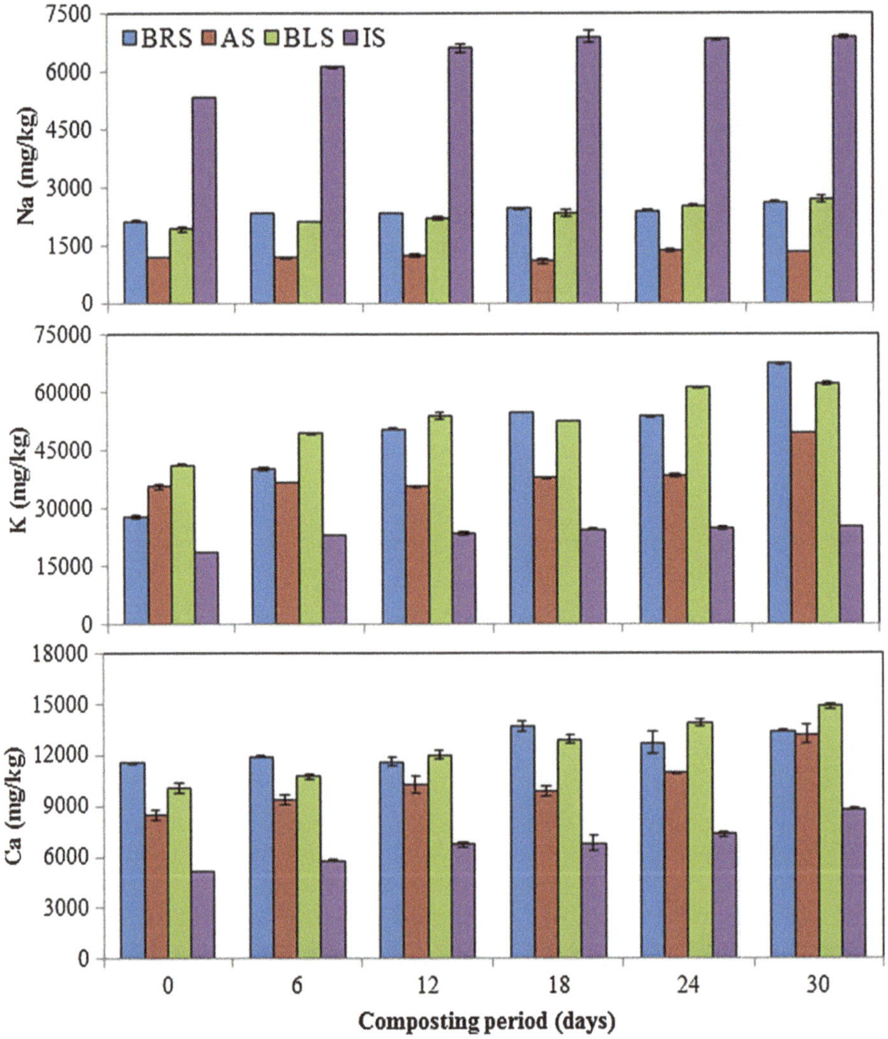

Fig. 3 Changes in Na, K and Ca during composting process

Biological analysis

Respiration of organic matter during degradation is measured as BOD, COD, OUR and CO_2 evolution rate. These parameters reduced with organic matter degradation, consequentially decreased the emission of carbon dioxide, ultimately indicating stabilization of the compost (Kalamdhad and Kazmi 2009). Figure 4a, b demonstrates that the soluble BOD and COD were reduced significantly ($P < 0.05$). The higher reduction in BOD was found in BRS (65.3 %) followed by BLS (63.3 %), AS (61.6 %) and IS (59.3 %); however, higher reduction in COD was found in IS (77.5 %) followed by BRS (69.7 %), BLS (63.2 %) and AS (62.3 %).

The CO_2 evolution directly correlates to aerobic respiration, the measurement of respiration and consequently aerobic biological activity (Kalamdhad et al. 2009). The OUR is the most accepted method for the determination of biological activity of a materials; it measures compost stability by assessing the amount of easily biodegradable organic matter still present in the compost through its carbonaceous oxygen demand (Singh et al. 2009). Significantly higher reduction ($P < 0.05$) in CO_2 evolution rate was found in BRS (78.7 %) followed by IS (77.0 %), BLS (67.8 %) and AS (66.8 %) (Fig. 4c), while higher reduction in OUR was found in BRS (74.4 %) followed by AS (71.2 %), BLS (71.1 %) and IS (68.6 %) during the process ($P < 0.05$) (Fig. 4d). The presence of coliform bacteria is generally used as an indicator of overall sanitary quality of the compost. Most of the pathogens and parasites are destroyed at high temperatures during the composting process. The fecal coliform was within the recommended limit of density 5×10^2 MPN/g dry weights for the compost hygienization (Kalamdhad et al. 2009). The total coliform levels declined from 2.3×10^{12} to 5.3×10^2, 1.5×10^8 to 2.1×10^6, 9.3×10^{10} to 4.6×10^7 and 1.1×10^{10} to 1.5×10^6 MPN/g wet weight in the agitated pile of BRL, AS, BLS and IS, respectively. The fecal coliform levels were also reduced from 3.6×10^8 to 23, 4.3×10^2 to 28, 3.6×10^2 to 1.2×10^2 and 2.4×10^2 to

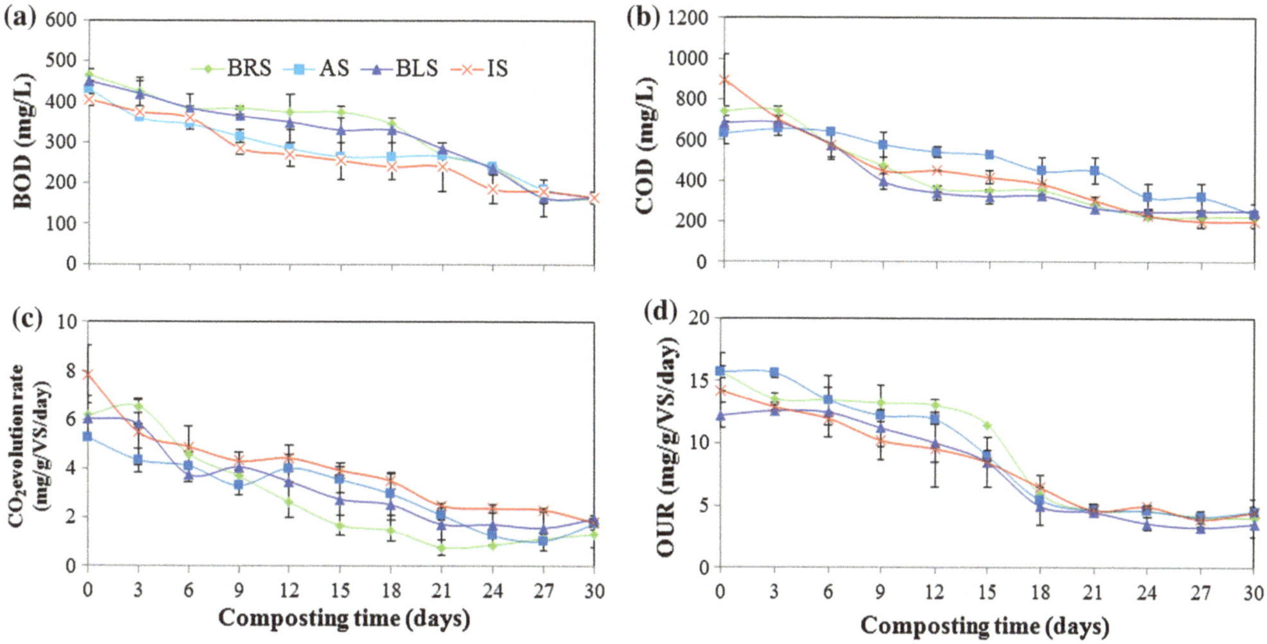

Fig. 4 Variation of BOD, COD, CO_2 evolution rate and OUR during composting process

15 MPN/g wet weight in the agitated pile of BRL, AS, BLS and IS, respectively, during the composting process. The results of present study were consistent with the results reported by Prasad et al. (2013) during the agitated pile composting of water hyacinth performed with different ratios of cattle manure, sawdust and water hyacinth.

The reduction of total and fecal coliform can be attributed to pathogens killed at high temperatures during thermophillic phase (Hassen et al. 2001)

Conclusion

Addition of optimum amount of cattle manure provides easily available carbon sources for composting microorganisms. The pH was changed from slightly acidic to neutral at the end of composting process. Highest reduction of moisture content, organic matter, ammonical nitrogen, CO_2 evolution rate and oxygen uptake rate was observed in the agitated pile of Bharalu River site as compared to industrial site, agricultural site and Boragaon landfill site during the composting process. The nutrient (nitrogen, available phosphorus, Na, K and Ca) content was increased significantly in the final compost of all agitated piles but higher increment was found in the final compost of water hyacinth collected from Bharalu River site. Total and fecal coliform were reduced significantly, which are pathogen indicators in the compost. Therefore, the present study concluded that the addition of optimum amount of cattle manure was very efficient for composting water hyacinth collected from four different sites, but better compost quality was found in water hyacinth collected from Bharalu River site.

Acknowledgments The authors gratefully acknowledge the financial support of the Department of Science and Technology (DST), Government of India.

References

Amir S, Hafidi M, Merlina G, Revel JC (2005) Sequential extraction of heavy metals during composting of sewage sludge. Chemosphere 59:801–810

APHA (2005) Standard methods for the examination of water and wastewater, 21st edn. American Public Health Association, Washington

Balasubramanian D, Arunachalam K, Arunachalam A, Das AK (2013) Water hyacinth [*Eichhornia crassipes* (Mart.) Solms.] engineered soil nutrient availability in a low-land rain-fed rice farming system of north-east India. Ecol Eng 58:3–12

Chen YX, Huang XD, Han ZY, Huang X, Hu B, Shi DZ, Wu WX (2010) Effects of bamboo charcoal and bamboo vinegar on nitrogen conservation and heavy metals immobility during pig manure composting. Chemosphere 78:1177–1181

Gajalakshmi S, Ramasamy EV, Abbasi SA (2002) High-rate composting–vermicomposting of water hyacinth (*Eichhornia crassipes*, Mart. Solms). Bioresour Technol 83:235–239

Hassen A, Belguith K, Jedidi N, Cherif A, Cherif M, Boudabous A (2001) Microbial characterization during composting of municipal solid waste. Bioresour Technol 80(217):225

Haug RT (1993) The practical handbook of composting engineering. Lewis publishers Boca Raton, Landon

Huang GF, Wong JWC, We QT, Nagar BB (2004) Effect of C/N on composting of pig manure with sawdust. Waste Manage 24:805–813

Jouraiphy A, Amir S, El Gharous M, Revel J-C, Hafidi M (2005) Chemical and spectroscopic analysis of organic matter transformation during composting of sewage sludge and green plant waste. Int Biodeterior Biodegrad 56:101–108

Kalamdhad AS, Kazmi AA (2009) Effects of turning frequency on compost stability and some chemical characteristics in a rotary drum composter. Chemosphere 74:1327–1334

Kalamdhad AS, Pasha M, Kazmi AA (2008) Stability evaluation of compost by respiration techniques in a rotary drum composter. Resour Conserv Recy 52:829–834

Kalamdhad AS, Singh YK, Ali M, Khwairakpam M, Kazmi AA (2009) Rotary drum composting of vegetable waste and tree leaves. Bioresour Technol 100:6442–6450

Malik A (2007) Environmental challenge vis a vis opportunity: the case of water hyacinth. Environ Int 33:122–138

Parkinson R, Gibbs P, Burchett S, Misselbrook T (2004) Effect of turning regime and seasonal whether conditions on nitrogen and phosphorus losses during aerobic composting of cattle manure. Bioresour Technol 91:171–178

Prasad R, Singh J, Kalamdhad AS (2013) Assessment of nutrients and stability parameters during composting of water hyacinth mixed with cattle manure and sawdust. Res J Chem Sci 3(4):1–4

Sarika D, Prasad R, Singh J, Vishan I, Varma VS, Kalamdhad AS (2014) Study of physico-chemical and biochemical parameters during rotary drum composting of water hyacinth. Int J Recycl Org Waste Agric 3(63):1–10

Singh J, Kalamdhad AS (2012) Concentration and speciation of heavy metals during water hyacinth composting. Bioresour Technol 124:169–179

Singh J, Kalamdhad AS (2013a) Assessment of bioavailability and leachability of heavy metals during rotary drum composting of green waste (Water hyacinth). Ecol Eng 52:59–69

Singh J, Kalamdhad AS (2013b) Effect of lime onbioavailability and leachability of heavy metals during agitated pile composting of water hyacinth. Bioresour Technol 138:148–155

Singh J, Kalamdhad AS (2014) Effects of natural zeolite on speciation of heavy metals during agitated pile composting of water hyacinth. Int J Recycl Org Waste Agricult 3(55):1–17

Singh YK, Kalamdhad AS, Ali M, Kazmi AA (2009) Maturation of primary stabilized compost from rotary drum composter. Resour Conserv Recyc 53:386–392

Tiquia SM, Tam NFY (2000) Fate of nitrogen during composting of chicken litter. Environ Pollut 110:535–541

Use of different organic fertilizers on soil fertility improvement, growth and head yield parameters of cabbage (*Brassica oleraceae* L)

Emmanuel Ibukunoluwa Moyin-Jesu[1]

Abstract

Purpose Field experiment was carried out in Akure in the rainforest zone of Nigeria to determine the effect of poultry manure, wood ash and rice bran on the soil fertility improvement, growth and head yield of cabbage (*Brassica oleraceae* L) in 2011 and 2012 cropping seasons.

Methods The three organic fertilizer treatments were applied each at 6 t/ha with a reference treatment NPK 15-15-15 applied at 300 kg/ha and a control treatment arranged in a randomized complete block design with three replicates.

Results The results showed significant increases ($P < 0.05$) in the growth and head yield parameters of cabbage crop under the different organic fertilizers compared to the control treatment. The application of poultry manure resulted in the highest values of cabbage plant height, stem girth, leaf number, leaf area, head weight, head length and head girth followed by NPK 15-15-15, wood ash and rice bran, respectively. Cabbage head weight, head girth, head length, plant height, stem girth, leaf number and leaf area increased by 17, 18, 8,17, 19, 10 and 16 %, respectively, with application of poultry manure compared to NPK fertilizer. Also, these parameters increased by 23, 21, 13 29, 25, 17 and 45 % compared to the wood ash treatment. The highest values of soil pH, Ca and Mg were obtained under wood ash application, while the highest values of soil O.M and moderate values of soil P, K, Ca and Mg were obtained under poultry manure application. Moderate values of soil pH, K, Ca and low % N were also recorded under rice bran treatment. However, the highest K/Ca, K/Mg and P/Mg ratios of 93.1, 74.1 and 572.1, respectively, were obtained under NPK 15-15-15 fertilizer application compared to 2:1 K/Ca, 2:1 K/Mg and 16:1 P/Mg in rice bran treatment.

Conclusion The poultry manure applied at 6 t/ha gave the best results in improving soil fertility, growth and head yield of cabbage, and this was because of its balanced nutrient composition and the least C/N ratio.

Keywords Cabbage (*Brassica oleraceae*) · Head yield and growth parameters · Organic fertilizers and soil fertility improvement

Introduction

Cabbage is a leafy vegetable belonging to the Brassicaceae family which is originated from California. Hague (2006) cabbage is also an herbaceous biennial plant that produces "heads", and it has short stem upon which is crowned with a mass of green leaves (John 1997).

Cabbage is consumed raw or cooked with other vegetables; it contains 93 ml water, 15 g protein, 0.2 g fat, 4.0 g carbohydrate, 4 g calcium and 0.5 g iron/100 g sample (Moamogwe 1995). In addition, it has high medicinal value because of its rich content in vitamins C, E and K as well as its antioxidant components which suppressed cancerous cell growth (Tindall 2000). It is one important vegetable in healthy diets of people around the world.

In spite of the economic, nutritional and health importance of cabbage to the national development, its optimum production has not been achieved due to its difficult agronomic characteristics, decline in soil fertility, lack of

✉ Emmanuel Ibukunoluwa Moyin-Jesu
moyinjesu2004@yahoo.com

[1] Agronomy Department, Federal College of Agriculture, Akure, Ondo, Nigeria

techniques of growing the crop in both the nursery and field. However, efforts made to increase the soil fertility through the use of inorganic fertilizers are limited due to high cost of purchase and continuous use which destroyed soil properties (Moyin-Jesu 2008).

Besides, the climatic condition for growing cabbage in temperate region is between 20 and 25 °C and adaptation of the crop to Nigeria, especially south-west, is difficult because temperature is between 29 and 37 °C. In other parts of the country (Northern region), the temperature ranged between 36 and 41 °C except the Jos Plateau where temperature is about 24–27 °C.

Therefore, there is justification to adapt the cultivation of cabbage to the climatic condition in Nigeria, especially in south-west, as well as adopting the use of low-cost organic fertilizers for the crop cultivation. Kwari et al. (1991) found that the incorporation of grass cow dung into the soils in Borno state improved the cation exchange capacity of the soils. Solomon and Ogeh (1995) indicated that leguminous materials and rice husk supplied mainly N, P, K, Zn, Fe, Cu, Mn and B to the soil which NPK 15-15-15 fertilizer did not possess.

Ojeniyi et al. (2001) also reported that application of wood ash to the vegetable crops improved the yield of crops significantly and increased the soil nutrients such as N, P, K, Ca, Mg in a trial conducted in south-west Nigeria. Awodun (2007) also reported that sawdust applied to soil significantly increased soil and leaf N, P, K Ca and Mg contents and number of pods, pod weight, number of branches, number of leaves and grain yield of cowpea.

Moyin-Jesu (2013a) reported that the modified neem extracts (neem leaf + wood ash extracts) increased significantly the garden egg fruit weight, fruit length and fruit diameter by 42, 24 and 13 % compared to NPK 15-15-15 fertilizer. He also observed that continuous application of NPK 15-15-15 fertilizer decreased soil pH, Ca, Mg and O.M as well as having highest K/Ca, K/Mg and P/Mg ratio caused soil nutrient imbalance.

Moyin-Jesu and Ogochukwu (2014) also reported that poultry manure applied at 6 t/ha increased the soil N, P, K, Ca and Mg concentrations and growth parameters of coconut seedlings. He also reported that the organic materials applied (wood ash, rice bran and so firth) have beneficial residual effects on soil properties which are in line with growing concern of using environment friendly fertilizer.

Furthermore, Obi and Ofonduro (1997) and Moyin-Jesu (2007) also reported that problems associated with continuous use of chemical fertilizers included nutrient imbalance, increased soil acidity, degradation in soil physical properties and loss of organic matter. Hence, the tendency to supply all plant nutrients through chemical fertilizer should be reconsidered in the future because of the deleterious effect on soil productivity on a long-term basis.

Finally, Afshar et al. (2014) reported that soil organic amendments improved the seed yield and flavonolignan production of milk thistle (*Silyburn marianum* L, Gaertin) compared to chemical fertilizers. Critical review of literature revealed that there is scarcity of research report on the use of wood ash, rice bran and poultry manure in growing cabbage except the works of Moyin-Jesu and Odewande (2013b); Ijoyah and Sophie (2009) and Mohammed and Solaiman (2012) working on varietal trial of cabbage and tillage influence on cabbage yield.

The choice of wood ash, rice bran and poultry manure as source of organic materials in the research study was because of their availability and abundance in large quantities from the experimental location and its surrounding communities where majority of the farmers are predominantly cassava, rice growers with associated processing mills and poultry producers. Moyin-Jesu and Ojeniyi (2006) also reported that application of rice bran wood ash and spent grain significantly reduced soil bulk density and increased porosity. Hence, the use of wood ash and rice bran as organic amendments would also have strong beneficial improvement on soil physical properties and base saturation.

Purpose of the experiment

The following research questions would be answered for the experiment: (1) Is there any significant difference between applied organic fertilizers on growth and head yield parameters of cabbage? (2) Is there any significant difference between the applied organic fertilizers and post-cropping soil properties?

The objectives of the research were to determine the effect of wood ash, rice bran and poultry manure on the growth and yield parameters of cabbage and on post-cropping soil properties.

Materials and methods

Description of the study area

Field and nursery experiments were carried out at Akure in the rainforest zone of Nigeria in 2011 and were repeated in 2012 to validate the results. The soil of the experimental site is loamy sand, skeletal, kaolinitic, isohyperthermic Oxic paleustalf (Alfisol) Soil Survey Staff (1999). The annual rainfall of the area is between 1000 and 2060 mm and the annual temperature is between 29–37 °C as presented in Table 1.

Table 1 Monthly rainfail and temprature data for Akure in 2011 and 2012

Months	Rainfall (mm)		Temperature (0 °C)		Modified temp. (0 °C) of the environment under shade structure during the experiment (March–July)	
	2011	2012	2011	2012	2011	2012
January	1.90	–	33.70	36.30	–	–
February	41.10	34.80	36.10	35.60	–	–
March	127.10	40.00	35.00	37.10	22.10	22.60
April	230.00	81.80	29.10	32.20	23.10	23.40
May	222.30	156.60	30.30	29.30	22.80	22.60
June	162.00	195.80	32.20	31.20	22.50	22.30
July	169.00	423.00	29.10	29.20	22.30	22.00
August	169.00	682.90	31.40	30.10	22.90	22.70
September	378.00	240.90	30.70	32.30	–	–
October	141.90	111.60	33.10	31.10	–	–
November	51.30	93.20	33.60	35.30	–	–
December	–	–	32.70	34.60	–	–

Source: meteorological station Federal College of Agriculture, Akure

Pre-cropping soil sampling and analysis

Thirty core samples were collected from 0 to 15 cm depth, bulked together, air-dried, sieved with 2 mm and ready for routine analysis.

The soil pH (1:1 soil/water and 1:2 soil/0.01 M $Cacl_2$) was read on pH meter (Crockford and Nowel 1956), while organic matter was determined using wet oxidation method through chromic acid digestion (Walkley and Black 1934). Soil P was extracted by Bray P1 extractant, and the extract was developed on Murphy blue colouration and determined on a spectronic 20 (Murphy and Riley 1962).

Soil K, Ca, Mg and Na were extracted with 1 M NH_40AC pH7, and their contents K, Ca and Na were read on the flame photometer (Jackson 1958), while Mg content was determined using the atomic absorption spectrophotometer.

The soil nitrogen was determined using the micro-Kjeldahl method (Jackson 1964), while the micronutrients (Fe, Cu, Zn and Mn) were extracted with 0.1 M, HCl and read on atomic absorption spectrophotometer. Particle size analysis was carried out using hydrometer method (Bouycous 1951).

Table 2 shows the soil chemical composition before planting. The soil P 5.20 mg/kg is lower than 10 mg/kg P recommended for crop production in south-west Nigeria (Agboola and Corey 1973). The organic matter was also below 3 % critical level for crop production (Agboola and Corey 1973).

Soil N content was 0.08 % which is lower than 0.15 % N critical level for crops recommended by Sobulo and Osiname (1981). The exchangeable bases (K, Ca, Mg and Na) were below 0.20 mmol/kg recommended by Folorunso et al. (2000). The soil pH 5.82 showed that the soil is slightly acidic. The soil textural class is loamy sand.

Table 2 Soil properties before planting

Soil properties	Values
Soil pH (H_2O)	5.82
Soil pH (0.01 M $CaCl_2$)	5.34
Organic matter (%)	0.60
Nitrogen (%)	0.08
Available P (mg/kg)	5.20
Exchangeable bases	
K^+ (mmol/kg)	0.11
Ca^{2+} (mmol/kg)	0.13
Mg^{2+} (mmol/kg)	0.08
Na^{2+} (mmol/kg)	0.11
Exchangeable acidity	
H^+ (mmol/kg)	4.10
Al^{3+} (mmol/kg)	1.30
Micronutrients	
Fe (mg/kg)	8.30
Zn (mg/kg)	3.83
Cu (mg/kg)	2.20
Mn (mg/kg)	1.85
Particle size analysis	
Sand (%)	79.30
Silt (%)	14.70
Clay (%)	6.00

Sources and processing of organic fertilizers used for the experiments

Rice bran and wood ash were obtained from the large-scale processing unit of Federal College of Agriculture, Akure, which processed cassava tubers from 10,000 ha of cassava farm and paddy rice from 5000 hectares of rice plantation, while the poultry manure was also obtained from the 10,000 poultry birds in the livestock unit of the same institution. NPK 15-15-15 was purchased from Agricultural Development Programme, while the imported cabbage seeds (Benelli cultivar) were purchased from a certified seed company Agro-Farm Enterprises, Akure.

The organic materials were processed, wood ash was sieved with 2 mm sieve to remove pebbles, while rice bran was partially composted for four weeks to reduce C/N ratio. Poultry manure was cured before application.

Chemical analysis of the organic materials used

Two grammes each of the processed forms of the organic materials was analysed. The N content was determined by the Kjeldahl method (Jackson 1964), while the determination of other nutrients such as P, K, Ca, Mg was done using the wet digestion method based on 25-5-5 ml of HNO_3-H_2SO_4-$HClO_4$ acids (AOAC 1970).

Table 3 shows the chemical analysis of the organic fertilizer used for the experiment. Wood ash had the highest values of K, Ca and Mg followed by poultry manure and rice bran, respectively. Rice bran had the least values of N, P, K, Ca and Mg and highest value of C/N ratio. Poultry manure also had the highest value of N, P and least C/N ratio compared to others.

Nursery establishment for cabbage seedlings

Land clearing and packing of debris were carried out followed by shade construction made of erected bamboo poles and top covered with moderate number of palm fronds. Twelve nursery beds with a size of 1 × 2 m^2 each were prepared. The cabbage seeds were sown on March manually into rows spaced at 30 cm apart, irrigated twice a day (morning and evening) to ensure good germination and establishment. Germination of the seeds occurred 8 days after planting and seedlings were nursed for 2 weeks before transplanting. The nursery establishment is important because directly sown seeds will not germinate; hence, the erection of shade structure in the experiment was to control the prevailing temperature of the environment for seeds germination.

Field experiment

The experimental site was cleared, ploughed, harrowed and divided into different plots. Each plot size was 4 × 4 m (16 m^2). There were three organic fertilizers treatments, namely poultry manure, wood ash and rice bran, applied at 6 t/ha with a reference treatment NPK 15-15-15 fertilizer applied at 300 kg/ha and a control treatment (no fertilizer). The experiment was arranged in a randomized complete block design and replicated three times. The choice of 6 t/ha for wood ash, rice bran and poultry manure in the experiment was based on the works of Folorunso (1999) and Moyin-Jesu (2008) which carried out extensive research on determination of soil critical levels for N, P, K, Ca and Mg using organic amendments such as rice bran, cocoa husk, poultry manure, wood ash and pig manure for vegetables. They reported that application of these organic fertilizer materials at 6 t/ha was the best critical level for optimum yield of crop in the study area.

The organic amendments were incorporated into the soil using hand trowel one week before transplanting the seedlings. Twenty-day-old seedlings were transplanted to the experimental plots at a spacing of 50 × 50 cm. After transplanting, irrigation was done every mornings and evenings until the rain was steady to allow full establishment.

A low-cost shade made of erected bamboo poles and the top covered with palm fronds was constructed over the transplanted cabbage seedlings to reduce partially direct impact of sun radiations and enhanced the prevailing environment of the site to favour sustainable cabbage production.

Weeding was done manually 10 days after transplanting and continued at 2 weeks interval until harvesting. Cabbage seedlings were sprayed with Avesthrin (Cypermethrin 10 EC) at 10 ml/litre of water at 3 weeks interval until 10 weeks after harvesting (WAT) to control leaf defoliating beetles and caterpillars. Growth parameters such as height, stem girth (cm), leaf area (cm^2) and number of leaves were measured starting from 15 days after transplanting (DAT) until 80 days after transplanting. The leaf area was estimated using the non-destructive and accurate method of Jamal et al. (2009) based on determination of individual leaf area using a predictive equation constructed from leaf length (*L*) and width (*W*) and their combination.

Table 3 Chemical analysis of the organic fertilizer used for the experiment

Treatments	N (%)	C (%)	C/N	P (mg/kg)	°K (%)	Ca (%)	Mg (%)
Poultry manure	4.32	30.00	6.93	385.00	9.70	3.30	4.10
Wood ash	1.53	18.00	11.76	86.00	23.02	9.40	8.52
Rice bran	0.60	14.00	23.33	56.00	7.93	0.12	1.80

Harvest of matured cabbage head started at 82 DAT and continued as they attain maturity in each experimental plot on treatment basis, and yield parameters such as cabbage head weight, cabbage head length and cabbage head girth were measured. In addition, it was observed that new shoots began to grow at the base of the harvested stem.

Soil analysis after harvesting cabbage

After harvesting, soil samples were taken from each plot using soil auger, bagged, air-dried and sieved for routine laboratory analysis. Soil pH, organic matter (O.M), N, P, K, Ca and Mg were determined as described for the pre-cropping soil analysis.

Statistical analysis

The average data of 2011 and 2012 cropping seasons on cabbage height, stem girth, leaf area, number of leaves, head weight, head girth and head length were analysed using ANOVA F test, and the treatment means were compared using Duncan multiple range test at 5 % level of significance (Gomez and Gomez (1984).

Results

Effect of organic fertilizers on the growth parameters of cabbage between 15 and 80 days after transplanting (DAT)

There were significant increases ($P < 0.05$) in the plant height, stem girth, number of leaves/plant and leaf area of cabbage under different organic fertilizer treatments compared to the control treatment (Table 4).

The highest values of cabbage plant height, stem girth, number of leaves and leaf area were obtained with poultry manure followed by NPK 15-15-15, wood ash and rice bran, respectively. The application of poultry manure treatment increased cabbage plant height, stem girth, leaf number/plant and leaf area by 17, 19, 10 and 16 % compared to NPK 15-15-15 fertilizer. The cabbage plant height, stem girth, number of leaves/plant and leaf area increased by 34, 26, 18 and 16 % with application of poultry manure compared to rice bran. The least values of growth parameters in cabbage were also obtained under the control treatment.

Effect of organic fertilizers on the head yield parameters of cabbage plant

There were significant increases ($P < 0.05$) in the head yield parameters (head weight t/ha, head girth and head length) of cabbage under different organic fertilizers compared to the control treatment (Table 5).

The highest values of cabbage head yield parameters were obtained with poultry manure followed by wood ash and rice bran, respectively. When compared to NPK 15-15-15, application of poultry manure also increased cabbage head weight, head girth and head length by 17, 18 and 8 %, respectively.

The application of poultry manure also increased cabbage head weight, head girth and head length by 23, 21 and 13 % compared to the wood ash treatment, respectively. The least values of cabbage yield parameters were obtained under the control treatment.

Soil chemical analysis after harvesting cabbage

Significant increases ($P < 0.05$) occurred in the soil N, P, K, Ca and Mg under different organic fertilizers compared to the control treatment (Table 6).

The soils which received NPK 15-15-15 fertilizer had the highest values of soil N, P, K, and the lowest contents of O.M, Ca and Mg. The highest values of soil organic matter (O.M) and moderate values of soil P, K, Ca and Mg

Table 4 Effect of different organic fertilizers on the growth parameters of cabbage plants

Treatments	Plant height (cm)			Stem girth (cm)			Leaf number			Leaf area (cm^2)		
	2011	2012	Mean	2011	2012	Mean	2011	2012	Mean	2011	2012	Mean
Control	4.00e	4.00e	4.00e	2.94e	2.70a	2.82e	4.64e	4.70e	4.67e	25.73e	26.11e	25.92e
NPK 15-15-15	10.53b	10.75b	10.64b	5.66b	5.30b	5.48b	8.10b	8.16b	8.13b	74.90b	73.60b	74.75b
Rice bran	8.50d	8.30d	8.40d	4.90d	5.18c	5.04d	7.46d	7.48d	7.47d	37.36d	36.20d	36.73d
Wood ash	8.70c	8.90c	8.80c	5.04c	5.14d	5.09c	7.50c	7.56c	7.53c	49.42c	48.30c	48.86c
Poultry manure	12.88a	12.40a	12.69a	6.84a	6.70a	6.77a	8.96a	9.10a	9.03a	88.88a	87.90a	88.39a

Treatment means within each group followed by the same letters are not significantly different from each other using Duncan multiple range test at 5 % level of significance

Table 5 Cabbage head yield parameters under different organic fertilizer treatments

Treatments	Head weight (t/ha)			Head girth (cm)			Head length (cm)		
	2011	2012	Mean	2011	2012	Mean	2011	2012	Mean
Control	0.90e	0.94e	0.92e	12.36e	12.30e	12.33e	9.10e	8.90e	9.0e
NPK 15-15-15	29.30b	28.90b	29.02b	31.80b	32.20b	32.00b	19.20b	18.30b	19.0b
Rice bran	22.60d	22.20d	22.40d	24.87d	24.57d	24.67d	16.80d	17.20d	17.00d
Wood ash	27.20c	27.40c	27.30c	29.90c	30.76c	30.33c	18.40c	17.60c	18.00c
Poultry manure	35.30a	34.90a	35.10a	38.56a	38.10a	38.33a	20.30a	20.70a	20.50a

Treatment means within each group followed by the same letters are not significantly different from each other using Duncan multiple range test at 5 % level of significance

Table 6 Effect of different organic fertilizers on the soil chemical composition after harvesting cabbage

Treatment	Soil pH			O.M (%)			N (%)			P (mg/kg)		
	2011	2012	Mean	2011	2012	Mean	2011	2012	Mean	2011	2012	Mean
Control	5.30d	5.10e	5.20de	0.19e	0.21e	0.20e	0.029e	0.031e	0.03e	3.30e	3.10e	3.20e
NPK 15-15-15	5.12e	5.14d	5.13e	0.27d	0.23d	0.25de	0.39a	0.35a	0.37a	28.40a	28.80a	28.60a
Rice bran	7.16a	7.12a	7.14a	1.20b	1.22b	1.21b	0.25c	0.27c	0.26c	15.90c	15.52c	15.71c
Wood ash	6.60c	6.60c	6.62c	1.15c	1.11c	1.13c	0.15d	0.19d	0.17de	13.71d	13.53d	13.65d
Poultry manure	6.92b	7.06b	6.99b	2.30a	2.40a	2.53a	0.36b	0.34b	0.35b	25.55b	25.47b	25.51b

Treatment	Kmmol/kg			Ca (mmol/kg)			Mg (mmol/kg)		
	2011	2012	Mean	2011	2012	Mean	2011	2012	Mean
Control	0.03e	0.05e	0.04e	0.06d	0.04e	0.05d	0.032e	0.028e	0.03d
NPK 15-15-15	3.74b	6.68a	3.71a	0.03e	0.05d	0.04e	0.048d	0.052d	0.05e
Rice bran	3.90a	3.30b	3.10b	1.54a	1.52a	1.53c	1.28a	1.26a	1.27a
Wood ash	1.85d	1.53d	1.69d	0.93c	0.97c	0.95b	0.80c	0.86c	0.83c
Poultry manure	2.45c	2.65c	2.40c	1.24b	1.22b	1.23a	1.13b	1.11b	1.12b

Treatment means within each group followed by the same letters are not significantly different from each other using Duncan multiple range test at 5 % level of significance

were obtained with the application of poultry manure. The application of wood ash resulted in the highest values of soil pH, Ca and Mg compared to other treatments, while moderate values of soil pH, K, Ca and low % N were obtained with rice bran.

The soil N, P and K increased by 54, 53 and 55 % with application of NPK 15-15-15 compared to rice bran treatment. Also, it had higher K/Ca, K/Mg and P/Mg ratio of 93:1, 74:1 and 572:1 P/Mg, respectively, compared to 2:1 K/Ca, 2:1 K/Mg and 16:1 P/Mg in rice bran treatment.

The soil O.M, % N and P increased by 49, 26 and 39 %, respectively, with application of poultry manure compared to wood ash. However, soil pH, K, Ca and Mg were increased by 3, 30, 20 and 12 %, respectively, under wood ash treatment compared to poultry manure. The least values of soil pH, N, P, K, Ca and Mg nutrients were obtained under control treatment.

Discussion

The least values of soil, growth and head yield parameters (number of leaf, leaf area, stem height, head length, head girth and head weight) obtained under the control treatment compared to other treatments could be as a result of initial low soil nutrients status that often characterize continuous cultivation of land without fertilizer application. This finding was supported by Mohammed and Solaiman (2012) who worked on the efficacy of fertilizers on the growth and yield of cabbage and reported that nutrients supply was an important input for realizing higher cabbage yield. Hence, the application of fertilizers, especially organic fertilizers, enhanced both soil and crop productivity in the tropics.

The best cabbage head weight, head length, head girth, number of leaf, leaf area, stem girth and plant height obtained from the application of poultry manure could be attributed to its balanced nutrient contents. The least value of C/N ratio of poultry manure also encouraged faster decomposition and quick release of nutrients for crop uptake and higher head yield parameters. This observation was supported by Ijoyah and Sophie (2009) who reported that the application of poultry manure increased cabbage yield. However, the head yield of 35.10 t/ha of cabbage obtained in this experiment was higher than 25.76 t/ha reported by Ijoyah and Sophie (2009). The prior processing and stacking of poultry manure enhanced quick decomposition and reduction in C/N ratio. This could be responsible for quick nutrient release and uptake by cabbage crop and hence higher cabbage head yield.

The increased soil pH obtained by the application of wood ash was attributed to its high K, Ca and Mg contents which served as base saturation agents and consequently increased soil buffering capacity (Moyin-Jesu and Ogochukwu 2014). The soil pH influenced nutrient availability and uptake as reported by Obatolu (1995) that oil palm bunch ash, wood ash and cocoa pod husk improved soil K, Ca and Mg nutrients and corrected soil acidity in an Alfisol under cultivation of coffee and corn.

The lower values of cabbage growth and head yield parameters under rice bran treatment compared to wood ash and poultry manure could be adduced to the fact that higher C/N ratio (23:3:1) of rice bran compared to poultry manure decelerated the residue decomposition rate and subsequently it had the least values of % N, P, K, Ca and Mg. The observation was supported by Moyin-Jesu (2007) who reported that the growth and yield performance of Okra were very low under the application of rice bran and sawdust. However, the processing of the organic fertilizer (rice bran) by composting was still responsible for reduction of C/N ratio and the results would have been extremely different, if the rice bran was not processed because of its high C/N ratio of 135:1. This was corroborated by Adebayo and Olayinka (1984) who used the unprocessed forms of oil palm bunch ash, sawdust and rice bran to grow maize; hence, prior processing of the organic fertilizers before application is important for good crop yield. Nevertheless, the inclusion of rice bran as source of fertilizer was important because it has been reported by Moyin-Jesu and Ojeniyi (2006) that rice bran, wood ash and spent grain were good in the reduction of soil bulk density and improved % porosity which are good soil fertility indicators (i.e. soil physical properties).

The application of NPK 15-15-15 fertilizer significantly increased the cabbage head yield and growth parameters more than that of rice bran, and this could be traced to the supply of readily available nutrients from the NPK fertilizer to the plant. This observation agreed with that of Makinde (2013) who reported that an increase in the readily available nitrate from the NPK fertilizer unlike the organic manure which must be mineralized before being utilized by crops. Moyin-Jesu (2013a) also reported that nitrogen increased the vegetative and yield performance of garden egg crop.

However, the high content of nitrogen in the NPK fertilizer and its continuous use might be responsible for excessive vegetative growth of cabbage and subsequently delayed maturity. This could explain the lower values of cabbage head yield and growth parameters compared to the poultry manure treatment. Also, Tong et al. (1997) reported higher rate of nitrate nitrogen leaching which polluted underground water particularly with the continuous use of chemical fertilizers compared to the organic fertilizers. Furthermore, Moyin-Jesu (2012) reported that the high K/Ca, K/Mg and P/Mg ratios would cause nutrient imbalance and made availability of nutrients such as K, Ca, and Mg difficult for both immediate and subsequent cropping of cabbage.

Conclusion

This experiment showed that the use of the various organic fertilizers (poultry manure, wood ash and rice bran) applied at 6 t/ha increased the growth, head yield of cabbage and soil N, P, K, Ca, Mg, pH and organic matter (O.M). It is recommended that poultry manure be applied at 6 t/ha for increased nutrient availability, growth and head yield of cabbage as well as enhancing sustainable cultivation of cabbage on commercial basis. The use of poultry manure would replace the application of large quantity of NPK 15-15-15 fertilizer.

The recommendation is important because the inorganic fertilizers are becoming very expensive to be purchased by small-scale and commercial farmers of cabbage. In addition, these organic fertilizers appear to have beneficial secondary effects on soil properties and ensuring environmental sustainability.

References

Adebayo A, Olayinka A (1984) The effect of methods of application of sawdust on plant growth, nutrient uptake and soil chemical properties. Ife J Agric 2(1):36–44

Afshar RK, Chaichi MR, Assareh MH, Hashemi M, Liaghat A (2014) Interactive effect of deficit irrigation and soil organic

amendments on seed yield and flavonolignan production of milk thistle (*Silybum marianum* L, Gaertn). Ind Crop Prod 58:166–172
Agboola AA, Corey RB (1973) Soil testing N, P, K for maize in the soils derived from metamorphic and igneous rocks of Western State of Nigeria. J West Afr Sci. Assoc 17:93–100
AOAC (1970) Official method of analysis 12th eds. Association of official Analytical Chemists Arlington, VA
Awodun MA (2007) Effect of sawdust ash on nutrient status, growth and yieldof Cowpea (*Vigna unguiculata* L. Walp). Asian J Agric Res 1:92–96. doi:10.3923/ajar
Bouycous H (1951) Mechanical analysis of soil using hydrometer method analytical chem. Acta 22:32–34
Crockford L, Nowel P (1956) Laboratory manual of physical chemistry. Exp 31 and 32. John Wiley and Sons, New York
Folorunso OO (1999) Use of plant residues for improving soil fertility and yield of Okra (*Abelmuschus esculentum* L Moench) and Amaranthus (*Amaranthus viridis* L). PhD thesis, School of Agriculture and Agriculture Technology, Federal University of Technology, Akure, Nigeria: 45–50
Folorunso OO, Agboola AA, Adeoye GO (2000) Evaluation of three fertilizer models for P and K recommendation in maize (Zea Mays L). J Tech Educ 2:237–253
Gomez KA, Gomez AA (1984) Statistical procedures for agricultural research, 2nd edn. John Wiley and Sons, New York
Hague KMF (2006) Yield and nutritional quality of cabbage as affected by nitrogen and phosphorous fertilization. Bangladeshi J Sci Med Res 41:41–46
Ijoyah MO, Sophie VL (2009) Effect of different levels of decomposed poultry manure on the yield of cabbage (*Brassica Oleraceae* L). Agro Sci J Trop Agric Food Environ Ext 8(1):20–23
Jackson ML (1958) Soil chemical analysis. N.J Prentice Hall, Englewood Cliffs, pp 57–69
Jackson ML (1964) Soil chemical analysis. N.J Prentice Hall, Englewood Cliffs, pp 86–92
Jamal Ali Olfati, Peyvast G, Samavi M, Salehi M, Mahdipour M, Nosratie-Rad Z (2009) Comparison of leaf area estimation from linear measurement of Red cabbage. Int J Veg Sci 15:185–192. doi:10.1080/19315260802770
John J (1997) Drug and food interaction. J Nutr Fla 9(2):7–9
Kwari JD, Bababe B, Shehu HE (1991) Effect of cow dung on CEC and accumulation of mineral nitrogen, phosphorus and sulphur in soils of Borno state. In: Ojeniyi S.O (eds). Proceedings 19th Annual conference of Soil Sci Soc, Nigeria, Nov 17–21, 1991. University of Maiduguri: 154–158
Makinde A (2013) Effect of inorganic fertilizer on the growth and nutrient composition of moringa. J Emerg Trends Eng Appl Sci (JETEAS) 4(2):34–343
Moamogwe M (1995) Adaptation trial of introduced cabbage cultivars ARP training reports between 1993–1995. AVRDC Afr Reg Program Arusha Tanzan 4(1):27–29
Mohammed RH, Solaiman AHM (2012) Efficacy of organic and inorganic fertilizers on the growth of cabbage (*Brassica Oleraceae* L). Int J Agric Crop Sci 4(3):128–138. ISBN 2227-670X
Moyin-Jesu EI (2007) Use of plant residues for improving soil fertility, pod nutrients, root growth and pod weight of Okra (*Abelmuschus esculentum* L). Bioresour Technol 98(2007):2057–2064
Moyin-Jesu EI (2008) Determination of soil nutrient levels for maximum yield of Okra using sole and amended plant residues. J Trop Agric Sci 31(2):233–245 **(Malaysia)**
Moyin-Jesu EI (2012) Simple and blended organic fertilizers improve soil fertility of degraded nursery soil for production of Kolanut (*Cola acuminate*) seedlings in Nigeria. In: Tech Open Publishers. Chapter book 15. EdsJoann K. Whalen. ISBN: 978-953-307-945-5. p 209–306. http://www.intechweb.org
Moyin-Jesu EI (2013a) Evaluation of modified neem leaf and wood ash extracts on the soil fertility improvement on growth and yield of garden egg plant. (*Solanium melongena* L.) Colombia International Publishing. Am J Agric Sci Tech 1(3):77–76. doi:10.776/ajast.2013.1007
Moyin-Jesu EI, Odewande A (2013) Comparative evaluation of different tillage practices and method of planting on soil properties, growth and head yield of cabbage. Am J Res Commun 1:1–20
Moyin-Jesu EI, Ogochukwu AI (2014) Comparative evaluation of different organic fertilizer effects on soil fertility, leaf chemical composition and growth performance of coconut (*Cocos nucifera*) seedlings. Int J Plant Soil Sci 3(6):737–750. doi:10.9734/IJPSS/2014/4338
Moyin-Jesu EI, Ojeniyi SO (2006) Effects of sole and amended plant residues on soil nutrient contents and yield of Okra (*Abelimoschus esculentum* L). Discov Innov J 18(4):318–326
Murphy J, Riley JP (1962) A modified single solution method for determination of phosphate in natural waters. Anal Chem Acta 27:31–36
Obatolu CR (1995) Nutrient balance sheet of Alfisol grown to coffee and maize using organic fertilizers. In: Agboola AA (eds) Proceedings 3rd Annual Conference of All African Soil Sci Society, University of Ibadan, Ibadan, August 20–23,1995: 250–256
Obi ME, Ofonduro CO (1997) The effects of soil amendments on soil physical properties of a severely degraded sandy loam soil in south Eastern Nigeria. Usman Danfodio University, Sokoto, pp 30–35 **(March 2–7)**
Ojeniyi SO, Oso OP, Arotolu AA (2001) Response of vegetables to wood ash fertilizer. In: Obigbesan OO (ed) Proceedings 30th annual conference of agric society of nigeria. University of Agriculture, Abeokuta, pp 147–150
Sobulo RA, Osiname OA (1981) Soils and fertilizer use in Western Nigeria. Tech Bull no. 11. Institute of Agricultural Research and Training, University of Ife, Nigeria
Soil Survey Staff (1999) Soil taxonomy. A basic system for soil classification for making and interpreting soil surveys agric handbook no. 436. USDA, Washington DC
Solomon MG, Ogeh JS (1995) Use of leguminous plants and rice husk as fertilizer materials. In: Agboola AA (eds). Proceedings 3rd All African Soil Science Conference August 20–23, 1995, University of Ibadan. 327–333
Tindall M (2000) Mineral and organic fertilizing in cabbage and their residual effect for commercial cultivation on yield and quality performance of cabbage. Hort Brass 6:15–20
Tong Y, Ove E, Dianquing L, Harald G (1997) Effect of organic manure and chemical fertilizers on nitrogen uptake and nitrate leaching in a Eum orthic anthrosols profile. Nutr Cycl Agroecosyst 48(3):225–229
Walkley A, Black IA (1934) An examination of degtajroff method for determining soil organic acid filtration method. Soil Sci 37:29–38

24

Sewage sludge composting in a rotary drum reactor: stability and kinetic analysis

Ashish Kumar Nayak[1] · Ajay S. Kalamdhad[1]

Abstract

Background Composting is one of the sustainable practices to convert sewage sludge into useful agricultural product because it is rich in organic matter, micro- and macronutrients, which is essential for plants growth and soil fauna to live. Therefore, the present study was to recycle the sewage sludge by rotary drum composter. Five sets of experiments (C/N 15, 20, 25, 30 including control) were carried out in the reactor, where sewage sludge was coupled with sawdust and cattle manure to attain the optimal feedstock. The process was monitored with respect to physico-chemical characteristics and stability parameters during 20 days of composting period. In addition, the kinetic parameters k_m and r_m have been evaluated in the present work.

Results Higher reductions in CO_2 evolution and oxygen uptake rate (OUR) observed in C/N 30 showed the stability, resulting in the total biodegradable ingredients to be stabilized. Similarly, higher percentage reduction in carbon-nitrogen ratio was observed in C/N 30 followed by C/N 20, C/N 25, C/N 15 and control experiments, respectively, indicates maximum degradation was achieved; as it corroborate higher temperature regime occurred in C/N 30. The composting parameters such as moisture content, volatile solids, CO_2 evolution, OUR and C/N ratio were reduced significantly ($p < 0.05$). The kinetic model used illustrated the comparable variations in Michaelis constant (k_m) and maximum rate constant (r_m) under varying proportions of the initial feedstock. The overall assessment of kinetic data for C/N 30 experiment showed enhancement of degradation during the composting process.

Conclusions The results suggested that composting of sewage sludge with optimal proportion of cattle manure and saw dust, especially in C/N 30, can produce stable compost within 20 days of composting.

Keywords Sewage sludge · Sawdust · Rotary drum · C/N ratio · Stability · Kinetic analysis

Introduction

Sewage sludge, a by-product generated in enormous quantity from various biological, chemical and physical treatment processes, may create environmental concerns in its disposal due to the presence of soil contaminants such as organic compounds, heavy metals and human pathogens (Singh and Agrawal 2008). The problem of efficient disposal and management of sewage sludge has become more rigorous owing to rapidly rising of population, economic growth and increasing numbers of treatment plants. Hence, it is necessary that we find approaches to effectively reuse the sewage sludge and reduce their impact on the environment. Composting is one of the most promising biological techniques to stabilize the organic matters contained in the sewage sludge (Doublet et al. 2010). During the composting process, pathogens may be killed due to the heat generated during the thermophilic phase and the organic fractions in sewage sludge will be converted into stabilized humic substances through mineralization and humification with a significant reduction in volume (Tuomela et al. 2000; Gouxue et al. 2001; Khwairakpam and Bhargava 2009). The reason is that compost made from sludge returns useful resources to the environment and also

✉ Ashish Kumar Nayak
coco.ashish@gmail.com; nashish.iitg@gmail.com

[1] Department of Civil Engineering, Indian Institute of Technology Guwahati (IITG), Guwahati 781039, India

has a number of beneficial effects such as accelerating the plant growth, improving moisture retention capacity in the soil, increasing organic matter in the soil and improving erosion control (Liang et al. 2003; Cheng et al. 2007; Wang et al. 2014).

Rotary drum composting is an efficient and promising technology compared to conventional outdoor technologies such as windrows and static piles, as it provides adequate agitation, aeration and proper compost mixing to produce consistent and uniform end product without any nuisance odor or leachate-related problems (Kalamdhad et al. 2009; Rodríguez et al. 2012). Different types of wastes (cattle manure, swine manure, municipal biosolids, brewery sludge, chicken manure, animal mortalities and food residues) can be effectively composted in rotary drum reactors (Vuorinen and Saharinen 1997; Mohee and Mudhoo 2005; Aboulam et al. 2006; Kalamdhad and Kazmi 2008; Rodríguez et al. 2012). Until recently, reported data about the use of rotary drum in sewage sludge composting are rather scarce.

Many studies have reported that sewage sludge alone produces poor quality of compost due to its high moisture content and low air permeability. It is necessary to mix with other ingredients including bulking agents such as rice straw, sawdust, grass and leaves. These bulking agents are used to adjust moisture content and maintain inter-particle void dispersion, which provides adequate air and water exchange within the composting mass (Eftoda and McCartney 2004; Petric and Selimbasic 2008; Iqbal et al. 2010), and also provides optimal initial carbon-to-nitrogen (C/N) ratio to enhance the decomposition rate (Kalamdhad and Kazmi 2009). The fact that the application of immature and unstable compost may inhibit the growth of vegetables by competing for oxygen or causing phytotoxicity to plants, due to insufficient biodegradation of organic matter (Khwairakpam and Bhargava 2009). In addition, cattle manure having the tendency to bind the substrates together and also an important resource to supplement additional organic fraction, which improves the soil conditions. Furthermore, other composting factors such as temperature and nutritional requirements should also be controlled to attain good-quality compost (Kalamdhad and Kazmi 2009). Stability prevents nutrient from becoming tied up in rapid microbial growth, allowing them to be available for plant needs. In this regard, different techniques are widely used for measuring stability, based on physical (temperature, aeration demand, odor, color, etc.), chemical (organic matter loss, C/N ratio, chemical oxygen demand, etc.) and biological (respiration measured either as O_2 consumption, CO_2 production or heat generation, etc.) characteristics of composts (Wu et al. 2000; Brewer and Sullivan 2003; Matteson and Sullivan 2006; Gomez et al. 2006; Kalamdhad et al. 2008), but none has found universal acceptance. But, the respirometric indices (either oxygen uptake rate (OUR) or CO_2 evolution) have been proven to be useful for the prediction of the stability of final compost products in the recent years (Barrena et al. 2011; Varma et al. 2014). Most of the literatures reported on compost stability are generally dealt with traditional composting techniques such as windrows and static pile types for various kinds of wastes. The study of stability analysis in rotary drum composter for sewage sludge in combination with optimal mixtures of cattle manure and saw dust is rarely present. Hence, it is essential to check the stability in the final compost which correlates with the degree of decomposition.

Besides the above facts, it is observed that most published information on the composting process is qualitative. Very few studies have been performed based on quantitative knowledge to design the operating conditions of the composting process, but they possessed less systematic. From the engineering point of view, kinetic study is essential to describe the controlling factors in the optimization of the process (Whang and Meenaghan 1980; Agamuthu et al. 2000; Bari et al. 2000). Keeping issues in mind, the scope of the present study was to evaluate the effect of process variables such as temperature, moisture content, ash content, volatile solids, total nitrogen, CO_2 evolution, OUR and initial C/N ratio, during 20 days of composting period using pilot-scale rotary drum reactor. In addition, the dynamics of the composting process were also examined by simulating the results of organic matter degradation on the Monod equation, to determine the kinetic parameters like Michaelis–Menten constant k_m and maximum velocity constant r_m.

Materials and methods

Feedstock materials and its preparation

Undigested sewage sludge having high percentage of organic matter was procured from the sewage treatment plant of the IIT Guwahati campus. Fresh cattle manure and sawdust were obtained and/or purchased from nearby Amingaon village and saw mill, respectively. The initial characteristics of waste are detailed in Table 1. Prior to composting, sewage sludge (C/N = 11.2), cattle manure (C/N = 26.4) and sawdust (C/N = 135.9) were mixed with five different proportions (C/N = 15, 20, 25, 30 and control), as compiled in Table 2. To prepare the initial feedstock, C/N ratio of the compost mixture was calculated as given in Eqs. (1), 2 and (3) (CWMI 1996).

Table 1 Initial characteristics of waste materials

Parameters	Sewage sludge	Cattle manure	Sawdust
Moisture content (%)	79.2 ± 2.0	80.8 ± 0.1	10.3 ± 0.3
Ash content (%)	61.5 ± 0.3	29.9 ± 3.9	2.4 ± 0.1
Total organic carbon (TOC) (%)	21.4 ± 0.2	39.0 ± 2.1	54.2 ± 0.1
Total nitrogen (%)	1.9 ± 0.2	1.5 ± 0.2	0.40 ± 0.3
C/N ratio	11.2 ± 1.2	26.4 ± 2.5	135.9 ± 7.3
CO_2 evolution (mg CO_2/g VS/day)	12.1 ± 0.5	17.6 ± 0.5	10.8 ± 0.1
Oxygen uptake rate (mg O_2/g VS/day)	17.9 ± 0.2	21.8 ± 1.3	12.5 ± 0.7

All values are means of three replicates ± SD ($n = 3$)

Table 2 Waste proportion of mixtures and its physico-chemical characteristics

Feedstock materials	C/N 15	C/N 20	C/N 25	C/N 30	Control
Sewage sludge (kg)	130	104	98	87	150
Cattle manure (kg)	16	39	39	45	0
Sawdust (kg)	4	7	13	18	0
Total weight of mixture (kg)	150	150	150	150	150
Initial moisture content (%)	70.9 ± 0.2	68.7 ± 0.5	64.9 ± 0.2	62.3 ± 0.2	68.1 ± 0.1
Initial ash content (%)	51.7 ± 0.4	50.5 ± 0.2	48.7 ± 0.4	46.5 ± 0.3	57.5 ± 0.2
Initial total organic carbon (%)	26.8 ± 0.2	27.5 ± 0.1	28.5 ± 0.2	29.7 ± 0.2	23.6 ± 0.1
Initial total nitrogen (%)	1.54 ± 0.03	1.20 ± 0.02	1.12 ± 0.02	1.05 ± 0.03	1.69 ± 0.08
Initial CO_2 evolution (mg CO_2/g VS/day)	9.9 ± 0.1	10.0 ± 0.1	9.4 ± 0.2	9.5 ± 0.3	10.5 ± 0.2
Initial oxygen uptake rate (mg O_2/g VS/day)	20.0 ± 0.3	17.4 ± 0.3	13.1 ± 0.1	16.1 ± 0.3	17.9 ± 0.2

All values are means of three replicates ± SD ($n = 3$)

$$G = \frac{M_1 \times Q_1 + M_2 \times Q_2 + M_3 \times Q_3}{Q_1 + Q_2 + Q_3} \quad (1)$$

where Q_n nis the mass of material n (wet weight basis); G the moisture goal (%); M_n the moisture content (%) of material n.

$$R = \frac{Q_1(C_1 \times (100 - M_1)) + Q_2(C_2 \times (100 - M_2)) + Q_3(C_3 \times (100 - M_3)) + \cdots}{Q_1(N_1 \times (100 - M_1)) + Q_2(N_2 \times (100 - M_2)) + Q_3(N_3 \times (100 - M_3)) + \cdots} \quad (2)$$

where R is the C/N ratio of compost mixture; Q_n the mass of material n (wet weight basis); C_n the carbon (%) of material n; N_n the nitrogen (%) of material n; M_n the moisture content (%) of material n.

By simplifying and rearranging the above equation, the mass of the third material required would be Eq. (3):

$$Q_3 = \frac{RQ_1N_1(100 - M_1) + RQ_2N_2(100 - M_2) - Q_1C_1(100 - M_1) - Q_2C_2(100 - M_2)}{C_3(100 - M_3) - RN_3(100 - M_3)} \quad (3)$$

Rotary drum composter design

Figure 1 shows the schematic diagram of a pilot-scale rotary drum composter of 550 L capacity operated by batch-mode operation. The composter is made up of 1.22 m in length and 0.76 m in diameter, fabricated by a 4-mm-thick metal sheet. The inner side of the drum is covered with anti-corrosive coating. The drum is mounted on four rubber rollers attached to a metal stand and is rotated manually with its handle. To ensure appropriate mixing and agitation of the wastes during rotation,

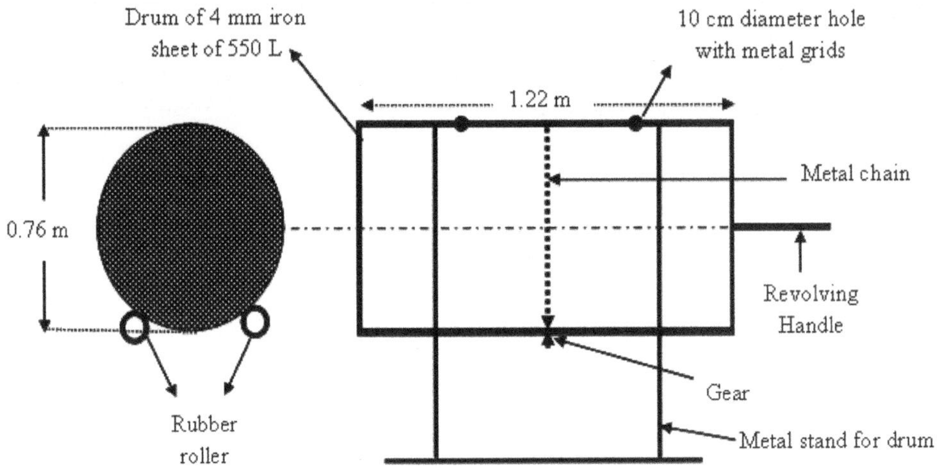

Fig. 1 Pilot-scale design of rotary drum composter

40 × 40 mm angles are welded longitudinally inside the drum, resulting in tumbling action inside the drum generally moves material through the drum for 20 days. In addition, two adjacent holes of 10 cm each are made on top of the drum to drain out the excess water. Rotary drum composter contained approximately 150 kg of different waste combinations and was manually turned on every 24 h through one complete rotation, to ensure that the material on the top portion moved to the central portion where it was subjected to higher temperature. Aerobic condition was maintained by opening the top half side doors of the two circular faces.

Sampling and parameter analysis

100 g of samples was collected through grab sampling from different locations, mainly from middle portion and terminal ends of the composter without disturbance of the adjacent materials. All the grab samples were thoroughly mixed together to make a homogenized sample. Triplicate homogenized samples were collected periodically and stored at 4 °C for stability analysis of the wet samples within 2 day. The sub-samples were air dried immediately, and pulverized to pass through 0.2 mm sieve and stored for physico-chemical analysis.

Temperature was monitored every 6 h using a digital thermometer throughout the composting period. Each air dried and grounded sample was analyzed for the following parameters: moisture content (105 °C for 24 h) and volatile solids/organic matter (loss ignition at 550 °C for 2 h). The total organic carbon (TOC) was calculated as follows in Eqs. (4), (5) and (6) (Mohee et al. 2008).

$$\% \text{ Ash} = \frac{\text{Weight}_{\text{Initial}} - \text{Weight}_{\text{Final}}}{\text{Weight}_{\text{Initial}}} \times 100 \quad (4)$$

$$\% \text{ Volatile solid} = 100 - \% \text{ Ash} \quad (5)$$

$$\% \text{ Total organic carbon} = \frac{\% \text{ Volatile solid}}{1.8} \quad (6)$$

The percentage of nitrogen content was calculated using Kjeldahl method following Eq. (7) (Codell and Verderame 1954).

$$\% \text{ Total nitrogen} = \frac{1.4(R - S)N}{W} \quad (7)$$

where R is the volume of H_2SO_4 to titrate boric acid (ml); S the volume of H_2SO_4 to titrate blank (ml), W the weight of sample (g); and N the normality of H_2SO_4.

Finally, the carbon-to-nitrogen ratio was determined by dividing the total organic carbon content to the total nitrogen content.

OUR was performed according to the method described by Kalamdhad et al. (2008). A liquid suspension of compost (8 g of compost in 500 mL of distilled water added with $CaCl_2$, $MgSO_4$, $FeCl_3$ and phosphate buffer at pH 7.2) incubated at room temperature (24 ± 2 °C) was placed in the sample bottle. A DO sensor (Eutech instruments, ECD11001K) was placed in the sample bottle at a depth of 5–7 cm below the water surface. The suspension was continuously stirred by means of a magnetic stirrer. Then, O_2 concentration (expressed in mg/L) was recorded every 30 s for approximately 15–20 min. Finally, OUR was calculated in terms of VS as mg O_2/g VS/day.

CO_2 evolution rate was measured using static measurement method (Knoepp and Vose 2002; Kalamdhad et al. 2008). Approximately, 10 g of sample was sealed in a 0.5 L vessel along with a beaker containing a known weight of oven-dried soda lime (105 °C, 1.5–2.0 sieve mesh) to trap CO_2. The samples were incubated at room temperature (24 ± 2 °C). Blank experiment was performed without putting the sample in a vessel, needs for initial CO_2 calculation. The soda lime traps were removed after 24 h, oven dried and reweighed to determine CO_2 absorbed. The CO_2 and OUR values were used to determine the Solvita maturity index on a scale of 1–8, which represents the maturity level of the compost samples.

All the results reported are the means of three replicates. The results were statistically analyzed at 0.05 levels using two-way analysis of variance (ANOVA) and Tukey's HSD test was used as a post hoc analysis to compare the means (STATISTICA package).

Results and discussion

Physico-chemical characteristics

Temperature profiles for the four C/N ratios (15, 20, 25 and 30) including control are shown in Fig. 2. It justifies the optimal conditions at which the biological process proceeds and also plays an important role in evolution and succession of the microbiological communities (Hassen et al. 2001). Due to early decomposition of the soluble, readily degradable compounds (especially in C/N 30), the substrate passed from an initial mesophilic phase (<40 °C) to a thermophilic stage on 2nd day of the composting period caused by metabolic heat, indicating quick establishment of microbial activities (Nolan et al. 2011). Elevated temperatures (>50 °C) were maintained in the reactor for 12–18 h which is sufficient time for the sanitation of the substrate from potential pathogenic microorganisms that exists in sewage sludge (Chroni et al. 2009). Thereafter, the amount of degradable organic matter in the reactor decreased over the course of the composting process until the maturation process takes place. The fact that heat generation decreased leads to reduction of the

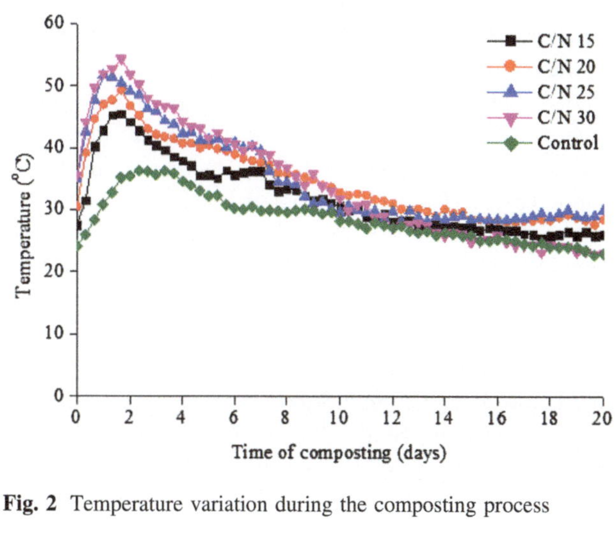

Fig. 2 Temperature variation during the composting process

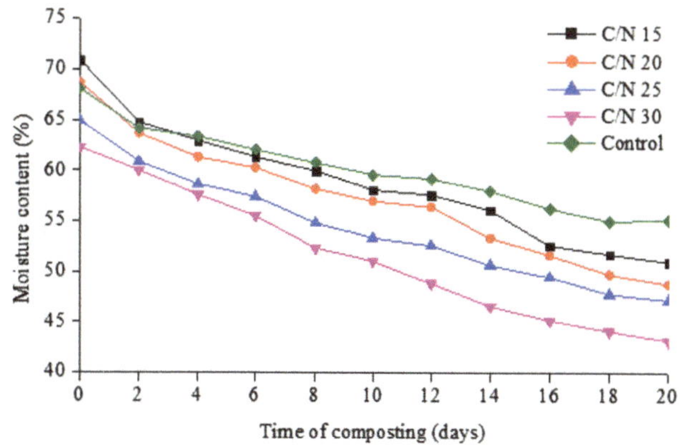

Fig. 3 Variation of moisture content with composting time during different sets of experiments

temperature in the composter. In addition, lower temperature profiles observed in control, C/N 15 and 20 experiments compared to others may be due to higher amount of sewage sludge which reduces porosity between substrates, indicates that sewage sludge without appropriate amendments is less favorable for growth and biological activity of microorganisms (Nakasaki et al. 1987). All of the experiments accomplished ambient temperature at the end of 20 day show the reflective image of stability. Similar results have been observed by Yañez et al. (2009), where maximum temperature was recorded as 62 °C.

Moisture loss during the composting process can be examined as an index of decomposition rate, since the heat generation which accompanies decomposition drives vaporization (Liao et al. 1996). From the results obtained (Fig. 3), it is seen that the C/N 15 achieved higher initial moisture content (70.91 %) due to huge quantity of sewage sludge, which is further dropped to 51.01 % resulting in only 28.06 % moisture loss, whereas in C/N 30, moisture content at the beginning was 62.30 % which is reduced to 43.23 % within 20 days. On the other hand, moisture contents were reduced from 68.74, 64.93 and 68.10 % to 48.80, 47.18 and 55.16 % for C/N 20, 25 and control experiments, respectively. Higher moisture loss was observed in C/N 30 (30.61 %) followed by others, which is directly correlated with its evaporation rate of moisture leading to large heat generation. Leachate formation was not observed during the composting period. The reduction in the percentage of moisture content was significantly varied in all the experiments ($p < 0.05$).

The contents of TOC decreased with composting period especially in thermophilic phase, leading to higher microbial activity. The variation of organic carbon during composting is shown in Fig. 4. Thereafter, the degradation rate of OM was reduced gradually. TOC loss during composting is mainly attributed by mineralization of organic matter resulting in the evolution of CO_2 and heat (Hassouneh et al.

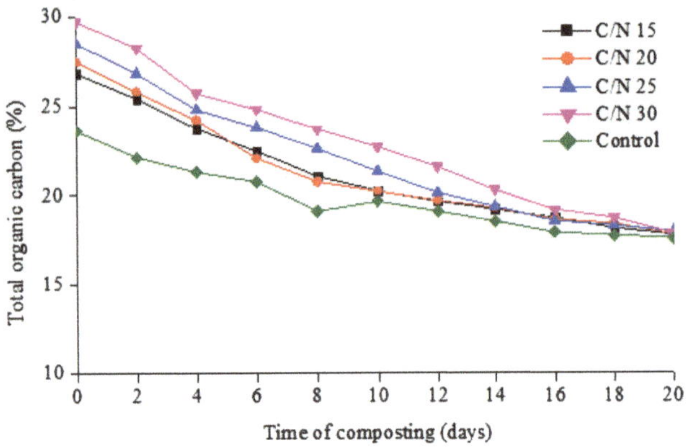

Fig. 4 TOC variation with composting time for all C/N ratios including control

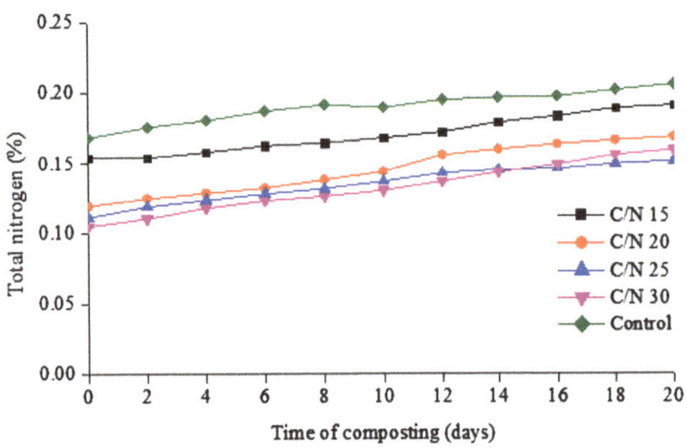

Fig. 5 Total nitrogen variation with composting time for all C/N ratios including control

1999). In addition, a large fraction (~30 %) of organic carbon was lost as utilized by microorganisms in the form of energy source (Kalamdhad and Kazmi 2008). Higher TOC reduction was observed in C/N 30 (39.97 %), followed by C/N 25 (37.08 %), C/N 20 (35.37 %), C/N 15 (33.81 %) and control (25.62 %), due to presence of readily degradable compounds as decomposed by microbes resulting in higher temperature regime in C/N 30. Significant variations in TOC were observed between all the reactors ($p < 0.05$).

Biodegradability was calculated from the initial and final organic matter contents, according to the following equation (Haug 1993; Külcu and Yaldiz 2004):

$$K_b = \frac{[OM_i(\%) - OM_f(\%)] \times 100}{OM_i(\%) \times [100 - OM_f(\%)]} \tag{8}$$

where OM_i is the organic matter content at the beginning of the process and OM_f is the organic matter content at the end of the process. The degradability coefficient (K_b) values for all the reactors are shown in Table 5. Higher values of biodegradability were observed in C/N 30 ($K_b = 0.5887$) and 25 (0.5478). As the added sewage sludge increases, lower values of biodegradability were observed in C/N 15 (0.4972) and C/N 20 (0.5199) including control (0.3841).

Figure 5 shows that total nitrogen was higher in final composts compared to initial substrate, as increased from 1.54 to 1.91 %, 1.20–1.69 %, 1.12–1.52 %, 1.05–1.60 % and 1.69–2.06 in C/N 15, 20, 25, 30 and control, respectively. Higher percentage in nitrogen content was observed in C/N 30 (34.38 %) followed by C/N 20 (28.99 %), C/N 25 (26.32 %), C/N 15 (19.37 %) and control (17.96 %) during the composting period. This enhancement was occurred due to the net loss of dry mass in terms of CO_2 as well as water loss by evaporation during mineralization of OM by microorganisms (Yamada and Kawase 2006; Kalamdhad and Kazmi 2009). Alternatively, nitrogen fixing bacteria might also contribute to increase in total nitrogen during later phase of composting (Zucconi et al. 1985). ANOVA analysis showed significant changes in nitrogen content among the experiments ($p < 0.05$).

Stability analysis

Compost stability has been evaluated by CO_2 evolution because it measures carbon derived directly from the compost being tested, caused by mineralization of the organic matter (Bernai et al. 1998; Kalamdhad et al. 2008). The CO_2 evolution rates of the C/N 15, 20, 25, 30 and control decreased from initial values of 9.94, 10.03, 9.40, 9.54 and 10.46–2.12, 1.41, 1.35, 0.51 and 3.60 mg CO_2/g VS/day, respectively (Table 3). The variation of CO_2 evolution among the reactors was statistically significant ($p < 0.05$). Results suggested that higher reduction in CO_2 evolution rate was observed in C/N 30 (94.7 %) followed by C/N 20 (85.9 %), C/N 25 (85.6 %), C/N 15 (78.7 %) and control (65.6 %). The reduction in CO_2 evolution rate was very low after 16 days of composting in all the experiments, indicating the stability of finished compost. The reason is CO_2 was mainly emitted during the thermophilic period because of the degradation of easily degradable carbon under vigorous bacterial and fungal activity. During the curing period, CO_2 emissions are related to the degradation of complex organic molecules such as lignin and lignocelluloses by some fungi and actinomycetes. The Solvita® maturation index based on CO_2 evolution increased from 3 to 8 for all the experiments. Solvita® results proved that the composts from all the reactors enter into the well stable condition after 20 days. Similarly, Wang et al. (2004) suggested that the composts become more stable when the respiration rates were below 1 mg CO_2/g VS/day.

As composting begins, large organic molecules are broken down to smaller soluble structures and temporarily more substrate may become available. Furthermore, the

Table 3 Variation in CO_2 evolution, OUR and C/N ratio during drum composting over time

Days	CO_2 evolution (mg CO_2/g VS/day)				
	C/N 15	C/N 20	C/N 25	C/N 30	Control
0	9.9 ± 0.1abc	10.0 ± 0.1ac	9.4 ± 0.2b	9.5 ± 0.3ab	10.5 ± 0.2c
2	9.5 ± 0.2a	9.5 ± 0.3a	8.6 ± 0.2b	8.3 ± 0.1b	9.8 ± 0.3a
4	8.6 ± 0.1a	8.9 ± 0.1b	6.9 ± 0.1c	7.7 ± 0.1d	8.8 ± 0.4ab
6	7.5 ± 0.1a	8.3 ± 0.1b	6.0 ± 0.1c	6.3 ± 0.5c	8.1 ± 0.3b
8	6.6 ± 0.2a	6.1 ± 0.2a	5.4 ± 0.1b	5.4 ± 0.3b	7.6 ± 0.1c
10	5.5 ± 0.2a	5.4 ± 0.3a	4.4 ± 0.2b	4.5 ± 0.2b	6.6 ± 0.3c
12	4.8 ± 0.2a	4.4 ± 0.4b	3.7 ± 0.2c	3.9 ± 0.1c	5.7 ± 0.1d
14	3.5 ± 0.1a	3.4 ± 0.1ab	3.0 ± 0.4b	3.3 ± 0.2ab	4.7 ± 0.2c
16	2.9 ± 0.3a	2.6 ± 0.2a	2.5 ± 0.2a	2.9 ± 0.1a	4.2 ± 0.2b
18	2.5 ± 0.2a	1.9 ± 0.3b	1.7 ± 0.2b	1.6 ± 0.1b	3.9 ± 0.1c
20	2.1 ± 0.3a	1.4 ± 0.2b	1.4 ± 0.1b	0.5 ± 0.1c	3.6 ± 0.1d
Oxygen uptake rate (OUR) (mg O_2/g VS/day)					
0	19.9 ± 0.3a	17.4 ± 0.3b	13.1 ± 0.2c	16.1 ± 0.3d	17.9 ± 0.2b
2	18.1 ± 0.3a	15.2 ± 0.2bde	10.9 ± 0.1c	13.9 ± 0.2d	14.5 ± 1.1e
4	16.37 ± 0.36a	13.1 ± 0.3bde	8.9 ± 0.7c	12.0 ± 0.2d	13.4 ± 0.3e
6	13.9 ± 0.4a	10.0 ± 0.2b	7.41 ± 0.3c	8.6 ± 0.4d	11.4 ± 0.1e
8	12.2 ± 0.3a	8.9 ± 0.3b	6.1 ± 0.5c	6.8 ± 0.1c	9.8 ± 0.4b
10	9.6 ± 0. 4a	7.8 ± 0.4b	5.3 ± 0.5c	5.4 ± 0.1c	8.5 ± 0.3b
12	8.5 ± 0.2a	7.1 ± 0.4b	4.1 ± 0.2c	3.6 ± 0.2c	7.7 ± 0.2b
14	7.0 ± 0.2a	6.6 ± 0.2a	2.9 ± 0.4b	2.8 ± 0.2b	7.0 ± 0.2a
16	5.9 ± 0.2ac	5.5 ± 0.1a	2.4 ± 0.4b	1.7 ± 0.6d	6.1 ± 0.2c
18	5.1 ± 0.1a	4.7 ± 0.1b	2.0 ± 0.1c	1.2 ± 0.9d	5.2 ± 0.1a
20	3.9 ± 0.2ac	3.7 ± 0.2a	1.8 ± 0.1b	0.8 ± 0.1d	4.4 ± 0.6c
C/N ratio					
0	17.4 ± 0.5a	23.2 ± 0.4b	25.6 ± 0.8c	28.3 ± 0.9d	14.0 ± 0.7e
2	16.5 ± 0.3a	20.6 ± 0.340b	22.6 ± 0.6c	25.6 ± 0.4d	12.6 ± 0.4e
4	15.2 ± 0.4a	18.9 ± 0.242b	20.1 ± 0.9b	21.8 ± 0.7c	11.8 ± 0.1d
6	13.8 ± 0.6a	16.7 ± 0.2b	18.6 ± 0.5c	20.2 ± 0.5d	11.1 ± 0.1e
8	12.8 ± 0.2a	15.3 ± 0.2b	17.2 ± 0.4c	18.8 ± 0.2d	9.9 ± 0.1e
10	12.0 ± 0.3a	14.4 ± 0.3b	15.6 ± 0.3c	17.4 ± 0.2d	10.4 ± 0.3e
12	11.4 ± 0.1a	12.6 ± 0.1b	14.1 ± 0.2c	15.7 ± 0.1d	9.8 ± 0.4e
14	10.7 ± 0.3a	12.4 ± 0.4b	13.3 ± 0.1c	14.1 ± 0.2d	9.4 ± 0.4e
16	10.2 ± 0.1a	11.4 ± 0.1b	12.7 ± 0.1c	12.9 ± 0.2c	9.1 ± 0.1d
18	9.6 ± 0.7a	11.0 ± 0.2b	12.2 ± 0.1c	11.9 ± 0.3c	8.8 ± 0.2d
20	9.3 ± 0.2a	10.5 ± 0.5b	11.9 ± 0.2c	11.2 ± 0.4d	8.5 ± 0.1e

All values are means of three replicates ± SD ($n = 3$)

Values followed by the same letter within each column are not significantly different (ANOVA; Tukey's test, $p < 0.05$)

microbial activity reduces with composting period as evidenced by the decrease in the values of OURs of all the experiments. It is observed that, a gradual decrease in OUR values from 19.98 to 3.96, 17.37–3.66, 13.12–1.77, 16.09–0.81 and 17.86–4.39 mg O_2/g VS/day in C/N 15, C/N 20, C/N 25, C/N 30 and control, respectively, after 20 days of composting period in the rotary drum. OUR values attained stability after the initial reduction in all the experiments, while after the 14th day of composting the drop was moderate indicating the compost approaches the maturation period. Higher reduction observed in C/N 30 (94.9 %) followed by C/N 25 (86.5 %), C/N 15 (80.2 %), C/N 20 (78.9 %) and control (75.4 %), respectively (Table 3), might be due to optimal amount of cattle manure and sawdust which provided a favorable condition for the growth and biological activity of microorganisms (Gomez et al. 2006). The percentage reduction in OUR value was significantly varied among the reactors ($p < 0.05$). Results indicated that the composts produced from C/N 25 and 30 experiments are considered to be very stable and possessed a Solvita® index value of 8, which indicated that compost was ready for usage of soil conditioner (Cabanas-Vargas et al. 2005). In addition, C/N 15, 20 and control experiments have a Solvita® maturity index of 6 which can be considered as immature and unstable compost.

The changes in the C/N ratios reflect the organic matter decomposition and stabilization achieved during composting. The microorganisms which decomposed the waste mixture require a supply of nutrients, mainly carbon as a source of energy and nitrogen for building cell structures; however, these nutrients should be present in optimal amount to perform the composting process efficiently (Kalamdhad and Kazmi 2009). Moreover, if the C/N ratio of compost is more, the excess carbon tends to utilize nitrogen in the soil to build cell protoplasm resulting in loss of nitrogen which is known as robbing of nitrogen in the soil. While, if the C/N ratio is too low, the resultant product does not help improve the structure of the soil. In this study, the C/N ratio decreased rapidly from an initial value of 17.43–9.30, 23.02–10.52, 25.60–11.85, 28.33–11.19 and 14.04–8.53 in C/N 15, 20, 25, 30 and control, respectively, after 20 days of composting (Table 3). Hirai et al. (1983) and CPHEEO (2000) have reported that C/N ratio less than or equal to 20 is considered as a satisfactory value for maturity when the initial value is in between 25 and 30. Results suggested that higher reduction was observed in C/N 30 (60.5 %) followed by C/N 20 (54.3 %), C/N 25 (53.7 %), C/N 15 (46.6 %) and control (39.2 %) experiment, which led to rapid decrease in organic carbon due to the oxidation of the readily available organic matter. C/N ratio was reduced significantly ($p < 0.05$) in all experiments during the composting process.

Kinetic analysis

Composting is a biochemical process where enzyme kinetics concept could applicable with some theoretical considerations. In this regard, the enzyme kinetics approach is carried out by considering following assumptions: (1) each microbe in the rotary drum is uniformly dispersed in a solution of soluble organic matter to maintain homogeneity, (2) under controlled aerobic conditions, the microbes are amended with organic matters and the moisture required for microbial growth is optimized with respect to readily available organic carbon, (3) These biocomplex systems can be analyzed by Monod's kinetics to describe its process. Various researchers have illustrated the comprehensive models for kinetic degradation (Finger et al. 1976; Whang and Meenaghan 1980; Haug 1993; Agamuthu et al. 2000; Kulcu and Yaldiz 2004; Yamada and Kawase 2006; Mason 2006). Among those models, the incorporation of organic matter or volatile solids' decomposition to predict the kinetic degradation becomes a successful and easy way for sewage sludge composting (Whang and Meenaghan 1980). In this sense, the composting of organic ingredients or decomposition of organic matter is regarded as enzymatic-related microbe system (Haug 1993; Agamuthu et al. 2000; Bari et al. 2000; Nelson and Cox 2000; Kumar et al. 2009).

In enzyme kinetics concept, enzyme concentration forms an intermediate complex with the organic matter concentration leading to complex formation represented as EC. The decomposition of EC is carried out in two ways as given in Eq. (9):

$$[E] + [C] \underset{k_2}{\overset{k_1}{\longleftrightarrow}} [EC] \overset{k_3}{\rightarrow} [E] + [P] \tag{9}$$

where E is the microbial concentration (%); C is the limiting organic matter concentration (%); P is the by-product generated by endogenous reaction (%); k_1, k_2, k_3 are the specific reaction rates.

In equilibrium conditions,

$$k_1[E][C] = k_2[EC] + k_3[EC] \tag{10}$$

By solving, EC is given by Eq. (10)

$$[EC] = \frac{[E_T][C]}{[C] + \frac{(k_2+k_3)}{k_1}} \tag{11}$$

where E_T is the total microbial concentration (%). On solving, we get

$$(r) = \frac{r_m[C]}{k_m + [C]} \tag{12}$$

where $r_m = k_2\ [E_T]$, represented as maximum rate of enzymatic reaction (day^{-1}).

Equation (12) is the projected kinetic rate equation for the composting process in accordance with Michaelis–Menten equation. To determine kinetic parameters, Eq. (12) is re-transformed into Lineweaver–Burk plot as follows:

$$\frac{1}{r} = \frac{k_m}{r_m}\left(\frac{1}{C}\right) + \frac{1}{r_m} \tag{13}$$

The kinetic parameters k_m and r_m can be graphically determined from the slope and intercept of a Lineweaver–Burk plot (Whang and Meenaghan 1980) using Eq. (13) incorporating $1/r$ and $1/C$ data. It correlates the initial rate of reaction (r) and the organic matter concentration (C) results into a linear relationship. In the Michaelis–Menten equation, the value of k_m (i.e., dissociation constant) is inversely proportional to the chemical affinity of the enzyme for the utilization of organic matter. The value of r_m (i.e., maximum or limiting velocity) gives the extent of reaction rate which directly correlated with the

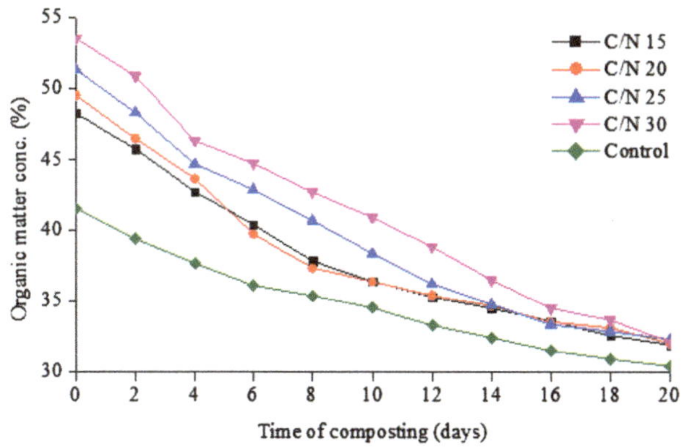

Fig. 6 Variation of organic matter (%) during the composting period

operational parameters of the process such as temperature, moisture, aeration and chemical conditions. Higher the value of r_m, the faster is the rate of degradation of organic matter.

The organic matter was evaluated over the 20 days of composting period (Fig. 6) and the consumption rate of organic matter due to microbial activities (i.e., r) can be obtained on each sampling day by plotting the tangent of the curve organic matter (on ash basis) vs. time, as shown in Table 4. The reciprocals of reaction rate (consumption rate, i.e., $1/r$) and organic matter content ($1/C$) were compiled by plotting a Lineweaver–Burk plot ($1/r$ vs. $1/C$), to determine the kinetic constants k_m and r_m for all sets of experiments as summarized in Table 5. It is observed that C/N 30 shows the least k_m value (0.4482) compared to other reactors, which reveals that 100 % ingredients act as feedstock to be satisfied for the enzyme–substrate affinity. Subsequently, C/N 30 yielded highest r_m value (114.9425) concludes the maximum rate of degradation which might be due to regular aeration and optimal nutrients balance occurred. Results showed that the reduction in temperature was constant after maintaining the peak value may caused due to lack of biodegradable organic matter, results the microbial deactivation. It is seen that C/N 30 showed higher R^2 value as compared to other experiments, which is well suited for the Lineweaver–Burk plot (Fig. 7).

While highest k_m value (98.5349) was observed in C/N 25 compared to other reactors, showed that the microorganisms does not decomposed the organic matter efficiently, might be due to insufficient oxygen as the air could not penetrate inside the waste mixture. It also correlated with lesser degradability coefficient ($K_b = 0.5478$) and limiting constant ($r_m = 1.1609$) values observed in C/N 25. The comparative analysis between C/N 15 ($k_m = 68.1105$), C/N 20 ($k_m = 80.1805$) and control ($k_m = 62.4675$) experiments revealed that cattle manure which contains readily available biodegradable materials, when coupled with sewage sludge and sawdust in the desired proportions; it does not satisfied the affinity of the

Table 4 Computed values of reaction rates during composting period

Days	Organic matter concentration (%)					r (day^{-1})				
	C/N 15	C/N 20	C/N 25	C/N 30	Control	C/N 15	C/N 20	C/N 25	C/N 30	Control
0	48.30	49.51	51.35	53.50	41.56	0.988	1.129	1.085	0.996	0.566
2	45.74	46.44	48.30	50.85	39.41	0.955	1.088	1.053	0.975	0.551
4	42.72	43.59	44.64	46.29	37.66	0.89	1.006	0.99	0.933	0.521
6	40.42	39.72	42.85	44.68	36.10	0.826	0.924	0.926	0.891	0.491
8	37.86	37.33	40.68	42.67	35.38	0.761	0.842	0.863	0.849	0.461
10	36.40	36.38	38.36	40.87	34.55	0.696	0.760	0.799	0.807	0.431
12	35.32	35.40	36.20	38.82	33.32	0.631	0.678	0.735	0.765	0.401
14	34.52	34.67	34.78	36.49	32.45	0.566	0.596	0.672	0.723	0.371
16	33.63	33.52	33.36	34.48	31.52	0.502	0.514	0.608	0.681	0.341
18	32.6	33.1	32.89	33.67	30.95	0.437	0.432	0.545	0.639	0.311
20	31.96	32.01	32.31	32.12	30.46	0.404	0.391	0.513	0.618	0.296

Table 5 Computed values of kinetic constants using Lineweaver–Burk plots for different C/N ratios

Parameters	C/N 15	C/N 20	C/N 25	C/N 30	Control
Correlation coefficient (R^2)	0.859	0.787	0.914	0.982	0.928
r_m	0.5214	0.8297	1.1609	114.9425	0.3330
k_m	68.1105	80.1805	98.5349	0.4482	62.4675
Degradability coefficient (K_b)[a]	0.4972	0.5199	0.5478	0.5887	0.3841

[a] Calculated on dry mass basis using Eq. (8)

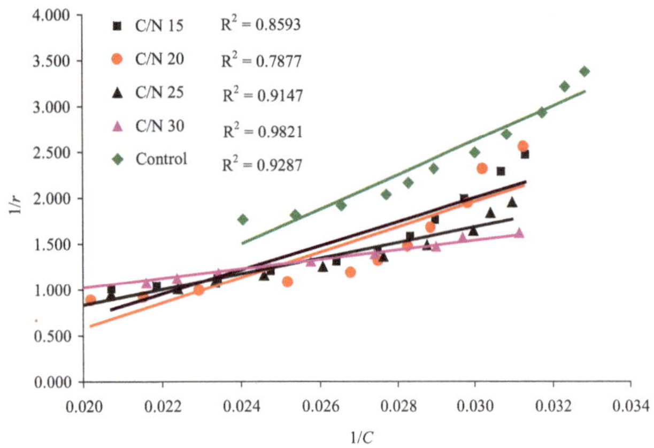

Fig. 7 Lineweaver–Burk plot for C/N 15, 20, 25, 30 and control experiments

compost mixture for microbial activities, and was further as evidenced with lesser r_m values (C/N 15: 0.5214; C/N 20: 0.8297; control: 0.3330). As a result, these waste proportions do not suit well for the degradation process due to lesser availability of the carbon source as compared to C/N 25 and 30. It also might be due to insufficient aeration which resembles lesser r_m values observed in C/N 15 (0.5214), C/N 20 (0.8297) and control (0.3330) experiments. The results concluded that higher degradation was achieved in C/N 30 as compared to other experiments, due to their optimal nutrient balance for the excellent microbial activities, which further correlated with its higher temperature regime.

Conclusions

Stability studies carried out in rotary drum composter for various C/N ratios reveal that C/N 30 produced more stable compost as compared to others after 20 days, implying that rigorous decomposition was occurred. Higher percentage reductions in CO_2 evolution (94.7 %), OUR (94.9 %) and C/N ratio (60.5 %) were observed in C/N 30 experiment, which showed that total biodegradable ingredients are stabilized. In addition, the degradation became decreased with increasing the amount of sewage sludge, caused by insufficient free spaces between the substrate mass for feasible microbial activities. Moreover, the decomposition process was analyzed by Monod kinetics which resembles Michaelis–Menten equation, and the following results (i.e., $k_m = 0.4482$ and $r_m = 114.9425$) suggested that the optimal nutrient balance has been occurred in C/N 30 experiment. Lower k_m value concluded that 100 % ingredients act as feeding material was to be satisfied for the enzyme–substrate affinity. Hence, it is concluded that the rotary drum composting of sewage sludge with initial C/N ratio of 30 can produce mature and stable compost after 20 days of the composting period.

Acknowledgments The authors are thankful to Indian Institute of Technology, Guwahati, India for the financial support to carry out this research work.

References

Aboulam S, Morvan B, Revel JC (2006) Use of a rotating-drum pilot plant to model the composting of household waste on an industrial scale. Compost Sci Util 14(3):184–190

Agamuthu P, Choong LC, Hasan S, Praven VV (2000) Kinetic evaluation of composting of agricultural wastes. Environ Technol 21(2):185–192

Bari QH, Koenig A, Guihe T (2000) Kinetic analysis of forced aeration composting. I. Reaction rates and temperature. Waste Manag Res 18:303–312

Barrena R, Turet J, Busquets A, Farrés M, Font X, Sánchez A (2011) Respirometric screening of several types of manure and mixtures intended for composting. Bioresour Technol 102(2):1367–1377

Bernai MP, Paredes C, Sanchez-Monedero MA, Cegarra J (1998) Maturity and stability parameters of composts prepared with a wide range of organic wastes. Bioresour Technol 63(1):91–99

Brewer LJ, Sullivan DM (2003) Maturity and stability evolution of composted yard trimmings. Compost Sci Util 11(2):96–112

Cabanas-Vargas DD, Sanchez-Monedero MA, Urpilainen ST, Kamilaki A, Stentiford EI (2005) Assessing the stability and maturity of compost at large-scale plants. Ingenieria 9(2):25–30

Cheng H, Xu W, Liu J, Zhao Q, He Y, Chen G (2007) Application of composted sewage sludge (CSS) as a soil amendment for turfgrass growth. Ecol Eng 29(1):96–104

Chroni C, Kyriacou A, Manios T, Lasaridi KE (2009) Investigation of the microbial community structure and activity as indicators of compost stability and composting process evolution. Bioresour Technol 100(15):3745–3750

Codell M, Verderame FD (1954) The determination of nitrogen in copper–titanium alloys. Anal Chim Acta 11(1):40–47

CPHEEO (2000) Manual on municipal solid waste management. Central public health and environmental engineering organization, New Delhi

CWMI (Cornell Waste Management Institute) (1996) Calculate C/N ratio for three materials. Cornell Waste Management Institute, Ithaca

Doublet J, Francou C, Poitrenaud M, Houot S (2010) Sewage sludge composting: influence of initial mixtures on organic matter evolution and N availability in the final composts. Waste Manag 30(10):1922–1930

Eftoda G, McCartney D (2004) Determining the critical bulking agent requirement for municipal biosolids composting. Compost Sci Util 12(3):208–218

Finger SM, Hatch RT, Regan TM (1976) Aerobic microbial growth in semisolid matrices: heat and mass transfer limitation. Biotechnol Bioeng 18:1193–1218

Gomez RB, Lima FV, Ferrer AS (2006) The use of respiration indices in the composting process: a review. Waste Manag Res 24(1):37–47

Gouxue L, Zhang F, Sun Y, Wong JWC, Fang M (2001) Chemical evaluation of sewage sludge composting as a mature indicator for composting process. Water Air Soil Pollut 132:333–345

Hassen A, Belguith K, Jedidi N, Cherif A, Cherif M, Boudabous A (2001) Microbial characterization during composting of municipal solid waste. Bioresour Technol 80(3):217–225

Hassouneh O, Jamrah A, Qaisi K (1999) Sludge stabilization by composting: a Jordanian case study. Bioprocess Eng 20:413–421

Haug RT (1993) The practical handbook of compost engineering. Lewis Publishers, Boca Raton

Hirai MF, Chanyasak V, Kubota H (1983) A standard method for measurement of compost maturity. Bio Cycles 24:54–56

Iqbal MK, Shafiq T, Ahmed K (2010) Characterization of bulking agents and its effects on physical properties of compost. Bioresour Technol 101(6):1913–1919

Kalamdhad AS, Kazmi AA (2008) Mixed organic waste composting using rotary drum composter. Int J Environ Waste Manag 2(1–2):24–36

Kalamdhad AS, Kazmi AA (2009) Effects of C/N ratio on mixed organic waste composting in a rotary drum composter. Int J Environ Eng 1(2):187–207

Kalamdhad AS, Pasha M, Kazmi AA (2008) Stability evaluation of compost by respiration techniques in a rotary drum composter. Resour Conserv Recycl 52(5):829–834

Kalamdhad AS, Singh YK, Ali M, Khwairakpam M, Kazmi AA (2009) Rotary drum composting of vegetable waste and tree leaves. Bioresour Technol 100(24):6442–6450

Khwairakpam M, Bhargava R (2009) Vermitechnology for sewage sludge recycling. J Hazard Mater 161(2–3):948–954

Knoepp JD, Vose JM (2002) Quantitative comparison of in situ soil CO_2 flux measurement method, USDA Forest Service, Southern Research Station, Research Paper SRS-28

Kulcu R, Yaldiz O (2004) Determination of aeration rate and kinetics of composting some agricultural wastes. Bioresour Technol 93(1):49–57

Kumar S, Sakhale A, Mukherjee S (2009) Simplified kinetic analysis for composting of municipal solid waste. Pract Period Hazard Toxic Radioact Waste Manag 13(3):179–186

Liang C, Das K, McClendon C (2003) The influence of temperature and moisture contents regimes on the aerobic microbial activity of a biosolids composting blend. Bioresour Technol 86:131–137

Liao PH, Jones L, Lau AK, Walkemeyer S, Egan B, Holbek N (1996) Composting of fish wastes in a full-scale in-vessel system. Bioresour Technol 59:163–168

Mason IG (2006) Mathematical modelling of the composting process: a review. Waste Manage 26(1):3–21

Matteson TL, Sullivan DM (2006) Stability evolution of mixed food waste composts. Compost Sci Util 14(3):170–177

Mohee R, Mudhoo A (2005) Analysis of the physical properties of an in-vessel composting matrix. Powder Technol 155(1):92–99

Mohee R, Driver MF, Sobratee N (2008) Transformation of spent broiler litter from exogenous matter to compost in a sub-tropical context. Bioresour Technol 99(1):128–136

Nakasaki K, Nakano Y, Akiyama T, Shoda M, Kubota H (1987) Oxygen diffusion and microbial activity in the composting of dehydrated sewage sludge cakes. J Ferment Technol 65(1):43–48

Nelson DL, Cox MM (2000) Lehninger principles of biochemistry, 3rd edn. Worth Publishers, New York

Nolan T, Troy SM, Healy MG, Kwapinski W, Leahy JJ, Lawlor PG (2011) Charaterization of compost produced from separated pig manure and a variety of bulking agents at low initial C/N ratios. Bioresour Technol 102(14):7131–7138

Petric I, Selimbasic V (2008) Composting of poultry manure and wheat straw in a closed reactor: optimum mixture ratio and evolution of parameters. Biodegradation 19(1):53–63

Rodríguez L, Cerrillo MI, García-Albiach V, Villaseñor J (2012) Domestic sewage sludge composting in a rotary drum reactor: optimizing the thermophilic stage. J Environ Manag 112:284–291

Singh RP, Agrawal M (2008) Potential benefits and risks of land application of sewage sludge. Waste Manag 28(2):347–358

Tuomela M, Vikman M, Hatakka A, Itavaara M (2000) Biodegradation of lignin in a compost environment: a review. Bioresour Technol 72(2):169–183

Varma VS, Chatuphale M, Kalamdhad A (2014) Effects of bulking agent in composting of vegetable waste and leachate control using rotary drum composter. Sustain Environ Res 24(4):245–256

Vuorinen AH, Saharinen MH (1997) Evolution of microbiological and chemical parameters during manure and straw co-composting in a drum composting system. Agr Ecosyst Environ 66(1):19–29

Wang P, Changa CM, Watson ME, Dick WA, Chen Y, Hoitink HAJ (2004) Maturity indices for composted dairy and pig manures. Soil Biol Biochem 36:767–776

Wang K, Li X, He C, Chen C, Bai J, Ren N, Wang J (2014) Transformation of dissolved organic matters in swine, cow and chicken manures during composting. Bioresour Technol 168:222–228

Whang DS, Meenaghan GF (1980) Kinetic model of composting process. Compost Sci 21(3):44–46

Wu L, Ma LQ, Martinez GA (2000) Comparison of methods for evaluating stability and maturity of biosolids compost. J Environ Qual 29:424–429

Yamada Y, Kawase Y (2006) Aerobic composting of waste activated sludge: kinetic analysis for microbiological reaction and oxygen consumption. Waste Manag 26(1):49–61

Yañez R, Alonso JL, Díaz MJ (2009) Influence of bulking agent on sewage sludge composting process. Bioresour Technol 100:5827–5833

Zucconi F, Monaco A, Forte M (1985) Phytotoxins during the stabilization of organic matter. In: Grasser JKR (ed) Composting of agricultural and other wastes. Elsevier, New York, pp 73–88

25

Influence of vermi-fortification on chickpea (*Cicer arietinum L.*) growth and photosynthetic pigments

Anoop Yadav[1] · V. K. Garg[1]

Abstract

Purpose The aim of this study was to investigate the effect of different vermicomposts and their dosages on chickpea (*Cicer arietinum* L.) plants in pot culture experiment.

Methods A total of five potting media were prepared containing soil and vermicompost. Soil fortified with 10 and 20 % vermicompost was used as potting media. The fertility status of soil and vermicomposts was quantified. In these potting media, growth, yield and biochemical parameters of chickpea plants were studied up to 90 days.

Results The results showed that the fortification of soil with vermicompost significantly stimulated the chickpea plant growth. The plant height, plant shoot biomass, number of pods and photosynthetic pigments were significantly higher in vermicompost-fortified experiments, whereas vermicompost fortification had no significant effect on chickpea seed germination as it was $\approx$100 % in all experiments. Total chlorophyll content in chickpea leaves was in the range of 0.437–1.07 mg/g. Similarly, carotenoid content was minimum in control and maximum in 20 % vermicompost containing potting media.

Conclusion It was concluded that if soil is fortified with appropriate quantities of vermicompost, the chickpea production per unit area could be enhanced significantly.

✉ V. K. Garg
vinodkgarg@yahoo.com

Anoop Yadav
yadavanoop@rediffmail.com

[1] Department of Environmental Science and Engineering, Guru Jambheshwar University of Science and Technology, Hisar 125001, India

Keywords Chickpea · Organic manures · Vermicompost · Chlorophyll · Carotenoids · Vermi-fortification

Abbreviations

CD	Cow dung
HC-3	Haryana channa-3
FIS	Food industry sludge
DAS	Days after sowing
TCa	Total calcium
TNa	Total sodium
NPK	Nitrogen phosphorous and potassium
EC	Electric conductivity
TOC	Total organic carbon
TKN	Total Kjeldhal nitrogen
TK	Total potassium
TP	Total phosphorus
CN	Carbon nitrogen ratio

Introduction

Conventional agricultural systems have been characterized by high input of chemical fertilizer and pesticides leading to soil health deterioration and poor product quality due to a reduction in soil organic matter content. Continuous addition of organic matter to soil is necessary to ensure a balanced supply of plant nutrients (Gupta et al. 2014). In recent years, organic farming has gained attention as it maintains soil health, prevents soil erosion, uses natural pest controlling agents and farm products have better nutritive quality. It is evident from the available literature that organic fraction of solid wastes contains significant quantities of NPK.

However, organic wastes cannot directly be applied to the agricultural fields since these may cause phytotoxicity or may destroy the natural fertility of the soil. Therefore, their stabilization is essential prior to their application in agricultural fields (Gupta et al. 2014).

Vermicomposting of non-toxic biodegradable matter produces stabilized humus like product known as vermicompost, which has a great potential as soil amendment (Arancon et al. 2003). Vermicompost is a very good soil conditioner that is rich in NPK, micronutrients and growth hormones. Vermicompost application to soil also increases microbial populations and activities that further influence nutrient cycling, production of plant growth-regulating materials and build up plant resistance or tolerance to pathogen and nematode attacks (Gopalakrishnan et al. 2011).

Organic amendments application enhances soil microbial activities as compared to mineral fertilizer and unamended treatments (Gopinath et al. 2008). Ferreras et al. (2006) reported that soil structure stability, organic carbon and microbial activity in soil were improved after two applications of vermicomposts. Jouquet et al. (2010) reported that both compost and vermicompost led to an improvement in soil properties with an increase in the pH, soil organic matter and nutrient content, compared to soil fertilized with chemical fertilizers. Khan et al. (2015) reported that application of vermicompost caused a substantial increase (54–83 %) in the soil organic carbon pool of the soil.

Chickpea (*Cicer arietinum* L.) is an important food legume, which ranks third among the world's pulse crops, and is grown in more than 50 countries (89.7 % area in Asia, 4.3 % in Africa, 2.6 % in Oceania, 2.9 % in Americas and 0.4 % in Europe) (Lake and Sadras 2014). India is the largest chickpea producing country accounting for 67 % (8.8 million metric tonnes) (FAO 2014). Chickpea is also a vital source of protein for millions of people in India and other developing countries, predominantly in South Asia, who are largely vegetarian. In addition to having high protein content (20–22 %), it is rich in fiber, P, Ca, Mg, Fe, Cu and β-carotene. Farmers have notion that chickpea, being a legume crop, does not need nutrients and usually cultivate it on marginal lands without applying chemical fertilizer. The yield gap of chickpea can be abridged, by adopting the advanced production technology accompanying with use of inoculums, balanced nutrition, weed management and high yielding varieties (Hakoomat et al. 2004).

A literature survey has indicated that vermicompost produced from different wastes has different physico-chemical properties and in turn has different effects on the crops' growth and yield (Table 1). Based on this it is hypothesized that vermicomposts with different NPK contents have different effects on crops and results of a vermicompost on a crop cannot be generalized. Keeping this view effect of food industry sludge and cow dung-based vermicompost on growth and productivity of chickpea plants has been investigated (Table 2).

Materials and methods

The soil used in potting media was collected from the agricultural fields of Guru Jambheshwar University of Science and Technology, Hisar. The soil was sandy–loam in nature. It was deficit in nitrogen, average in phosphorus and rich in potassium content. Vermicomposts used in potting media were prepared in the laboratory using (a) cow dung and (b) wastewater treatment plant sludge of food industry spiked cow dung (Yadav and Garg 2011). The physico-chemical characteristics of soil and vermicomposts are given in Table 3. The Haryana channa-3 (HC-3) variety of chickpea was procured from the certified dealer. This variety of chickpea was chosen because of its suitability for irrigated lands, blight and wilt disease suffering areas. The main characteristics of this species are hard and bold stem and light brown grain.

The soil was fortified with 10 and 20 % (w/w, on dry weight basis) vermicompost to prepare the pot culture media. Four vermicompost-fortified potting media were prepared along with one control (soil only). The potting media were well mixed and sieved through 2 mm sieve before use. Each pot was filled with 3.0 kg of potting media and three replicates were maintained for each treatment. The composition of soil and vermicompost in different treatments is given in Table 2.

After filling potting media, the pots were maintained as such for 2 weeks at 60–70 % moisture content. After that 10 chickpea seeds were drilled in each pot at a depth of approximately 2 cm in the third week of October. The temperature is in range of 18–27 °C during this period. The chickpea seeds were treated with systemic fungicide, viz., carbendazim (50 % WP @ 1.5 g kg^{-1} seed) to protect from seedling diseases. The germination was recorded in all pots 7 days after sowing (DAS) and two healthy seedlings pot^{-1} were chosen for further study. The pots were kept in open in a pot house without controlling temperature. Plant height was recorded periodically at 30, 60 and 90 DAS by measuring the height from the ground level to the furthest extremity of the longest branch. Fresh and fully mature leaves were used to determine the chlorophyll, carotenoid and protein contents at 30 DAS. At the end of experiment, plants were removed carefully from the pots and separated in roots and shoots. Then fresh root and shoot biomass was recorded. No additional fertilizers were added during the study. The irrigation was done periodically as per requirement.

Germination percentage was calculated using the below given formula:

Table 1 Effects of different vermicompost on physico-chemical properties of soil and crop

Raw material used for vermicomposting	Ratio of mixing/dosage	Duration of study	Crop grown	Conclusion	References
Vermicompost–compost (mix) + bone meal	Vermicompost–compost mix 2 kg m^2 + bone meal, dosage 0.15 kg m^2	132 days	Beet	An increment in extractable P and stimulated microbial respiration in soil was found in all the treatments. N and P concentration in beet aerial biomass was also recorded in treatments with amendment application	González et al. (2010)
Buffalo manure	Soil + vermicompost (equivalent to 20 t ha^{-1} of organic substrate)	1 year	Maize and tomato	Vermicompost amendment modified soil chemical properties leading to higher C and N, higher pH and CEC, and lower available P, NH_4^+ and NO_3^- than in the control	Doan et al. (2013)
Buffalo manure	At the rate of 20 t ha^{-1}	3 years	Maize	CEC, N, P and K content was significantly higher at the end of the experiment in vermicompost-amended soils than control soil. The combination of vermicompost and biochar improves plant productivity and reduce the negative impact of agriculture on water quality	Doan et al. (2015)
Temple, farmyard and kitchen wastes	5, 10 and 25 % vermicompost–soil (v/v)	–	Chickpea	Vermicompost–water extract (VCE) stimulates seed germination, seedling growth and plant growth of chickpea	Singh et al. (2013)
Vegetable peels and leaf litter	10, 25 and 50 % of vermicompost–soil (v/v)	–	Chickpea	Substitution of soil with 10, 25, and 50 % VC significantly increased growth and biomass production of chickpea seedlings progressively with the increase in vermicompost substitution compared with control	Sahni et al. (2008a)
–	Vermicompost application @ 0, 1, 2 and 3 t/ha	3 years	Chickpea	The application of vermicompost improves plant growth, yield attributes and grain yield	Singh et al. (2012)
Vegetable peels and leaf litter	Garden soil substituted with different amounts of vermicompost (10, 25 and 50 % w/w)	30 days	Chickpea	Substituting the soil with different amounts of vermicompost showed significant reduction in mortality of chickpea compared to control. Maximum reduction in plant mortality was noticed in 50 % vermicompost substitution	Sahni et al. (2008b)
Paddy straw (*Oriza sativa*) and *Ageratum conyzoides* residues	1 t/ha	60 days	Zea mays, *Phaseolus vulgaris* and *Abelmoschus esculentus*	Vermicompost amendment of soil increased the nutrient supply and plant productivity	Roy et al. (2010)
Cassava peels, poultry dropping, cow dung and guava leaves	8 kg/plot (1 × 6 m)	4 months	Cowpea	This field study observed that cassava peel vermicompost enhanced cowpea aerial biomass production but acidified the soil. The effect of various vermicompost on soil physico-chemical and biological properties is very pronounced	Mba (1996)
Cattle manure, market food waste and recycled paper waste	10 or 20 t ha-1 (d.w.) to tomatoes and peppers, 5 or 10 t ha-1 (d.w.) to strawberries	For one season	Tomato, pepper and strawberry	The marketable tomato, pepper and strawberry yields in all vermicompost-treated plots were consistently greater than yields from the inorganic fertilizer-treated plots	Arancon et al. (2003)

$$\text{Percent germination} = \frac{[\text{No. of seeds germinated}]}{[\text{Total no. of seed drilled}]} \times 100$$

Chlorophyll, carotenoid and protein contents were determined at 30 DAS. Chlorophyll '*a*' and '*b*' contents and total chlorophyll were determined as reported by

Table 2 Composition of soil and vermicompost in different treatments

Treatment	Soil (%)	Vermicompost
T_1	100	0 %
T_2	90	10 % Vermicompost (cow dung)
T_3	80	20 % Vermicompost (cow dung)
T_4	90	10 % Vermicompost (30 % FIS + 70 % cow dung)
T_5	80	20 % Vermicompost (30 % FIS + 70 % cow dung)

Table 3 Physico-chemical characteristics of experimental soil and vermicomposts

Parameters	Soil	Vermicompost I (100 % cow dung)	Vermicompost II (70 % CD + 30 % FIS)
pH	8.5 ± 0.4	6.8 ± 0.1	6.6 ± 0.2
EC (dS/m)	0.4 ± 0.02	1.6 ± 0.04	1.8 ± 0.06
TOC (g/kg)	24 ± 1.7	280 ± 5	310 ± 8
TKN (g/kg)	1.6 ± 0.1	28.8 ± 0.35	26 ± 0.5
OM (%)	4.13 ± 0.3	48.2 ± 1.3	52.5 ± 1.15
TAP (g/kg)	0.6 ± 0.05	10.5 ± 0.4	9.75 ± 0.3
TK (g/kg)	2.0 ± 0.2	10 ± 0.2	7.6 ± 0.4
C:N ratio	15 ± 0.4	9.72 ± 0.54	11.9 ± 0.3

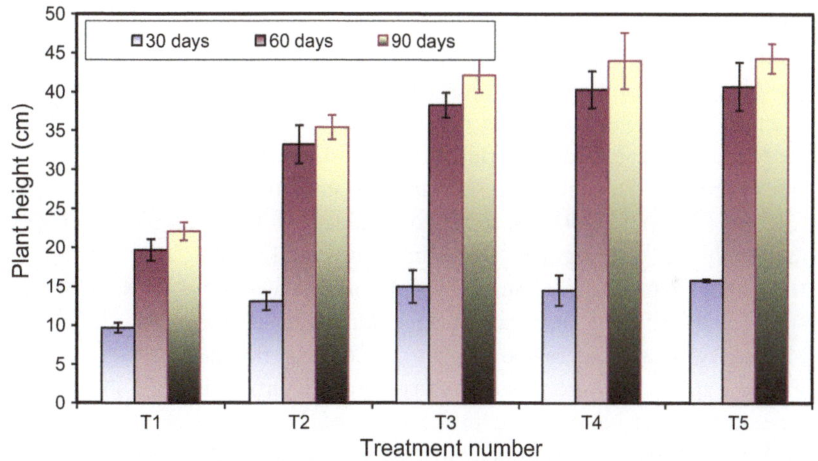

Fig. 1 Height of chickpea plants grown in different vermicompost treatments

Arnon (1949). Carotenoid was determined by the method given by Ikan (1969). Protein content in leaf tissues was estimated using the method of Bradford (1976).

The reported results are the mean of three replicates. One-way analysis of variance (ANOVA) was used to analyze the significant differences among different treatments for studied parameters. All statistical tests were evaluated at the 95 % confidence level using the SPSS software.

Results and discussion

Effects of vermicomposts on chickpea plants

Seed germination is an important and decisive phase in growth cycle of a plant since it determines plant establishment and yield. Seed germination was 96.6–100 % in different potting media. This shows that vermicomposts amendment had no antagonistic effect on chickpea seed germination. Several other authors have reported moderate or no effect of vermicompost on seed germination (Singh et al. 2013; Roy et al. 2010; Alves and Passoni 1997). In contrast, Ievinsh (2011) has reported that vermicompost addition inhibits seed germination depending on crop species and cultivar tested even at moderate concentrations.

The plant height increased progressively throughout the study period in all the plants (Fig. 1). It varied significantly in different treatments. At 30 DAS, the plant height was in the ranges of 9.7–15.8 cm in different treatments. On 30 DAS minimum plant height was in treatment T_1 and maximum was in treatment T_5. At 90 DAS plant height was

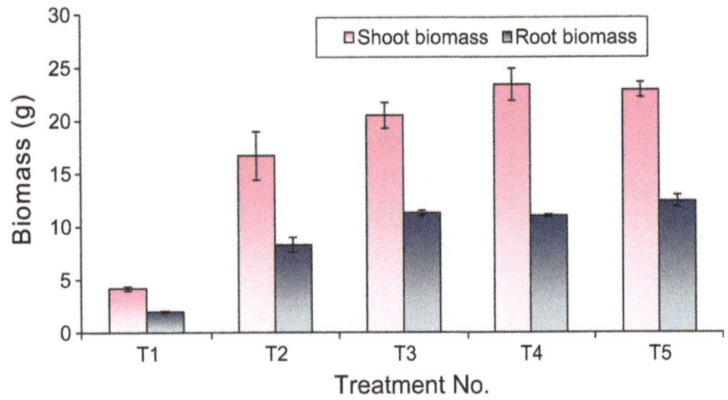

Fig. 2 Shoot biomass and root biomass in different vermicompost treatments

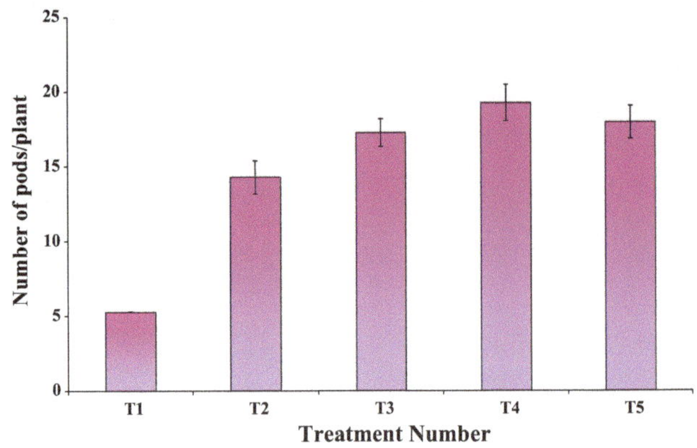

Fig. 3 Total numbers of pods in chickpea plants grown in different vermicompost treatments

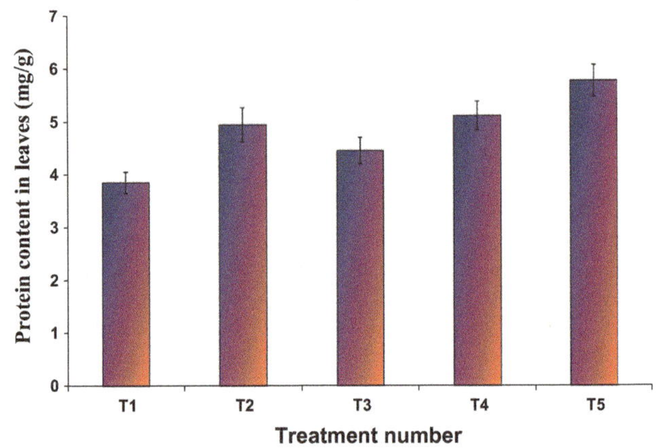

Fig. 4 Protein content in the leaves of chickpeas grown in different vermicompost treatments

in the range of 22.1 (T_1) to 43.3 cm (T_5) in different treatments. Statistically, the plant height in vermicompost-fortified treatment was significantly different from control (T_1) ($F = 52.7$, $P < 0.001$). This increase in plant height may be due to enhanced supply of nutrients in the vermicompost-amended soils. Roy et al. (2010) have reported the synergistic effect of vermicompost on the plant height of maize, French bean and okra. Gopinath et al. (2008) reported that height of *Triticum aestivum* L. was significantly higher in all vermicompost and farmyard manure compost amended treatments, than the control.

Fresh shoot biomass (above ground plant weight) was measured at the end of experiment. The results of fresh shoot biomass are given in Fig. 2. Maximum shoot biomass (23.4 ± 1.7 g per plant) was recorded in treatment T_4 and minimum (4.2 ± 0.2 g per plant) in treatment T_1. The fresh shoot biomass in treatment T_3, T_4 and T_5 was not significantly different from each other. Similarly fresh root biomass was also measured at the end of experiment. Figure 2 shows the fresh root biomass recorded in different treatments. Root biomass was in the range of 2.0 (T_1) to 12.4 g per plant (T_5). The root biomass was significantly different in different treatments ($F = 98.7$, $P < 0.001$). Bachman and Metzger (2008) have also reported that the plants cultivated in media mixes containing up to 20 % pig manure vermicompost had higher shoot biomass as compared to controls.

The number of pods per plant is an important yield contributing parameter. The number of pods per plant was different in different plants grown in vermicompost spiked soil that had more pods per plant as compared to control. The number of pods per plant in different treatments varied from 5.3 to 19.3 (Fig. 3). Maximum number of pods per plant was recorded in treatment T_4 and minimum was in control. The difference in numbers of pods per plant was non-significant within different vermicompost-fortified treatments, whereas the numbers of pods per plant were significantly different in control and vermicompost-fortified treatments ($F = 12.42$, $P < 0.001$).

Protein content in plant leaves was measured at 30 DAS and it varied from 5.83 to 3.85 mg/g in different treatments (Fig. 4). It was minimum in the leaves of chickpea plants grown in control. Maximum protein content was in the leaves of the plants grown in treatment T_5 followed by T_4. This was in agreement with reported literature that protein contents in onion grown in photo-catalytically treated dyeing industry effluent were higher than control (Ameta et al. 2003). The protein contents in treatment T_4 and T_5 were significantly different from other treatments ($F = 34.4$, $P < 0.001$).

Photosynthetic pigments

Determination of chlorophyll content in plant leaves is an indirect method of estimating the crop productivity. Chlorophyll *a*, chlorophyll *b* and total chlorophyll content measured in all the treatments on 30 DAS (Fig. 5). The chlorophyll *a* was significantly higher than chlorophyll *b* in all the treatments. Maximum chlorophyll *a* content was in

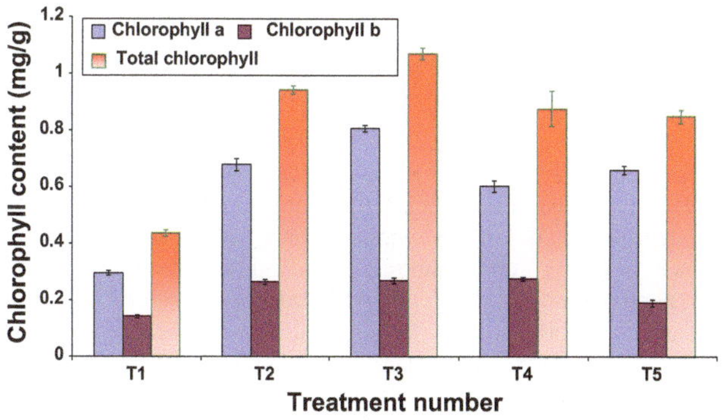

Fig. 5 Chlorophyll content in the leaves of chickpeas grown in different treatments

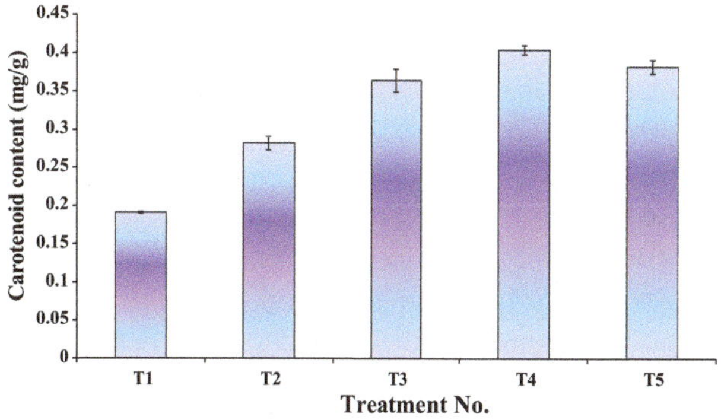

Fig. 6 Carotenoid content in the leaves of chickpeas grown in different treatments

those plants grown in treatment T_3 followed by T_5 and minimum was in control. Chlorophyll *a* content ranged from 0.296 to 0.807 mg/g in different treatments. The differences in Chlorophyll *a* content after 30 DAS between the treatments were significantly different ($F = 693$, $P < 0.001$).

Chlorophyll *b* content in plant leaves was minimum in control and maximum in treatments T_4 (Fig. 5). The chlorophyll *b* content in different treatments ranged from 0.141 to 0.275 mg/g. The difference in Chlorophyll *b* content between the treatment T_2, T_3 and T_4 was non-significant ($P = 0.65$). Total chlorophyll content ranged from 0.437 to 1.07 mg/g. Total chlorophyll content was maximum in treatment T_3 and minimum in control. The results indicated that spiking of vermicomposts with soil has a synergistic effect on chlorophyll content of chickpea leaves. The total Chlorophyll contents after 30 DAS were not significantly different from other treatments ($P = 0.99$), except control (T_1) ($P < 0.001$) Supply of more nutrients by vermicomposts as compared to soil may be the reason of enhanced leaf chlorophyll content in chickpea plants. Nitrogen is the most important mineral element in the process of chlorophyll biosynthesis, adding nitrogen to the soil may have a positive effect, which leads to increase of leaf chlorophyll content (Cecchin and Terezinha 2004). This may be the cause for higher total chlorophyll content in treatment T_3. Atiyeh et al. (2002) have reported that plants grown in 10–20 % vermicomposted food wastes contained more chlorophyll than the plants grown on control. Sangwan et al. (2010) have also reported that plants grown in cow dung vermicompost had higher total chlorophyll content than control.

Carotenoids play important role in photosynthesis process. Biosynthesis of these pigments in plants is a genetic characteristic, but some time environmental conditions also play a vital role. Figure 6 represents the carotenoid content in fresh leaf of plants measured at 30 DAS in all treatments. Carotenoid content was minimum in control and maximum was in treatments T_4. It varied from 0.191 to 0.404 mg/g in different treatments. The Carotenoid content in different treatments showed statistical significant difference ($F = 320.8$, $P < 0.001$). The more availability of nutrients in FIS containing vermicomposts may be the reason of higher carotenoid content in treatments (T_4 and T_5).

Conclusion

The results indicate that the vermicompost application has increased plant height, plant shoot biomass, seedpods, and photosynthetic pigments. Substitution of soil with 10 and 20 % vermicompost significantly increased growth in chickpea plants progressively with the increase in vermicompost substitution compared with those grown in soil (control). The increased growth may be due to enhanced capacity of plants to produce more leaves and the greater photosynthetic activity in leaves. Besides nutrient supply, the increased soil microbial activity and biomass in amended soils may indirectly have contributed to the enhanced growth and yield of plants.

Author contribution AY carried out the experiment and physico-chemical analysis and drafts the manuscript. VKG planned this study, supervision on the data analysis and revised the manuscript. Both authors read and approved the final manuscript.

Compliance with ethical standards

Conflict of interest The authors declare that they have no competing interests.

References

Alves WL, Passoni AA (1997) Compost and vermicompost of urban solid waste in *Licania tomentosa* (Benth) seedlings production to arborization. Pesqui Agropecu Bras 32:1053–1058

Ameta SC, Punjabi PB, Kothari S, Sancheti A (2003) Effect of untreated and photo catalytically treated dyeing industry effluent on growth and biochemical parameters of *Allium cepa* (Onion). Poll Res 22(3):389–392

Arancon NQ, Lee S, Edwards CA, Atiyeh R (2003) Effects of humic acid derived from cattle, food and paper-waste vermicomposts on growth of green–house plants. Pedobiologia 47:741–744

Arnon DL (1949) A copper enzyme is isolated chloroplast polyphenols oxidase in *Beta vulgaries*. Plant Physiol 24:1–15

Atiyeh RM, Arancon NQ, Edwards CA, Metzger JD (2002) The influence of earthworm-processed pig manure on the growth and productivity of marigolds. Bioresour Technol 81:103–108

Bachman GR, Metzger JD (2008) Growth of bedding plants in commercial potting substrate amended with vermicompost. Bioresour Technol 99:3155–3161

Bradford M (1976) A rapid and sensitive method for the quantitation of microgram quantities of protein utilizing the principle of protein dye binding. Anal Biochem 77:248–254

Cecchin I, Terezinha FF (2004) Effect of nitrogen supply on growth and photosynthesis of sunflower plants grown in the greenhouse. Plant Sci 166:1379–1385

Doan TT, Ngo PT, Rumpel C, Nguyen BV, Jouquet P (2013) Interactions between compost, vermicompost and earthworms influence plant growth and yield: a one-year greenhouse experiment. Sci Hortic 160:148–154

Doan TT, Tureaux TH, Rumpel C, Janeau JL, Jouque P (2015) Impact of compost, vermicompost and biochar on soil fertility, maize yield and soil erosion in Northern Vietnam: a three year mesocosm experiment. Sci Total Environ 514:147–154

FAO (2014) Production of chickpea by countries, statistical database. Food and Agriculture Organization of the United Nations, Rome, Italy. Accessed 21 July 2014

Ferreras L, Gomez E, Toresani S, Firpo I, Rotondo R (2006) Effect of organic amendments on some physical, chemical and biological properties in a horticultural soil. Bioresour Technol 97:635–640

González M, Gomez E, Comese R, Quesad M, Conti M (2010) Influence of organic amendments on soil quality potential indicators in an urban horticultural system. Bioresour Technol 101:8897–8901

Gopalakrishnan S, Pande S, Sharma M, Humayun P, Kiran BK, Sandeep D, Vidya MS, Deepthi K, Rupela O (2011) Evaluation of actinomycete isolates obtained from herbal vermicompost for biological control of Fusarium wilt of chickpea. Crop Prot 30:1070–1078

Gopinath KA, Saha S, Mina BL, Pande H, Kundu S, Gupta HS (2008) Influence of organic amendments on growth, yield and quality of wheat and on soil properties during transition to organic production. Nutr Cycl Agroecosys 82:51–60

Gupta R, Yadav A, Garg VK (2014) Influence of vermicompost application in potting media on growth and flowering of marigold crop. Int J Recycl Org Waste Agricult 3:47

Hakoomat A, Muhammad AK, Shakeel AR (2004) Interactive effect seed inoculation and phosphorus application on growth and yield of chickpea (*Cicer arietinum* L.). Int J Agricul Biol 6:110–112

Ievinsh G (2011) Vermicompost treatment differentially affects seed germination, seedling growth and physiological status of vegetable crop species. Plant Growth Regul 65:169–181

Ikan R (1969) Natural products. A laboratory guide. Academic Press, New York

Jouquet P, Plumerea T, Doan TT, Rumpel C, Duc TT, Orange D (2010) The rehabilitation of tropical soils using compost and vermicompost is affected by the presence of endogeic earthworms. Appl Soil Ecol 46:125–133

Khan K, Pankaj U, Verma SK, Gupta AK, Singh RP, Verma RK (2015) Bio-inoculants and vermicompost influence on yield, quality of *Andrographis paniculata*, and soil properties. Ind Crop Prod 70:404–409

Lake L, Sadras VO (2014) The critical period for yield determination in chickpea (*Cicer arietinum* L.). Field Crop Res 168:1–7

Mba CC (1996) Treated–cassava peel vermicomposts enhanced earthworm activities and cowpea growth in field plots. Resour Conserv Recycl 17:219–226

Roy S, Arunachalam K, Dutta BK, Arunachalam A (2010) Effect of organic amendments of soil on growth and productivity of three common crops viz. *Zea mays*, *Phaseolus vulgaris* and *Abelmoschus esculentus*. Appl Soil Ecol 45:78–84

Sahni S, Sarma BK, Singh DP, Singh HB, Singh KP (2008a) Vermicompost enhances performance of plant growth-promoting rhizobacteria in *Cicer arietinum* rhizosphere against *Sclerotium rolfsii*. Crop Prot 27:369–376

Sahni S, Sarma BK, Singh KP (2008b) Management of Sclerotium rolfsii with integration of non-conventional chemicals, vermicompost and *Pseudomonas syringae*. World J Microbiol Biotechnol 24:517–522

Sangwan P, Kaushik CP, Garg VK (2010) Growth and yield response of marigold to potting media containing vermicompost produced from different wastes. Environmentalist 30:123–130

Singh G, Sekhon HS, Kaur H (2012) Effect of farmyard manure, vermicompost and chemical nutrients on growth and yield of chickpea (*Cicer arietinum* L.). Int J Agric Res 7(2):93–99

Singh A, Jain A, Sarma BK, Abhilash PC, Singh HB (2013) Solid waste management of temple floral offerings by vermicomposting using *Eisenia fetida*. Waste Manag 33:1113–1118

Yadav Anoop, Garg VK (2011) Recycling of organic wastes by employing *Eisenia fetida*. Bioresour Technol 102:2874–2880

Application of spent *Agaricus subrufescens* compost in integrated production of seedlings and plants of tomato

Raul Xavier Lopes[1] · Diego Cunha Zied[2] · Emerson Tokuda Martos[1] · Rovilson José de Souza[3] · Romildo da Silva[1] · Eustáquio Souza Dias[1]

Abstract
Purpose Spent mushroom compost (SMC) is a major solid waste product of the mushroom industry and is the material which remains at the end of a mushroom crop.
Methods Different proportions of the SMC from *Agaricus subrufescens* cultivation were tested to produce seedlings and plants of tomato. A commercial substrate was used, both as a control and in combination with the spent compost in different proportions. So two experiments were carried out, the first was the production and evaluation of tomato seedlings and the second was tomato cultivation from seedlings produced in the first experiment.
Results The use of different proportions of spent *A. subrufescens* compost resulted in a decreasing trend of all the parameters in the production of the tomato seedlings. However, in tomato cultivation for all periods of harvest, a positive effect was observed in fruiting when the seedlings were produced with spent *A. subrufescens* compost in comparison to the commercial control.
Conclusion It was concluded that the use of spent *A. subrufescens* compost for seedling production led to a higher total tomato production compared to previously reported production levels in organic cultivation systems with green, organic and other types of fertilization. These results demonstrated the great potential of spent *A. subrufescens* compost for use in organic tomato production because of the better quality of harvested fruit.

Keywords Almond mushroom · *Agaricus blazei* · Reuse of compost · Waste

Introduction

Agaricus subrufescens, which is synonymous with *Agaricus blazei* ss. Heinemann and *Agaricus brasiliensis*, is a basidiomycete fungus that has aroused great interest in various parts of the world because of its medicinal properties, which is used in the prevention or treatment of diabetes, atherosclerosis, hepatitis, hypercholesterolemia, and heart disease (Dias et al. 2013; Zied et al. 2013).

Cultivation of mushrooms generates a large amount of spent mushroom compost (SMC) that must be rapidly discarded by the grower to avoid the contamination of new cultivation cycles and infestation by flies or other insects (Marques et al. 2014). Furthermore, after cultivation, uncovered SMC should not be stockpiled due to the risk of contaminating groundwater by leaching after rain events (Ribas et al. 2009).

Depending on the type of mushroom and substrate materials, the SMC can be directly used as animal feed because it is nutritionally rich in proteins, easy to digest and a stimulator of the immune system of animals (Sánchez 2004; Machado et al. 2007; Kwak et al. 2008; Azevedo et al. 2009).

Apart from use in animal feeds, compost produced by mushroom cultivation could be used directly as an organic fertilizer (Rinker 2002). However, under these conditions, the compost is not yet fully stable and can

✉ Diego Cunha Zied
dczied@gmail.com

1 Departamento de Biologia, Universidade Federal de Lavras (UFLA), CP 3037, Lavras, MG 37200-000, Brazil

2 Universidade Estadual Paulista (UNESP), Câmpus de Dracena, Rod. Cmte João Ribeiro de Barros, km 651, Bairro das Antas, Dracena, SP 17900-000, Brazil

3 Departamento de Agricultura, Universidade Federal de Lavras (UFLA), CP 3037, Lavras, MG 37200-000, Brazil

undergo changes if it is stored wet over long periods of time and may affect plant growth because of the presence of heavy metals, poor physical properties and excessive salts, that result in high electrical conductivity (Abad et al. 2001; Bustamante et al. 2008; Medina et al. 2009).

In addition to fertility and plant nutrition, SMC that is produced using the appropriate technology also has the advantage of containing a rich microbiota that can ensure the necessary balance to guarantee the phytosanitary requirements of plant cultures (Elo et al. 2000; Viji et al. 2003). It is also possible that some species of microorganisms have the potential to induce resistance in seedlings pre-grown on this substrate.

The use of microorganisms to induce plant resistance to pathogens of root and leaves is now well documented, showing the great potential of this strategy (Ishida et al. 2008; Fontenelle et al. 2011). The microbial community that is present in organic compost and in the substrate after cultivation of mushrooms might play an important role in controlling the microbiota in the rhizosphere region.

Therefore, this type of SMC could be an important source of microorganisms of different species with the potential to induce resistance against diseases. In addition, the mushroom species itself produce biotic resistance inducers, substances that elicit plant defense responses (Silva et al. 2013). Considering all this, the use of SMC can have a much greater potential than simply an organic fertilizer or soil conditioner.

In the present manuscript, tomato plants were chosen to evaluate the potential use of spent mushroom compost after cultivation of the mushroom *A. subrufescens* in different proportions in a mixture with a commercial substrate as growing media for tomato production in an organic system.

Materials and methods

Location of the experiment

The experiment was conducted under greenhouse conditions for the production of tomato seedlings and under field conditions for the production of tomatoes (*Solanum lycopersicum*) in the area located at 21°14′S and 45°00′W and at 920 m of altitude. The average annual temperature is 19.3 °C, and the average annual rainfall is 1411 mm, with a concentration of rainfall during summer. The average relative humidity is 78 %. The soil was an oxisol, and the soil showed good depth, a clayey texture, good water retention capacity, good aeration and a flat topography. The analysis of the soil is described in Table 1.

Table 1 Physicochemical characteristics of the soil at the experimental location

Characteristics	Values
pH in water (v:v; 1:2.5)	5.9
P (mg dm^{-3})	9.3
K (mg dm^{-3})	116
Ca^{2+} ($cmol_c$ dm^{-3})	3.4
Mg^{2+} ($cmol_c$ dm^{-3})	1.4
Al^{2+} ($cmol_c$ dm^{-3})	0.1
SB ($cmol_c$ dm^{-3})	5.1
CEC ($cmol_c$ dm^{-3})	8.0
Clay (g kg^{-1})	600
Silt (g kg^{-1})	220
Sand (g kg^{-1})	180
H + Al ($cmol_c$ dm^{-3})	2.9

SB sum of bases, *CEC* cation exchanged capacity

Procurement of substrates for the production of seedlings

Bioplant© a commercial product[1] was used as a substrate for the production of tomato seedlings, both as a control and in combination with the spent *A. subrufescens* compost (SASC) in different proportions, depending on each treatment. The SASC was obtained from the study carried out by Figueiredo et al. (2013), who used the CS10 strains (University of Lavras, Brazil), compost based on sugarcane bagasse, and coastcross hay with a casing layer of soil. In total, there were six treatments: Treatment 1–100 % Bioplant; Treatment 2–80 % Bioplant +20 % SASC; Treatment 3–60 % Bioplant +40 % SASC; Treatment 4–40 % Bioplant +60 % SASC, Treatment 5–20 % Bioplant +80 % SASC; Treatment 6–100 % SASC. The moisture contents of both substrates were determined after drying the samples in an oven at 60 °C for 48 h. The dry weight was determined by the weight difference between dry and wet weight of each sample (Bioplant and SASC) used in each treatment. The analyses of the SASC and chemical properties of the mixtures were done with colorimetric determination methods and are described in Tables 2 and 3 (Ansorena 1994; Pardo-Gimenez et al. 2010).

Production and evaluation of tomato seedlings

The seedlings were grown in a greenhouse covered with a plastic sheet and sided with white screens. The seeds were

[1] Composition: *Pinus* bark, aggregating agents, vermiculite, NPK and micronutrients with pH of 8.2 and e Electrical conductivity 0.96 mS cm^{-1}.

Table 2 Chemical composition of the spent *A. subrufescens* compost (SASC)

pH	P (mg dm^{-3})	K (mg dm^{-3})	Ca^{+2} (mg dm^{-3})	Mg^{+2} (mg dm^{-3})	Al^{+3} (mg dm^{-3})	H + Al^{+3} (mg dm^{-3})	CEC (mg dm^{-3})	*M* (mg dm^{-3})	*V* (%)	OM (%)	*P* (mg L^{-1})
5.0	0.9	624	2.0	0.7	0.0	2.1	6.4	0.0	67.2	63.0	1.5

CEC cation exchanged capacity, *M* percentage of Al saturation, *V* percentage of base saturation, *OM* Organic matter

Table 3 Chemical properties of the growing medium SASC based on tomato seedlings production

Treatment	pH	Electrical conductivity (mS cm^{-1})
T1: 0 % SASC	8.2	0.96
T2: 20 % SASC	7.6	1.5
T3: 40 % SASC	6.6	3.1
T4: 60 % SASC	6.5	4.13
T5: 80 % SASC	6.1	5.7
T6: 100 % SASC	6.0	5.51

sown in plastic trays with 200 cells, each containing the substrate of each of the six treatments, totaling 1200 seedlings. The seedlings were irrigated daily and were transferred to the field after 17 days, when they showed four true leaves.

At the transfer to the field, ten seedling samples from each of the six treatments were randomly selected to evaluate and quantify the level of growth of the tomato seedlings based on measurements and fresh weight of the seedlings. The following characteristics were analyzed: mass of the aerial portion, height of aerial portion, mass of root, length of root and total fresh weight.

Tomato cultivation from seedlings produced on different substrates

The tomato plants were cultivated with a spacing of 45 × 60 cm in three rows with staking. The experiment was a randomized block design with six repetitions, in which each plant is considered as a repetition. The buds were thinned when they were 3–5 cm long, and the tip was pruned when the last bunch had sprouted, reaching the desired amount of 4 bunches per plant.

To analyze the production of tomatoes, all the fruits from each plant were harvested in each of the four production flushes. The fruits were harvested when the "breaker" stage was attained, that is, when each tomato showed a reddish color on the apex in the weekly harvest. Only the commercially acceptable production was evaluated, and all damaged fruits were discarded.

The fruits were then evaluated with regard to the number of fruits per plant, the size according to the classification of the Ministry of Agriculture of Brazil (Table 4) and the total weight (tomato production, g plant^{-1}).

Regarding culture management, insecticides and fungicides were not used to control pests and diseases. Nutrients were also not supplemented through the use of fertilizers. Likewise, the weed growth was controlled manually rather than through the use of herbicides. Irrigation was performed daily from the moment of transplanting until the harvest period.

Table 4 Classification of tomato fruits with regard to size, according to the norms of the Ministry of Agriculture of Brazil

Classification	Fruit diameter (mm)
Extra-AA (coarse)	>52
Extra-A (medium)	47 < 52
Extra (small)	40 < 47
Especial (extra small)	33 < 40

Statistical analysis

All the parameters were subjected to regression analysis using the program SISVAR.

Results and discussion

Production of the tomato seedlings

In the production of the tomato seedlings, the use of different proportions of SASC resulted in a decreasing tendency of all the parameters (Figs. 1, 2, 3).

Considering the treatments with SASC, the best results were generally obtained with the treatments using 20 and 40 % SASC. The fresh mass of roots and size of the root were the parameters most negatively affected by the use of SASC, sharply decreasing with the addition of SASC, which may indicate a deficiency in the physical characteristics of the SASC.

Costa et al. (2012) used different substrates for the production of tomato seedlings. The results reported for all the tested parameters (plant height, stem diameter, total weight, mass of the shoots and mass of the root) were worse than those shown here. Steffen et al. (2010) tested substrates of rice husk and cow manure and also obtained worse results for plant height and fresh weight of the aerial portion than those obtained in this work with the use of SASC.

Medina et al. (2009) evaluated the use of SMC of *Agaricus bisporus* and *Pleurotus ostreatus* for the seedlings' production of different vegetables species. None of the treatments containing SMC matched the peat (control) as growing medium for tomato seedlings production, but SMC from *A. bisporus* was better than *P. ostreatus* as component of the growing media. In the present study, although the best results were obtained with the control substrate (100 % Bioplant), all the treatments with different ratios of SASC achieved results as good as or even better than those obtained with other substrates in other studies.

Cultivation of tomatoes from seedlings grown on substrates based on SASC

In all flushes of tomato production, a positive effect was observed when the seedlings were produced with SASC in comparison to the control (Fig. 4). For each flush, variations among the treatments regarding the best production were observed. However, the treatment with 100 % SASC generally led to the best production in two flushes and the second-best production in the remaining two flushes. Considering the total production, the seedlings produced in the substrate with 100 % SASC had the greatest tomato production (Fig. 5). It is worth noting that, in addition to the treatment with 100 % of SASC, all the other treatments

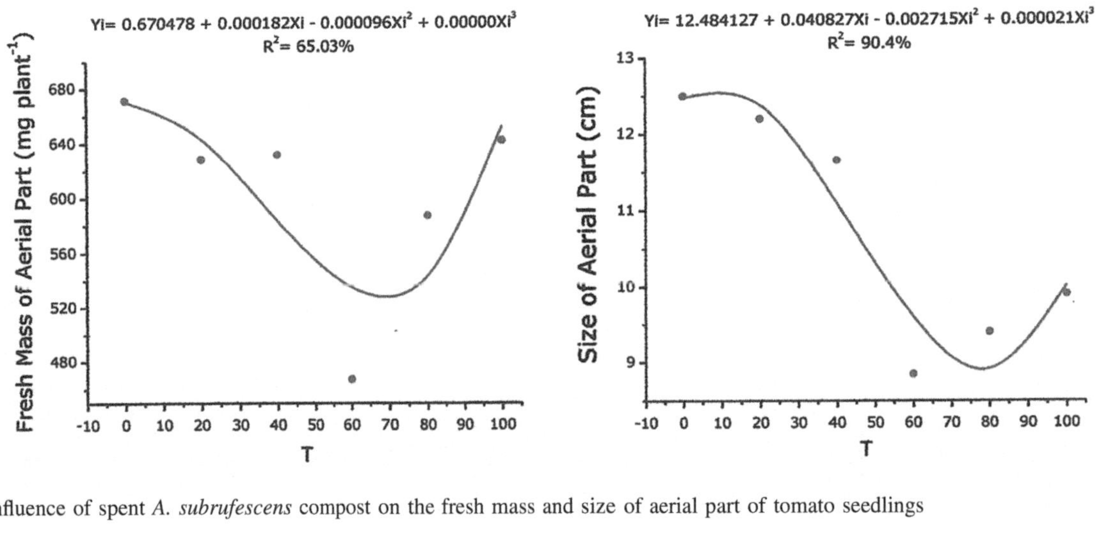

Fig. 1 Influence of spent *A. subrufescens* compost on the fresh mass and size of aerial part of tomato seedlings

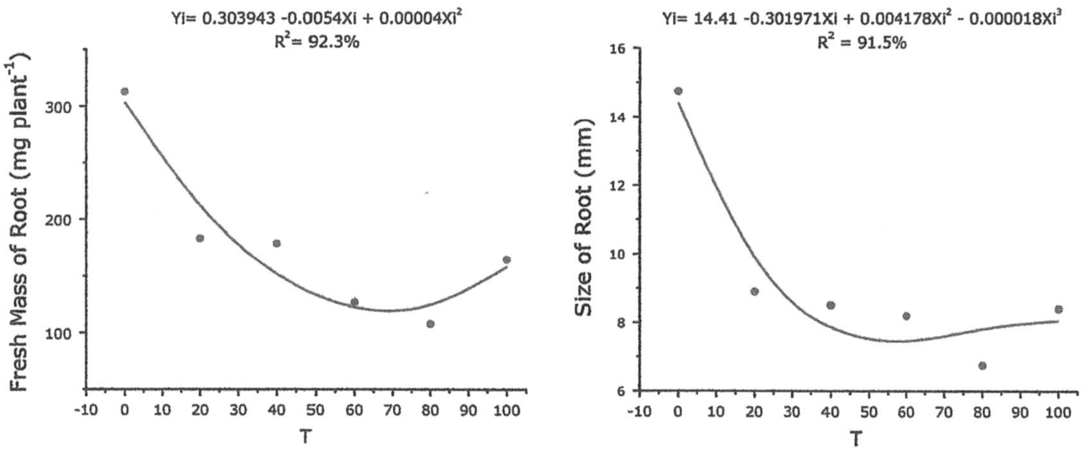

Fig. 2 Influence of spent *A. subrufescens* compost on the fresh mass and length of roots of tomato seedlings

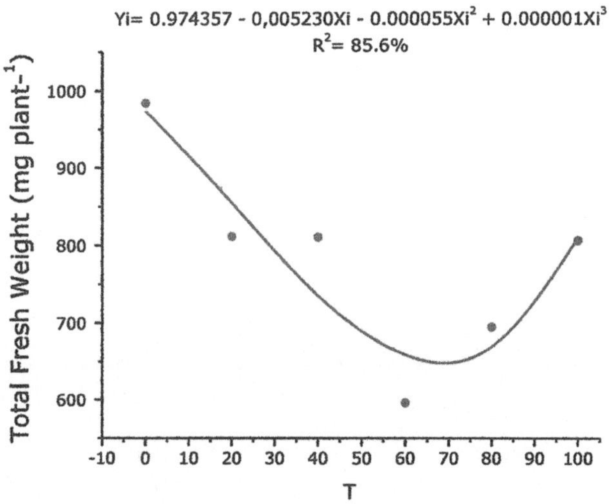

Fig. 3 Influence of spent *A. subrufescens* compost on the total fresh weight of tomato seedlings

led to a larger production than the control, showing the positive effect of this substrate.

In comparison to the results obtained with conventional crops, the tomato production was lower in this study because pesticides and fertilizers were not used. The maximum production (53 t ha^{-1}) was obtained when treating with 100 % SASC resulting in a yield of 1.43 kg of tomatoes per plant. When comparing organic and conventional cultivation, Bettiol et al. (2004) concluded that organic farming production corresponds to only 36.5 % of conventional production. Wamser et al. (2012) studied the productivity of tomato hybrids in super-high-density cultivation fields (44,444 plants ha^{-1}), reporting a maximum yield of 170.7 t ha^{-1}, whereas in traditional cultivation (22,222 plants ha^{-1}), the maximum yield was 128.5 t ha^{-1}. Assuming that the proportion of 36.5 % proposed by Bettiol et al. (2004) is correct, yields of 62.2 and 46.9 t ha^{-1} should be expected for super-dense and traditional cultivation, respectively, under an organic cultivation system. Therefore, the maximum yield obtained in the present work, 53 t ha^{-1}, falls in the expected range for organic system.

In hydroponic cultivation, Albuquerque Neto and Peil (2012) reported a maximum total production of 54.6 t ha^{-1}. The highest total yield reported by these authors is therefore practically the same as the maximum yield obtained in the present work. In addition to the numbers that were presented above, it is also important to emphasize that no fertilization procedures of any kind were used in the present work, either before or during the crop, because we aimed to evaluate the behavior of seedlings in the simplest crop management conditions possible. This information is important because despite the need to avoid chemical fertilizers in organic farming, the use of organic compost, green manure, bone meal and so forth is recommended.

Costa et al. (2011) evaluated the productivity of tomato cultivars tolerant to high temperature in an organic farming system using organic manure, bone meal and biofertilizer, with a maximum total production of 36.3 t ha^{-1}. Marouelli et al. (2011) evaluated the effect of this type of irrigation system (dripping vs. sprinkler) on organic tomato cultivation using green fertilization with Crotalaria (*Crotalaria pumila*), applying organic compost and thermophosphate as well as applying Bordeaux mixture and a biopesticide (*Bacillus thuringiensis*). Under these conditions, these authors obtained a total production of 58.8 and 64.8 t ha^{-1} for the dripping and the sprinkler systems, respectively. These results were better than those obtained in this study (53 t ha^{-1}). Nonetheless, it is important to emphasize that the authors used green fertilization, organic fertilizer and thermophosphate.

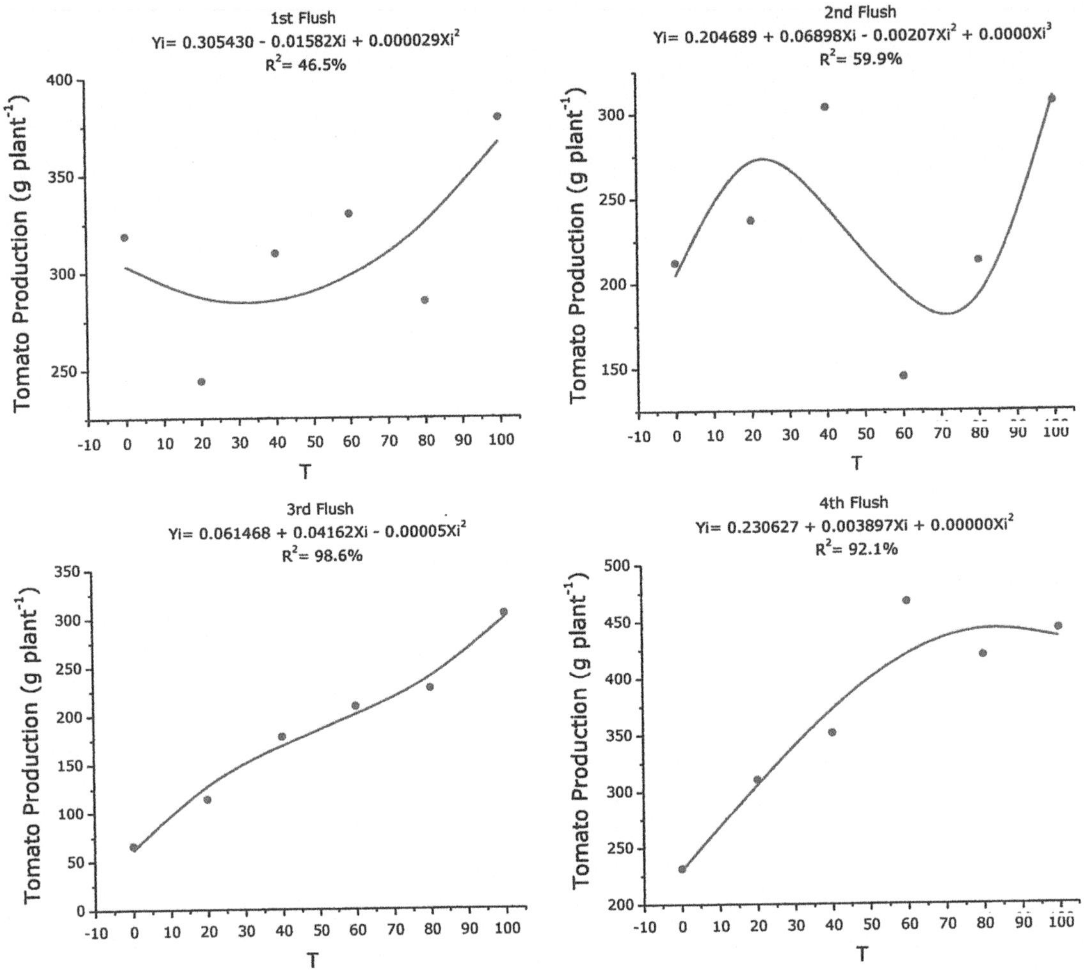

Fig. 4 Flushes of tomato production from the seedlings grown using different proportions of spent *A. subrufescens* compost

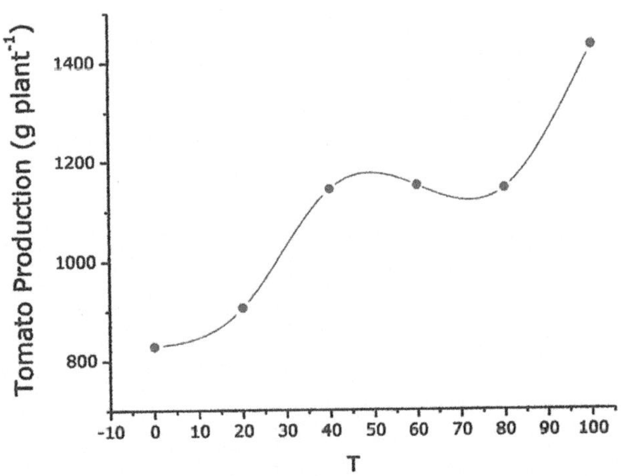

Fig. 5 Total tomato production from seedlings grown using different proportions of spent *A. subrufescens* compost

Spent mushroom compost and *Trichoderma harzianum* were effective when used against *Meloidogyne javanica* in tomato under field conditions in Dargai in Malakand Agency, Pakistan (Irfan-ud-Din et al. 2005). The author conclude that SMC stimulated flowering, plant height, fresh and dry shoot weight and increased yield by 101.9, 61.5, 27.9, 38.3 and 102.5 %, respectively; moreover, it reduced the number of galls/plant by 73.5 %. Modolon et al. (2012) used homeopathic preparations for pest control during the organic cultivation of tomatoes, also employing green fertilization procedures and applying compost. Under these conditions, the largest total production obtained was 15.5 t ha^{-1} in field cultivation and 29.8 t ha^{-1} in greenhouse cultivation.

The treatment without SASC induced a higher proportion of small and very small fruit in the tomato production, whereas the treatment with 100 % SASC led to a higher proportion of large and medium-sized fruit (Table 5).

Therefore, the use of this substrate in the production stage of the seedlings resulted in more vigorous plants, leading not only to greater yields but also to fruits with greater size compared to the commercial control. Interestingly, the treatment with 80 % SASC led to the highest

Table 5 Influence of spent *A. subrufescens* compost on the classification of the total tomato production regarding the fruit size

Treatment	Classification of tomatoes								Total number of fruits
	Extra small		Small		Medium		Coarse		
	N[a]	%[b]	*N*	%	*N*	%	*N*	%	
Control	12	12.0	49	49.0	35	35.0	4	4.0	100
20 % SMS	2	2.1	51	53.7	37	38.9	5	5.3	95
40 % SMS	0	0	55	43.3	72	56.7	0	0	127
60 % SMS	0	0	55	44.7	68	55.3	0	0	123
80 % SMS	3	2.1	73	51.4	66	46.5	0	0	142
100 % SMS	0	1.5	40	31.0	73	56.6	14	10.8	129

The values are presented as a percentage of the total number of fruits harvested

[a] Number of fruits

[b] Percentage of fruits with size classification

number of fruits among all the treatments. However, compared to the treatment with 100 % SASC, the production consisted almost exclusively of small and very small fruit, with not a single large fruit harvested.

Conclusions

We can conclude that the use of SASC for seedling production led to a higher, or at least equal, total tomato production compared to previously reported production levels in organic cultivation systems with green, organic and other types of fertilization. These results demonstrate the great potential of SASC for use in organic tomato production because of the better quality of harvested fruit.

Acknowledgments We would like to thank the Fundação de Amparo à Pesquisa do Estado de Minas Gerais (FAPEMIG) and Fundação de Amparo à Pesquisa do Estado de São Paulo (FAPESP-2014/12129-0) and Coordenação de Aperfeiçoamento de Pessoal de Nível Superior (CAPES) for financial support.

References

Abad M, Noguera P, Bures S (2001) National inventory of organic wastes for use as growing media for ornamental potted plant production: case study in Spain. Bioresour Technol 77:197–200

Albuquerque Neto AAR, Peil RMN (2012) Biological productivity of tomato plant genotypes in hydroponic systems during fall/winter. Hortic Bras 30(4):613–619

Ansorena J (1994) Sustratos. Propiedades y caracterización. Mundi-prensa S.A., Madrid, p 172

Azevedo RS, Ávila CLS, Bertechini G, Schwan R (2009) Utilization of the spent substrate of *Pleurotus sajor caju* mushroom in broiler chicks ration and the effect on broiler chicken performance. Acta Scient Anim Sci 31(2):139–144

Bettiol W, Ghini R, Galvão JAH, Siloto RC (2004) Organic and conventional tomato cropping systems. Sci Agric 61(3):253–259

Bustamante MA, Paredes C, Moral R, Agulló E, Pérez-Murcia MD, Abad M (2008) Composts from distillery wastes as peat substitutes for transplant production. Resour Conserv Recy 52:792–799

Costa CA, Silva AC, Sampaio RA, Martins ER (2011) Productivity of determinate growth tomato lines tolerant to heat under the organic system. Hortic Bras 29(4):590–593

Costa E, Leal PAM, Benett CGS, Benett KS, Salamene LCP (2012) Production of tomato seedlings using different substrates and trays in three protected environments. Eng Agric 32(5):822–830

Dias ES, Zied DC, Rinker DL (2013) Physiologic response of *Agaricus subrufescens* using different casing materials and practices applied in the cultivation of *Agaricus bisporus*. Fungal Biol 117(7):569–575

Elo S, Maunuksela L, Salkinoja-Salonen M, Smolander A, Haahtela K (2000) Humus bacteria of Norway spruce stands: plant growth promoting properties and birch, red fescue and alder colonizing capacity. FEMS Microbiol Ecol 31:143–152

Figueiredo VR, Martos ET, Siqueira FG, Maciel WP, Silva R, Rinker DL, Dias ES (2013) Microbial inoculation during composting improves productivity of sun mushroom (*Agaricus subrufescens* Peck). Afr J Microbiol Res. doi:10.5897/AJMR2013.5944

Fontenelle ADB, Guzzo SD, Lucon CMM, Harakawa R (2011) Growth promotion and induction of resistance in tomato plant against *Xanthomonas euvesicatoria* and *Alternaria solani* by *Trichoderma* spp. Crop Prot 30(11):1492–1500

Irfan-ud-Din Saifullah, Khan H, Khattak B (2005) Biological control of *Meloidogyne javanica* (Treub) Chitwood with *Trichoderma harzianum* Rifai and spent mushroom compost in tomato under field conditions. Pak J Phytopathol 17(2):144–145

Ishida AKN, Souza RM, Resende MLV, Cavalcanti FR, Oliveira DL, Pozza EA (2008) *Rhizobacterium* and acibenzolar-S-methyl (ASM) in resistance induction against bacterial blight and expression of defense responses in cotton. Trop Plant Pathol 33(1):27–34

Kwak WS, Jung SH, Kim YI (2008) Broiler litter supplementation improves storage and feed-nutritional value of sawdust-based spent mushroom substrate. Bioresour Technol 99(8):2947–2955

Machado AMB, Souza Dias E, Santos EC, Freitas RTF (2007) Spent mushroom substrate of *Agaricus blazei* in broiler chicks diet. R Bras Zootec 36(4):1113–1118

Marouelli WA, Medeiros MA, Souza RF, Resende FV (2011) Production of organic tomato irrigated by sprinkling or dripping in single cultivation or in consortium with coriander. Hortic Bras 29(3):429–434

Marques ELS, Martos ET, Souza RJ, Silva R, Zied DC, Dias ES (2014) Spent mushroom compost as a substrate for the production of lettuce seedlings. J Agric Sci 6(7):138–143

Medina E, Paredes C, Pérez-Murcia MD, Bustamante MA, Moral R (2009) Spent mushroom substrates as component of growing media for germination and growth of horticultural plants. Bioresour Technol 100(18):4227–4232

Modolon TA, Boff P, Boff MIC, Miquelluti DJ (2012) Homeopathic and high dilution preparations for pest management to tomato crop under organic production system. Hortic Bras 30(1):51–57

Pardo-Gimenez A, Zied DC, Pardo-Gonzalez JE (2010) Utilización de compost agotado de champiñón como capa de coberturas en nuevos ciclos de producción. Pesq agropec bras 45(10):1164–1171

Ribas LCC, Mendonça MM, Camelini CM, Soares CHL (2009) Use of spent mushroom substrates from *Agaricus subrufescens* (syn. *A. blazei, A. brasiliensis*) and *Lentinula edodes* productions in the enrichment of a soil-based potting media for lettuce (*Lactuca sativa*) cultivation: growth promotion and soil bioremediation. Bioresour Technol 100(20):4750–4757

Rinker DL (2002) Handling and using "spent" mushroom substrate around the world. Mushroom biology and mushroom products. Universidad Autónoma del Estado de Morelos, Cuernavaca, pp 43–60

Sánchez C (2004) Modern aspects of mushroom culture technology. Appl Microbiol Biot 64(6):756–762

Silva RF, Pascholati SF, Bedendo IP (2013) Induced resistance in tomato plants to *Clavibacter michiganensis* subsp. *Michiganensis* by *Lentinula edodes* and *Agaricus subrufescens* (syn. *Agaricus brasiliensis*). J Plant Pathol 95(2):285–297

Steffen GPK, Antoniolli ZI, Steffen RB, Machado RG (2010) Rice husk and cow manure as substrates for worm multiplication and production of tomato and lettuce seedlings. Acta Zool Mex 26(2):333–343

Viji G, Uddin W, Romaine CP (2003) Suppression of gray leaf spot (blast) of perennial ryegrass turf by *Pseudomonas aeruginosa* from spent mushroom substrate. Biol Control 26:233–243

Wamser AF, Mueller S, Suzuki A, Becker WF, Santos JP (2012) Productivity of tomato hybrids subjected to super-dense cultivation. Hortic Bras 30:168–174

Zied DC, Pardo Giménez A, Pardo González JE, Souza Dias E, Carvalho MA, Minhoni MTA (2013) Effect of cultivation practices on the β-Glucan content of *Agaricus subrufescens* basidiocarps. J Agric Food Chem 62(1):41–49

27

Foliar application of liquid biofertilizer of brown alga *Stoechospermum marginatum* on growth, biochemical and yield of *Solanum melongena*

S. Sivasangari Ramya[1] · N. Vijayanand[2] · S. Rathinavel[3]

Abstract

Background Seaweed extracts are used as nutrient supplements, biostimulants and or biofertilizers as an alternative to chemical fertilizers in agriculture. The study was set up to evaluate the biofertilizing efficiency of liquid extracts of brown marine alga *Stoechospermum marginatum* on growth, biochemical and yield of brinjal, a vegetable crop. To achieve the objectives, liquid extracts at different concentrations were prepared and applied as foliar spray on the brinjal seedlings raised in experimental pots maintained under natural conditions.

Results After 30 and 180 days, the growth and biochemical parameters and yield attributes were monitored, respectively. The results exhibited that shoot and root length, total fresh and dry weight, leaf area and the content of moisture, photosynthetic pigments, protein, amino acids, reducing sugar, ascorbic acid and nitrate reductase activity were found to be enhanced in the leaves of brinjal plants which received 1.5 % of *Stoechospermum marginatum* extracts. Similarly, number of fruits and fruit weight were also increased at lower concentration only (1.5 %). In contrast, liquid extracts at high concentration (5 %) was found to have inhibitory effect on brinjal plants as compared to the control sprayed with water.

Conclusions The study evidence that lower concentrations were found to show promoting effect on growth and productivity of brinjal plants. The fertilizing efficiency of liquid extracts of testing marine alga might me due to the presence of micro and macro nutrients, growth hormones and vitamins at preferential levels. It can be concluded that seaweed extracts could be used as eco-friendly liquid biofertilizer to substitute chemical fertilizer and also play a pivotal role in organic farming practices toward sustainable agriculture.

Keywords Seaweed extract · Liquid biofertilizer · *Stoechospermum marginatum* · Growth · Biochemical · Yield · Brinjal

Electronic supplementary material The online version of this article (doi:10.1007/s40093-015-0096-0) contains supplementary material, which is available to authorized users.

✉ N. Vijayanand
vijayapsac@gmail.com

[1] Department of Biochemistry, Sri Sarada Niketan College for Women Amaravathipudur, Karaikudi, India

[2] Department of Botany, Arumugam Pillai Seethai Ammal College, Tiruppattur, India

[3] Centre for Research in Botany, Saraswathi Narayanan College (Autonomous), Madurai, India

Introduction

In modern agriculture, chemical fertilizers have degraded the fertility of soil making it acidic and rendering it unsuitable for raising crop plants. The intensive use of inputs has led to severe health and environmental hazards viz., soil erosion, water contamination, pesticide poisoning, falling ground water table, water logging and depletion of biodiversity. The practice of chemical farming has also put the long-run sustainability of Indian agriculture and the survival of the farming community at risk. In recent years, the use of natural seaweed as fertilizer has allowed for partial substitution of conventional synthetic fertilizer (Khan et al. 2009; Zodape et al. 2010). In addition, a number of commercial seaweed extract products are available for use in agriculture and horticulture. A number of seaweeds used a liquid fertilizer by applying them as foliar spray, soil drench, or in granular/powder form as soil conditioners and manure (Thirumaran et al. 2009). These extracts are marketed as liquid biofertilizers because the chemical analyses of seaweeds and their extracts have revealed the presence of a wide variety of plant growth-promoting substances such as auxins, cytokinins and betaines (Khan et al. 2009).

Seaweeds are one of the important marine bio-resources which are nowadays termed as fantastically promising plants. Seaweeds and their derivatives are used in agriculture as potential plant growth regulators. Moreover, manures of seaweeds are used as a soil amendment in agriculture in many parts of the world (Eyras et al. 1998) and are an inexpensive local resource in coastal agricultural areas. Seaweed contains all major and minor plant nutrients, trace elements, vitamins, auxins and other bioactive substances. The growth promoting efficiency of fertilizing efficiency of extracts of several marine algae was evaluated in the cereals, pulses and vegetable crops (Kalidass et al. 2010; Sasikumar et al. 2011; Zodape et al. 2011; Bai et al. 2013; Parthiban et al. 2013; Kalaivanan et al. 2012; Hernández-Herrera et al. 2014). An attempt has been made to assess the biofertilizing potential of liquid extracts of marine brown alga *Stoechospermum marginatum* on important vegetable crop *Solanum melongena.*

Materials and methods

Preparation of liquid extracts of seaweeds

The marine alga *Stoechospermum marginatum* (C.Agardh) Kützing 1843:339 was collected from Mandapam (Lat 9° 45′ N; Long 79° 15′ E) located in South east coast of Tamil Nadu (Supplementary Fig. 1). The alga was brought to the laboratory and washed thoroughly in tap water for 3 or 4 times to remove all epiphytes, sand particles and associated fauna. The wet weight of sample of collected algal samples was taken and shade dried. After shade drying, it was cut into small pieces and kept in hot air oven for one day at 60 °C. Then the dried sample of seaweed (1 kg) was further made into small pieces and powdered. The powdered sample of seaweed was mixed with water (20 litres) in the proportion of 1:20. Then it was boiled for one hour. After one hour, the mixture was squeezed and filtered through muslin cloth (Bhosle et al. 1975). The obtained extract was designated as stock solution and was used to prepare different concentrations viz; 0.5, 1.0, 1.5, 2.0, 2.5 and 5.0 % by mixing appropriate proportions of seaweed liquid extract (LE) with sterilized distilled water.

Preparation of pot study

Healthy seeds of brinjal plants (Variety CO-2, accession number—EG 203) were purchased from Agriculture College and Research Institute, Madurai. They were surface sterilized with 0.1 % mercuric chloride and then sown in earthenware pots (9 cm dia) filled with a sterilized standard soil mix supplemented with sufficient quantity of NPK. Ten seeds were sowed in each pot. The seed to seed distance in pot was maintained at 3–5 cm and pots were watered regularly. After 10 days, potted plants were treated with different concentrations of SLE in the form of foliar spray. About 50 ml of different concentrations of extracts was applied at intervals of 3 days. In treating plants, growth parameters viz., shoot length, root length, total height, total fresh and dry weight, leaf area, moisture content and relative water content were determined. Photosynthetic pigments (Arnon 1949), protein content (Lowry et al. 1951), reducing sugar (Nelson 1944), ascorbic acid (Roe 1954) and nitrate reductase activity (Jaworski et al. 1971) were assessed in the leaves of treated plants. Similarly, yield parameters such as number of fruits and fruit weight were also observed. Growth and biochemical parameters were recorded in 30-day-old treated and control plants. After 70 days, yield characters were observed. Routinely, the brinjal plants irrigated with water alone served as control. All pot experiments were done in four replicates each under natural uniform conditions.

Physico-chemical and hormone analyses of seaweed extracts of *Stoechospermum marginatum*

The physical observations such as colour and pH were made using standard methods. The presence of elements such as copper, manganese, zinc, iron, potassium, magnesium, cobalt and sodium were estimated using Atomic Absorption Spectrophotometer (Humpshires 1956). Further, liquid extract of seaweed was also subjected for estimation of auxin (Gordon and Paleg 1957), gibberellin (Graham and Henderson 1961) and cytokinin (Syono and Torrey 1976) using standard methods.

Statistical analysis

Data were subjected to one-way ANOVA and means were separated by Duncan's test ($P < 0.05$, $n = 5$). Statistical analysis was carried out using IRRISTAT ver. 4.0 (IRRI, Manila, Phillipines) (Duncan 1965).

Results and discussion

Physio chemical analysis of liquid extracts of *S. marginatum*

The mineral analyses of liquid extract of our experimental brown marine alga revealed the presence of potassium

Table 1 Physio-chemical and hormone analyses of liquid extract of *S. marginatum*

Physical parameters	
Colour	Brown
pH	6.3
Chemical parameters	
Copper	3.014
Manganese	1.53
Zinc	1.58
Iron	0.500
Potassium	1.070
Magnesium	17.31
Cobalt	0.103
Sodium	5.77
Growth hormones	
Auxin	3.5
Cytokinin	9.2
Gibberellin	5.5

All the parameters given are in mg/L except colour and pH

(1.070 mg/l), copper (3.014 mg/l), manganese (1.53 mg/l), zinc (1.58 mg/l), iron (0.50 mg/l), cobalt (0.103 mg/l), sodium (5.77 mg/l) and magnesium (17.31 mg/l) in appreciable level. Among the elements estimated, magnesium (17.31 mg/l) was found to be abundant in the extract. Similarly, in case of phytohormone analysis, cytokinin (9.2 mg/l) was found to be more when compared to auxin (3.5 mg/l) and gibberellins (5.5 mg/l) (Table 1).

The presence of phytohormones is in accordance with the earlier findings that reported auxins in the extracts of *Ascophyllum nodosum* (Sanderson and Jameson 1986) and cytokinins in the extracts of *Ulva sp.* (Sekar et al. 1995).

Growth parameters of brinjal plants

In next experiments, we applied liquid extracts of *S. marginatum* on brinjals. The results have shown that extracts generally increased the rate of growth and physiology of plant objects. Growth parameters such as total height (30 %), total fresh (150 %) and dry weight (125 %), leaf area (61 %) and moisture content (56 %) were enhanced when 1.5 % of *S. marginatum* was given to brinjal plants. Higher concentrations (above 1.5 %) were found to show retarding effect on all the growth parameters (Table 2; Figs. 1 and 2).

Plants treated with different seaweed extracts exhibited wide range of responses that have been well documented in a number of reviews. In particular, seaweed extract concentrations were found to be effective in *Brassica nigra* (Kalidass et al. 2010), *Abelmoschus esculentus* (Sasikumar et al. 2011), *Lycopersicon esculentum* (Zodape et al. 2011), *Vigna radiata* (Bai et al. 2013; Parthiban et al. 2013), *Vigna mungo* (Kalaivanan et al. 2012), *Solanum lycopersicum* (Hernández-Herrera et al. 2014), *Mangifera indica* (Ahmed et al. 2013) and *Fagopyrum esculentum* (Anisimov et al. 2013). On the contrary, it has also been reported that concentration at 20 % of brown alga *Sargassum wightii* (Jothinayagi and Anbazhagan 2009) and red alga *Rosenvingea intricata* (Thirumaran et al. 2009) promoted shoot length, root length, fresh and dry weight of *Abelmoschus esculenthus* and *Cyamopsis tetragonolaba,* respectively. Growth enhancement by seaweed extracts may be due to components such as macro and micro elements, amino acids, vitamins, cytokinins, auxins and abssisic acid (ABA)-like growth substances which affect cellular metabolism in treated plants leading to enhanced growth and crop yield (Ordog et al. 2004; Durand et al. 2003). Similarly, in our study, the increased growth of brinjal plants could be associated with the occurrence of some growth promoting substances present in the seaweed extract of *S. marginatum* as in other macro algal extract (Mooney and Van Staden 1986). The growth hormones play an imperative role in enhancement of cell size and cell division and together they complement each other as cytokinins are effective in

Table 2 Influence of liquid extracts of *Stoechospermum marginatum* on growth characteristics of brinjal

Seaweed extract treatments	Shoot length (cm)	Root length (cm)	Total height (cm)	Total fresh wt (mg)	Total dry wt (mg)	Leaf Area (mm^2)	Moisture content (%)
Control	8.725b	12.250a	20.00ab	1.015a	0.400a	42.115b	41.533a
0.5 %	9.325b	13.050ab	21.400bc	1.130bc	0.492b	44.183b	54.685bc
1.0 %	10.000c	13.400bc	23.33bc	1.620de	0.620c	50.790c	63.645bc
1.5 %	11.400cd	15.250d	26.65bc	2.595f	0.962d	68.165d	64.825c
2.0 %	10.300c	14.450cd	24.750c	1.800e	0.605c	58.400d	64.808c
2.5 %	8.950b	14.200cd	23.150bc	1.365cd	0.515b	43.425b	53.720b
5.0 %	6.425a	12.300a	18.725a	0.915ab	0.400a	35.450a	55.785bc

Means sharing within the rows are significantly different ($P \leq 0.05$ level)

Different letters followed in each row are statistically significant based on DMRT

Fig. 1 Influence of liquid extract of *Stoechospermum marginatum* on the total plant height of brinjal

Fig. 2 Influence of liquid extract of *Stoechospermum marginatum* on the leaf area of brinjal

shoot formation and auxin in root development, while micronutrients improve soil health.

Biochemical characteristics of brinjal plants

In brinjal plants, there was a noticeable increase in biochemical parameters when 1.5 % of seaweed liquid extract of *S. marginatum* applied to brinjal plant. The content of total chlorophyll pigments, (77 %), protein (38 %), reducing sugar (201 %) and ascorbic acid (36 %) and nitrate reductase activity (159 %) was enhances when the brinjal plants were treated with SLE at 0.5, 1.0 and 1.5 % concentrations of *S. marginatum* (Table 3). However, steep decline was recorded in plants that received 2.0, 2.5 and 5.0 % of *S. marginatum* extracts.

This is in accordance with the earlier reports that lower concentrations of seaweed extracts enhanced the biochemical constituents in *Cajanus cajan* (Erulan et al. 2009), *Brassica nigra* (Kalidass et al. 2010), *Citrullus lanatus* (Abdel-Mawgoud et al. 2010), *Trigonella foenum-graecum* (Pise and Sabale 2010), *Solanum melongena* (Bozorgi 2012) and *Abelmoschus esculentus* (Sasikumar et al. 2011). In our study, reducing sugar was found to be increased twofold when 1.5 % of *S. marginatum* was given to brinjal plants. The increase in reducing sugar (201 %) may be due to the presence of magnesium (Table 1) which could have triggered chlorophyll synthesis and thereby subsequently increasing the photosynthetic rate in the plants. The increase in photosynthetic pigments may be also be due to the presence of betaines (Blunden et al.

Table 3 Influence of liquid extract of *Stoechospermum marginatum* on biochemical characteristics of brinjal

Seaweed extract treatments	Chl-a (mg/g/fr.wt)	Chl-b (mg/g/fr.wt)	Total chlorophyll (mg/g/fr.wt)	Protein (mg/g/fr.wt)	Reducing sugar (mg/g/fr.wt)	Ascorbic acid (mg/g/fr.wt)	NRA (μ moles No^{-2}/gm fr.wt/hr)
Con	0.5225a	0.4200a	0.9425a	19.200b	41.650a	0.568a	0.820a
0.5 %	0.6500a	0.4450b	1.095a	24.550b	55.850b	0.656ab	1.090b
1.0 %	0.6650ab	0.515c	1.180c	28.250c	62.650c	0.700c	1.635d
1.5 %	0.975d	0.710d	1.685d	33.30d	84.10d	0.775bc	2.125e
2.0 %	0.805c	0.530b	1.335b	26.625bc	66.80c	0.637ab	1.725d
2.5 %	0.650c	0.442b	1.192a	21.350a	51.50b	0.593a	1.345c
5.0 %	0.590b	0.3 15a	1.052a	19.90a	42.10a	0.582a	0.970ab

Means sharing within the rows are significantly different ($P \leq 0.05$ level)

Different letters followed in each row are statistically significant based on DMRT

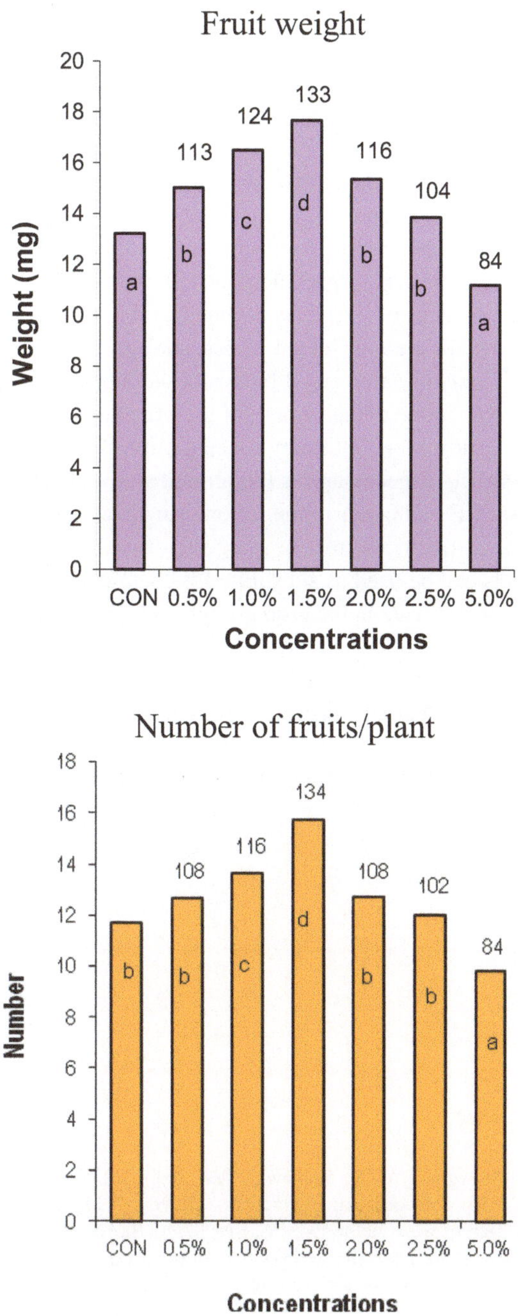

◀ **Fig. 3** Influence of liquid extracts of *Stoechospermum marginatum* on fruit weight and number of fruits/plant of brinjal. *Numbers* present above each *bar* are percent over control. *Different letters* inside each *bar* is statistically significant based on DMRT ($P \leq 0.05$ level)

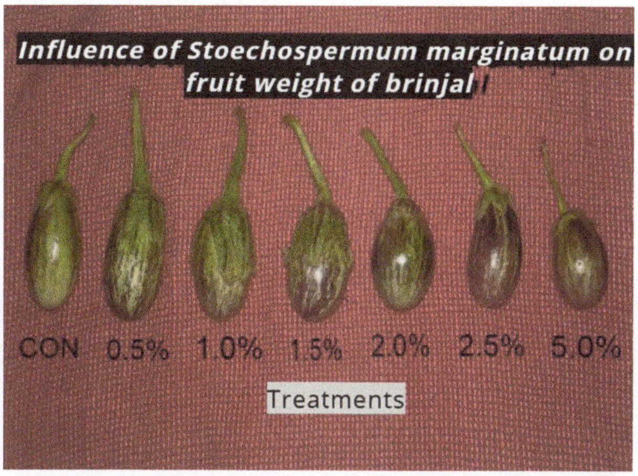

Fig. 4 Influence of liquid extract of *Stoechospermum marginatum* on the fruit weight of brinjal

1997), increase in number and size of the chloroplast and better grana development (Atzmon and Van Staden 1994) in the SLE-treated plants. Moreover, the increase in the protein content and nitrate reductase activity at lower concentrations of SLE confirmed the efficiency of foliar spray as it enhanced the absorption of most of the necessary elements by the seedlings. The increase in chlorophyll content could also be a result of reduction in chlorophyll degradation, which might be caused in part by betaines in the seaweed liquid extract (Whapman et al. 1993). In addition, in a study, 1 % *Ulva lactuca* extract along with 50 % recommended rate of chemical fertilizers enhanced the content of protein, carbohydrate and lipid in *Tagetus erecta* (Sridhar and Rengasamy 2010). However, it has been reported that seaweed liquid fertilizer at 10 % extracted from brown alga *Sargassum wightii* increased the content of photosynthetic pigments, protein and total

Table 4 Influence of liquid extracts of *Stoechospermum marginatum* on fruit weight and number of fruits/plant of brinjal

Seaweed treatments	Control	0.5 %	1.0 %	1.5 %	2.0 %	2.5 %	5.0 %
Fruit weight (g)	13.22 ± 0.14a	15.02 ± 0.13b	16.5 ± 0.34c	17.67 ± 0.34d	15.37 ± 0.30b	13.87 ± 0.09b	11.20 ± 0.30a
Number of fruits	11.72 ± 0.25b	12.70 ± 0.26b	13.62 ± 0.10c	15.75 ± 0.26d	12.75 ± 0.20b	12.0 ± 0.14b	9.8 ± 0.10a

Results are mean ± SD ($n = 5$). Means sharing within the rows are significantly different ($P \leq 0.05$ level)

Different letters followed in each row statistically significant based on DMRT

sugars in *Vigna radiata* (Sivasankari et al. 2006) and SLE of *Rosenvingea intricata* at 20 % enhanced the photosynthetic pigments and carotenoids in *Cyamopsis tetragonoloba* (Thirumaran et al. 2009).

Yield attributes of brinjal plants

In many crops, yield is associated with the number of flowers at maturity. As the onset and development of flowering and the number of flowers produced are linked to the developmental stage of plants, seaweed extracts probably encourage flowering by initiating robust plant growth (Khan et al. 2009). In our experiments, different concentrations of *S. marginatum* extracts showed differential responses in the yield characters also. Significant increase in number of flowers (34 %) and fruit weight (33 %) was noticed when 1.5 % of *S. marginatum* was given as foliar spray. (Fig. 3a, b). However, decrease in number and weight of the fruits was observed in higher concentrations (above 1.5 %) Fig. 4.

Seaweed extract at low concentrations not only increase the vegetative growth of the plant but also triggers the early flowering and fruiting in crops. In our study, reports showed that treatment of *Ablemoschus esculentus* with seaweed extract increased the length (31.7 %), diameter (18.2 %) and yield (37.4 %) of fruits as compared to (Zodape et al. 2008). The liquid extract of *Hypnea musciformis, Laurencia obtusa, Padina tetrastromatica and Stoechospermum mariginatum* acted as a biostimulant to increase the productivity and quality of tea (Thevanathan et al. 2005). Zodape et al. (2011) observed that foliar application of liquid extracts of *Kappaphycus alvarezii* promoted the yield potency of *Lycopersicon esculentum*, a vegetable crop. Bai et al. (2013) applied the liquid extract of *Padina pavonia* and got maximum yield of pulses. Moreover, in potato, Siddagangaiah et al. (2010) recorded that per cent tuber yield, tuber dry matter content, dry weight of haulms and tubers were enhanced when potato plants received 0.5 % and 0.4 % of Phyton-T, a commercial seaweed extracts. Yield increase in seaweed-treated plants is thought to be associated with the hormonal substances present in the extracts, especially cytokinins (Featonby-Smith and Van Staden 1984). Cytokinins in vegetative plant organs are associated with nutrient partitioning, whereas in reproductive organs, high levels of cytokinins may be linked with nutrient mobilization. It may also be attributed with the process of fruit ripening that generally causes an increase in transport of nutrient resources. Moreover, the phytohormones play a significant role in retention of flowers and fruits during the reproductive phase of crops and thus promoted crop productivity (Table 4).

Conclusions

Our results show that enhancement of growth, biochemical and yield parameters of brinjal plant might be due to the presence of differential level of micro and macro elements, growth hormones, trace elements, vitamins, etc., in the liquid extracts of *S. marginatum.* Cytokinin and magnesium, which are considered as essential growth promoting constituents in chlorophyll biosynthesis might have played a key role in the enhancement of growth and physiology of brinjal. However, optimum concentration of seaweed liquid extracts is necessary as the present study exhibited that 1.5 % SLE had better influence on growth and productivity of brinjal plants. It may be suggested that liquid seaweed extracts can be used as substitute for chemical fertilizers to improve the sustainable crop growth and yield. Further, the study also emphasizes that the application of seaweed liquid extracts of seaweeds can be effectively used as eco-friendly approach to organic farming.

Conflict of interest The authors declare that they have no competing interests.

References

Abdel-Mawgoud AMR, Tantaway AS, Hafez MM, Habib HAM (2010) Seaweed extract improves growth, yield and quality of different watermelon hybrids. Res J Agri Biol Sci 6(2):161–168

Ahmed MY, Sehrawy EL (2013) Effect of seaweed extract on fruiting of Hindy Bisinnara mango trees. J Amer Sci 9(6):539–544

Anisimov MM, Chaikina EL, Klykov AG, Rasskazov VA (2013) Effect of seaweeds extracts on the growth of seedling roots of Buckwheat (*Fagopyrum esculentum* Moench) is depended on the season of algae collection. Agric Sci Dev 2(8):67–75

Arnon DI (1949) Copper enzymes in isolated chloroplasts, polyphenol oxidase in *Beta vulgaris*. Plant Physiol 24:1–15

Atzmon N, Van Staden J (1994) The effect of seaweed concentrate on the growth of *Pinus pinea* seedlings. New Forest 8(3):279–288

Bai NR, Mary Christi R, Christy Kala T (2013) Growth and yield characteristics of *Dolichos biflorus* Linn as influenced by seaweed liquid fertilizer. Plant Arch 13(1):163–166

Bhosle NB, Untawale AG, Dhargalkar VK (1975) Effect of seaweed extract on the growth of *Phaseolus vulgaris* L. Indian J Mar Sci 4:209–210

Blunden G, Jenkins T, Wan LY (1997) Enhanced leaf chlorophyll levels in plants treated with seaweed extracts. J Appl Phycol 8(6):535–543

Bozorgi HR (2012) Effects of foliar spraying with marine plant *Ascophyllum nodosum* extract and nano iron chelate fertilizer on fruit yield and several attributes of eggplant (*Solanum melongena* L.). ARPN J Agri Bio Sci 7(5):200–206

Duncan BD (1965) Multiple range and multiple test. Biometric 11:1–42

Durand N, Briand X, Meyer C (2003) The effect of marine bioactive substances (NPRO) and exogenous cytokinins on nitrate reductase activity in *Arabidopsis thaliana*. Physiol Plant 119:489–493

Erulan V, Soundrapandian P, Thirumaran G, Ananthan G (2009) Studies on the effect of *Sargassum polycystum* (C. Agardh. 1824) extract on the growth and biochemical composition of *Cajanus Cajan* (L.) Mill sp. Amer Eur J Agri Environ Sci 6:392–399

Eyras MC, Rostagno CM, Defosse GE (1998) Biological evaluation of seaweed composting. Comp Sci Util 6:74–81

Featonby-Smith BC, Van Staden J (1984) Identification and seasonal variation of endogenous cytokinin in *Ecklonia maxima* (Osbeck) Papen F. Bot Mar 17:527–531

Gordon SA, Paleg LG (1957) Quantitative measurement of indole acetic acid. Physiol Plant 10:37–48

Graham HD, Henderson JHM (1961) Reaction of gibberellic acid and gibberellins with Folin-Wu Phosphomolybdic acid reagent and its use for quantitative assay. Plant Physiol 36:405–408

Hernández-Herrera RM, Santacruz-Ruvalcaba F, Alberto Ruiz-López M, Norrie J, Hernández-Carmona H (2014) Effect of liquid seaweed extracts on growth of tomato seedlings (*Solanum lycopersicum* L). J Appl Phycol 26:619–628

Humpshires EC (1956) Mineral components and ash analysis. In: Rach K, Tracer MV (eds) Modern methods of Plant analysis. Springer-Verlag, Berlin, pp 468–502

Jaworski EG (1971) Nitrate reductase in intact plant tissue. Biochem Biophys Res Commun 43:1274–1279

Jothinayagi N, Anbazhagan C (2009) Effect of seaweed liquid fertilizer of *Sargassum wightii* on the growth and biochemical characteristics of *Abelmoschus esculentus* (L.) Medikus. Recent Res Sci Tech 1:155–158

Kalaivanan C, Chandrasekaran M, Venkatesalu V (2012) Effect of seaweed liquid extract of *Caulerpa scalpelliformis* on growth and biochemical constituents of black gram (*Vigna mungo*) (L.) Hepper. Phykos 42(2):46–53

Kalidass C, Jayarani S, Glory M (2010) Effect of seaweed liquid fertilizers on growth and biochemical constituents of *Brassica nigra* (L.). Intl J Agri Environ Biol 3:307–311

Khan W, Usha RP, Sowmyalakshmi S, Jithesh MN, Rayorath P, Hodges DM, Critchley AT, Craigie JS, Norrie J, Prithiviraj B (2009) Seaweed extracts as biostimulants of plant growth and development. J Plant Growth Reg 28:386–399

Lowry OH, Rosebrough H, Farr AL, Randall RJ (1951) Protein measurement by folin-phenol reagent. J Biol Chem 193:265–275

Mooney PA, Van Staden J (1986) Algae and cytokinins. J Plant Physiol 123:1–21

Nelson N (1944) A photometric adaptation of Somogyis method for the determination of reducing sugar. Anal Chem 31:426–428

Ordog V, Stirk WA, van Staden J, Novak O, Strnad M (2004) Endogenous cytokinins in the three genera of macroalgae from the chlorophyta. J Phycol 40:88–95

Parthiban C, Saranya C, Hemalatha A, Kavitha B, Anantharaman P (2013) Effect of seaweed liquid fertilizer of *Spatoglossum asperum* on the growth and pigment content of *Vigna radiata*. Int J Recent Sc. Res 4(9):1418–1421

Pise NM, Sabale AB (2010) Effect of seaweed concentrates on the growth and biochemical constituents of *Trigonella foenum-graecum*. J Phytol 2(4):50–56

Roe JH (1954) Chemical determination of ascorbic, dehydroascorbic and diketogluconic acids. Meth Biochem Anal 1:115–139

Sanderson KJ, Jameson PE (1986) The cytokinins in a liquid seaweed extract: could they be the active ingredients? Acta Hort (ISHS) 179:113–116

Sasikumar K, Govindan T, Anuradha C (2011) Effect of seaweed liquid fertilizer of *Dictyota dichotoma* on growth and yield of *Abelmoschus esculentus* (L). Eur J Exp Biol 1:223–227

Sekar R, Thangaraju N, Rengasamy R (1995) Effect of seaweed liquid fertilizer from *Ulva lactuca* on *Vigna unguiculata* L. (Walp). Phykos 34:49–53

Siddagangaiah K, Raveesha A, Vasanth Kumar T (2010) Effect of foliar application of phyton– a seaweed extract on growth and yield of potato. Potato J 37(1–2):44–47

Sivasankari S, Chandrasekaran M, Kannathasan K, Venkatesalu V (2006) Studies on the biochemical constituents of *Vigna radiata* Linn treated with seaweed liquid fertilizer. Seaweed Res Utiln 28(1):151–158

Sridhar S, Rengasamy R (2010) Effect of seaweed liquid fertilizer on the growth, biochemical constituents and yield of *Tagetes erecta* under field trial. J. Phytol 2(6):61–68

Syono K, Torrey JG (1976) Identification of cytokinins of root nodules of the Garden Pea, *Pisum sativum* L. Plant Physiol 57:602–606

Thevanathan R, Anjana D, Dinamani DS, Bhavani ILG (2005) Studies on the impact of application of marine algal manure and liquid fertilizer in the linear growth of the seedlings of some pulses. Seaweed Res Utiln 27(1&2):125–133

Thirumaran G, Arumugam M, Arumugam R, Anantharaman P (2009) Effect of seaweed liquid fertilizer on growth and pigment concentration of *Cyamopsis tetrogonolaba* (L.) Taub. Am Euras J Agron 2:50–56

Whapman CA, Blunden G, Jenkins T, Hankins SD (1993) Significance of betaines in the increased chlorophyll content of plants treated with seaweed extract. J Appl Phycol 5:231–234

Zodape ST, Kawarkhe VJ, Patolia JS, Warade AD (2008) Effect of liquid seaweed fertilizer on yield and quality of okra (*Abelmoschus esculentus* L.). J Sci Ind Res 67:1115–1117

Zodape ST, Mukhopadhyay S, Eswaran K, Reddy MP, Chikara J (2010) Enhanced yield and nutritional quality in green gram (*Phaseolus radiata* L.) treated with seaweed (*Kappaphycus alvarezii*) extract. J Sci Ind Res 69:468–471

Zodape ST, Gupta A, Bhandari SC (2011) Foliar application of seaweed sap as biostimulant for enhancement of yield and quality of tomato. J Sci Ind Res 67:215–219

28

Assessment of changes in different fractions of the organic carbon in a soil amended by nanozeolite and some plant residues: incubation study

Milad Mirzaei Aminiyan[1] · Ali Akbar Safari Sinegani[1] · Mohsen Sheklabadi[1]

Abstract

Introduction Organic carbon (OC) fractions play important roles in soil and many ecosystem processes. This study focuses on changing of OC in density and soluble fractions in a soil amended by nanozeolite and plant residues that incubated in lab condition for 90 days.

Results The results showed that amounts of OC in light fraction (LF) and heavy fraction (HF) increased with the increasing percentage of nanozeolite and plant residues in the soil. The highest amounts of LF (7.54 g LF. $kg^{-1}{}_{Soil}$) and HF (11.10 g $kg^{-1}{}_{Soil}$) were found when 30 % nanozeolite and 5 % wheat and alfalfa straws were added to the soil. Accordingly, wheat and alfalfa straws were effective on increasing the LF and HF, respectively. However, they decreased with declining the OC from the 1st day of experiment until the 90th day of experiment. Soluble OC in hot (2.22 g $kg^{-1}{}_{Soil}$) and cool (1.54 g $kg^{-1}{}_{Soil}$) water fractions increased by addition of 30 % nanozeolite and 5 % plant residues particularly alfalfa straw in comparison with control. Although these soluble fractions increased after initial 30 days of incubation, they decreased in the continuation of the experiment.

Conclusion In fact, OC contents in density and soluble fractions increased by addition of 30 % nanozeolite and 5 % plant residues into the soil; however, they decreased in initial 30 days of incubation with declining the OC. The findings of this research revealed the application of nanozeolite and plant residues improved carbon pools in density and soluble fractions and carbon sequestration increased by increasing OC contents in soil.

Keywords Alfalfa straw · Wheat straw · Light fraction · Heavy fraction · Hot water · Cool water

Introduction

It has been recognized in last decades that the amounts of carbon stored in soils have important effects on a global scale. Therefore, land management practices affecting soil organic carbon (SOC) content may have a global impact, if they are applied over large areas (Bronick and Lal 2005). Therefore, the small fluctuations of SOC may potentially alter the atmospheric carbon dioxide (CO_2) concentration and the global climate (Mahmoodabadi and Heydarpour 2014). The carbon storage and rate of CO_2 sequestration in soils depend on climate, soil properties and management. Soils as a sink for atmospheric CO_2 play a key role in the global carbon budget as well as in the global carbon cycle (Eshel et al. 2007). Soils are known as one of the largest active carbon pools after the hydrosphere and the lithosphere. The role of soils as either a source or a sink for greenhouse gases, in general, and that of CO_2, in particular, has been a focus of recent studies (Majumder et al. 2008; Bhattacharyya et al. 2009).

Whereas the largest terrestrial pool of carbon is located in the soils (Bhattacharyya et al. 2009), there are many factors that influence on carbon retention and release in soil and also carbon exchange between soil and atmosphere (Majumder et al. 2008). Storage of organic carbon (OC) in

✉ Milad Mirzaei Aminiyan
miladmir67@yahoo.com

Ali Akbar Safari Sinegani
safari_sinegani@yahoo.com

Mohsen Sheklabadi
sheklabadi@yahoo.com

1 Soil Science Department, College of Agriculture, Bu-Ali Sina University, Azadegan Street, Hamedan 6517833131, Iran

agricultural systems is a balance between carbon additions from non-harvested portions of crops (Wang et al. 2011), organic sources (Thelen et al. 2010) and carbon losses, primarily through organic matter decomposition and release of respired CO_2 to the atmosphere (Bird et al. 2002). Last researches have shown that organic substances improve soil aggregation and consequently soil structure and also diminish soil compaction and surface crusting, enlarge carbon sequestration and nutrient availability, and increase infiltration and water-holding capacity (Balashov et al. 2010; Aminiyan et al. 2015).

Density fractionation

Physical fractionation, in contrast to chemical fractionation, allows for isolation of fractions, can isolate as intact as possible the SOC associated predominantly with soil minerals (primary organo-mineral complexes) and also the SOC protected within aggregates (secondary organo-mineral complexes) due to their three-dimensional architecture (Straathof et al. 2014). OC fractions exhibit different rates of biochemical and microbial degradation (Stevenson 1994; Tan et al. 2007) as well as different accessibility and interactions (Sollins et al. 1999). Very little is known of the dynamics of soil organic matter (SOM) after agricultural abandonment. SOC contains fractions with a rapid turnover rate as well as fractions with a slower turnover rate (Aminiyan et al. 2015). The labile fractions of OC, such as microbial biomass carbon (MBC) and dissolved organic carbon (DOC), can respond rapidly to changes in C supply. The dynamics of SOC are usually described by dividing SOM into two or more fractions. Density fractionation, that is a laboratory procedure, physically separates soil into light (LF) and heavy fractions (HF) (Wander and Traina 1996; Sollins et al. 1999). The procedure is useful method for studying labile pools of SOC that are more sensitive to cropping practice than is the total SOC pool in temperate soils (Janzen et al. 1992). Generally, sodium polytungstate (SPT) (1.85 g mL^{-1}) and NaI (1.3 g cm^{-3}) solutions are often used for density fractionation (Golchin et al. 1994; Magid et al. 1996; Six et al. 2002). Although LF is commonly referred to a plant-like and less stable fraction with high OC concentration (Gregorich et al. 2003), HF is more stable and high-density organo-mineral fractions having lower C concentrations (Golchin et al. 1995a, b).

Light fraction of SOM not only is sensitive to changes in management practices (Cambardella and Elliott 1992; Bremer et al. 1994) but also correlates well with the rate of nitrogen (N) mineralization (Hassink 1995; Barrios et al. 1996). The importance of LF (including free and occluded organic C within aggregates) is widely recognized for its role in formation and stability of soil structure, especially in stabilization of soil macroaggregates (>250 mm) (Miller and Jastrow 1990; Kay 1998). Janzen et al. (1992) concluded that LF in surface soil (0–7.5 cm) accounted for 2–17 % of SOC, depending principally on cropping systems. However, there are few studies about these two fractions and their contributions to OC storage as related to changes soil management systems.

Water-soluble organic carbon

Dissolved organic carbon (DOC) is defined as a mixture of organic molecules produced by the decomposition of SOM and plant material and by root exudation (Strobel et al. 2001). Although it has been assumed that dissolved organic matter (DOM) represents a labile part of SOM and that total DOC concentration and especially its easily degradable part resembles soil microbial activity, it has been suggested that a great part of DOM in soil represents a relatively stable by-product of microbial activity (Zhao et al. 2008; Wang et al. 2013). The movement of DOM is significant to the cycling and distribution of nutrients and carbon within and between ecosystems and contributes to soil forming processes (Kalbitz et al. 2003). Plant residue and humus are the most significant sources of soluble organic matter in soil. Gregorich et al. (2000) hypothesized that, although the water-soluble carbon pool was small, it had a high turnover and was in equilibrium with soil humus. Liang et al. (2012) suggested that the difference in water-soluble and biodegradable C in agricultural soils was greater than forest soils due to increases in soluble humic materials in agricultural soils. Therefore, humus is probably the main source of DOC because of the relatively large amount of humus present in soil relative to that contributed by the microbial biomass or recently deposited plant residues. DOC inputs to soil solution originate from biological decomposition, throughfall or litter leaching, root exudates and from deposition of soot and dust (Wang et al. 2013).

Laboratory studies (Smolander and Kitunen 2011; Liang et al. 2012; Kiikkilä et al. 2014) have shown that microorganisms can decompose many amounts of water-soluble organic matter fraction. These studies, which ranged in duration from hours to months, indicated that 10–40 % of the water-soluble OC was decomposable under laboratory conditions. Liang et al. (2012) reported that SOC had significantly positive correlations with labile organic C fractions in the 0–20 cm depth. Smolander and Kitunen (2011) observed positive correlations between DOC concentration extracted from soil and the rate of C and net N mineralization and amount of C and N in microbial biomass, which were used to assess soil microbial activity. High temperature is known to hydrolyze organic structures, lyse cells and dissociate organic materials from inorganic colloids (Kiikkilä et al. 2014). Kim et al. (2014) reported strong correlations between organic C extracted with hot water and mineralizable carbon. This easily degradable DOM seemed to decrease fairly consistently during the degradation of organic matter as observed previously (Don and Kalbitz 2005; Sanderman et al. 2008).

During last years, great strides have been made in a number of research topics including characterizing the spatial and temporal variations in the concentration and flux of DOC reviewed by Kalbitz et al. 2000 and Aitkenhead-Peterson et al. 2003, quantifying its role in soil chemistry and pedogenesis (e.g., Kaiser and Zech 1998; Jansen et al. 2003; Cances et al. 2003), describing the chemical composition of DOC (Guggenberger and Zech 1994; Kaiser and Zech 1998; Strobel et al. 2001), and quantifying the availability of DOC to soil microflora (Zsolnay and Gorlitz 1994; Yano et al. 2000; Kalbitz et al. 2003; Marschner and Kalbitz 2003). There are few studies about assaying of zeolitic materials effects on SOC. Zeolitic materials are extensively used to improve soil physical environment, particularly, in sandy and clay poor soils (Abdi et al. 2006). The assessment of nanozeolite effects on SOC showed that the addition of higher percentage of nanozeolite with alfalfa straw into the soil increased SOC pools and improved soil aggregation stability (Aminiyan et al. 2015). The main objectives of this study were to determine water-soluble and density fractions of OC in soils that treated by different percentage of nanozeolite and some plant residues and incubation them in field capacity for 90 days.

Materials and methods

Study area

This study was conducted on agricultural soil in Azandariyan, Hamedan province, the west of Iran. This area was located between longitudes 47°42′ and 48°45′ E and latitudes 33°28′ and 34°29′ N. The climate of the region is semiarid with a mean annual precipitation of 300 mm and a mean annual temperature of 10 °C. The soil of the area is mostly classified as Typic Haplocalcids (Aminiyan et al. 2015).

Sampling, treatment and analysis of soil

The methods used for soil sampling, treatment and analysis were reported in Aminiyan et al. (2015). The treated and moistened soils were incubated in laboratory condition (20–25 °C) for 90 days. After 1, 5, 10, 20, 30, 45, 60, 75 and 90 days of incubation, a portion of each soil was taken for the study of in density (light and heavy) and soluble (hot water and cool water) OC fractions.

Density fractionation

About 10 g dried sample was transferred to a 20-ml graduated centrifuge tube, and 50 ml of NaI solution ($d = 1.3$ g cm^{-3}) was added. Suspensions were immediately centrifuged at 3000 rpm for 10 min. The supernatant containing the LF was decanted onto Whatman no. 50 filters (2.7-p.m. retention) and vacuum-filtered. The HF residue was re-suspended twice in fresh NaI solution, and the LFs were combined. LF and HF were then washed four times into pre weighed tins with deionized water, afterward dried at 55 °C for 24 h in the oven, and weighed (Sollins et al. 1999). Then OC content in HF was determined by Walkley and Black (1934) method.

Soluble water organic carbon fractions

The soluble water OC in the whole soil and the three aggregate fractions were extracted using cold water followed by hot water. Soluble organic matter in cold water was extracted from soils by adding 150 ml of distilled/deionized water to a tube containing 15 g of air-dried whole soil or aggregate fraction. The soil water suspension was shaken for 30 min and centrifuged at 4500 rpm for 20 min. The supernatant solutions were decanted and passed through a 0.45-μm cellulose nitrate filter. The weight of extraction tubes with remaining wet soil was recorded in order to calculate the amount of cold water extract remaining. Hot water-soluble organic matter was extracted from these soils by adding water to the wet soil remaining in each tube to return the water volume to 150 ml, then by placing the tubes in a water bath at 80 °C for 16 h. After this period of time, the samples were centrifuged, decanted and filtered as above. Filtered solutions were stored in a refrigerator (4 °C) prior to incubation (Gregorich et al. 2003). Then OC content in HF was determined by Walkley and Black (1934) method.

Table 1 Some of chemical and physical properties of applied soil

EC (dS m^{-1})	pH	CEC (Cmol^{+} kg^{-1} $_{Soil}$)	Total organic C (g kg^{-1})	CCE[a] (%)	Sand (%)	Clay (%)	Silt (%)
1.1	7.2	4.80	3.41	1.79	69	12	19

[a] Carbonate calcium equivalent

Table 2 Some properties of applied plant residues in this study

	pH	EC (dS m^{-1})	Total organic carbon (g kg^{-1})	Total nitrogen (g kg^{-1})	Total phosphorous (g kg^{-1})	C/N	C/P
Alfalfa straw	6	9.5	511	22	5.98	23.30	85.20
Wheat straw	7.97	4.3	532	7	4.31	90.75	123.50

Table 3 Some properties of applied nanozeolite

EC (dS m^{-1})	pH	Organic C (g kg^{-1})	CEC (Cmol^{+} kg^{-1})
0.98	7.17	1.03	400.39

Table 4 Nanozeolite compositions and their %wt

Composition	%wt
SiO_2	69.44
Al_2O_3	11.87
Fe_2O_3	1.30
Ti_2O	0.18
K_2O	1.31
Na_2O	0.68
CaO	3.28
MgO	0.99

Statistical data analysis

The experiment was a completely randomized factorial design with three replicates. The factors applied were alfalfa straw (0 and 5 %), wheat straw (0 and 5 %), nanozeolite (0, 10 and 30 %) and incubation time (1, 5, 10, 20, 30, 45, 60, 75 and 90 days). All statistical analyses were performed in the SAS ver.9.2 statistical framework; to obtain the main differences between the treatments, the Duncan's ($\alpha = 0.01$) test was applied.

Results and discussion

Table 1 shows some of chemical and physical properties of applied soil. According to the sand, clay and silt contents, the soil texture was loamy sand. Table 2 presents some properties of applied plant residues. Alfalfa and wheat straw had neutral pH, high OC values and C/N and C/P ratios. Some of applied nanozeolite properties are given in Table 3. Also nanozeolite compositions and their weight percentage are shown in Table 4; according to this table, SiO_2 and Al_2O_3 (69.44 and 11.87 %), respectively, had much higher portion than the other compositions.

The effect of nanozeolite and plant residues on OC in density fractions

Table 5 shows the analysis of variance of the effects of nanozeolite, plant residues application, incubation time and their interaction on LF, HF, soluble OC in cool water and hot water fractions in soil ($p < 0.01$). However, the

Table 5 Analysis of variance (mean square) of the effects of nanozeolite, plant residues application, incubation time and their interaction on LF and HF, soluble OC in cool water and hot water fractions in soil

Source	DF	Organic carbon fractions			
		LF (g LF. kg^{-1}_{Soil})	OC in HF (g kg^{-1}_{Soil})	Soluble OC in cool water (g kg^{-1}_{Soil})	Soluble OC in hot water (g kg^{-1}_{Soil})
Nanozeolite	2	643.18**	1495.55**	6.70**	28.16**
Plant residues	3	49.63**	165.98**	0.14**	1.48**
Time	8	134.34**	354.49**	3.20**	9.42**
Nanozeolite × time	16	42.76 ns	2.24 ns	1.53 ns	5.08 ns
Plant residues × time	24	34.96 ns	0.95 ns	0.74 ns	2.28 ns
Nanozeolite × plant residues	6	130.61**	2.63**	8.26**	5.63**
Nanozeolite × plant residues × time	48	23.94 ns	0.59 ns	0.37 ns	1.28 ns
Error	216	8.02	0.6	0.3	0.1

** Mean square of the treatment is significant at the 0.01 level

ns Mean square of the treatment is not significant

Table 6 Light fraction and HF content in all of treatment

Treatment	LF (g LF. kg^{-1}_{Soil})	OC in HF (g kg^{-1}_{Soil})
Control	16.32 ± 1.30* g	17.60 ± 3.00 e
N0A5	17.08 ± 1.27 fg	20.50 ± 3.01 d
N0W5	18.04 ± 1.28 ef	19.70 ± 3.02 de
N10PR0	16.50 ± 2.15 ef	17.98 ± 3.02 c
N10A5	19.20 ± 2.13 cd	25.70 ± 3.01 ab
N10W5	19.52 ± 2.14 de	24.60 ± 3.02 bc
N30PR0	16.63 ± 2.10 bc	18.10 ± 3.00 bc
N30A5	21.78 ± 2.12 ab	28.70 ± 3.02 a
N30W5	23.86 ± 2.11 a	27.00 ± 3.01 b

N0A5 (0 % nanozeolite + 5 % alfalfa straw), N0W5 (0 % nanozeolite + 5 % wheat straw), N10PR0 (10 % nanozeolite + 0 % plant residue), N10A5 (10 % nanozeolite + 5 % alfalfa straw), N10W5 (10 % nanozeolite + 5 % wheat straw), N30PR0 (30 % nanozeolite + 0 % plant residue), N30A5 (30 % nanozeolite + 5 % alfalfa straw), N30W5 (30 % nanozeolite + 5 % wheat straw)

The same letters are not significantly different at $p < 0.01$ using Duncan's LSD test

* Mean ± standard deviation

interactions between nanozeolite and incubation time, plant residues and incubation time and nanozeolite, plant residues and incubation time did not have significant effects on mentioned OC fractions in the soil.

Table 6 reveals the OC content in LF and HF increased by the addition of nanozeolite and plant residues ($p < 0.01$). LF value in 30 % nanozeolite plus 5 % wheat straw (N30W5) treatment was greater than the other treatments; as its value 7.54 (g LF. kg^{-1}_{Soil}) was greater than control, because C/N ratio in wheat straw was higher than alfalfa straw, and thus subsequently wheat straw had lower stage of biodegradation by microorganisms in soil. This fraction of OC decreased from the 1st day (20.61 g LF. kg^{-1}_{Soil}) to the 90th day (16.99 g LF. kg^{-1}_{Soil}) during soil incubation (Fig. 1). The HF value in N30A5 treatment was significantly increased (11.1 g kg^{-1}_{Soil}) in comparison with the control treatment (Table 6). Also this table investigates that N30A5 treatment increased HF (3 g kg^{-1}_{Soil}) and (8.2 g kg^{-1}_{Soil}) in comparison with the N10A5 and N0A5 treatments, respectively.

It is known that the alfalfa straw was more efficiency due to increasing OC in HF than wheat straw in all of the treatments with the similarity percentage of nanozeolite (Table 6). SOC in the LF plays an important role in retaining of cellulase molecule from washing out and nutrition of soil microorganisms and subsequently humus production. Thus SOM quality is an important factor in its disintegration rate (Schmidt et al. 2002; Beheshti et al. 2012). According to Fig. 2, OC in HF had a distinct downward trend from the 1st day (27 g kg^{-1}_{Soil}) until the 90th day (18.4 g kg^{-1}_{Soil}). The recent research on OC decay dynamics showed that LF and HF were decreased during soil incubation (Hassink et al. 1995; Creamer et al. 2012; Aminiyan et al. 2016). The results of Rovira and Vallejo' studies (2003) were in line with those of the present study.

The effect of nanozeolite and plant residues on water-soluble organic carbon fractions

As shown in Table 7, soluble OC contents in hot water and cool water increased by the addition of nanozeolite and plant residues especially alfalfa straw. The results showed that soluble OC in hot water in N30A5 treatment was greater 2.22, 1.36 and 2.06 (g kg^{-1}_{Soil}) than control, N10A5 and N0A5 treatments, respectively (Table 7). Working on

Fig. 1 Light fraction changes with the passage of time

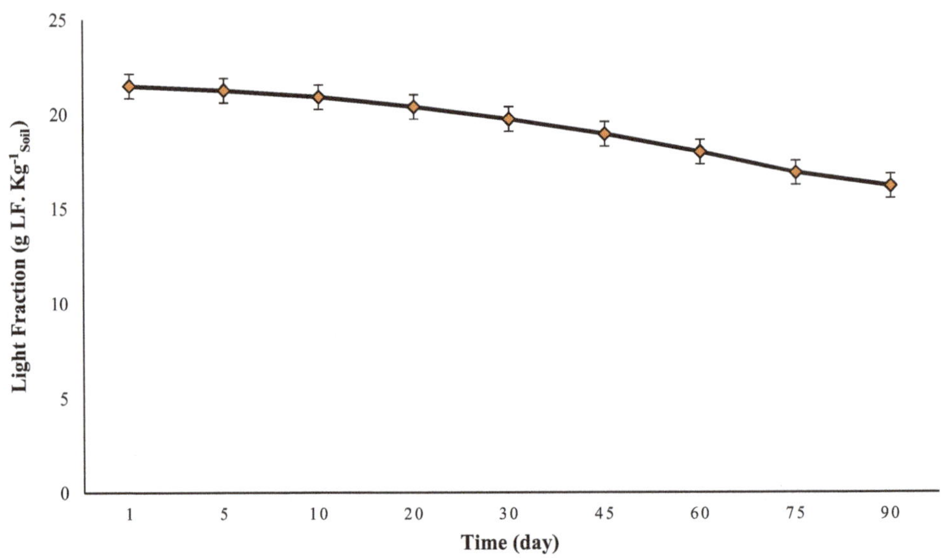

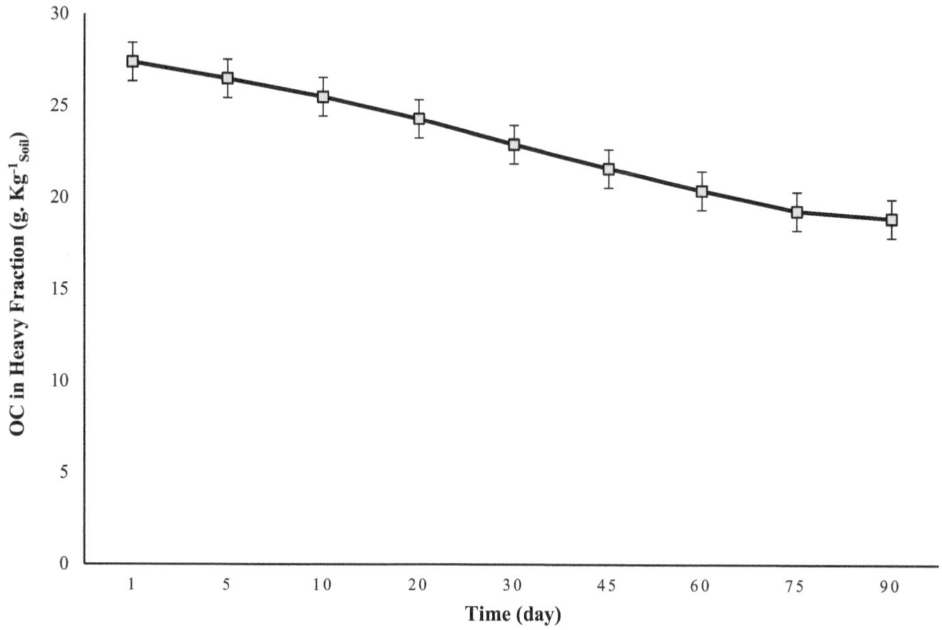

Fig. 2 Organic carbon changes in HF with the passage of time

Table 7 Soluble OC in hot and cool water content in all of treatments

Treatment	Soluble OC in cool water (g $kg^{-1}{}_{Soil}$)	Soluble OC in hot water (g $kg^{-1}{}_{Soil}$)
Control	1.93 ± 0.358* c	2.34 ± 0.552 e
N0A5	2.01 ± 0.360 c	2.50 ± 0.546 de
N0W5	1.94 ± 0.357 c	2.42 ± 0.561 de
N10PR0	1.98 ± 0.255 b	2.52 ± 0.454 cd
N10A5	2.33 ± 0.251 ab	3.20 ± 0.461 ab
N10W5	2.29 ± 0.248 ab	3.10 ± 0.477 bc
N30PR0	2.00 ± 0.255 ab	2.63 ± 0.480 ab
N30A5	3.47 ± 0.251 a	4.56 ± 0.491 ab
N30W5	2.51 ± 0.247 ab	3.44 ± 0.473 a

N0A5 (0 % nanozeolite + 5 % alfalfa straw), N0W5 (0 % nanozeolite + 5 % wheat straw), N10PR0 (10 % nanozeolite + 0 % plant residue), N10A5 (10 % nanozeolite + 5 % alfalfa straw), N10W5 (10 % nanozeolite + 5 % wheat straw), N30PR0 (30 % nanozeolite + 0 % plant residue), N30A5 (30 % nanozeolite + 5 % alfalfa straw), N30W5 (30 % nanozeolite + 5 % wheat straw)

The same letters are not significantly different at $p < 0.01$ using Duncan's LSD test

* Mean ± standard deviation

the chemical composition of DOC suggested that most DOC is an end product of microbial metabolism (Guggenberger and Zech 1994); However, short-term experimental manipulations of organic matter sources showed that fresh litter also contributes significantly to the production of DOC (Park et al. 2002). These two views are not necessarily mutually exclusive, but they do point out the considerable difficulty in determining the influence of substrate (litter, SOM), microbial community composition (Muller et al. 1999) and abiotic factors such as temperature and water flux on DOC production and flux (Brooks et al. 1999). Aminiyan et al. (2015) reported that the addition of 30 % nanozeolite with 5 % alfalfa straw to the soil redounded increasing OC in different aggregate particle size classes.

Based on Table 7, the same results were achieved similar to the results of hot water to cool water; accordingly, soluble OC in cool water increased with the greater percentage of nanozeolite and plant residues particularly alfalfa straw. Soluble OC in cool water value increased 1.54 and 0.58 (g $kg^{-1}{}_{Soil}$) into the control in N30A5 and N30W5 treatments, respectively. Thus N30A5 treatment more effective to increasing soluble OC in cool water than N30W5 treatment. Soluble OC content in hot water was

Fig. 3 Soluble OC changes in hot and cool water with the passage of time

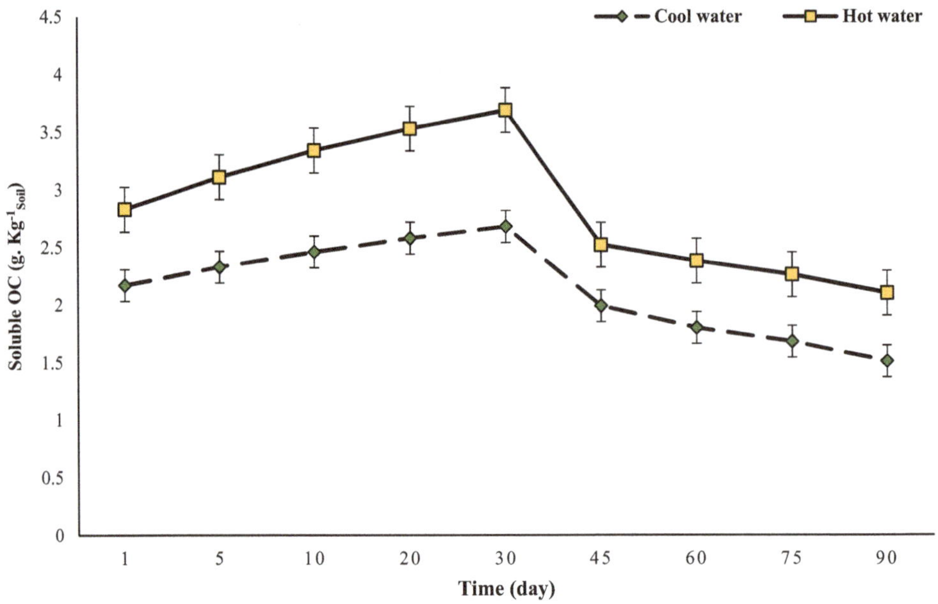

greater than soluble OC content in cool water (Table 7), because hot water had greater ability to extract of lysis microbial cells and extractable soluble organic matter may be adsorbed to clay or complexed with other organic material produced by plants or decomposing organic matter than cool water (Guggenberger and Zech 1994; Muller et al. 1999). According to the Gregorich et al. (2003) findings, high temperature is known to hydrolyze organic structures, lyse cells and dissociate organic materials from inorganic colloids. The plant residues with lower C/N ratio are a readily decomposable substrate for microorganisms, and they have additional soluble OC content than plant residues with higher C/N ratio (Gregorich et al. 2003). Also these researchers found that the quantity of biodegradable soluble organic matter was related to the extraction procedure and the quantity of organic matter present in the soil.

Figure 3 indicates that soluble OC in both hot water and cool water increased with over time from 1st day until the 30th day of incubation period, but then decreased by the end of the experiment. Accordingly, the soluble OC in cool water increased from 2.17 (g $kg^{-1}{}_{Soil}$) in the 1st day to 2.68 (g $kg^{-1}{}_{Soil}$) in the 30th day and then it decreased by the end of experiment 1.92 (g $kg^{-1}{}_{Soil}$). As shown in Fig. 3, Soluble OC in hot water value increased from 2.83 (g $kg^{-1}{}_{Soil}$) in the 1st day to 3.69 (g $kg^{-1}{}_{Soil}$) in the 30th day, and finally it decreased by the 90th day 2.39 (g $kg^{-1}{}_{Soil}$). Since soluble OC was increased with the development and promoting plant residues biodegradation in the initial 30 days and when the growth and development of microbial communities were increased with the passage of time and subsequently soluble OC decreased with the passage of time. Kalbitz et al. (2003) observed that soluble OC increased with the passage of time, but in another study soluble OC decreased by over the time (Gregorich et al. 2003). Alfalfa straw had greater soluble OC than wheat straw, and thus its degradation rate and OC content decreasing was done by higher rate in this fraction (Swanston et al. 2002; Preston and Schmidt 2006; Aminiyan et al. 2016). It is known in recent reviews that the organic matter quality is particularly important for SOC stabilization (Amelung et al. 2008; Schmidt et al. 2011).

Conclusion

Organic carbon fractions in soils play important roles in many ecosystem processes. OC fractions exhibit different rates of biochemical and microbial degradation. Density fractionation is a laboratory procedure that separates SOC into LF and HF. Also DOC is affected by the extraction procedure used. Extraction procedures involving higher temperatures extract a greater amount of soluble organic matter than extractions carried out at room temperature. The results of this study showed that LF and HF and water-soluble OC was increased by the addition of greater percentage of nanozeolite and plant residues into the soil. The results of this study showed that LF was greater in N30W5 treatment than in the other treatments. But OC in HF and soluble OC in hot and cool water had maximum amounts in (N30A5) than in the other treatments. LF and HF decreased with the passage of time from the 1st day until the 90th day. Soluble OC in hot and cool water increased from 1st day until the 30th day, and then they decreased by the end of the experiment. In fact, OC content increased by application and addition of nanozeolite and plant residues into the soil, but these pools decreased with the passage of time. Finally, it can be said that the application of nanozeolite and plant residues improve carbon sequestration process and increase carbon pools in soil.

Acknowledgments The authors sincerely thank editor and the anonymous reviewers for reviewing the manuscript and providing critical comments to improve the paper. Thanks so much for Fadak institute in Isfahan and Afrazand factory in Tehran.

References

Abdi GH, Khui MK, Eshghi S (2006) Effects on natural zeolite on growth and flowering on straw berry. Int J Agric Resour 1:384–389

Aitkenhead-Peterson JA, McDowell WH, Neff JC (2003) Sources, production, and regulation of allochthonous dissolved organic matter. In: Findlay S (ed) Dissolved organic matter sources, transport, and transformation in aquatic ecosystems. Academic Press, New York

Amelung W, Brodowski S, Sandhage-Hofmann A, Bol R (2008) Combining biomarker with stable isotope analyses for assessing the transformation and turnover of soil organic matter. In: Sparks DL (ed) Advances in Agronomy. Academic Press, Burlington, pp 155–250

Aminiyan MM, Sinegani AAS, Sheklabadi M (2015) Aggregation stability and organic carbon fraction in a soil amended with some plant residues, nanozeolite, and natural zeolite. Int J Recycl Org Waste Agric 4(1):11–22. doi:10.1007/s40093-014-0080-0

Aminiyan MM, Sinegani AAS, Sheklabadi M (2016) The effect of zeolite and some plant residues on soil organic carbon changes in density and soluble fractions: Incubation study. Eurasian J Soil Sci 5(1). doi:10.18393/ejss.2016.1.074-083

Balashov E, Kren J, Prochazkova B (2010) Influence of plant residue management on microbial properties and water-stable aggregates of two agricultural soils. Int Agrophys 24(1):9–14

Barrios E, Buresh RJ, Sprent JI (1996) Nitrogen mineralization in density fractions of soil organic matter from maize and legume cropping systems. Soil Biol Biochem 28:1459–1465

Beheshti A, Raiesi F, Golchin A (2012) Soil properties, C fractions and their dynamics in land use conversion from native forests to croplands in northern Iran. Agri Eco Environ 148:121–133

Bhattacharyya R, Prakash V, Kundu S, Srivastva AK, Gupta HS (2009) Soil aggregation and organic matter in a sandy clay loam soil of the Indian Himalayas under different tillage and crop regimes. Agric Ecosyst Environ 132(1):126–134

Bird M, Santruckova H, Lloyd J, Lawson E (2002) The isotopic composition of soil organic carbon on a north–south transect in western Canada. Eur J Soil Sci 53(3):393–403

Bremer E, Jansen HH, Johnston AM (1994) Sensitivity of total, light fraction and mineralizable organic matter to management practices in a Lethbridge soil. Can J Soil Sci 74:131–138

Bronick CJ, Lal R (2005) Soil structure and management: a review. Geoderma 124(1):3–22

Brooks PD, McKnight DM, Bencala KE (1999) The relationship between soil heterotrophic activity, soil dissolved organic carbon (DOC) leachate, and catchment-scale DOC export in headwater catchments. Water Resour Res 35:1895–1902

Cambardella CA, Elliott ET (1992) Particulate soil organic matter changes across a grassland sequence. Soil Sci Soc Am J 56:777–783

Cances B, Ponthieu M, Castrec-Rouelle M, Aubry E, Benedetti MF (2003) Metal ions speciation in a soil and its solution: experimental data and model results. Geoderma 113:341–355

Creamer C, Filley T, Boutton T (2012) Long-term incubations of size and density separated soil fractions to inform soil organic carbon decay dynamics. Soil Biol Biochem 51:8–11

Don A, Kalbitz K (2005) Amount and degradability of dissolved organic carbon from foliar litter at different decomposition stages. Soil Biol Biochem 37:2171–2179

Eshel G, Fine P, Singer MJ (2007) Total soil carbon and water quality: an implication for carbon sequestration. Soil Sci Soc Am J 71(2):397–405

Golchin A, Clarke P, Oades JM, Skjemstad JO (1995a) The effects of cultivation on the composition of organic matter and structural stability of soils. Aust J Soil Res 33:975–993

Golchin A, Oades JM, Skjemstad JO, Clarke P (1994) Study of free and occluded particulate organic matter in soils by solid-state 13C NMR spectroscopy and scanning electron microscopy. Aust J Soil Res 32:285–309

Golchin A, Oades JM, Skjemstad JO, Clarke P (1995b) Structure and dynamic properties of soil organic matter reflected by ^{13}C natural abundance, pyrolysis mass spectrometry and solid-state ^{13}C NMR spectroscopy in density fractions of an Oxisol under forest and pasture. Aust J Soil Res 33:59–76

Gregorich EG, Beare MH, Stoklas U, St-Georges P (2003) Biodegradability of soluble organic matter in maize-cropped soils. Geoderma 113:237–252

Gregorich EG, Liang BC, Mackenzie AF, Drury CF, McGill WB (2000) Elucidation of the source and turnover of water soluble and microbial biomass carbon in agricultural soils. Soil Biol Biochem 32:581–587

Guggenberger G, Zech W (1994) Composition and dynamics of dissolved organic carbohydrates and lignin degradation products in two coniferous forests, N.E. Bavaria. Germany. Soil Biol Biochem 26:19–27

Hassink J (1995) Density fractions of soil macro organic matter and microbial biomass as predictors of C and N mineralization. Soil Biol Biochem 27:1099–1108

Jansen B, Nierop KGJ, Verstraten JM (2003) Mobility of Fe(II), Fe(III) and Al in acidic forest soils mediated by dissolved organic matter: influence of solution pH and metal/organic carbon ratios. Geoderma 113:323–340

Janzen HH, Campbell CA, Brandt SA, Lafond GP, Townley-Smith L (1992) Light-fraction organic matter in soils from long-term crop rotations. Soil Sci Soc Am J 56:1799–1806

Kaiser K, Zech W (1998) Rates of dissolved organic matter release and sorption in forest soils. Soil Sci 168:714–725

Kalbitz K, Schmerwitz J, Schwesig D, Matzer E (2003) Biodegradation of soil-derived dissolved organic matter as related to its properties. Geoderma 113:273–291

Kalbitz K, Solinger S, Park JH, Michalzik B, Matzner E (2000) Controls on the dynamics of dissolved organic matter in soils: a review. Soil Sci 165:277–304

Kay BD (1998) Soil structure and organic carbon, a review. In: Lal R, Kimble JM, Follett RF, Stewart BA (eds) soil processes and the carbon cycle. CRC Press, Boca Raton, pp 169–197

Kiikkilä O, Kanerva S, Kitunen V, Smolander A (2014) Soil microbial activity in relation to dissolved organic matter properties under different tree species. Plant Soil 377(1–2):169–177

Kim Y, Ullah S, Moore TR, Roulet NT (2014) Dissolved organic carbon and total dissolved nitrogen production by boreal soils and litter: the role of flooding, oxygen concentration, and temperature. Biogeochemistry 118(1–3):35–48

Liang Q, Chen H, Gong Y, Fan M, Yang H, Lal R, Kuzyakov Y (2012) Effects of 15 years of manure and inorganic fertilizers on soil organic carbon fractions in a wheat-maize system in the North China Plain. Nutr Cycl Agroecosyst 92(1):21–33

Magid J, Gorissen A, Giller KE (1996) In search of the elusive ''active'' fraction of soil organic carbon: three size-density fractionation methods for tracing the fate of homogeneously ^{14}C labeled plant materials. Soil Biol Biochem 28:89–99

Mahmoodabadi M, Heydarpour E (2014) Sequestration of organic carbon influenced by the application of straw residue and farmyard manure in two different soils. Int Agrophys 28(2):169–176

Majumder B, Mandal B, Bandyopadhyay PK (2008) Soil organic carbon pools and productivity in relation to nutrient management in a 20-year-old rice–berseem agroecosystem. Biol Fertil Soils 44(3):451–461

Marschner B, Kalbitz K (2003) Controls of bioavailability and biodegradability of dissolved organic matter in soils. Geoderma 113:211–235

Miller RM, Jastrow JD (1990) Hierarchy of root and mycorrhizal fungal interactions with soil aggregation. Soil Biol Biochem 22:579–584

Muller J, Miller M, Kjuller A (1999) Fungal–bacterial interaction on beech leaves: influence on decomposition and dissolved organic carbon quality. Soil Biol Biochem 31:367–374

Park JH, Kalbitz K, Matzner E (2002) Resource control on the production of dissolved organic carbon and nitrogen in a deciduous forest floor. Soil Biol Biochem 34:813–822

Preston CM, Schmidt MWI (2006) Black (pyrogenic) carbon: a synthesis of current knowledge and uncertainties with special consideration of boreal regions. Biogeosciences 3:397–420

Rovira P, Ramón Vallejo V (2003) Physical protection and biochemical quality of organic matter in Mediterranean calcareous forest soils: a density fractionation approach. Soil Biol Biochem 35:245–261

Sanderman J, Baldock JA, Amundson R (2008) Dissolved organic carbon chemistry and dynamics in contrasting forest and grassland soils. Biogeochemistry 89:181–198

Schmidt MWI, Skjemstad JO, Jager C (2002) Carbon isotope geochemistry and nanomorphology of soil black carbon: black chernozemic soils in central Europe originate from ancient biomass burning. Global Biogeochem Cycles 16(4):1–70

Schmidt MWI, Torn MS, Abiven S, Dittmar T (2011) Persistence of soil organic matter as an ecosystem property. Nature 478:49–56

Six J, Callewaert P, Lenders S, De Gryze S (2002) Measuring and understanding carbon storage in afforested soils by physical fractionation. Soil Science Society of American Journal 66:1981–1987

Smolander A, Kitunen V (2011) Comparison of tree species effects on microbial C and N transformations and dissolved organic matter properties in the organic layer of boreal forests. Appl Soil Ecol 49:224–233

Sollins P, Glassman C, Paul EA, Swanston C, Lajtha K, Heil JW, Elliott EA (1999) Soil carbon and nitrogen: pools and fractions. In: Robertson GP, Bledso CS, Coleman C, Sollins P (eds) Standard soil methods for long-term ecological research. Oxford University Press, New York, pp 89–105

Stevenson FJ (1994) Humus chemistry: genesis, composition, reaction, 2nd edn. Wiley, New York, p 496

Straathof AL, Chincarini R, Comans RN, Hoffland E (2014) Dynamics of soil dissolved organic carbon pools reveal both hydrophobic and hydrophilic compounds sustain microbial respiration. Soil Biol Biochem 79:109–116

Strobel BW, Hansen HCB, Borggaard OK, Andersen MK, Raulund-Rasmussen K (2001) Composition and reactivity of DOC in forest floor soil solutions in relation to tree species and soil type. Biogeochemistry 56:1–26

Swanston CW, Caldwell BA, Homann PS, Ganio L, Sollins P (2002) Carbon dynamics during a long-term incubation of separate and recombined density fractions from seven forest soils. Soil Biol Biochem 34:1121–1130

Tan Z, Lal R, Owens L, Izaurralde RC (2007) Distribution of light and heavy fractions of soil organic carbon as related to land use and tillage practice. Soil Tillage Res 92:53–59

Thelen KD, Fronning BE, Kravchenko A, Min DH, Robertson GP (2010) Integrating livestock manure with a corn–soybean bioenergy cropping system improves short-term carbon sequestration rates and net global warming potential. Biomass Bioenergy 34(7):960–966

Walkley A, Black IA (1934) An examination of digestion method for determining soil organic matter and a proposed modification of the chromic acid titration. Soil Sci 37:29–38

Wander MM, Traina SJ (1996) Organic matter fractions from organically and conventionally managed soils. 1. Carbon and nitrogen distribution. Soil Sci Soc Am J 60:1081–1087

Wang X, Cammeraat LH, Wang Z, Zhou J, Govers G, Kalbitz K (2013) Stability of organic matter in soils of the Belgian Loess Belt upon erosion and deposition. Eur J Soil Sci 64:219–228

Wang Y, Xu H, Wu X, Zhu Y, Gu B, Niu X, Liu A, Peng Ch, Ge Y, Chang J (2011) Quantification of net carbon flux from plastic greenhouse vegetable cultivation: a full carbon cycle analysis. Environ Pollut 159(5):1427–1434

Yano Y, McDowell WH, Aber JD (2000) Biodegradable dissolved organic carbon in forest soil solution and effects of chronic nitrogen deposition. Soil Biol Biochem 32:1743–1751

Zhao M, Zhou J, Kalbitz K (2008) Carbon mineralization and properties of water extractable carbon in soils of the south Loess Plateau in China. Eur J Soil Biol 44:158–165

Zsolnay A, Gorlitz H (1994) Water-extractable organic matter in arable soils: effects of drought and long-term fertilization. Soil Biol Biochem 26:1257–1261

29

Obtaining commercial humic products from uncomposted manures: previous acid hydrolysis to enhance yields

Carla Bellera[1] · Issa Abaalkheel[2] · Pere Rovira[1] · Jamal Alrefai[2]

Abstract

Purpose In Saudi Arabia, owing to the abundant livestock, manures are a source of commercial humic amendments. Climate (hot weather plus scarcity of water sources) makes composting very difficult and compulsory to obtain these humic amendments without any previous composting step. A substantial fraction of manures, cellulosic in nature, are unextractable by alkaline solutions. Our aim is to verify whether a sequential combination of acid hydrolysis and KOH extraction could result in higher yields and better quality of the humic extract.

Methods Acid hydrolysis was applied to manures either (1) in a single step, with HCl or H_2SO_4, or (2) in two steps (Saeman method). In both cases, acid hydrolysis was followed by extraction with hot KOH.

Results Acid hydrolysis before KOH extraction increases total extracted organic matter. If acid hydrolysis is performed, the proportion of humic fractions in the KOH extract alone is strongly increased, reaching very high values (up to 80 % of C in the KOH extract after Saeman acid hydrolysis). Acid hydrolysis enhances some optical indicators of humification (E_4E_6, aromaticity, E_{BTZ} ratio) in the KOH extract. Nevertheless, because increasing acid concentration enhances carbohydrate extraction but decreases that of compounds other than carbohydrates, the overall result may be a drop in the proportion of humic fractions in the whole extracted matter (acid hydrolysate + KOH extract).

Conclusions Acid hydrolysis improves yields of commercial humic extracts from manures. However, care must be taken to avoid a decrease in the richness of humic fractions in the obtained products.

Keywords Manures · Commercial humic amendments · Humic substances · Acid hydrolysis · Humification

Introduction

Liquid humic amendments are used in many countries to enhance crop production. Though doubtful or even negative results may appear sometimes (for instance, Pinheiro et al. 2010; Verlinden et al. 2010), recent reviews state their general usefulness for crops under field conditions (Rose et al. 2014). The use of humic amendments, usually added to irrigation water, is almost compulsory in arid countries, where soils are often extremely poor in organic matter, which translates in a general way to a poor soil biochemistry.

At a commercial scale, humic amendments are most commonly obtained from fossil sources such as leonardite, but other sources such as peat, composted lignin-rich residues, or livestock residues (manures) may be envisaged. A consequence of this heterogeneity in the sources is the heterogeneity with regard to the physico-chemical characteristics of the commercial humic products. Such a heterogeneity is enhanced by the fact that the features of the extracted humic compounds depend on the concentration and the nature of the chemical extractive agent (Stevenson 1982; Tan 2003; Zaccone et al. 2007).

The Kingdom of Saudi Arabia (KSA) is not rich in peat or leonardite, but there is abundance of feedlot livestock

✉ Pere Rovira
pere.rovira@ctfc.cat

[1] CEMFOR, Forest Science Centre of Catalonia (CTFC), Carretera de Sant Llorenç de Morunys, km 2, 25280 Solsona, Spain

[2] Petrochemicals Research Institute, KACST (King Abdulaziz City for Science & Technology), Riyadh, Saudi Arabia

(especially, cow and sheep), which makes the collection of manures in great amounts very easy—an essential condition if extraction at a commercial scale is to be applied. Livestock manure, available in impressive amounts in some countries, may be a source for commercial humic products comparable to fossil sources such as leonardite (Tan 2003).

Fresh manures do not contain true humic substances (HS), which appear during decomposition or as a result of composting. The compost-derived humic products seem to give the best results in the enhancement of plant production (Rose et al. 2014). Information about the processes associated with composting animal manures is abundant (Bernal et al. 2009); actually, the hormonal-like activity of humic extracts from composted manures has been known since the 1970s (Hernando et al. 1969a, b, 1977). In KSA, however, composting on a mass scale is extremely difficult owing to the hot climate and the scarcity of water resources. The problem is worsened by the very long times often needed to attain compost maturity, even for manures (Mondini et al. 1996; Som et al. 2009). Thus, obtaining these humic amendments without any previous composting is almost compulsory.

In a previous work (Alrefai et al. 2015), we performed extraction trials from uncomposted sheep and cow manures, using KOH solutions at a range of concentrations (0.25–2 M), at different temperatures (100–150 °C). The extracts produced in this way have properties similar to those of humic substances, but the yield was not too big and the amount of unextracted dung remained high. By solid-state ^{13}C-NMR, we showed the dominance in the residue of *O*-alkyl C forms, usually attributed to polysaccharides. It is reasonable to assume that they are mostly cellulose and other cell wall polysaccharides. These forms, barely extractable by hot KOH, are prone to be hydrolyzed by acid. Therefore, an obvious strategy to drop the amount of unextracted residue is to perform an acid attack before the alkaline extraction.

Here, we show the results of our experiences of obtaining humic alkaline extracts from uncomposted manures, with or without a previous acid hydrolysis step. The aim of our work is not only to increase the total yield (or, in other words, to decrease the amount of unextracted residue), but also to verify whether the humic character of the obtained alkaline extracts is enhanced as a result of acid hydrolysis.[1]

[1] We must state here that, since the manures used are fresh and uncomposted, in the alkaline extracts obtained from them strictly there will be no humic nor fulvic acids, but humic-like and fulvic-like acids. Nevertheless the terms 'humic' and 'fulvic' will be used throughout this paper, in order to avoid terminological complexities and make easier its reading.

Materials and methods

Manures

Sheep and cow manures were obtained from farms near Riyadh. Manures were air dried for several days in the open air and then passed through a 2 mm sieve. Subsamples were finely ground for chemical analyses. Table 1 summarizes their most relevant properties.

Alkaline extraction and acid hydrolysis

A flow diagram of the applied fractionation method is given in Fig. 1. Essentially, we applied the classic alkaline fractionation method to obtain humic and fulvic acids, either after a previous acid hydrolysis or without any previous acid hydrolysis step. All procedures were performed in triplicate.

(a) *Direct extraction with KOH* Dried ground manures (500 mg) were extracted with 20 ml of 1N KOH, for 1 h at 105 °C, in sealed Pyrex tubes placed in a heating block. Alkaline extracts were recovered by centrifugation. Unextracted residues were washed with water; the washings were recovered also by centrifugation and combined with the alkaline extracts. Extracts were filtered through sintered glass filters, acidified to pH 7.5 and stored at 4 °C until analyses.

(b) *Extraction with KOH after single-step acid hydrolysis* Dried ground manures (500 mg) were hydrolyzed with 20 ml of HCl (0.1, 0.5 or 1N), or with 20 ml of H_2SO_4 (0.1, 0.5 or 1N), for 1 h at 105 °C in sealed Pyrex tubes placed in a heating block. The hydrolysates were recovered by centrifugation. Unhydrolyzed residues were washed with water; the washings were recovered also by centrifugation and combined with the hydrolysates. Hydrolysates were filtered through sintered glass filters. Unhydrolyzed residues were then extracted with 20 ml of 1N KOH, as explained above. The two consecutive liquid extracts (acid hydrolysate and KOH extract) were not combined; each one was studied separately. Both were stored at 4 °C until analysis.

(c) *Extraction with KOH after two-step Saeman acid hydrolysis* Saeman acid hydrolysis (Saeman et al. 1945) is the common method to hydrolyze cellulose and leave the so-called Klason lignin; it has ben also applied to the quantification of lignin in animal excreta (e.g., Kraus and Marlett 1990). Briefly, dried ground manures (500 mg) were hydrolyzed in a Pyrex tube, with 2 ml of 72 % H_2SO_4 at room temperature overnight in an end-over-end shaker. Water was then added to the tube to drop acid concentration to 1N and manures were hydrolyzed at 105 °C 1 h in a heating block. Recovery of hydrolysates and further

Table 1 Main characteristics of the studied manures

	Sheep manure	Cow manure
Organic carbon (%)[a]	37.3	35.2
Total nitrogen (%)	1.51	1.66
Forms of carbon[b]		
Alkyl C (0–46 ppm)	19.9	17.8
Methoxyl C (46–59 ppm)	5.3	4.6
O-alkyl C (59–91 ppm)	44.6	45.2
Di-*O*-alkyl C (91–110 ppm)	8.4	8.7
Aromatic C (110–142 ppm)	3.3	5.1
O-aromatic C (142–156 ppm)	3.2	3.5
Carbonyl C (156–186 ppm)	9.4	10.1
Carboxyl C (186–220 ppm)	5.7	4.9

Data from Alrefai et al. (2015)

[a] Carbon and nitrogen analyzed by dry combustion in a ThermoQuest CHN analyzer

[b] By solid-state ^{13}C–CP–MAS NMR. Areas are in % of total integrated area. Identification of areas followed Song et al. (2008)

extraction with KOH were done as in (b). Again, the acid hydrolysates and the KOH extracts were not combined, but studied separately.

None of these steps was performed under inert atmosphere (N_2 or Ar). The reasons and implications of this choice are discussed further (see "Methodological constraints"). Both acidic and alkaline extracts were analyzed for carbon in a TOC analyzer. In all cases, the unextracted solid residue was recovered, dried, weighed and analyzed for carbon in a ThermoQuest CHN analyzer. From these, and taking into account the total C in the initial manure, the amount of C lost in the process was obtained:

Unaccounted C = Initial C − C in acid hydrolysate − C in the KOH extract.

Carbohydrates

In all acid hydrolysates and alkaline extracts, carbohydrates were quantified by the phenol–sulfuric method of Dubois

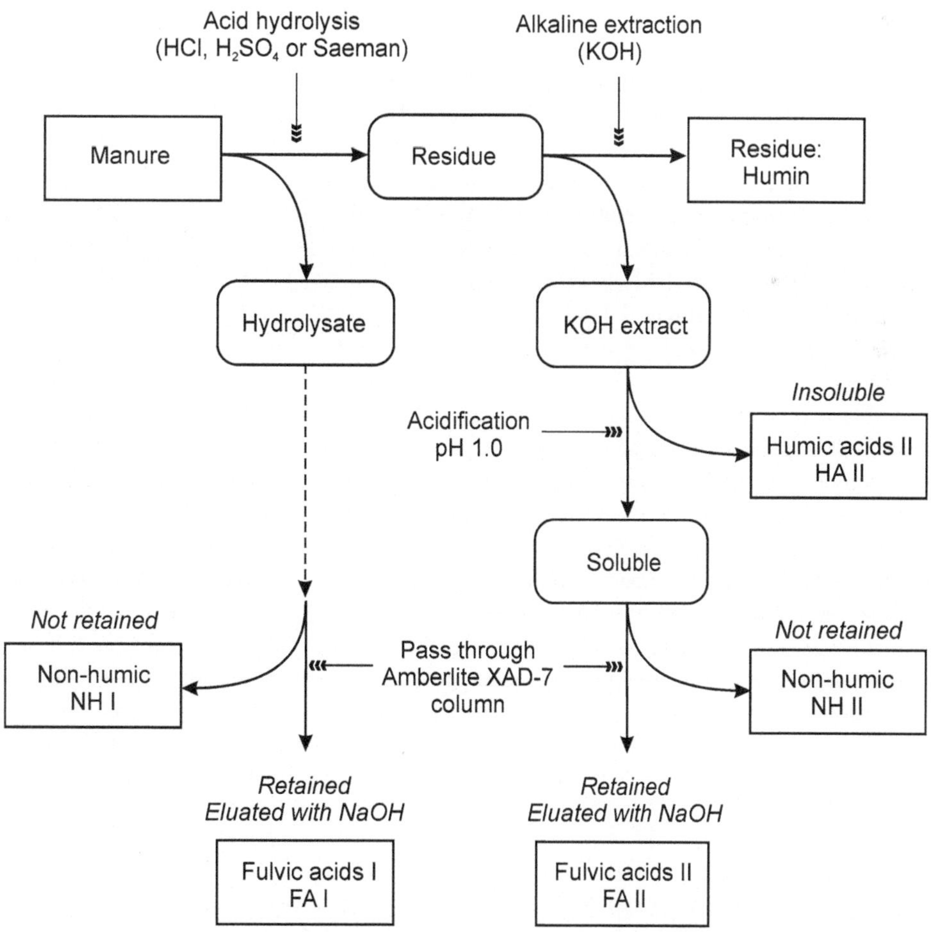

Fig. 1 Flow diagram of the fractionation method

et al. (1956), using glucose as standard. In the KOH extracts, to account for the possibility that a substantial part of carbohydrates were still in polysaccharide form, the aliquot to be analyzed was hydrolyzed previously with 1N H_2SO_4 (proportion 1:1 v/v) at 105 °C for 1 h. Carbohydrates were obtained as glucose equivalents, and its carbon was estimated by multiplying the obtained result by 0.4.

Fractionation of the hydrolysates and the KOH extracts

The KOH extracts were fractionated into humic acids, fulvic acids and non-humic compounds. Briefly, 50 ml of the extract was acidified to pH 1.0 with 2N H_2SO_4 and left to stand for 4 h. The insoluble fraction (humic acids: HA) was recovered by centrifugation and the soluble fraction (fulvic fraction) was saved. The humic acids were redissolved with 0.1N KOH, acidified again to pH 1.0 and centrifuged. The soluble fraction was saved and combined with the previous one. Humic acids were redissolved with a minimum amount of 0.1N KOH, its pH settled to 7.5, made up to 50 ml in a volumetric flask and stored at 4 °C.

The fulvic fraction (acid soluble) was fractionated through a column of Amberlite XAD-7 resin. The not retained, colorless material (non-humic fraction: NH) was recovered. The brown material, retained by the resin (fulvic acids: FA) was recovered by elution with NaOH (0.1N first and 1N thereafter), collected at the bottom of the column, acidified to pH 7.5 with 2N H_2SO_4, made up to 50 ml in a volumetric flask and stored at 4 °C.

Acid hydrolysates were also fractionated in this way, except for the first step (HA precipitation), because acid-insoluble compounds are by definition absent in them.

Both HA and FA (in the acid hydrolysate: FA I, and in the alkaline extract: FA II and HA II) were analyzed for carbon in a TOC analyzer. In the NH fractions (NH I and II), carbon was not analyzed directly, for they had become largely diluted during the fractionation. Their carbon contents were quantified by the difference: total alkaline extract − (HA II + FA II) in the alkaline extract; total hydrolysate − FA I in the acid hydrolysates.

From the C quantified in each fraction, we obtained the degree of humification (DH), according to Ciavatta et al. (1988):

$$DH(\%) = [(HA + FA)/TEC] \times 100,$$

where TEC is the total extracted carbon. This ratio measures how rich a given extract in humic fractions is and may be applied to the hydrolysate, the KOH extract or to the sum of both, i.e., the overall extracted matter. Ciavatta's index has been preferred to other indicators, because all commercial humic products must declare (i.e., in the label) its relative richness in humic substances, which is considered by the farmers to be an indicator of the quality of the purchased product.

Optical properties

Optical properties are often used as indicators of the humification degree of humic substances, and they have been suggested as a criterion to classify them (Kumada 1987). To evaluate the apparent humification degree of the obtained fractions, three optical properties were studied:

(a) The ratio between absorptions at 465 and 665 nm (E_4E_6 ratio), in solutions of 150 μg C ml^{-1} and pH 7.5, as recommended by Chen et al. (1977).
(b) In a second solution of 15 μg C ml^{-1} and pH 7.5, absorptions at 280, 253 and 203 nm were measured. The ratio between absorptions at 253 and 203 nm (Korshin et al. 1997) was then calculated (henceforth, E_{BTZ} ratio). In addition to the E_{BTZ} ratio, the E_{253}/E_{220} ratio (Fuentes et al. 2006) was also applied to the obtained humic fractions. The latter is the ratio between absorptions at 253 and 220 nm, and it was proposed as an alternative to the former, because the absorption at 203 nm may be affected by the presence of nitrate, which absorbs also at this wavelength. However in our work, the original E_{BTZ} ratio (253/203) gave more consistent results and allowed detecting more differences between the humic fractions depending on their origin (sheep or cow manure). Thus, only the results for the E_{BTZ} ratio will be shown here.

Optical properties were studied in all the humic fractions: FA I, FA II and HA II. The non-humic fractions (NH I and NH II) were not included in this study. All absorptions were measured in 1-cm quartz cuvettes, in a T80 UV/vis spectrometer (PG Instruments). The E_4E_6 ratio, a classical index of humification, has been largely considered an optical indicator of condensation of the humic molecules (Stevenson 1982). Chen et al. (1977), in a detailed study, showed that it is related to the apparent size of the molecule, but not to the concentration of condensed aromatic rings. The specific UV absorption at 280 nm (SUVA), normalized per 100 μg C ml^{-1} cm^{-1}, is widely used as an indicator of abundance of aromatic C in the solution (Chin et al. 1994; Kalbitz et al. 2003). With regard to the absorptions at 253 and 203 nm, they are related to the electron-transfer bands of benzene and to the benzenoid band of benzene rings, respectively (Korshin et al. 1997; Fuentes et al. 2006).

As shown by Fuentes et al. (2006), all these indices increase with humification and under this point of view

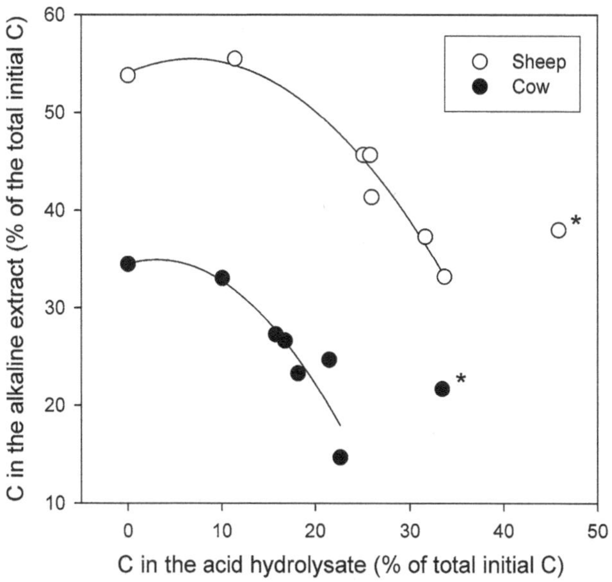

Fig. 2 Relationship between the amounts of C recovered in the acid hydrolysate and those recovered in the KOH extract. Data have been fitted to polynomial curves. Results obtained with Saeman hydrolysis, with *asterisk*, have not been included in the fitting, as they were clearly outside the main trend

were applied in our experiment, i.e., as a criterion to detect increases or decreases in the apparent degree of humification in the humic fractions obtained.

Results

Carbon balance: extracted and lost

Table 2 summarizes the amounts of total C extracted either in the acid hydrolysis (with HCl, H_2SO_4 or by the Saeman hydrolysis) or in the subsequent alkaline extraction, or the sum of both. The yields obtained by direct alkaline extraction with KOH, without any previous hydrolysis step, are given also for comparison.

For both sheep and cow manure, acid hydrolysis increases the total extracted carbon. The highest yields are always obtained through the Saeman hydrolysis. Irrespective of the experimental condition, yields are much higher for sheep than for cow manure.

For a given acid (either HCl or H_2SO_4), increasing acid concentration results in an increased extraction of carbon in the acid hydrolysate, but a decrease of carbon in the subsequent alkaline extract (Fig. 2). It is noteworthy that for both sheep and cow manures, the maximum yield in the acid hydrolysate is obtained by the Saeman hydrolysis; but the yield of the KOH extraction after this hydrolysis is one of the lowest. Thus, the amount of C extracted in the acid hydrolysis and in the KOH extraction are inversely related (Table 2). As a result, increasing acid concentration from 0.1 to 1.0N does not always increase the total amount of

Table 2 Amount of carbon extracted at each step (acid hydrolysis, KOH extraction and the sum of both) and unaccounted C, all given in g C per 100 g of initial C in the manure

Acid	Conc (N)	Hydrolysate	KOH extract	Total extracted	Unaccounted
(a) Sheep manure					
None	–	–	53.8 ± 0.4d	53.8 ± 0.4a	9.5 ± 1.5abc
HCl	0.1	11.4 ± 1.0a	55.5 ± 1.5d	67.0 ± 0.8b	3.0 ± 1.3a
	0.5	25.1 ± 1.2b	45.6 ± 1.2c	70.8 ± 1.3b	7.4 ± 1.8abc
	1.0	26.0 ± 0.5b	41.4 ± 2.0bc	67.3 ± 2.9b	12.7 ± 3.4c
H_2SO_4	0.1	25.8 ± 0.5b	45.6 ± 1.7c	71.5 ± 1.7b	5.6 ± 1.4ab
	0.5	31.7 ± 0.8c	37.3 ± 3.6ab	69.0 ± 2.9b	10.1 ± 3.9bc
	1.0	33.7 ± 1.2c	33.2 ± 9.8a	66.8 ± 8.6b	11.6 ± 1.5bc
	Saeman[a]	45.9 ± 7.3d	38.0 ± 2.7ab	83.9 ± 5.5c	6.7 ± 2.7abc
(b) Cow manure					
None	–	–	34.5 ± 1.8d	34.5 ± 1.8a	32.5 ± 11.2
HCl	0.1	10.1 ± 0.3a	33.0 ± 0.7d	43.1 ± 1.1bc	28.5 ± 6.7
	0.5	16.8 ± 0.4bc	26.7 ± 0.9c	43.4 ± 1.0bc	24.2 ± 7.6
	1.0	18.2 ± 0.7bc	23.3 ± 1.0bc	41.5 ± 1.6abc	26.1 ± 10.9
H_2SO_4	0.1	15.8 ± 0.6 ab	27.3 ± 2.2 c	43.1 ± 2.2 bc	35.4 ± 12.0
	0.5	21.5 ± 1.0bc	24.7 ± 3.1bc	46.2 ± 4.0c	26.7 ± 1.6
	1.0	22.6 ± 0.7c	14.7 ± 4.3a	37.3 ± 4.8ab	30.9 ± 2.0
	Saeman	33.4 ± 9.2d	21.7 ± 1.0b	55.2 ± 9.0d	27.5 ± 4.9

Data are averages of $n = 3$ ± standard deviations. For a given manure, data in columns followed by the same lowercase letter do not differ, at $p = 0.05$. No letters are given when no significant differences are detected

[a] In the Saeman procedure, hydrolysis is performed in two steps, involving changes in temperature and acid concentration

extracted C. Actually in one case (Table 2: cow manure, hydrolysis with 1N H_2SO_4), the total extracted C does not significantly differ from that extracted with KOH without any previous acid hydrolysis.

Carbon losses in the process (acid hydrolysis plus/or KOH extraction) are small for sheep manure, but quite high for cow manure (Table 2). Losses by volatilization (due to decarboxylation reactions or to generation of volatile organic compounds) may partly explain the low yields obtained from cow manure and are always smaller than those obtained from sheep manure.

Extracted carbohydrates

Acid hydrolysis clearly increases carbohydrate extraction (Table 3). Whatever the conditions for acid hydrolysis, the amounts of extracted carbohydrates are much higher for sheep—than for cow manure. At high acid concentrations (1N), the effectiveness of HCl and H_2SO_4 seem similar, but at lower concentrations H_2SO_4 is more effective for polysaccharide hydrolysis. For both sheep and cow manure, carbohydrate extraction is highest by the Saeman hydrolysis method: the carbohydrates extracted from sheep manure by this procedure account for more than 40 % of the total initial C, and more than 30 % from cow manure.

Using 0.1N HCl for acid hydrolysis, the amount of carbohydrates extracted in the acid hydrolysate and in the alkaline extract is similar. In the rest of the cases, the acid hydrolysate accounts for most of the extracted carbohydrates. The amount of carbohydrates in the KOH extract is irrelevant.

Increasing the amount of carbohydrates released by acid hydrolysis does not automatically translate to a net increase in total extracted C (acid hydrolysate plus KOH extract). Even though acid hydrolysis improves substantially the extraction of carbohydrates, it does not improve (even decreases) the extraction of other compounds (Table 3). Hydrolysis with 0.1N HCl is the only exception to this rule.

Fractionation of acid hydrolysates and alkaline extracts

In absolute amounts (i.e., as unit of C extracted from unit of initial C in the extracted manure), increasing the concentration of acid in the previous acid hydrolysis results mainly in a net increase in NH I and, to a lesser amount, also FA I. In contrast, all fractions of the subsequent KOH extract tend to decrease in absolute amounts, leaving aside the increase in HA at the lowest acid concentration (Fig. 3).

Yields of all fractions are higher when extracted from sheep manure. Except for the FA I fraction, the kind of acid also affects the results. For extracting non-humic substances, hydrolysis with H_2SO_4 is more efficient than with

Table 3 Amount of carbohydrates extracted at each step (acid hydrolysis, KOH extraction and the sum of both), all given in g carbohydrate C per 100 g of initial C in the manure and amount of non-carbohydrate compounds extracted in g C per 100 g of initial C in the manure

Acid	Conc (N)	Hydrolysate	KOH extract	Total extracted	Non-carbohydrates
(a) Sheep manure					
None	–	–	3.6 ± 0.3e	3.6 ± 0.3a	50.2 ± 0.4b
HCl	0.1	3.5 ± 0.7a	3.1 ± 0.3d	6.7 ± 0.4b	60.3 ± 1.0c
	0.5	22.8 ± 0.6bc	0.7 ± 0.1b	23.5 ± 0.6c	47.2 ± 1.5ab
	1.0	27.0 ± 1.4d	0.4 ± 0.0ab	27.5 ± 1.4d	39.8 ± 3.2a
H_2SO_4	0.1	21.0 ± 0.5b	0.6 ± 0.1b	21.6 ± 0.6c	49.8 ± 2.2b
	0.5	25.9 ± 0.6cd	0.7 ± 0.1b	26.6 ± 0.6d	42.3 ± 3.3ab
	1.0	27.2 ± 1.4d	1.2 ± 0.3c	28.4 ± 1.6d	38.4 ± 7.3a
	Saeman[a]	43.5 ± 4.1e	0.2 ± 0.2a	43.7 ± 3.9e	40.2 ± 2.6a
(b) Cow manure					
None	–	–	2.1 ± 0.1e	2.1 ± 0.1a	32.4 ± 1.8b
HCl	0.1	1.8 ± 0.6a	1.7 ± 0.0d	3.5 ± 0.6a	39.6 ± 1.5c
	0.5	10.9 ± 0.7b	0.9 ± 0.0c	11.5 ± 0.7b	31.9 ± 1.3b
	1.0	13.8 ± 0.4b	0.3 ± 0.0b	14.2 ± 0.4b	27.3 ± 1.4ab
H_2SO_4	0.1	9.2 ± 0.4b	0.6 ± 0.0c	9.8 ± 0.4b	33.3 ± 2.1 bc
	0.5	14.3 ± 0.7b	0.3 ± 0.0b	14.5 ± 0.7b	31.6 ± 3.3b
	1.0	15.2 ± 1.3b	0.7 ± 0.2c	15.9 ± 1.5b	21.4 ± 3.3a
	Saeman	27.1 ± 9.5c	0.1 ± 0.1a	27.2 ± 9.4c	27.9 ± 2.4b

Data are averages of $n = 3 \pm$ standard deviations. For a given manure, data in columns followed by the same lowercase letter do not differ, at $p = 0.05$

[a] In the Saeman procedure, hydrolysis is performed in two steps involving changes in temperature and acid concentration

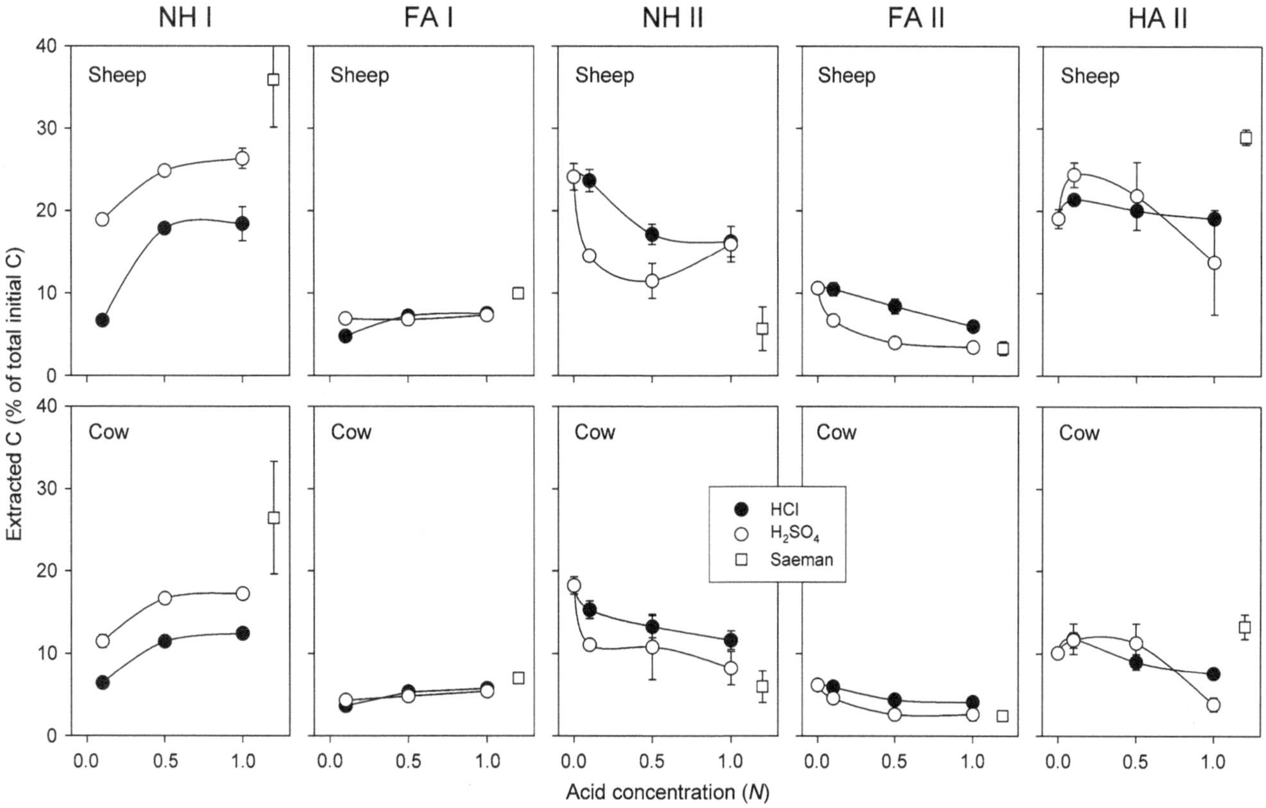

Fig. 3 Amounts of C in each extracted fraction. Points are averages of $n = 3$, and *vertical bars* show standard deviations. Data for Saeman hydrolysis have been plotted together with the other conditions, but out of scale, since in this procedure acid concentration is not constant

HCl; in the subsequent KOH extraction, this results in lower amounts of non-humic substances (NH II) and fulvic acids (FA II). In contrast, yields of humic acids (HA II) are higher using sulfuric acid in the hydrolysis. Only when using the highest concentration (H_2SO_4 1N), a drop in the yield of humic acids is observed.

Saeman hydrolysis gives results outside the general trends. It yields very high amounts of non-humic substances in the hydrolysate (NH I) and also of humic acids in the KOH extract (HA II). In contrast, the yield of non-humic substances in the KOH extract (NH II) is remarkably low for sheep-derived manure. In the rest of the fractions, Saeman hydrolysis yields amounts similar to those obtained using the highest concentration of H_2SO_4 (1N).

The overall result of the above-described behavior is a general decrease in the humic character of the total extracted carbon with acid concentration (Fig. 4, right graphs): the higher the acid concentration (for both HCl and H_2SO_4), the lower is the humification index. However, if only the KOH extract is considered (Fig. 4, middle graphs), the humification index is overall higher than for the whole C and tends to increase with acid concentration, except for the highest concentration of H_2SO_4 (1N).

For both sheep and cow manures, the highest humification indices (both I and II) are obtained when acid hydrolysis is carried out following Saeman' procedure. The humification index is particularly high for the KOH extract obtained from sheep manure: >80 % of the KOH-extracted C is found in the humic fractions (FA II and HA II).

Optical indicators of humification

E_4E_6 ratios

In the FA I fraction obtained from cow manure, the E_4E_6 ratio is not affected by acid concentration, both using HCl and H_2SO_4. However, this was the exception; in the rest of the studied fractions, the E_4E_6 was largely affected by the acid used for hydrolysis and its concentration. The results were inconsistent, depending on the kind of manure (Fig. 5).

Using HCl for hydrolysis, increasing acid concentration increases the E_4E_6 ratio of the studied fractions: the effect is clear for sheep manure and less for cow manure. In contrast, using H_2SO_4, the behavior of E_4E_6 is largely inconsistent. In humic acids (HA II), the E_4E_6 ratios

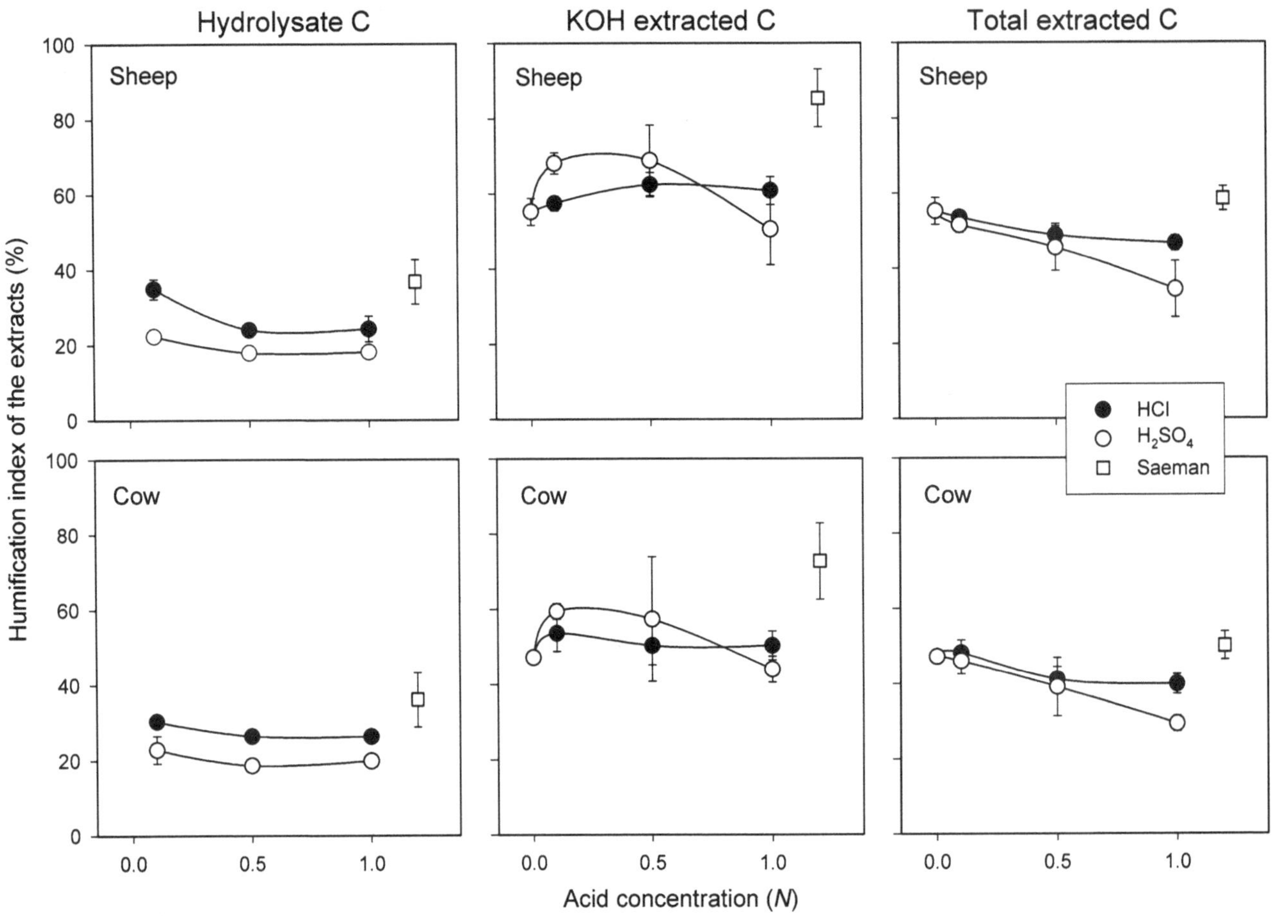

Fig. 4 Humification index (humic fractions/total C) in the acid hydrolysate (*left*), KOH extract (*middle*) and the sum of both (*right*). Points are averages of $n = 3$, and *vertical bars* show standard deviations. Data for Saeman hydrolysis have been plotted together with the other conditions, but out of scale, since in this procedure acid concentration is not constant

increase with acid concentration. For fulvic acids, a two-phase behavior is observed: E_4E_6 ratios increase for low acid concentrations, but drop again beyond a given threshold (0.5N for sheep-derived FA II, 0.1N for cow derived). In sheep-derived FA I obtained using H_2SO_4 for hydrolysis, increasing acid concentration consistently decreases E_4E_6 ratios.

Aromaticity of humic fractions (absorption at 280 nm)

The aromaticity of fulvic acids in the hydrolysates (FA I) is the lowest of all humic fractions (Fig. 6). Aromaticity seems also little affected by the acid used (HCl or H_2SO_4) or its concentration and is very similar in sheep and cow manure-derived hydrolysates.

The aromaticity of the humic fractions extracted in the subsequent KOH extraction, in contrast, is strongly affected by the previous acid hydrolysis. Aromaticities are usually higher if acid hydrolysis has been performed with H_2SO_4. If HCl is used, the aromaticity of both FA II and HA II consistently increases with the concentration of acid, for both sheep- and cow-derived manures. If H_2SO_4 is used, the aromaticity drops beyond a given threshold (either 0.1 or 0.5N), due probably to artifact generation. In HA II, the aromaticity increases with the concentration of acid; only for the highest H_2SO_4 concentration the trend seems to drop.

E_{BTZ} ratio

In contrast to the E_4E_6 ratio, or the absorption at 280 nm, the behavior of the E_{BTZ} ratio in the humic fractions was homogeneous (Fig. 7). The ratio tended to increase with the concentration of the acid used in the hydrolysis step. The trend is very consistent; only in the FA I fraction obtained with HCl, an initial drop was observed from 0.1 to 0.5N. As a general trend, the E_{BTZ} ratios are higher if acid hydrolysis has been performed with H_2SO_4. Increases are noteworthy (from about 0.22 to more than 0.40) in humic acids (HA II) when acid hydrolysis has been performed with H_2SO_4.

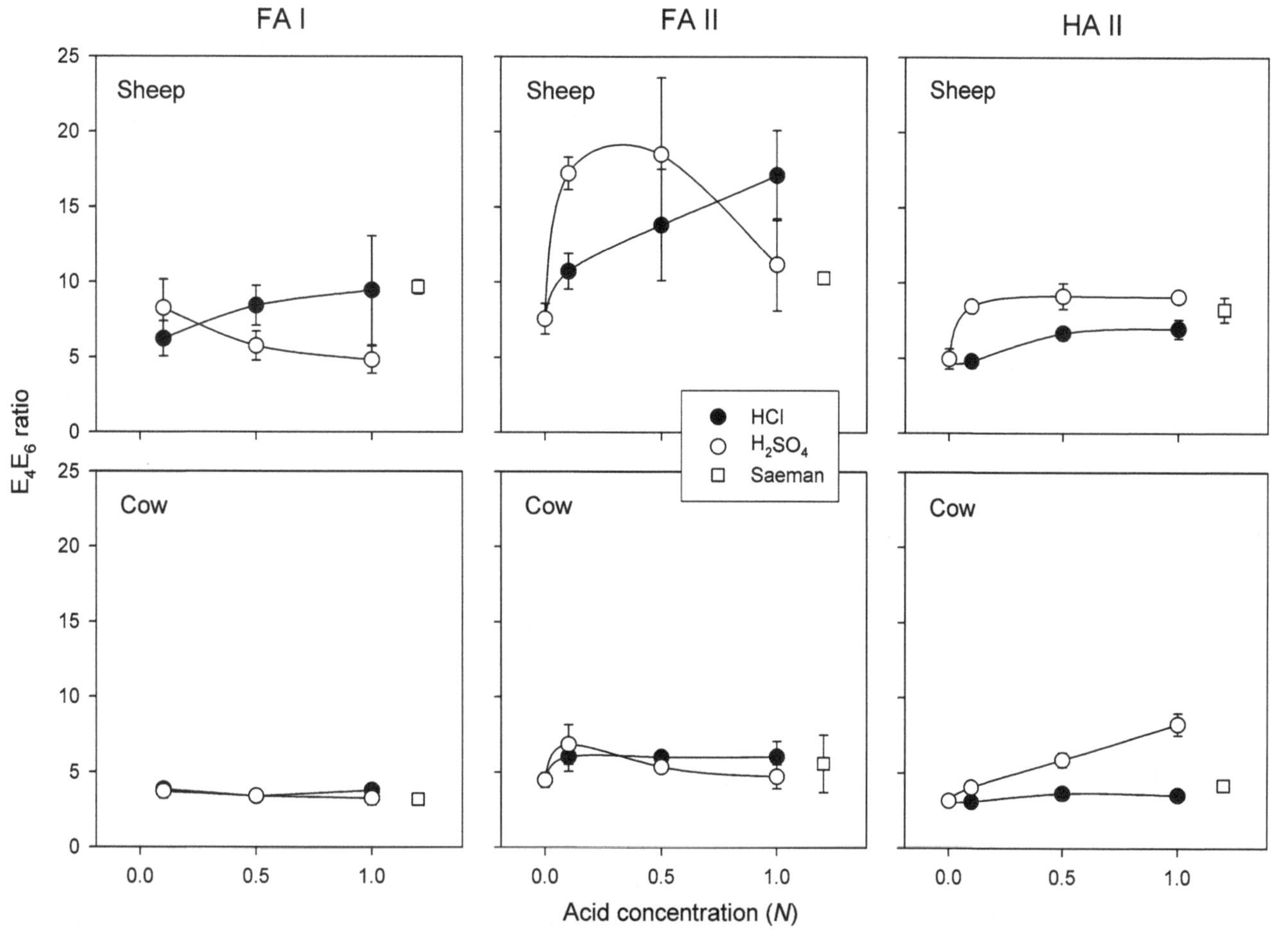

Fig. 5 E_4E_6 ratios in the three humic fractions obtained. Points are averages of $n = 3$, and *vertical bars* show standard deviations. Data for Saeman hydrolysis have been plotted together with the other conditions, but out of scale, since in this procedure acid concentration is not constant

Discussion

Methodological constraints

In comparison to other studies on obtaining humic substances from organic sources, our work shows three peculiarities: (1) KOH is used throughout, instead of the more common NaOH; (2) extractions are performed under hot conditions and short times, instead of long extractions at room temperature; and (3) an inert atmosphere (of either N_2 or Ar) was not applied in any of the extraction steps. All these peculiarities reflect the aim of applying extraction conditions as similar as possible to those expected when the extraction of humic products from manures will be applied at an industrial scale.

In scientific studies on humic substances, avoiding the generation of artifacts during the extraction is a main concern, and mild conditions are the common choice (extractions at room temperature and low NaOH concentrations, balanced by very long extraction times, often overnight): yield considerations are secondary. When humic amendments are to be obtained at a commercial scale, in contrast, yield considerations may never be secondary. Thus, harsh extraction conditions are usual. For instance, KOH is used at high concentrations (1 M or higher) to reach high extractions of humic matter, at the price of obtaining highly saline humic products, even though this is a minor problem because commercial humic products are largely diluted when applied to crops. With some exceptions (e.g., Garcia et al. 1994), it is usually observed that yields of extracted humic matter increase with temperature, at the price of shifts in the chemical characteristics of the extracted humic matter (Yefimov and Vasil'kova 1970; Cegarra et al. 1974; Garcia et al. 1994). Under these conditions, the obtained alkaline extracts may be partly artifactual, as well as their humic character.

We observed (Alrefai et al. 2015) that increasing extraction temperature of KOH concentration may mimic in the alkaline extracts the effect of composting, as deduced from their optical properties (Fuentes et al. 2006).

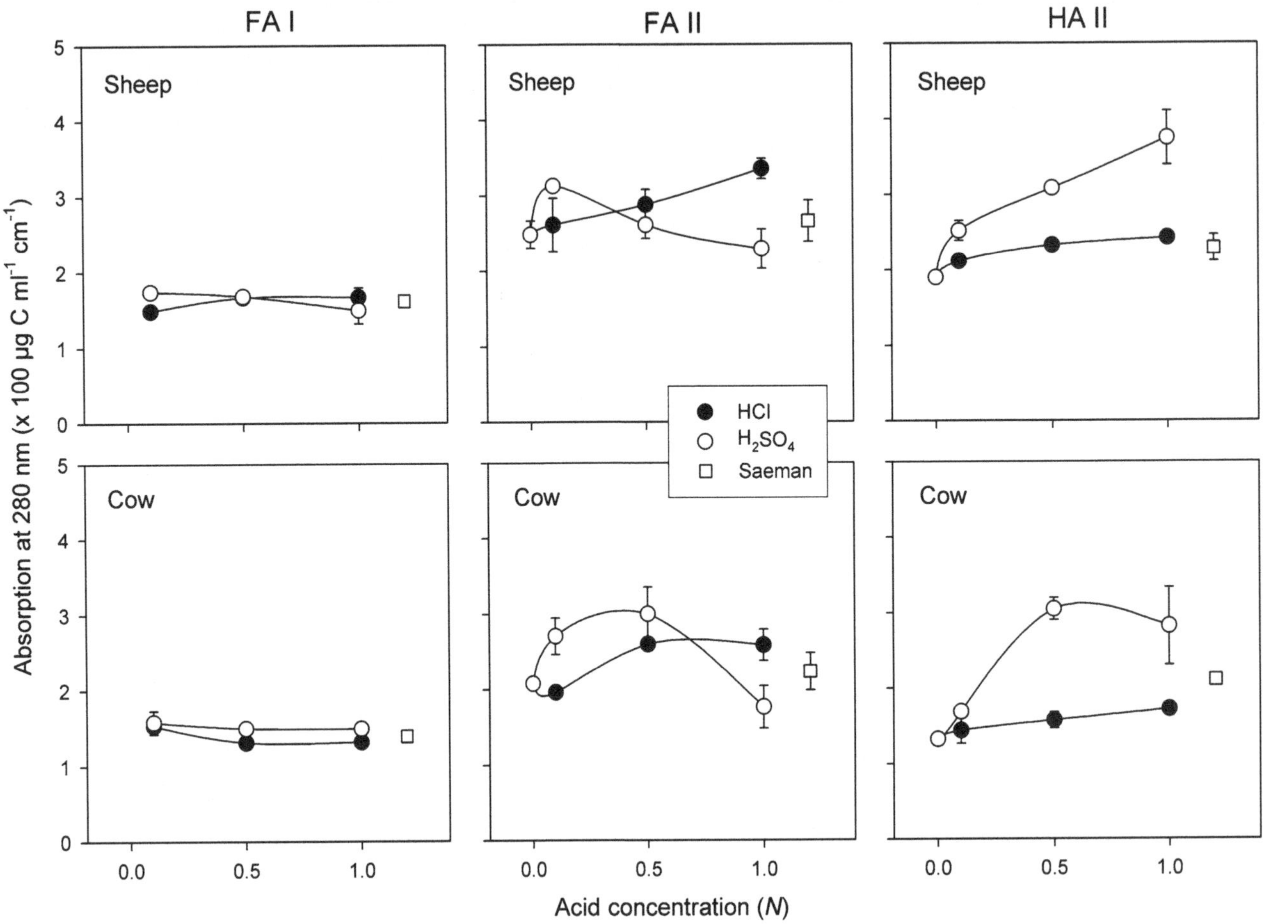

Fig. 6 Specific UV absorption (SUVA) values, normalized as per solutions of 100 µg C per ml, measured in 1 cm quartz cuvettes. Points are averages of $n = 3$; *vertical bars* show standard deviations. Data for Saeman hydrolysis have been plotted together with the other conditions, but out of scale, since in this procedure acid concentration is not constant

The results of the current paper agree with this trend: most optical indicators of humification in the humic fractions (FA I and II, HA II) tend to increase with the concentration of acid used in the previous acid hydrolysis.

This principle may be also applied to the use of inert atmospheres (N_2 or Ar) for humus extraction, a recommended practice in scientific studies (Swift 1996). There are detectable differences between the humic fractions extracted under inert atmosphere and those extracted in the presence of atmospheric O_2 (Tan et al. 1991). However, applying an inert atmosphere for industrial-scale extraction of humus compounds would increase their cost notably, thus making them largely unattractive for farmers.

The unavoidable consequence of these harsh conditions (hot temperature, short time, lack of inert atmosphere) is the generation of artifacts during the hydrolysis and the further KOH extraction. However, we must accept that a substantial fraction of most commercial humic products are artefactual in nature. Artifact generation is the most reasonable explanation for most inconsistencies observed in our work about the behavior of humic fractions when increasing acid concentration in the hydrolysis step, such as erratic behaviors for E_4E_6 ratios (Fig. 5) or sudden drops in aromaticity in FA II (Fig. 6).

Usefulness of acid hydrolysis

To obtain commercial humic products at an industrial scale, an acid hydrolysis step before the standard KOH extraction implies an added cost, which translates to the final product. Acid hydrolysis may be useful only if this added cost is balanced by yield increases, improvements of the obtained product—increased humic character—or both.

Yield increases are not always satisfactory, owing to the generation of artifacts: acid hydrolysis substantially improves the extraction of carbohydrates, but often it does not improve or even decrease that of other compounds (Table 3). Increasing acid concentration increases the yield of the acid hydrolysate, but decreases the yield of the further KOH extract, particularly for sheep manure

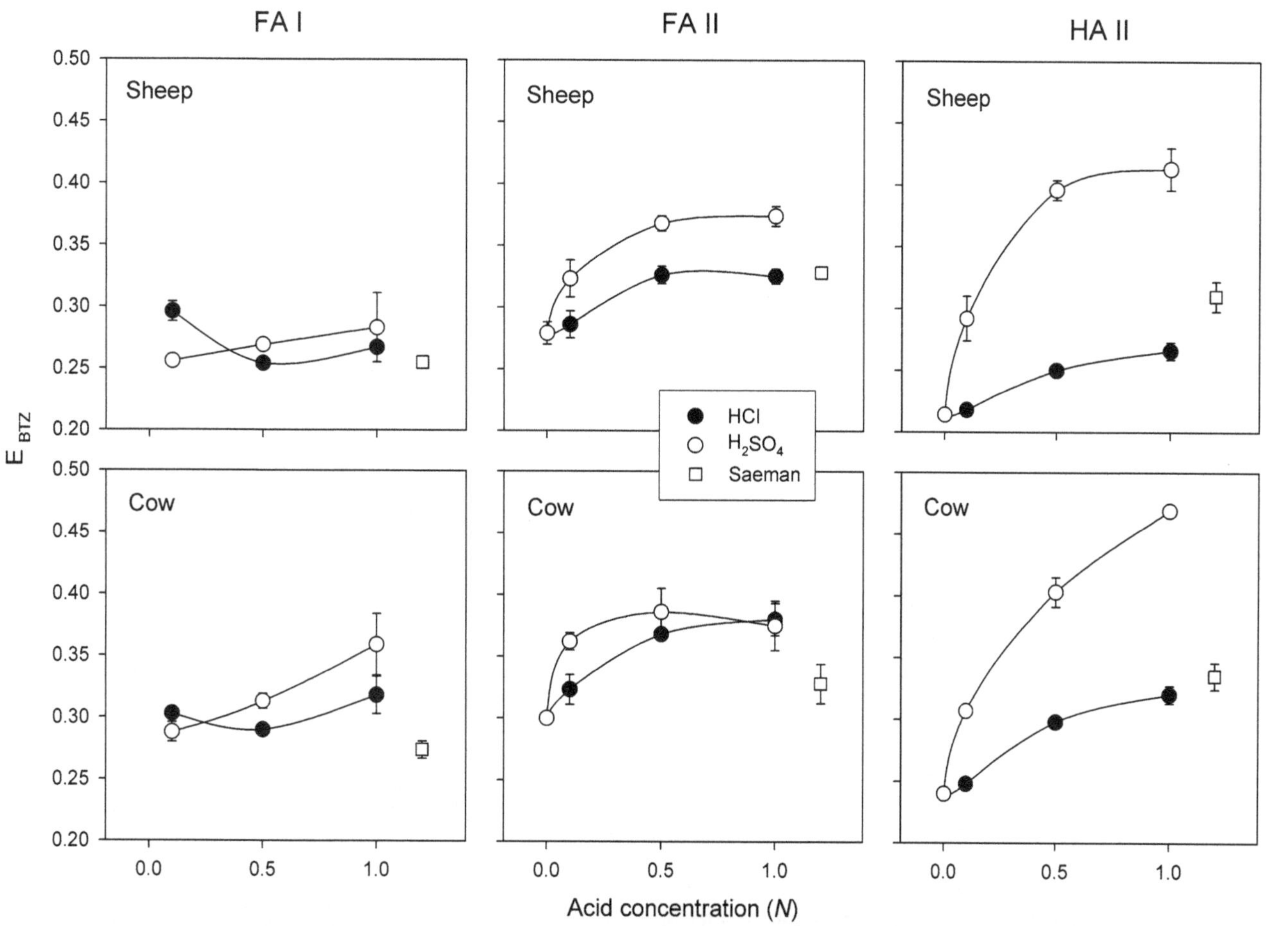

Fig. 7 Ratio between absorptions at 253 and 203 nm (E_{BTZ} ratio). Points are averages of $n = 3$, and *vertical bars* show standard deviations. Data for Saeman hydrolysis have been plotted together with the other conditions, but out of scale, since in this procedure acid concentration is not constant

(Table 2). The highest yields are obtained by the Saeman hydrolysis, but this procedure is longer and more complex than the rest and its applicability at a commercial scale is unclear.

As to the changes in the characteristics of the extracted matter, the proportion of humic fractions to total extracted matter (HD index) decreases with increasing acid concentration (Fig. 4). This is a relevant result, because commercial humic products are evaluated by their potential users (farmers) by their richness in humic or humic compounds, which in many countries must be stated in their labels, by law. Under such a point of view, acid hydrolysis may be a self-defeating practice. Nevertheless, it is not so simple. As shown in Fig. 6, the decrease in the humification index is largely due to the acid hydrolysate: on the contrary, in the KOH extract the HD may increase after acid hydrolysis. The apparent humification indices obtained in the KOH extract after Saeman's hydrolysis are very high, particularly for sheep manure. On the other hand, most optical indicators used here show that increasing acid concentration may improve the apparent humic character of the fulvic and humic acids in the KOH extract: E_{280}, E_{BTZ} and E_4E_6 in FA II and HA II tend to increase with acid concentration, though not always in a consistent manner.

There are several ways by which acid hydrolysis may be useful for manufacturers. The first and most obvious method is to apply the conditions that apparently give the best results. Our results suggest that acid hydrolysis with a dilute acid solution (0.1N) is the best option: it results in increased total yield without any decrease in the humification ratio. The drawback of this option is that the small increase in total yield obtained under these conditions could not justify the cost of a previous acid hydrolysis. In such a case, other options could be considered:

(a) Taking separately the acid hydrolysate (rich in carbohydrates, poor in humic matter) and the KOH extract (poor in carbohydrates, rich in humic matter), rather than joining them in a single humic extract, is an option that may be considered. In this way, only the KOH extract, highly humic in character (Fig. 4), would be retained for use in agriculture as humic amendment. The challenge of this option should be to find a different and specific use for the acid hydrolysate: if this goal is attained, obtaining two different and consecutive products from a single substrate could be commercially interesting.

(b) Enhancing the humic character of the overall product (at least, the acid hydrolysate) by further chemical treatments. This option would take advantage of the available information about obtaining humic substances by purely chemical methods, a matter of study in soil science for many years (e.g., among many, Arfaioli et al. 2003; Cataldo 1998; Datta et al. 2001; Mathur and Schnitzer 1978; Wang and Huang 2000). The genesis of chemical analogs of humic substances relies often on secondary reactions between carbohydrates and amino acids in the presence of catalyzers; since carbohydrates and amino acids or peptides are expected to be the main constituents of the acid hydrolysate, such an option should be technically possible. Whether or not it could be feasible at an industrial scale from these hydrolysates should be the matter of future research.

End comment

In spite of the promising results obtained in this paper and in our previous one (Alrefai et al. 2015), a further step must be the study of the effects on plant growth of the obtained humic extracts and verifying the lack of deleterious effects on crops. The toxicity of manures usually disappears with composting. Since composting was avoided in our experiments, we must expect that the strong acid and/or alkaline treatments applied would have a similar effect, in the sense of provoking in the liquid extracts chemical changes strong enough to suppress any toxic effect. This compulsory step must be undertaken before translating our findings to a commercial or industrial scale.

Acknowledgments This researched was financed by Grant Number 33-785, of the KACST (King Abdulaziz City for Science & Technology). The authors acknowledge also the CTFC (Forest Science Centre of Catalonia) for their technical support. Pere Rovira is a member of the Graccie-CONSOLIDER research network of the Spanish Ministry of Science and Innovation and also a member of the Consolidated Research Group CEMFOR (2014 SGR 1526), recognized by the Generalitat de Catalunya.

References

Alrefai J, Rovira P, Alsewailem F, Abaalkheel I, Alsolami A, Albishi H, Alfantokh S (2015) Un-composted sheep and cow manures as a source of commercial humic amendments: an exploratory study. Commun Soil Sci Plant Anal 46:1137–1156. doi:10.1080/00103624.2015.1019084

Arfaioli P, Ugolini FC, Bosetto M, Corti G (2003) Synthesis of humic-like substances under sterile and non-sterile conditions in presence of quartz and volcanic ash. Commun Soil Sci Plant Anal 34:285–298

Bernal MP, Alburquerque JA, Moral R (2009) Composting of animal manures and chemical criteria for compost maturity assessment. A review. Bioresr Technol 100:5444–5453

Cataldo F (1998) On the structure of macromolecules obtained by oxidative polymerization of polyhydroxyphenols and quinones. Polym Intern 46:263–268

Cegarra J, Reverte L, Lax A, Costa F (1974) Factores que influyen en la extracción y fraccionamiento de la materia orgánica del suelo. Anal Edafol Agrobiol 33:575–590

Chen Y, Senesi N, Schnitzer M (1977) Information provided on humic substances by E4/E6 ratios. Soil Sci Soc Am J 41:352–358

Chin YP, Aiken G, Loughlin EO (1994) Molecular weight, polydispersity, and spectroscopic properties of aquatic humic substances. Environ Sci Technol 28:1853–1858

Ciavatta C, Vittori Antisari L, Sequi P (1988) A first approach to the characterization of the presence of humified materials in organic fertilizers. Agrochimica 32:510–517

Datta A, Sanyal SK, Saha S (2001) A study on natural and synthetic humic acids and their complexing ability towards cadmium. Plant Soil 235:115–125

Dubois M, Gilles KA, Hamilton JK, Rebers PA, Smith F (1956) Colorimetric method for the determination of sugars and related substances. Anal Chem 28:350–356

Fuentes M, González-Gaitano G, García-Mina JM (2006) The usefulness of UV-visible and fluorescence spectroscopies to study the chemical nature of humic substances from soils and composts. Org Geochem 37:1949–1959

Garcia D, Cegarra J, Roig A, Abad M (1994) Effects of the extraction temperature on the characteristics of a humic fertilizer obtained from lignite. Bioresr Technol 47:103–106

Hernando V, Sánchez Conde MP, Ortega BC (1969a) Acción del humato sódico sobre el desarrollo de diversas plantas (Action of sodium humate on the growth of several plants). Anal Edafol Agrobiol 28:791–809

Hernando V, Sánchez Conde MP, Ortega BC (1969b) Acción del ácido húmico sobre la planta de maíz cultivada en soluciones nutritivas equilibradas de concentración superior a la normal (Action of sodium humate on corn plants grown on nutrient solutions of concentration higher than usual). Anal Edafol Agrobiol 28:835–846

Hernando V, Ortega BC, Fortún C (1977) Estudio de la acción ejercida sobre la planta de maíz por dos tipos de ácido húmico [a study of the effect of two kinds of humic acid on corn plants]. In: Soil organic matter studies, vol 2, Rapport IAEA—SM—211/57. Pergamon Press: Oxford, pp 307–317

Kalbitz K, Schmerwitz J, Schwesig D, Matzner E (2003) Biodegradation of soil-derived dissolved organic matter as related to its properties. Geoderma 113:273–291

Korshin GV, Li CW, Benjamin MM (1997) Monitoring the properties of natural organic matter through UV spectroscopy: a consistent theory. Water Res 31:1787–1795

Kraus RJ, Marlett JA (1990) Measuring lignin in human and rat faeces. J Sci Food Agric 52:275–286

Kumada K (1987) Chemistry of soil organic matter. Elsevier, New York, p 241

Mathur SP, Schnitzer M (1978) A chemical and spectroscopic characterization of some synthetic analogues of humic acids. Soil Sci Soc Am J 42:591–596

Mondini C, Chiumenti R, da Borso F, Leita L, de Nobili M (1996) Changes during processing in the organic matter of composted and air-dried poultry manure. Bioresour Technol 55:243–249

Pinheiro GL, Silva CA, Furtini Neto AE (2010) Crescimento e nutriçâo de clone de eucalipto em resposta à aplicaçâo de concentraçôes de C-ácido húmico (Growth and nutrition of eucalyptus clone as affected by the application of C-humic acid doses). Rev Bras Ci Solo 34:1217–1229

Rose MT, Patti AF, Little KR, Brown AL, Jackson WR, Cavagnaro TR (2014) A meta-analysis and review of plant growth response to humic substances: practical implications for agriculture. Adv Agron 124:37–89

Saeman JF, Bubl JL, Harris EE (1945) Quantitative saccharification of wood and cellulose. Ind Eng Chem 17:35–37

Som MP, Lemée L, Amblès A (2009) Stability and maturity of a green waste and biowaste compost assessed on the basis of a molecular study using spectroscopy, thermal analysis, thermodesorption and thermochemolysis. Bioresour Technol 100:4404–4416

Song G, Novotny EH, Simpson A, Clapp CE, Hayes MHB (2008) Sequential exhaustive extraction of a Mollisol soil, and characterization of humic components, including humin, by solid and solution state NMR. Eur J Soil Sci 59:505–516

Stevenson FJ (1982) Humus chemistry. Genesis, composition, reactions. Wiley, New York

Swift RS (1996) Organic matter characterization. In: Sparks DL (ed), Methods of soil analysis, part 3, chemical methods. SSSA Book series no. 5, pp 1011–1070

Tan KH (2003) Humic matter in soil and the environment: principles and controversies. Marcel Dekker, New York

Tan KH, Lobartini JC, Himmelsbach DS, Asmussen LE (1991) Composition of humic acids extracted under air and nitrogen atmosphere. Commun Soil Sci Plant Anal 22:861–877

Verlinden G, Coussens T, de Vliegher A, Baert G, Haesaert G (2010) Effect of humic substances on nutrient uptake by herbage and on production and nutritive value of herbage from sown grass pastures. Grass Forage Sci 65:133–144

Wang MC, Huang PM (2000) Characteristics of pyrogallol-derived polymers formed by catalysis of oxides. Soil Sci 165:737–747

Yefimov VN, Vasil'kova MG (1970) Method of extracting humic substances from peat soils. Sov Soil Sci 2:335–345

Zaccone C, Cocozza C, D'Orazio V, Plaza C, Cheburkin A, Miano TM (2007) Influence of extractant on quality and trace elements content of peat humic acids. Talanta 73:820–830

A bibliometric analysis on the solid waste-related research from 1982 to 2013 in Iran

Alireza Mesdaghinia[1,2] · Amir Hossein Mahvi[2,3] · Simin Nasseri[1,2] · Ramin Nabizadeh Nodehi[2] · Mahdi Hadi[1]

Abstract

Background A bibliometric analysis was used to assess the trend of solid waste-related publications produced by Iran and indexed in Scopus from 1982 to 2013. The study aims to analyze the distribution of languages, countries, institutes, journals, author keywords, authorship pattern and co-authorship relationships.

Results The publication outputs showed an exponential trend ($R^2 = 0.98$) and increased by 45.3 % per year from 1982 to 2011. It was received with a decreasing trend to 44 papers at the end of 2013. *Journal of Environmental Studies* published the most papers. University of Tehran (25 %), Islamic Azad University (8.24 %), Tehran University of Medical Sciences (7.35 %) played active roles in the publication and Abduli MA was the most productive author (9.12 %) in this field. The English language was dominant language of all publications (88.24 %). The appearance of keywords namely "Heavy metals", "landfill leachate" and "site selection" reveals the importance of release of heavy metals through landfill leachates and site selection issues in Iran.

Conclusions A downward trend in the number of publications is likely to be continued in future years. The highest cooperation was found among the capital universities in Iran and other institutions had a minor role in the production of articles. It is hoped that transferring the experiences by main universities and starting up teamwork activities may increase the number of researches in this field. It is expected that this study could be the basis for a better understanding of development of researches related to solid waste in Iran.

Keywords Solid waste · Iran · Bibliometric analysis · Scopus

Introduction

The attention to the solid waste-related problems is paid by large amount of scientific articles originated from many countries around the world (Fu et al. 2010). So far, numerous studies have been conducted in the field of solid waste management and engineering in Iran. In one of the earliest studies in this field, Chokouhmand (1982) studied the energy recovery by incineration of municipal solid waste. He established a laboratory to obtain the physical and chemical compositions of the solid waste and analyzed the amount of collected solid waste data. The practical aspects associated with solid waste management systems in Tehran (the capital of Iran) such as equipment management, personnel, organizational structure, financing, cost accounting and budgeting were described by Abduli (Abduli 1995). The status of industrial waste management in the capital of Iran and the principle guidelines and policies regarding the collection, handling and safe disposal were also assessed by Abduli (Abduli 1996). In the recent years, municipal solid waste (MSW) has been one of the most important environmental concerns throughout Iran's

✉ Mahdi Hadi
hadi_rfm@yahoo.com

1 Center for Water Quality Research (CWQR), Institute for Environmental Research (IER), Tehran University of Medical Sciences, Tehran, Iran

2 Department of Environmental Health, School of Public Health, Tehran University of Medical Sciences, Tehran, Iran

3 Center for Solid Waste Research (CSWR), Institute for Environmental Research (IER), Tehran University of Medical Sciences, Tehran, Iran

regions and universities and research institutes around the country have produced several publications in this field. Assessing the research trend, identification of the most influential research institutions, authors, journals, and the collaboration between universities and between authors, in the field of solid waste management and technology, may help relevant researchers to realize the scope of solid waste more carefully and establish the further research direction in Iran.

The bibliometric methods have been used commonly in many disciplines of science and engineering to study the scientific production and research trends (Fu et al. 2013, 2014; Ho 2014; Tan et al. 2014). Fu et al. (2010) evaluated the distribution pattern of the publications of solid waste researches from 1993 to 2008. Today many studies have been conducted by focusing on the solid waste management and engineering but the trend of researches in the field of solid waste has not been studied yet in Iran. Thus, a systematic gathering and analyzing data to get a panoramic look at the solid waste researches is necessary to be conducted in Iran.

This study aims to find the most important authors in this field and to assess the interaction between universities and authors around the country. The bibliometric methods often evaluate the trend of the research by publication outputs of countries, research institutes, journals, and the research field analysis (Braun et al. 1995; Colman et al. 1995; Ugolini et al. 1997). The analysis of the author keywords, languages and co-authorship pattern are also going to be conducted in this study.

A research tool utilized by bibliometric practitioners includes the use of Scopus, a bibliographic database containing the abstracts and citations for academic journal articles. Scopus is owned by Elsevier B.V. Company (Elsevier B.V. Company 2014) and in this study the bibliometric data were based on the Scopus bibliographic database.

In this study, all solid waste-related researches produced during 1982–2013, indexed by Scopus, were analyzed by emphasizing on the article production, keywords analysis, authorship pattern analysis, co-authorship networks mapping, the distribution of countries, institutes, authors, journals, subject areas and languages, to provide a more comprehensive and complete assessment of the solid waste-related researches in Iran.

Methods

The data was based on the Scopus bibliographic database. Scopus is a database containing abstracts and citations for academic journal articles. It covers nearly 21,000 titles from over 5000 publishers, of which 20,000 are peer-reviewed journals in the scientific, technical, medical, and social sciences (including arts and humanities) (Elsevier B.V. Company 2014). Scopus offers about 20 % more coverage than Web of Science, whereas Google Scholar offers results of inconsistent accuracy (Falagas et al. 2008). In this study for bibliometric analysis, the Scopus was searched on 15 March 2014 with a search script (presented in "Appendix") to compile the bibliography of all publications related to the solid waste researches conducted in Iran.

The final number of publications fulfilling the search script criteria was 407. The collaboration type was determined by assessing the addresses of the authors. The term "Single Country" (SC) was assigned if the all authors of paper were from the same country and the term "Multiple Country" (MC) was assigned if the article was co-authored by researchers from different countries. The term "Single Institute" (SI) was assigned if researchers were from the same institute and the term "Multiple Institute" (MI) was assigned if the article was co-authored by researchers from different institutes. Collaboration type was also assessed in term of the names of authors. The term "Single Author" (SA) was assigned if the publication was produced by one author and the term "Multiple Author" (MA) was assigned the article was co-authored by at least two researchers (Chiu and Ho 2005).

In this study all articles referring to the researches on the solid waste, during the past 31 years (1982–2013), were assessed in term of following aspects: distribution of languages and publication year, distribution of countries, co-authorship relations among countries, distribution of journals and subject categories, distribution of author keywords, distribution analysis of authors and institutions, authorship pattern analysis and co-authorship relations among authors. All analyses referring to the document type, language, journal, country, institutes and author were performed using R programming language v.3.0.2 (R Core Team 2013). The VOSviewer v.1.5.5 was used for constructing and viewing bibliometric maps.

Results and discussion

In all publications met the search criteria ("Appendix"), four document types were included. The journal articles (published or in-press) were the most frequently used document type (340; 83.54 %). The three other types were the conference papers (51; 12.53 %), review papers (14; 3.44 %) and book chapters (2; 0.49 %). Because journal articles are peer-reviewed within this field and were dominant among the document types, only journal articles were selected for further analysis and all others were discarded. There was no marked change in the number of

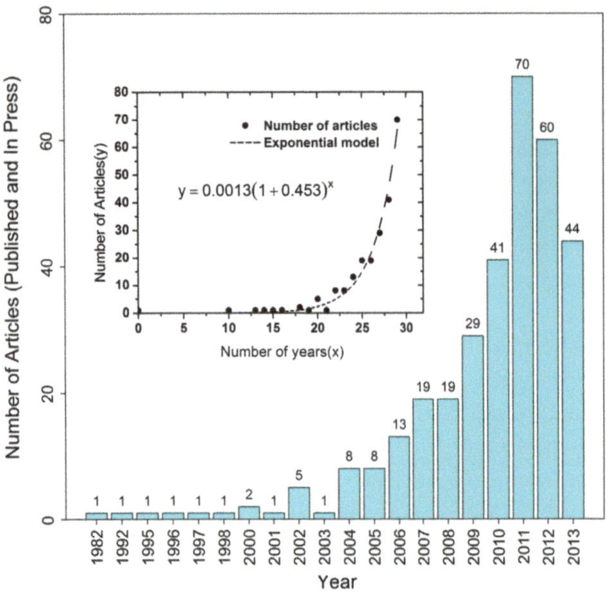

Fig. 1 Annual journal articles publication

conference papers during the period from 1977 to 2013. The distribution of annual journal article publications is shown in Fig. 1. A significant decreasing trend in the number of research articles from 2011 to 2013 is notable. It is obvious that there was a significant exponential increasing trend ($R^2 = 0.98$) in the number of article publications and they were increased by 45.3 % per year during the period from 1982 to 2011. From Fig. 1 it can be seen that the number of articles in this period increased from one paper in 1982 to 70 papers in 2011 and then received with a downward trend to 44 papers at the end of 2013. This sudden downward trend in the number of articles started in 2011, is likely to continue in the future. Given the importance of environmental issues and solid waste management in Iran, the policy makers must take some supportive actions to hinder this downward trend in the number of the research publications in this area of science.

In bibliometric analysis, it is important to note the language of publication. The Analysis of the language distribution revealed that the English was the predominant language of articles on solid waste-related studies from 1982 to 2013 in Iran. Out of the 340 records, English was occupied the first position with 300 article records (88.24 %). There were four other languages in addition to English, which were Persian (27; 7.94 %), Arabic (6; 1.76), Persian and English (6; 1.76 %), English and Spanish (1; 0.29 %). Therefore, the English language is the main language of article publications. English language is the dominant language in bibliometrics because it is official language in many countries and many articles and conference proceedings are published in English language only (Patra et al. 2006). The scientific bibliometric studies conducted so far (Fu et al. 2010; Ma et al. 2011; Yang et al. 2013), revealed that the English language is the predominant language of the publications in the field of solid waste.

The contribution analysis of the co-countries publications was based on journal articles in which the address and affiliation of at least one author were provided. Out of 340 articles that Iran are involved, 280 (82.35 %) were independent papers produced by Iran and 60 papers (17.65 %) were internationally collaborative publications between Iran and other countries (Table 1). Nineteen countries were

Table 1 Most productive institutions in research on solid waste from 1982 to 2013

Institute	TP (%) R	SI (%) R	MI (%) R	SI_P	MI_P
University of Tehran	85 (25) 1	42 (27.27) 1	43 (23.12) 1	49.41	50.59
Islamic Azad University/Science and Research Branch	28 (8.24) 2	5 (3.25) 4	23 (12.37) 2	17.86	82.14
Tehran University of Medical Sciences	25 (7.35) 3	7 (4.55)2	18 (9.68) 4	28	72
Tarbiat Modares University	24 (7.06) 4	5 (3.25) 4	19 (10.22) 3	20.83	79.17
Iran University of Science and Technology	18 (5.29) 5	5 (3.25) 4	13 (6.99) 5	27.78	72.22
Isfahan University of Technology	15 (4.41) 6	7 (4.55) 2	8 (4.3) 6	46.67	53.33
Amirkabir University of Technology	11 (3.24) 7	3 (1.95) 10	8 (4.3) 6	27.27	72.73
Shiraz University	10 (2.94) 8	2 (1.3) 16	8 (4.3) 6	20	80
Toosi University of Technology	9 (2.65) 9	5 (3.25) 4	4 (2.15) 18	55.56	44.44
Isfahan University of Medical Sciences	8 (2.35) 10	3 (1.95) 10	5 (2.69) 11	37.5	62.5
University of Guilan	8 (2.35) 10	5 (3.25) 4	3 (1.61) 25	62.5	37.5
University of Kurdistan	8 (2.35) 10	1 (0.65) 24	7 (3.76) 9	12.5	87.5
Ferdowsi University of Mashhad	7 (2.06) 13	2 (1.3) 16	5 (2.69) 11	28.57	71.43
Iran University of Medical Sciences	7 (2.06) 3	0 (0) 59	6 (3.23) 10	0	85.71
Sharif University of Technology	7 (2.06) 13	3 (1.95) 10	4 (2.15) 18	42.86	57.14

TP Total publication, *R* Rank, *MA* Multiple author publications, *SA* Single author publications, SA_P Percent of Single author publication of the total publications, MA_P Percent of multiple author publications of the total publications involved in the production of articles with Iran. The Malaysia contributes to produce of 5.88 % of total article publications (TP) or 33 % of the total articles were produced by international collaboration (MC). The United States is the second country has cooperated with Iran in the production of 3.24 % of total article publications (18.33 % of internationally collaborative publications) in the field of solid waste. The Canada (2.94 % TP and 16.67 % MC) and Germany (1.18 % TP and 6.67 % MC) were ranked third and fourth in terms of being involved in the producing of papers with Iran.

The top 15 institutes within 31 years period are displayed in Table 1. Tehran University ranked first among all institutes. Out of 340 articles, 85 (25 %) articles produced by Tehran University. 50.59 % of this university's publications produced by inter-institute collaboration and the rest (49.41 %) independently. 27.27 % of all "single institute" publications (154 articles) and 23.12 % of all "inter-institute publications" (186 articles) were attributed to the University of Tehran. Islamic Azad University (Science and Research Branch), Tehran University of Medical Sciences and Tarbiat Modares University with 28

Table 2 Most productive authors with at least 4 papers in research on solid waste from 1982 to 2013

Author	TP (%) R	MA (%) R	SA (%) R	SA_P	MA_P
Abduli MA	31 (9.12) 1	28 (8.81) 1	3 (13.64) 1	9.68	90.32
Safari E	9 (2.65) 2	7 (2.2) 3	2 (9.09) 2	22.22	77.78
Shariatmadari N	9 (2.65) 2	9 (2.83) 2	0 (0) 17	0	100
Mehrdadi N	7 (2.06) 4	7 (2.2) 3	0 (0)17	0	100
Nabizadeh R	7 (2.06) 4	7 (2.2) 3	0 (0) 17	0	100
Naddafi K	7 (2.06) 4	7 (2.2) 3	0 (0) 17	0	100
Farzadkia M	6 (1.76) 7	6 (1.89) 7	0 (0) 17	0	100
Noori R	6 (1.76) 7	6 (1.89) 7	0 (0) 17	0	100
Sabour MR	6 (1.76) 7	6 (1.89) 7	0 (0) 17	0	100
Jaafarzadeh N	5 (1.47) 10	5 (1.57) 10	0 (0) 17	0	100
Karbassi AR	5 (1.47) 10	5 (1.57) 10	0 (0)17	0	100
Mosaferi M	5 (1.47) 10	5 (1.57) 10	0 (0) 17	0	100
Olfati JA	5 (1.47) 10	5 (1.57) 10	0 (0) 17	0	100
Torabian A	5 (1.47) 10	5 (1.57) 10	0 (0) 17	0	100
Abbasi M	4 (1.18) 15	4 (1.26) 15	0 (0) 17	0	100
Ahmadi M	4 (1.18) 15	4 (1.26) 15	0 (0) 17	0	100
Alidadi H	4 (1.18) 15	4 (1.26) 15	0 (0) 17	0	100
Ashori A	4 (1.18) 15	3 (0.94) 34	1 (4.55) 6	25	75
Badv K	4 (1.18) 15	4 (1.26) 15	0 (0) 17	0	100
Baghvand A	4 (1.18) 15	4 (1.26) 15	0 (0) 17	0	100
Bahmanyar MA	4 (1.18) 15	4 (1.26) 15	0 (0)17	0	100
Ghiasinejad H	4 (1.18) 15	4 (1.26) 15	0 (0) 17	0	100
Jalili GM	4 (1.18) 15	4 (1.26) 15	0 (0) 17	0	100
Jorfi S	4 (1.18) 15	4 (1.26) 15	0 (0) 17	0	100
Kalbasi M	4 (1.18) 15	4 (1.26) 15	0 (0) 17	0	100
Mahvi AH	4 (1.18) 15	4 (1.26) 15	0 (0) 17	0	100
Moharamnejad N	4 (1.18) 15	4 (1.26) 15	0 (0) 17	0	100
Monavari SM	4 (1.18) 15	4 (1.26) 15	0 (0) 17	0	100
Nasseri S	4 (1.18) 15	4 (1.26) 15	0 (0) 17	0	100
Nourbakhsh F	4 (1.18) 15	3 (0.94) 34	1 (4.55) 6	25	75
Peyvast G	4 (1.18) 15	4 (1.26) 15	0 (0) 17	0	100
Taghipour H	4 (1.18) 15	4(1.26)15	0 (0) 17	0	100
Yaghmaeian K	4 (1.18) 15	4(1.26) 15	0 (0) 17	0	100
Yousefi Z	4 (1.18) 15	4(1.26) 15	0 (0) 17	0	100
Zazouli MA	4 (1.18) 15	4(1.26) 15	0 (0) 17	0	100

TP Total publication, *R* Rank, *MA* Multiple author publications, *SA* Single author publications, SA_P Percent of Single author publication of the total publications, MA_P Percent of multiple author publications of the total publications

(8.24 %), 25 (7.35 %) and 24 (7.06 %) publications were ranked second to forth, respectively. 72 % of the publications of Tehran University of Medical Sciences were published by inter-institute collaboration. Among these four top institutes, the Islamic Azad University (Science and Research Branch) is only university that produced the highest percentage (82.14 %) of its published articles through inter-institute collaborations.

Table 2 shows the productive authors identified with at least four publications. Abduli MA from the Faculty of the Environment of the University of Tehran is the most productive author in this field with 31 publications, which account about 9.12 % of the total article publications. Out of his 31 publications, 28 (90.32 %) and 3 (9.68 %) were "MA" and "SA" publications, respectively. 13.64 % of all single author publications (22 articles) and 8.81 % of all multiple authors publications (318 articles) were produced by Abduli MA. Two authors, Safari E and Shariatmadari N, ranked in the second position with nine publications. Mehrdadi N, Nabizadeh R and Naddafi K ranked in the third position with seven publications. These six top authors produced about 20 % of total publications in the solid waste field.

The distribution analysis of the journals showed that the journals namely *Journal of Environmental Studies*, *Waste Management and Research*, *Waste Management* and *International Journal of Environmental Research* with 34 (10 %), 23 (6.76 %), 17 (5 %) and 17 (5 %) publications, respectively, were the top journals each produced more than 10 publications on the field of solid waste. The distribution analysis of the subject categories also indicates that "environmental sciences", "Agricultural and Biological Sciences", "Medicine", "Engineering", "Earth and Planetary Sciences" and "Chemistry" are the top six most popular subject categories which have involved 197 (57.94 %), 75 (22.06 %), 31 (9.12 %), 29 (8.53 %), 21 (6.18 %) and 20 (5.88 %) of the total article publications, respectively.

The bibliometric methods concerning author keywords analysis in a specific period, could be found in recent years (Chiu and Ho 2007; Ho 2007). The useful information about the trend of researches that are concerned by researchers could be provided by analyzing the author keywords. The statistical analysis of author keywords might be aimed at discovering directions of science, and proved to be important for monitoring development of science (Hu et al. 2010). The assessment of author keywords in this study period revealed that 500 author keywords were used. Among them, 338 (67.60 %) keywords appeared only once, 94 (18.80 %) keywords appeared twice, 25 (5 %) keywords appeared thrice and only 43 author keywords (8.6 %) were used more than three times. The large number of once-only author keywords probably indicated a lack of continuity in research and a wide disparity in research focuses (Chuang et al. 2007). Table 3 shows top 20 most frequently used author keywords appeared in articles on the solid waste-related studies from 1982 to 2013. Except for eight top keywords, which were searching keywords in this study, the two frequently used keywords were "Heavy metals" and "Site selection". The appearance of keyword "Heavy

Table 3 Top 20 keywords used by the authors

Keyword	TP (%)	R
Landfill	29 (8.53)	1
Municipal solid waste	28 (8.24)	2
Solid waste	26 (7.65)	3
Leachate	25 (7.35)	4
Iran	23 (6.76)	5
Compost	19 (5.59)	6
Management	13 (3.82)	7
Waste management	13 (3.82)	7
Heavy metals	13 (3.82)	7
Environment	12 (3.53)	10
Landfill leachate	10 (2.94)	11
Tehran	8 (2.35)	12
Site selection	8 (2.35)	12
Vermicompost	8 (2.35)	12
Composting	8 (2.35)	12
Hazardous waste	7 (2.06)	16
GIS	7 (2.06)	16
Municipal solid waste compost	7 (2.06)	16
Biosolids	7 (2.06)	16
Solid waste management	7 (2.06)	16

TP Total publication, *R* Rank

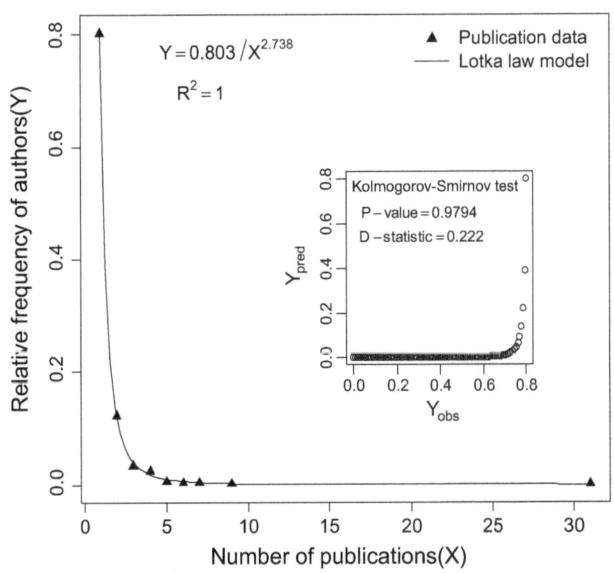

Fig. 2 Relative frequency of authors versus the number publications

metals" reveals the importance of release of heavy metals to environment from solid waste especially through landfill leachates. This is a concern that has been investigated in 3.82 % of articles. Given the importance of an appropriate selection of disposal site that is one of the major problems in waste management in Iran, "Site selection" is another keyword that appeared in 2.35 % of studied articles.

Table 4 Authorship pattern

X	No. of authors	Y	Y_{Lotka}
1	688	0.802	0.802
2	105	0.122	0.120
3	29	0.033	0.039
4	21	0.024	0.018
5	5	0.005	0.009
6	3	0.003	0.005
7	3	0.003	0.003
9	2	0.002	0.001
31	1	0.001	0.000

According to the publication data, it was revealed that 857 authors publish 340 articles, thus there is about 0.39 articles per author. It is clear that single authorship is common in this field. The numbers of 688 (80.28 %) authors have only one publication and 105 (12.25 %) authors have two publications. On the other hand, only one author produced the maximum numbers of publications (31 articles). The relationship between the relative frequency of authors and their corresponding publications is shown in Fig. 2. The Lotka's (1926) law, an inverse square law, describes the frequency of publication by authors in any given field. It states that "the number of authors making n contributions is about $1/n^a$ of those making one; and the proportion of all contributors, that make a single contribution, is about 60 percent". This means that out of all the authors in a given subject field, about 60 % publish only one article, 15 % publish two articles, 7 % publish three articles, and so on. According to Lotka's law of scientific productivity, only six percent of the authors in a given field will produce more than 10 articles. The generalized form of Lotka's Law can be expressed as:

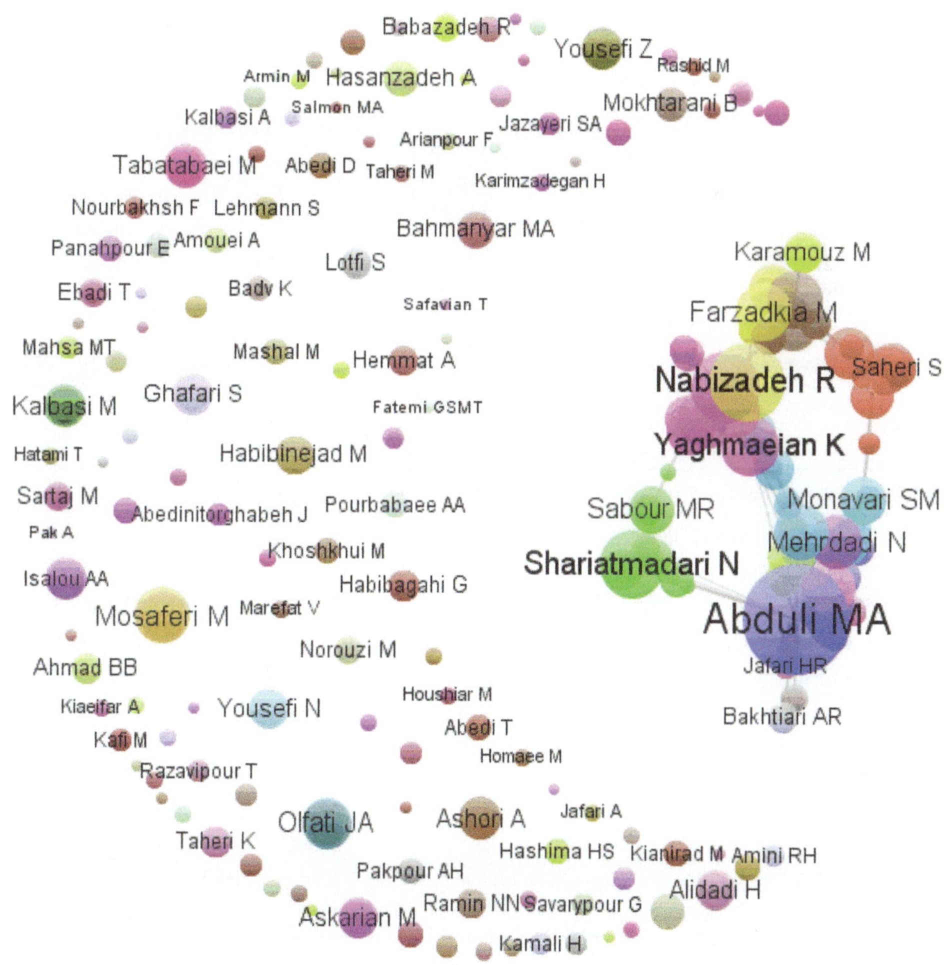

Fig. 3 Bibliometric network map of the co-occurrence of the authors

$$Y = \frac{C}{X^n} \quad (1)$$

where Y is the relative frequency of authors with X articles, the exponent n and constant C are parameters to be estimated from a given set of author productivity data.

Table 5 Author names with at least 3 co-authored publications

Author 1	Author 2	TP (%)	R
Mosaferi M	Taghipour H	4 (1.18)	1
Olfati JA	Peyvast G	4 (1.18)	1
Abduli MA	Noori R	4 (1.18)	1
Abduli MA	Baghvand A	4 (1.18)	1
Karimpourfard M	Shariatmadari N	3 (0.88)	5
Ashori A	Hamzeh Y	3 (0.88)	5
Ghanbarzadeh LM	Sabour MR	3 (0.88)	5
Alidadi H	Najafpoor AA	3 (0.88)	5
Mokhtarani B	Mokhtarani N	3 (0.88)	5
Nabizadeh R	Naddafi K	3 (0.88)	5
Bahmanyar MA	Pirdashti H	3 (0.88)	5
Nabizadeh R	Yaghmaeian K	3 (0.88)	5
Nabizadeh R	Yunesian M	3 (0.88)	5
Koolivand A	Yunesian M	3 (0.88)	5
Naddafi K	Yaghmaeian K	3 (0.88)	5
Koolivand A	Nabizadeh R	3 (0.88)	5
Machado SL	Shariatmadari N	3 (0.88)	5
Abbasi M	Abduli MA	3 (0.88)	5
Abduli MA	Safari E	3 (0.88)	5
Jafari AJ	Nabizadeh R	3 (0.88)	5
Jafari AJ	Koolivand A	3 (0.88)	5
Jafari AJ	Yunesian M	3 (0.88)	5
Karimpourfard M	Machado SL	3 (0.88)	5
Yousefi Z	Zazouli MA	3 (0.88)	5

TP Total publication, *R* Rank

Using above formula and the modifications given by Pao, Fang (Fang and Fang 1995; Nicholls 1986; Pao 1985) the values of C and n for the solid waste-related researches in Iran were determined to be 0.803 and 2.738, respectively. The Kolmogorov–Smirnov Goodness-of-Fit test shows that the literature does follow Lotka's law (P value = 0.979). It was found that for this study the maximum (absolute) deviation is 0.222, leading to the acceptance of the Lotka's law. Figure 2 shows the observed and theoretical relative frequencies of authors (Y) versus the number of corresponding publications (X). Table 4 shows the observed (Y) and theoretical (Y_{Lotka}) relative frequencies of authors and the number of corresponding publications.

Bibliometric mapping of co-authorship relations among authors allows for the representation of information in ways, which make relationships among them easier to understand. Figure 3 shows a co-occurrence network map generated from publications of the authors. Several different components including author nodes (circles), co-occurrence weight (circle size), networked relationship clustering (color and proximity), and name of authors (text) are included in a map. The paper co-authorship network is a network expressing existence of co-authorship relation between authors of scientific papers (Van Eck and Waltman 2011).

In Fig. 3 the circle's color indicates the cluster or group which the authors are associated. Clustering shows the dimension of similarity to other authors in the display. The co-authorship relations are relations representing whether an author have written a paper with another author. Typically, a paper is written by two or more authors. Analyzing co-authorship information on a larger database of scientific publications will assist in identifying groups of people who work closely together (Van Eck and Waltman 2011). The authors with at least three co-authored publications are shown in Table 5. The maximum number of co-

Table 6 Institute names with at least three co-instituted publications

Institute 1	Institute 2	TP (%)	R
Islamic Azad University/Science and Research Branch	University of Tehran	9 (2.65)	1
Tarbiat Modares University	University of Tehran	5 (1.47)	2
Tarbiat Modares University	Tehran University of Medical Sciences	4 (1.18)	3
Hamadan University of Medical Sciences	Tehran University of Medical Sciences	3 (0.88)	4
Isfahan University of Medical Sciences	Mashhad University of Medical Sciences	3 (0.88)	4
Islamic Azad University/Science and Research Branch	Sharif University of Technology	3 (0.88)	4
Islamic Azad University/Science and Research Branch	Universiti Kebangsaan Malaysia	3 (0.88)	4
Tehran University of Medical Sciences	University of Applied Science and Technology	3 (0.88)	4
Iran University of Science and Technology	University of Tehran	3 (0.88)	4
Federal University of Bahia	Iran University of Science and Technology	3 (0.88)	4
Iranian Research Organization for Science and Technology	University of Tehran/Karaj	3 (0.88)	4
Hamadan University of Medical Sciences	University of Applied Science and Technology	3 (0.88)	4

TP Total publication, *R* Rank

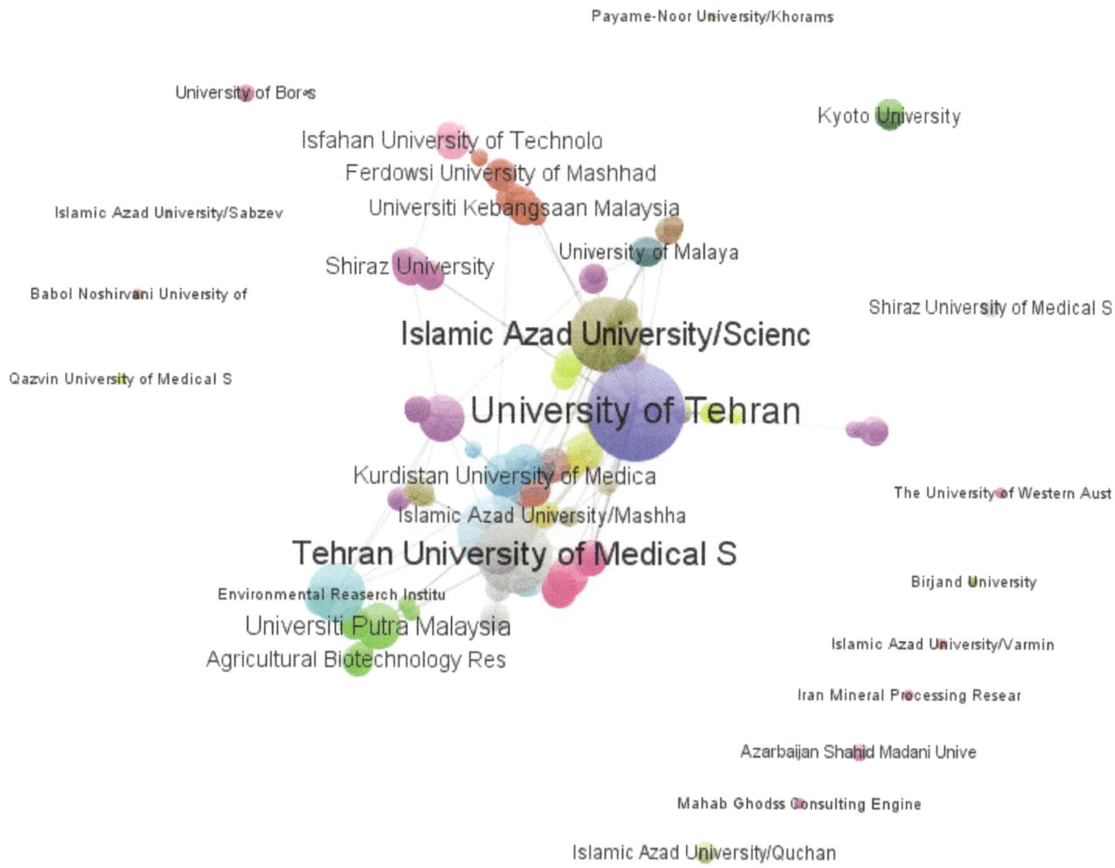

Fig. 4 Bibliometric network map of the co-occurrence of the institutions

authored publications per each pair of authors was four articles. Seven top authors in Table 5 involved in the production of 4.72 % of total publications. In other words, each pair of authors, Mosaferi M and Taghipour H, Olfati JA and Peyvast G, Abduli MA and Noori R and Abduli MA and Baghvand A, has been involved in the production of four articles (1.18 %). The mentioned pairs ranked first in terms of the number of co-authored publications.

Each circle in Fig. 3 is given a label derived from the name of author being analyzed. Visually, the size of the circles and labels show the co-occurrence weight. The authors, Abduli MA, Nabizadeh R, Naddafi K with 74, 37 and 31 co-occurrence weights, respectively, have most collaboration with other authors. The value of weight is the number of co-authored publications for each author. By default, for the authors with same co-occurrence weight, for example., Yaghmaeian K and Nasseri S, only one of them has been shown on the Fig. 3 to avoid overlapping author's names. Among all authors, Abduli MA with 74 co-occurrence weights published 31 papers. Thus, this author may be considered as the most influential author in the field of solid waste in Iran. This means, while Abduli MA produced the greatest number of papers, he has the most collaboration with other researchers.

Similar to what was shown for the case of authors, the Co-authorship relationships among the institutes can provide a useful representation of the relationships among the institutes and makes them easier to understand. Table 6 represents the names of the institutes with at least three co-instituted publications. The co-occurrence network map from the publications of these institutes is shown in Fig. 4 included the institution nodes, co-occurrence weight, networked relationship clustering, and the name of institutes. The institutions with at least three co-instituted publications are shown in Table 6. The Islamic Azad University (Science and Research Branch) with the University of Tehran have been involved in the production of nine articles (2.65 %). This pair ranked first in terms of the number of co-instituted publications.

Clustering methods have a long tradition in bibliometrics as tools for grouping bibliometric units on the basis of similarity properties measuring the distance between them (Callon et al. 1983). In Fig. 4 the similarity strength between institutes has been established, the institutes are represented as graph nodes and the similarity relationship between two institutes is represented as a weighted edge connecting them, where weights stand for the similarity intensity. Universities are colored based on the cluster to

Table 7 Institutions with at least ten co-occurrences weights

Institute	Weight	Cluster
University of Tehran	66	3
Tehran University of Medical Sciences	38	16
Islamic Azad University/Science and Research Branch	36	9
Tarbiat Modares University	36	18
Iran University of Science and Technology	23	6
Iran University of Medical Sciences	15	11
University of Kurdistan	15	14
Universiti Putra Malaysia	14	2
Kurdistan University of Medical Sciences	12	7
Bu-Ali Sina University	11	13
Isfahan University of Technology	10	10
Amirkabir University of Technology	10	4
Shiraz University	10	5
Ahvaz Jundishapur University of Medical Sciences	10	7

which they belong. As shown in Table 7 the most of universities with at least ten co-occurrences weights are belong to different clusters. In this way, universities that belong to the same cluster are very similar to each other, whereas universities belonging to different clusters are very different in term of bibliometric data. Each cluster is interpreted as providing a characterization of research activity by institutes, identifying both their strengths and weakness (Ibáñez et al. 2013). As is shown in Table 7 the University of Tehran with 66 co-occurrence weights have the most collaboration with other institutes. The weight parameter is the number of co-instituted publications for each institution. By default, for the institutions with same co-occurrence weights, for example., Islamic Azad University/Science and Research Branch and Tarbiat Modares University, only one of them has been shown in Fig. 4. Tehran University of Medical Sciences was ranked second with 38 co-occurrence weights.

Conclusions

To increase the understanding and to provide a clearer insight into the trend and contributions on the characteristics of the solid waste-related research activities in Iran, bibliometric methods were applied in this study. The results indicate that the number of annual article publications was increased by 45.3 % per year during the period from 1982 to 2011, but then it was continued with a decreasing trend up to the end of 2013 (37 % reduction). It may be a concern because this downward trend in the number of publications is likely to continue in future years. University of Tehran, Islamic Azad University (Science and Research Branch) and Tehran University of Medical Sciences as three main universities in the capital of Iran (Tehran) were the most productive universities with 25, 8.24 and 7.35 % of total article publications, respectively. Based on this finding, major research activities on the field of solid waste in Iran have been conducted by the universities mainly in the capital of Iran and other institutes had a minor role in the production of articles. The highest cooperations were also found among the main universities in Tehran.

The analysis of co-authorship network showed that the most influential author, Abduli MA, is from the University of Tehran, which is ranked first among other institutes. This might indicate the existence of a correlation between the ranks of universities and that for authors but which of them leads to another get promoted in ranks, is an issue needs further investigation.

It is foreseeable that to develop more research activities in the field of solid waste in Iran, other universities from other cities of Iran should also involve themselves in solid waste research works and cooperate with the main capital universities using their experiences in this field. Transferring the experiences by capital universities to other institutes and walking toward teamwork activities may increase the number of researches and publications in the field of solid waste engineering and management in Iran. Finally, it was expected the results of this study could be the basis for a better understanding of the international development of researches related to solid waste in Iran.

Acknowledgments This work was supported by the Center for Water Quality Research (CWQR) of Tehran University of Medical Sciences. This support is gratefully acknowledged.

Conflict of interest The authors declare that they have no competing interests.

Appendix: The search script used for the finding of the solid waste-related documents in Scopus database

((AUTHKEY("Municipal Solid Waste" OR "Sanitary Landfill" OR "urban solid waste" OR "Leachate Management" OR "Landraising" OR "Institutional Waste" OR "Household Waste" OR "Domestic Waste" OR "waste incineration" OR "waste incinerator" OR "Construction waste" OR "Demolition Waste" OR "Commingled Recyclables waste" OR "Commercial Waste" OR "waste collection" OR "municipal solid waste" OR "solid waste" OR "solid wastes" OR "solid waste disposal" OR "Curb-side Collection" OR "composting" OR "landfill" OR "hospital solid waste" OR "medical solid waste" OR "industrial solid waste" OR "solid waste management" OR "solid waste recycling" OR "solid wasteforms") OR TITLE("Municipal Solid Waste" OR "Sanitary Landfill" OR "urban solid waste" OR "Leachate Management" OR "Landraising" OR "Institutional Waste" OR "Household Waste" OR "Domestic Waste" OR "waste incineration" OR "waste incinerator" OR "Construction waste" OR "Demolition Waste" OR "Commingled Recyclables waste" OR "Commercial Waste" OR "waste collection" OR "municipal solid waste" OR "solid waste" OR "solid wastes" OR "solid waste disposal" OR "Curb-side Collection" OR "composting" OR "landfill" OR "hospital solid waste" OR "medical solid waste" OR "industrial solid waste" OR "solid waste management" OR "solid waste recycling" OR "solid wasteforms") OR ABS("Municipal Solid Waste" OR "Sanitary Landfill" OR "urban solid waste" OR "Leachate Management" OR "Landraising" OR "Institutional Waste" OR "Household Waste" OR "Domestic Waste" OR "waste incineration" OR "waste incinerator" OR "Construction waste" OR "Demolition Waste" OR "Commingled Recyclables waste" OR "Commercial Waste" OR "waste collection" OR "municipal solid waste" OR "solid waste" OR "solid wastes" OR "solid waste disposal" or "Curb-side Collection" OR "composting" OR "landfill" OR "hospital solid waste" OR "medical solid waste" OR "industrial solid waste" OR "solid waste management" OR "solid waste recycling" OR "solid wasteforms")) AND AFFIL("Iran")) AND NOT (KEY("adsorption" OR "sorption" or "desorption") OR TITLE("adsorption" OR "sorption" OR "desorption") OR ABS("adsorption" OR "sorption" OR "desorption")).

References

Abduli MA (1995) Solid waste management in Tehran. Waste Manag Res 13:519–531. doi:10.1006/wmre.1995.0050

Abduli MA (1996) Industrial waste management in Tehran. Environ Int 22:335–341. doi:10.1016/0160-4120(96)00020-7

Braun T, Glänzel W, Grupp H (1995) The scientometric weight of 50 nations in 27 science areas, 1989–1993. Part I. All fields combined, mathematics, engineering, chemistry and physics. Scientometrics 33:263–293

Callon M, Courtial J-P, Turner WA, Bauin S (1983) From translations to problematic networks: an introduction to co-word analysis. Soc Sci Inf 22:191–235

Chiu W-T, Ho Y-S (2005) Bibliometric analysis of homeopathy research during the period of 1991 to 2003. Scientometrics 63:3–23

Chiu W-T, Ho Y-S (2007) Bibliometric analysis of tsunami research. Scientometrics 73:3–17

Chokouhmand H (1982) Energy recovery from incineration of Tehran municipal solid waste and its air pollution effects. Energy Convers Manag 22:231–234. doi:10.1016/0196-8904(82)90047-4

Chuang K-Y, Huang Y-L, Ho Y-S (2007) A bibliometric and citation analysis of stroke-related research in Taiwan. Scientometrics 72:201–212

Colman AM, Dhillon D, Coulthard B (1995) A bibliometric evaluation of the research performance of British university politics departments: publications in leading journals. Scientometrics 32:49–66

Elsevier B.V. Company Scopus Content Overview (2014). http://www.elsevier.com/online-tools/scopus/content-overview. Accessed 15 Mar 2014

Falagas ME, Pitsouni EI, Malietzis GA, Pappas G (2008) Comparison of PubMed, Scopus, web of science, and Google scholar: strengths and weaknesses. FASEB J 22:338–342

Fang P, Fang JM (1995) A modification of Lotka's function for scientific productivity. Inf Process Manage 31:133–137

Fu H-Z, Ho Y-S, Sui Y-M, Li Z-S (2010) A bibliometric analysis of solid waste research during the period 1993–2008. Waste Manag 30:2410–2417

Fu H-Z, Wang M-H, Ho Y-S (2013) Mapping of drinking water research: a bibliometric analysis of research output during 1992–2011. Sci Total Environ 443:757–765

Fu H-Z, Long X, Ho Y-S (2014) China's research in chemical engineering journals in Science Citation Index Expanded: a bibliometric analysis. Scientometrics 98:119–136

Ho Y-S (2007) Bibliometric analysis of adsorption technology in environmental science. J Environ Prot Sci 1:1–11

Ho Y-S (2014) Classic articles on social work field in Social Science Citation Index: a bibliometric analysis. Scientometrics 98:137–155

Hu J, Ma Y, Zhang L, Gan F, Ho Y-S (2010) A historical review and bibliometric analysis of research on lead in drinking water field from 1991 to 2007. Sci Total Environ 408:1738–1744

Ibáñez A, Larrañaga P, Bielza C (2013) Cluster methods for assessing research performance: exploring Spanish computer science. Scientometrics 97:571–600

Lotka AJ (1926) The frequency distribution of scientific productivity. J Wash Acad Sci 16:317–323

Ma H, Ho Y-S, Fu H-Z (2011) Solid waste related research in Science Citation Index Expanded. Arch Environ Sci 5:89–100

Nicholls PT (1986) Empirical validation of Lotka's law. Inf Process Manage 22:417–419

Pao ML (1985) Lotka's law: a testing procedure. Inf Process Manag 21:305–320

Patra SK, Bhattacharya P, Verma N (2006) Bibliometric study of literature on bibliometrics. DESIDOC J Libr Inf Technol 26:27–32

R Core Team (2013) R: a language and environment for statistical computing. R Foundation for Statistical Computing, 3.0.2 edn., Vienna, Austria

Tan J, Fu H-Z, Ho Y-S (2014) A bibliometric analysis of research on proteomics in science citation index expanded. Scientometrics 98:1473–1490

Ugolini D, Parodi S, Santi L (1997) Analysis of publication quality in a cancer research institute. Scientometrics 38:265–274

Van Eck NJ, Waltman L (2011) Text mining and visualization using VOSviewer. ISSI Newsl 7(3):50–54

Yang L, Chen Z, Liu T, Gong Z, Yu Y, Wang J (2013) Global trends of solid waste research from 1997 to 2011 by using bibliometric analysis. Scientometrics 96:133–146

31

The effect of feeding fresh swine manure, poultry waste, urea, molasses and bakery by-products ensiled for lambs

Fernando Martínez San Pedro[1] · Ignacio A. Domínguez Vara[1] · José Luis Bórquez[1] · Manuel Gonzalez-Ronquillo[1]

Abstract

Introduction The use of by-products such as swine manure (SM), poultry waste (PW), urea (U), molasses (M) and bakery by-product (BB) is an alternative method for lamb feeding. The objective of the present study was to determine the chemical composition, dry matter intake and digestibility in growing lambs using ensiled PW combined with BB (PWBB), SM with BB (SMBB), SM with M (SMM) and U with M (UM) in their diets.

Methods Four silages—PWBB, SMBB, SMM and UM—were prepared, the chemical composition of the silages was determined ($n = 3$), silages and concentrate ratio (60:40) as fresh matter were given to growing lambs over 21 days as total period, and the collection of samples to determine intake and digestibility was carried out for the last 7 days, in a 4 × 4 Latin square design, with a significance level $P < 0.05$.

Results The organic matter (OM) was lower ($P < 0.05$) in silage with SMM (886 g/kg), and crude protein increased with UM (206 g/kg) followed by PWBB (170 g/kg). DM, OM, NDF and ADF intake was higher ($P < 0.05$) in UM diets compared with the rest of the treatments; ADF digestibility was lower ($P < 0.05$) for SMM (389 g/kg) than UM (417 g/kg) diets. N intake was higher ($P < 0.05$) in UM (51.3 g N/day) diets than the rest of the treatments (39.9 ± 1.3 g N/day).

Conclusions The use of PWBB, SM combined with BB or M is an alternative method for lamb feeding, provided that a proper formulation supplies the nutrient requirements of crude protein and diminish the N excretion compared with UM diet.

Keywords Poultry waste · Fresh pig manure · Bakery by-product · Silage · Lambs

Abbreviations

ADF	Acid detergent fiber
BB	Bakery by-product
DM	Dry matter
M	Molasses
ME	Metabolizable energy
N	Nitrogen
NDF	Neutral detergent fiber
NPN	Non-protein N
OM	Organic matter
PW	Poultry waste
SM	Swine manure
U	Urea

Introduction

The sustainability of animal diets is crucial in the development of livestock production systems, and feed efficiency can be improved by reusing food waste (i.e. bakery by-products, residues of the sugar industry, pineapple and citrus by-products) in ruminant diets, thus diminishing the use of food grains (Makkar and Ankers 2014).

The use of excretes from livestock is an alternative method of supplying non-protein nitrogen (NPN) in the feed of ruminants (Ortiz et al. 2007; Nasiru et al. 2014);

✉ Manuel Gonzalez-Ronquillo
mrg@uaemex.mx

1 Departamento de Nutrición Animal, Facultad de Medicina Veterinaria y Zootecnia, Universidad Autónoma del Estado de México, CP 50000 Toluca, Edo de México, México

however, they usually have low energy content for optimal rumen bacteria growth. The use of by-products from the baking and sugar industries provides an attractive energy source for ruminants. Increased cellulolytic activity of microorganisms in the rumen can increase the digestible energy due to better utilization of fibrous feeds, and otherwise improve the supply of microbial protein (Chandrasekharaiah et al. 2012). Several studies (Bórquez et al. 2009; Trujillo et al. 2014) have proved that the inclusion of up to 50 % silage manure as dry matter does not affect the intake and metabolic response in lambs, thus presenting an option as a sustainable resource in ruminant feeding.

The aim of this study was to evaluate the effects of swine manure (SM), poultry waste (PW) and urea (U) as nitrogen sources, with the inclusion of molasses (M) or bakery by-product (BB) as carbohydrate sources, on the chemical composition of silages and their intake and digestibility in lambs.

Materials and methods

Experimental silages and chemical analysis

Four silages were prepared using nitrogen (SM, PW and U) and energy sources (M and BB), and mixed with corn stover in different proportions as a fiber source (Table 1). The silages were prepared with PW combined with BB (PWBB), SM with BB (SMBB), SM with M (SMM) and U with M (UM). Each combination was ensiled with a bacterial additive (Sill-All 4 × 4 Alltech®, 10 mg/kg DM; *Streptococcus faecium*, *Lactobacillus plantarum*, *Pediococcus acidilactici* and *Lactobacillus salivarius* and enzymes cellulase, hemicellulase, pentosanase and amylase). The mixing process was performed by adding water (480 ml/kg fresh matter) to SM, PW and U, followed by M or BB. Once diluted, these combinations were mixed with corn stover in different proportions in plastic bags with a capacity of 50 kg, in three replications, compacted and sealed to prevent the ingress of air; for details see Mejía-Uribe et al. (2013).

After 60 days, the silage bags were opened (Serrano-García et al. 2008) and samples of 500 g were taken from each bag; pH was determined with a pH meter (Conductronic pH 130), samples were dried in a forced air oven (60 °C, 48 h) and grounded in a Willey mill (2 mm diameter). Silage samples were analyzed for dry matter (DM, #934.01), ash (#942.05) and N (#954.01) according to AOAC (1997). Neutral and Acid detergent fiber (NDF and ADF; Van Soest et al. 1991) were analyzed using an ANKOM200 Fiber Analyzer Unit (ANKOM Technology Corporation, Macedon, NY, USA), and lignin (AOAC 1997; #973.18). NDF was assayed with alpha amylase and sodium sulfite in the NDF. Both NDF and ADF are

Table 1 Proportion of the ingredients and chemical composition (g/kg DM) of the experimental silages

Ingredient/silage	PWBB	SMBB	SMM	UM	SEM	*P* value
Poultry waste	384					
Swine manure		250	274			
Urea				64		
Bakery by-product	231	281				
Molasses			214	262		
Corn stover	385	469	512	674		
Additive[a]	0.01	0.01	0.01	0.01		
Total	1000	1000	1000	1000		
Chemical composition						
pH	4.54a	4.72a	3.93a	6.52b	0.7	0.01
Dry matter[b]	359bc	357bc	338c	408a	5.1	0.01
Organic matter	913b	921a	886c	920ab	2.3	0.01
Crude protein	170bc	161c	147c	206ab	6.0	0.01
Neutral detergent fiber	529c	524cb	511b	485a	7.7	0.03
Acid detergent fiber	263	255	241	255	6.8	0.47
Lignin	39	39	38	39	1.5	0.58

abc Different letters indicate significance ($P < 0.05$)

PW poultry waste, *SM* swine manure, *U* urea, *BB* bakery by-product, *M* molasses, *SEM* standard error of mean

[a] Additive sill all 4 × 4® (10 mg/kg DM)

[b] Dry matter expressed as fresh matter

expressed without residual ash. Moisture content of the silages was determined through distillation with toluene (Haigh and Hopkins 1977).

Animals and diets

Four Hampshire lambs with live weight (LW) of 30 ± 3.0 kg and <1 year old, provided with ruminal cannulas, were fed with the four experimental silages using a 4 × 4 Latin square design. The animals were placed in metabolic cages. The diet consisted of the inclusion of silages (Table 1) and concentrate supplement (Table 2) in order to meet growth requirements (NRC 2007). At the beginning of the experiment, animals were dewormed (IVOMEC®; Ivermectin 1 ml 50 kg LW), supplemented with ADE complex (1 ml/head IM) and vaccinated (Bobact 8®, 2.5 ml/animal). The experimental diets were formulated to contain approximately 140 g/kg CP and 10.25 MJ ME/kg DM on average; treatments were administered ad libitum twice a day at 08.00 and 16.00 h. Each experimental period lasted 21 days, allowing 14 days for adaptation to the diet and 7 days for sample collection. Feed and ort samples were collected on days 14 to 21, weighed and composited daily, both for each individual sheep and across days. Total fecal matter and urine were weighed and sub-sampled (10 % of wet weight) for each lamb and period and stored at −20 °C for laboratory analysis.

On day 21, ruminal fluid samples (250 ml) were drawn using a suction strainer, obtained via ruminal cannula at 0 (previous ingestion), 3, 6, 9 and 12 h after feeding. Samples were filtered through a double layer of cheesecloth gauze, and the pH was recorded (Conductronic pH 130).

Table 2 Proportion of the ingredients used and chemical composition (g/kg DM) of the diets for growing lambs, with the inclusion of silages of poultry waste (PW), swine manure (SM) or urea (U) with the inclusion of bakery by-product (BB) or molasses (M)

Ingredients	PWBB	SMBB	SMM	UM
Experimental silages	376	374	360	406
Corn stover	178	260	221	241
Soybean meal 44 % CP	27	28	28	0
Corn grain	275	195	244	243
Wheat bran	71	71	73	54
Fish meal	30	30	31	23
Vitamin and mineral premix[a]	43	42	43	33
Chemical composition				
Dry matter	556	546	524	685
Organic matter	940	939	929	941
Crude protein	141	134	130	140
Rumen degradable protein	86	88	85	102
Neutral detergent fiber	417	471	435	452
Acid detergent fiber	212	250	222	245
Lignin	34	41	37	40
ME (MJ/kg DM)	10.25	10.25	10.25	10.25
Calcium	7.6	9.0	11.8	7.9
Phosphorus	6.5	6.0	6.2	6.1

DM is dry matter expressed as fresh matter, ME is metabolizable energy expressed as MJ/kg DM

[a] Content of vitamin/mineral premix/kg: 60 g phosphorus; 160 g calcium; 100 g sodium; 20 g potassium; 4 g sulfur; 2 g magnesium; 30 mg zinc; 0.6 mg copper; 1.8 mg iron; 2 mg manganese; 20 mg iodine; 6 mg cobalt; 12 mg selenium; 50 000 IU vitamin A; 10 000 IU vitamin D; 250 IU vitamin E

Statistical analysis

Chemical composition data were processed as a complete randomized design (Steel et al. 1997), using the following model:

$$Y_{ij} = \mu + T_i + \varepsilon_{ij}, \quad (1)$$

where Y_{ij} is the response variable, μ is the general mean, T_i is the effect due to diet and ε_{ij} is the random experimental error.

The in vivo experiment data were analyzed according to a 4 × 4 Latin square design, following the model:

$$Y_{ijk} = \mu + T_i + A_j + P_k + \varepsilon_{ijk} \quad (2)$$

where Y_{ij} is the response variable, μ is the general mean, T_i is the effect due to diet, A_j is the animal effect, P_k is the effect due to experimental period, and ε_{ijk} is the random experimental error. The GLM procedure of SAS (2002) was used. The means of treatments were compared by Tukey's test (Steel et al. 1997) where the effect was significant ($P \leq 0.05$).

Results

Chemical composition of silages

Silage pH was higher for UM ($P < 0.001$) compared with the rest of the treatments (Table 2); the OM content was higher ($P < 0.001$) for SMBB and UM, followed by PWBB compared with SMM. The CP concentration was higher ($P < 0.001$) for UM than SMBB and SMM. There were no differences ($P > 0.05$) for ADF and lignin among silages. The NDF content was higher ($P < 0.03$) for PWBB and SMBB than UM. The inclusion of experimental silages (Table 2) in the diets ranged from 360 to 406 g/kg DM, and the CP of the diets varied from 130 to 141 g/kg, being lower for SMBB and SMM. The NDF and ADF content in the diets were lower in PWBB and SMM compared with SMBB and UM diets.

Nutrient intake and digestibility

Table 3 shows the pH, intake, digestibility and nitrogen balance in growing lambs fed manure silage. By far the most effective source of nitrogen and minerals (ash) was provided by the livestock manure (swine and poultry), urea as we know, is only a source of nitrogen. Ruminal pH values in lambs were similar ($P > 0.05$). The DM, OM and ADF intakes were higher ($P < 0.05$) for silage based on UM compared with the rest of the treatments. The NDF intake was higher ($P < 0.05$) for UM, followed by PWBB and SMM, and lower for SMBB. Digestibility of DM, OM and NDF was similar among treatments ($P > 0.05$). The ADF digestibility was lower ($P < 0.05$) for SMM compared with UM treatment. Nitrogen intake (g N/day) was higher ($P < 0.05$) for UM compared with the rest of the treatments. Nitrogen excretion (feces and urine) and retention were similar among treatments ($P > 0.05$).

Discussion

Chemical composition of silages

All the silages had acceptable quality (Frenkel 1984), except UM, which showed higher pH values (Table 1). This might be due to the fact that BBP and M were a better carbohydrate sources to be fermented into desirable organic acids such as lactic, acetic and propionic acids, diminishing the pH. The low pH in silages in our study suggests that undesirable fecal microorganisms (Coliforms, Salmonella, Shigella, Proteus), yeasts and molds may had been eliminated (Pagán et al. 2014; Serrano-García et al. 2008), which have been reported for cattle manure 1 week after ensiling (Cornman et al. 1981; McCaskey and Wang 1983). Optimum moisture recommended for manure-blended silage is at least 600 g/kg (McCaskey and Wang 1983), whereas in our study moisture of silages ranged from 524 to 685 g/kg. Bórquez et al. (2009) using cattle manure silage with BB, the amount of DM was lower with respect to which was added M. The amount of water varies among the different ingredients added in the silages, thus varying among silages. The CP and NDF content can vary depending on the type and amount of bedding floor used by the livestock species (Tobia and Vargas 2000). Evans and Smith (1986) reported that the use of U leads to changes in cell wall components of forages treated, destroying the linkages of phenolic groups between hemicellulose and lignin, which solubilizes the hemicellulose and make it available to the cell wall unless UM silage includes more corn stover than the rest of the silages. Tobia and Vargas (2000) found a similar protein content in PW and NDF, but higher for ADF. Mthiyane et al. (2001) showed a lower content of CP and OM compared with the present study, but higher in NDF for PW. These variations in nutrient content depend on the type of food, floor and management of excreta that is offered to poultry and pigs (Tobia and Vargas 2000; Morales et al. 2002; Teixeira et al. 2015),

Table 3 Intake (g/kg $LW^{0.75}$), digestibility (g/kg) and N balance (g/day) in lambs fed with silages of poultry waste (PW), swine manure (SM) or urea (U) with the inclusion of bakery by-product (BB) or molasses (M)

Item	PWBB	SMBB	SMM	UM	SEM	*P* value
Rumen pH	6.63	6.46	6.62	6.83	0.09	0.37
Intake (g/kg $LW^{0.75}$)						
Dry matter	77.7b	77.9b	79.2b	88.0a	1.73	0.01
Organic matter	68.7b	70.0b	70.4b	78.6a	1.65	0.01
Neutral detergent fiber	40.7c	38.2c	43.0b	47.9a	1.65	0.01
Acid detergent fiber	34.2b	32.9b	34.6b	40.4a	1.70	0.01
Digestibility (g/kg)						
Dry matter	718	675	681	674	31.90	0.52
Organic matter	728	722	687	717	18.30	0.56
Neutral detergent fiber	525	538	501	537	8.50	0.48
Acid detergent fiber	402ab	396ab	389b	417a	5.70	0.05
N balance (g/day)						
N intake	40.7b	38.5b	40.6b	51.3a	1.36	0.05
N excretion						
N feces	13.0	12.9	15.3	13.8	1.32	0.35
N urine	11.7	10.8	10.9	14.9	1.61	0.42
N retention	16.1	15.7	14.4	22.5	1.84	0.34

abc Different letters indicate significance ($P < 0.05$)

SEM standard error of mean

effecting a wide variation in the chemical composition of PW and SM.

Nutrient intake and digestibility

The inclusion level of silage in the diet had no effect on ruminal pH in the lambs (Table 3), which could be due to silage inclusion which ranged around 60 % as fresh matter and there was a sufficient amount of NDF in the entire diet (ranging 417–471 g/kg DM). The intake of DM, OM, NDF and ADF was similar to Trujillo et al. (2014). The digestibility (g/kg) of DM and OM based on PW silage was lower than Morales and Egaña (1997), but similar for NDF digestibility; these variations depend on the chemical composition of the different poultry waste sources. In the present study, there were no differences ($P > 0.05$) among DM, OM and NDF digestibility, which was similar to Obeidat et al. (2011) and Trujillo et al. (2014), but ADF digestibility was higher ($P < 0.05$) for UM compared with SMM. This effect could be related with the lower ADF intake and higher ash content in the SMM diet, as was found by Jakhmola et al. (1988) and Iñiguez-Covarrubias et al. (1990).

The lower N intake in the silage diets compared with the UM diet is a direct response to the higher DM intake in the UM diet compared with the rest of the silages; the N retained in lambs fed diets with the inclusion of PW and SM silage was numerically lower ($P = 0.34$) than UM, which provided more efficient N retention in the animals. The inclusion level of silages in the diets of up to 37 % neither enhanced nor adversely affected animal performance as compared with UM, which are in agreement with the results of Zia-ul-Hassan et al. (2011) and Sarwar et al. (2011) in feeding traits with different levels of cattle manure in lactating Nili–Ravi buffaloes and growing cattle calves, respectively; and with Trujillo et al. (2014) who studied different levels of SM or PW silages in growing lambs.

Conclusion

UM silage showed higher pH and DM, OM, CP content, but lower NDF. Treatment with these silage also showed higher DM, OM, NDF and ADF intake. Nitrogen intake was higher with la inclusion of UM silage, but there was no difference in N retention. It is concluded that UM silage inclusion in complete diets for growing lambs can be as good as the use of livestock manures silages.

Acknowledgments Mr. Martinez San Pedro was granted for a CONACyT fellowship during his studies in the University Autonomous State of Mexico. This study was supported by the project UAEMex 2238/2006 and 2750/2009. Dr. González was granted for a CONACyT fellowship (Estacias Sabaticas en el Extranjero, 2014), also we thanks to Miss. Penelope Krumm for the critical review of this paper.

Author contribution All authors have made reasonable effort on all parts of the work necessary for the development of this manuscript in accordance with their expertise. All authors read and approved the final manuscript.

Compliance with ethical standards

Conflict of interest The authors declare they have no conflicts of interest with regard to the work presented in this report.

References

AOAC (Association of Official Analytical Chemists) (1997) Official methods of analysis, 16th edn. AOAC, Arlington

Bórquez JL, Gonzalez Muñoz SS, Pinos Rodriguez JM, Dominguez I, Barcena JR, Mendoza GD, Cobos MA, Bueno G (2009) Feeding value of ensiling fresh cattle manure with molasses or bakery by products in lambs. Lives Sci 122:276–280

Chandrasekharaiah M, Thulasi A, Sampath KT (2012) Effect of different rumen degradable nitrogen levels on microbial protein synthesis and digestibility in sheep fed on finger millet straw (*Eleucine coracana*) based diet. Small Rum Res 102:51–156

Cornman AW, Lamm WD, Webb KE, Fontenot JP (1981) Ensiling cattle waste with rye straw as a diet supplement for ruminants. J Anim Sci 52:1233–1239

Evans MR, Smith MPW (1986) Treatment of farm animal wastes. J Appl Bact 61(Suppl.):27S–41S

Frenkel MA (1984) Forage conservation: hays, silages, dehydration. Albatros, Buenos Aires

Haigh PM, Hopkins JR (1977) Relationship between oven and toluene dry matter in grass silage. J Sci Food Agric 28(6):477–480

Iñiguez-Covarrubias G, Cuarón-Ibargüengoitia JA, Pérez-Gavilán P, de la Torre-Martinez M, Magaña Plaza I (1990) Fermentation characteristics, digestibility and performance of ensiled swine waste, wheat straw and cane molasses fed to sheep. Biol Wastes 34(4):281–299

Jakhmola RC, Kundu SS, Punj ML, Singh Kiran, Kamra DN, Singh Rameshwar (1988) Animal excreta as ruminant feed-Scope and limitations under Indian conditions. Anim Feed Sci Technol 19:1–23

Makkar HPS, Ankers P (2014) Towards sustainable animal diets: a survey-based study. Anim Feed Sci Technol 198:309–322

McCaskey TA, Wang YD (1983) Microflora and acid development of rations ensiled with varied percents of bovine manure and moisture. J Dairy Sci 66:93–98

Mejía-Uribe LA, Bórquez JL, Salem AZM, Dominguez IA, González-Ronquillo M (2013) Effects of adding different protein and carbohydrates sources on chemical composition and in vitro gas production of corn stover silage. Spanish J Agric Res 11(2):427–430

Morales MS, Egaña JI (1997) Efecto del peletizado y ensilaje de las camas de broiler sobre su valor nutritivo para rumiantes. 1. Evaluación nutricional y productiva. Arch Zoot 46:159–167

Morales TH, Gutierrez HE, Bernal BH (2002) El uso de la cama de pollo de buena calidad mejora la productividad de bovinos en crecimiento en engorda intensiva. Tec Pec México 40:1–15

Mthiyane DMN, Nsahlai IV, Bonsi MLK (2001) The nutritional composition, fermentation, characteristics, in sacco degradation and fungal pathogen dynamics of sugarcane tops ensiled with broiler litter with or without water. Anim Feed Sci Technol 94:171–185

Nasiru A, Ibrahim MH, Ismail N (2014) Nitrogen losses in ruminant manure management and use of cattle manure vermicast to improve forage quality. Int J Recycl Org Waste Agric 3:57

NRC (2007) National Research Council: nutrient requirements of small ruminants: sheep, goats, cervids and new world camelids. National Academy Press, Washington, DC

Obeidat BS, Awawdeh MS, Abdullah AY, Muwalla MM, Abu Ishmais MA, Telfah BT, Ayrout AJ, Matarneh SK, Subih HS, Osaili TO (2011) Effects of feeding broiler litter on performance of Awassi lambs fed finishing diets. Anim Feed Sci Technol 165:15–22

Ortiz RMA, Orskov ER, Milne J, Galina HMA (2007) Effect of different sources of nitrogen on in situ degradability and feed intake of zebu cattle fed sugarcane tops (*Saccharum officinarum*). Anim Feed Sci Technol 139:143–158

Pagán S, Rodríguez AA, Valencia EM, Randel PF (2014) Pineapple and citrus silage as potential feed for small ruminant diets: fermentation characteristics, intake, nutrient digestibility, and aerobic stability. Rev Colomb Cienc Pecu 27:37–46

Sarwar M, Shahzad M, Nisa M, Afzal D, Sharif M, Saddiqi H (2011) Feeding value of urea molasses-treated wheat straw ensiled with fresh cattle manure for growing crossbred cattle calves. Trop Anim Health Prod 3(3):543–548. doi:10.1007/s11250-010-9745-5

SAS (2002) Statistical analysis system institute. User's guide: statistic version 9. SAS Institute Inc., Cary

Serrano-García E, Castrejón-Pineda F, Herradora-Lozano MA, Ramírez-Pérez AH, Angeles-Campos S, Buntinx SE (2008) Fungal survival in ensiled swine faeces. Bioresour Technol 99:3850–3854

Steel RGD, Torrie TH, Dickey DA (1997) Principles and procedures of statistics o biometrical approach, 2nd edn. Mc Graw Hill Book Co., New York

Teixeira AS, de Oliveira MC, Menezes JFS, Gouvea BM, Teixeira SR, Gomes AR (2015) Poultry litter of wood shavings and/or sugarcane bagasse: animal performance and bed quality. Rev Colomb Cienc Pecu 28(3):238–246

Tobia C, Vargas E (2000) Evaluación de las excretas de pollos de engorde (pollinaza) en la alimentación animal. 1. Disponibilidad y Composición química. Agronomia Costarricence 24(1):47–53

Trujillo GD, Bórquez GJL, Pinos-Rodríguez JM, Domínguez-Vara IA, Rojo RR (2014) Nutritive value of ensiled pig excreta, poultry litter or urea with molasses or bakery by-products in diets for lambs. South Afr J Anim Sci 44(2):114–122. doi:10.4314/sajas.v44i2.3

Van Soest PJ, Robertson JB, Lewis BA (1991) Methods for dietary fiber neutral detergent fibre and non starch polysaccharides in relation to animal nutrition. J Dairy Sci 74:3583–3597

Zia-ul-Hassan M, Aasif Shahzad, Nisa M, Sarwar M (2011) Nutrient utilization and milk yield response of early lactating Nili-Ravi buffaloes fed on urea-molasses treated wheat straw fermented with cattle manure. Livest Sci 139:271–276

Agricultural waste as growing media component for the growth and flowering of *Gerbera jamesonii* cv. hybrid mix

Atif Riaz[1] · Adnan Younis[1] · Imran Ghani[1] · Usman Tariq[1] · Muhammad Ahsan[1]

Abstract

Purpose This study was performed to investigate the impact of various agricultural wastes as potting media on growth, yield, and quality flower production of *Gerbera jamesonii* L. cv. hybrid mix, and to estimate hazardous impact of chemical fertilizers.

Method The effect of four agricultural substrates viz. farm yard manure, coconut coir dust, Lahore compost (local produce), and leaf compost combined with conventional media i.e., garden soil, sand, and silt in equal proportion was studied for commercial production of gerbera. All the treatment combinations were arranged into completely randomized design with three replicates.

Result Visible variations in morphological characters were evidenced among conventional medium and agricultural waste as substrates with significant superiority of flower quality and plant growth. The presented results showed that treatment combination of silt, coconut coir dust, and top soil (SCT) produced maximum plant height while mixture of silt, Lahore compost and top soil (SLT) gave maximum number of leaves per plant and maximum flower stalk thickness. The highest values for leaf area, number of roots, flower fresh to dry weight ratio and number of flowers per plant were achieved in combination of silt, farm yard manure, and top soil (SFT) whereas silt, sand, and top soil (SST) presented maximum flower diameter, maximum fresh, and dry weight of flowers. Chemical characteristics of growing media showed superiority of silt, farm yard manure, and top soil (SFT) and silt, sand, and top soil (STL) for available phosphorus and potassium with values of 27.0 and 500 ppm, respectively.

Conclusion It was concluded that the application of waste materials in combination with silt provided positive results for vegetative and reproductive growth of *Gerbera jamesonii* plants.

Keywords Agricultural waste · Farm yard manure · Flower quality · *Gerbera jamesonii* · Leaf compost

✉ Atif Riaz
atiff23@gmail.com

Adnan Younis
adnanyounis@uaf.edu.pk; adnanyounis1976@yahoo.com

Imran Ghani
imran_mulghani@yahoo.com

Usman Tariq
usmanflori@yahoo.com

Muhammad Ahsan
ahsan2065@gmail.com

1 Institute of Horticultural Sciences, University of Agriculture, Faisalabad 38040, Pakistan

Introduction

Gerbera (*Gerbera jamesonii*), is a herbaceous perennial plant with flowering stalks produced in between the leaves, ranging from 12 to 37 cm in height depending upon the cultivar (Weeraratne et al. 2012). Single daisy like flowers of gerbera are used for garden and home decoration in early stages but rarely used as cut flower because of their weak pedicle, bent neck, and poor vase life. The introduction of new varieties having double flowers (Torrey 1951) is considered better for cut purposes in all respects, encouraging the cut flower growers to enhance their production. It's potential as a cut flower is now widely accepted because of its ability to propagate through seed, rapid growth, and the demand for cheap labor supplies. For a better production of gerbera cut flowers, it is necessary to estimate all the growth factors like nutrition, growth environment, growth media, etc.

Growth media are considered among one of the most important factors affecting the production of Gerbera (Nowak and Strojny 2003; Sindhu et al. 2010; Khalaj et al. 2011; Ahmad et al. 2012). Natural soil along with peat are the most common growing substrates used for the container production of annual and perennial ornamental plants since many years (Guerrero et al. 2002; Hernandez-Apaolaza et al. 2005). In order to enhance quality and production of flowering plants chemical fertilizer are frequently applied to soils but the excessive applications of chemical fertilizers have harmful effects on soil flora, fauna, and enzymes. This ultimately leads to decrease their activity for maintaining the natural fertility of soil (Gupta et al. 2014) and increase the dependence on chemical fertilizers. As a result, there is a growing concern to reduce the application of chemical fertilizer to soils, and in order to use soil nutrients more efficiently and to overcome pollution hazards the application of organic matter is gaining acceptance among farming community.

In the past few years, usage of peat has been reduced due to environmental concerns for conservation (Abad et al. 2002). Therefore, researchers showed their interest in other organic materials like agricultural wastes which could be viable source of organic nutrients for the production of container grown plants. Agricultural waste such as rice hulls, coir, sawdust, and compost as growing media has been proved as an important substitute of simple soil for ornamental plant production (Younis et al. 2007; Wilson et al. 2009; Tariq et al. 2012) including gerbera. Furthermore, production of large sized gerbera flowers is entirely based on the structure, texture, pH, organic matter contents, as well as nitrogen, phosphorus, and potassium levels of the growth media (Rader 1998; Lang and Pannuk 1998). Organic media that are light in weight, rich in nutrients and well-drained considered ideal choices for growth and development of flowering plants like gerbera (Awang et al. 2009; Dubey et al. 2013), as these plants are notoriously ranked as heavy feeders and require abundant quantities of plant food for maximum development. It thrives best in a media with a pH slightly below or above the neutral point which helps the plants to face the seasonal hazards (Abad et al. 2002).

Currently, in Pakistan less work has been carried out on this flowering plant using different growth media sources. Most of the farmers are using chemical fertilizers and peat in order to attain maximum production of flowering plants. Both ways are eco destructive in several ways and growers in Pakistan are less familiar about the usage of other organic sources for better crop production. A less expensive and more adoptable growing media from agriculture waste not only benefits the environment but also attains meaningful production outcomes for Gerbera growers. Keeping in view, environmental concerns, economic constraints, and growers need, this study was performed to investigate the impact of various agricultural wastes as potting media on morphological characters as well as yield and quality flower production of Gerbera.

Materials and methods

The comparative study for evaluation of different potting media for *Gerbera jamsonii* cv. Hybrid Mix plants production was conducted at Floriculture Research Area, Institute of Horticultural Sciences, University of Agriculture, Faisalabad (latitude 31°30′N, longitude 73°10′E and altitude 213 m). Eight-week-old seedlings were purchased from local nursery in the month of November and 108 plants of uniform size (3 cm) having two sets of true leaves were selected to grow in pots having different combinations of FYM, coconut coir dust, Lahore compost (local produce), sand, leaf compost, silt, and garden soil. Six treatment combinations including control were made with the similar ratio by volume (Table 1).

All treatment combinations were arranged in completely randomized design (CRD) with three replicates and six plants were planted in each replication. Management practices like irrigation, weeding, and pesticide application remained same for the all treatments.

Plant growth characteristics

Following observations on plant growth characteristics were evaluated during the course of study.

Plant height The height of plants was measured in centimeters, from base to the top leaf of the plant.

Number of leaves per plant The total number of leaves per plant was counted for each media after every 15 days.

Leaf area The leaf area (cm^2) was measured with the help of LI-3000 Portable Leaf Area Meter, USA.

Days to flower emergence Number of days were counted from the date of sowing to emergence of first flower on the plant.

Number of flowers per plant Flowers per plant were counted regularly during blooming period of the crop and then average was calculated.

Flower diameter Diameter of all the flowers was measured in centimeters (cm) with the help of vernier caliper and average was estimated.

Flower stalk thickness Flower stalk thickness was measured in centimeters (cm) with the help of vernier caliper and average was calculated.

Table 1 Different treatment combinations of agricultural wastes used as growing media

Treatments	Combinations	Ratio by volume (*v*:*v*:*v*)
T_0	Control (garden soil = GS)	
T_1	Silt + farm yard manure + top soil (SFT)	1:1:1
T_2	Silt + coconut coir dust + top soil (SCT)	1:1:1
T_3	Silt + Lahore compost + top soil (SLT)	1:1:1
T_4	Silt + sand + top soil (SST)	1:1:1
T_5	Silt + top soil + leaf compost (STL)	1:1:1

Fresh weight of a flower One flower was collected at random from each plant and weighed on an electric balance (in grams) and average was computed.

Dry weight of a flower Flowers were dried in oven at 65 °C for 48 h, and then weighed again on electric balance. In this way, dry weight of flowers was recorded in grams and average was computed.

Fresh and dry weight ratio Fresh and dry weight ratio of a flower was calculated by dividing the fresh weight to dry weight of the flower.

Number of roots per plant The number of roots of plants was counted individually by uprooting the plant from the pots at mature stage.

Chemical characteristics of growing medium

The pH and Electrical conductivity (EC) in dS m^{-1} was determined by HI 9811-5 portable pH/EC/TDS/°C meter (Mclean 1982) and digital electrical conductivity (EC) meter (Lovibond Senso Direct 150). Organic matter was determined using the method of Walkley (1947), while, available phosphorus was estimated using Spectrophotometer, model spectronic 21 USA (Olsen et al. 1954). The flame photo metric method was used for estimation of Potassium (U.S. Salinity Lab. Staff 1954).

Statistical analysis

The collected data were subjected to Fisher's analysis of variance technique (Steel et al. 1997). The Duncan's multiple range (DMR) test at 0.05 % level of probability was employed to compare the differences among treatment means.

Results

Plant growth characteristics

Results regarding plant height showed significant superiority of SCT (T_2) treatment, which produced maximum plant height with 40.22 % increase as compared to control (T_0) followed by 30.50 and 20.66 % increase with SLT (T_3) and STL (T_5), respectively. Mean comparison of treatments in Table 2 regarding number of leaves per plant showed significant superiority of SLT (T_3) again with 34.85 % increase in number of leaves followed by STL (T_5) with 30.30 % more leaves, compared to minimum number of leaves (3.30) in garden soil (GS) (T_0). Therefore, maximum leaf area (39.24 cm^2) was produced in mixture of SFT (T_1) with 164.96 % increase followed by STL (T_5), SCT (T_2), and SST (T_4) with 160.50, 125.79, and 107.63 % rise, respectively as compared to conventional media i.e., GS (T_0). Growing media combination in the form of STL (T_5) presented the highest value for number of roots (6.70) followed by SST (T_4) and SCT (T_2) treatments and finally up to 148.15 % improvement over GS (T_0) was noted (Table 2).

Size of flower also improved in SST (T_4) significantly with maximum increases in flower diameter (20.75 %), in comparison to all growing media used in this study. Whereas, plants grown in SCT (T_2), SLT (T_3), and SFT (T_1) behaved similar in case of flower diameter (Fig. 1). The maximum flower stalk thickness was produced in SLT (T_3) growth media, while the least flower stalk thickness was attained in plants grown in SCT (T_2) after GS (Fig. 2). The utmost but statistically similar stalk thickness obtained in SST (T_4) and STL (T_5). There was 50 % improvement in flower stalk thickness in plants grown in SLT as compared to control treatment. On average, maximum fresh weight of flower with 31.87 % increase in weight was attained in SST (T_4) growing media compared to control (Fig. 3). The smallest count of fresh weight of flower was observed in SFT (T_1) as this growing media caused 5.49 % decrease in weight when compared with control. Similarly, results regarding dry weight of flowers showed that the SST (T_4) and SLT (T_3) gave the highest results without any significant difference. Whereas, the lowest dry weight of flowers was observed in STL (T_5) with 43.77 % decrease from control (GS) (Fig. 4). The STL (T_5) mixture possessed the highest fresh to dry weight ratio, which was 104.56 % greater than the control treatment, whereas minimum fresh/dry weight ratio was produced again by GS (T_0) (see Fig. 5).

However, difference for days to flower emergence was found to be non-significant but in general, STL (T_5) mixture attained least days to first flower emergence and showed 8.30 % earliness in this regard than control

Table 2 Effect of growing substrates on plant height, number of leaves per plant, leaf area and number of roots of *Gerbera jamesonii*

Treatments	Plant height (cm)	Number of leaves per plant	Leaf area (cm^2)	Number of roots
T_0	7.31 ± 0.81f	3.30 ± 0.88d	14.81 ± 2.06f	2.70 ± 0.45f
T_1	7.86 ± 1.36e	4.09 ± 0.34c	39.24 ± 1.79a	3.66 ± 0.41e
T_2	10.25 ± 0.75a	4.15 ± 0.40c	33.44 ± 1.69c	4.45 ± 0.20c
T_3	9.54 ± 1.04b	4.45 ± 0.70a	30.75 ± 2.00e	4.07 ± 0.04d
T_4	8.44 ± 0.94d	4.11 ± 0.36c	32.24 ± 1.49d	5.79 ± 0.03b
T_5	8.82 ± 1.32c	4.30 ± 0.75b	38.58 ± 2.08b	6.70 ± 0.57a
F value	1436**	104**	26,562**	512**
p value	3.16×10^{-16}	1.23×10^{-11}	7.97×10^{-24}	2.13×10^{-13}

NS non significant, ±SE

* Significant ($P < 0.05$)

** Highly significant ($P < 0.01$)

Letters (a–f) exhibit significant difference among means at $P < 0.05$

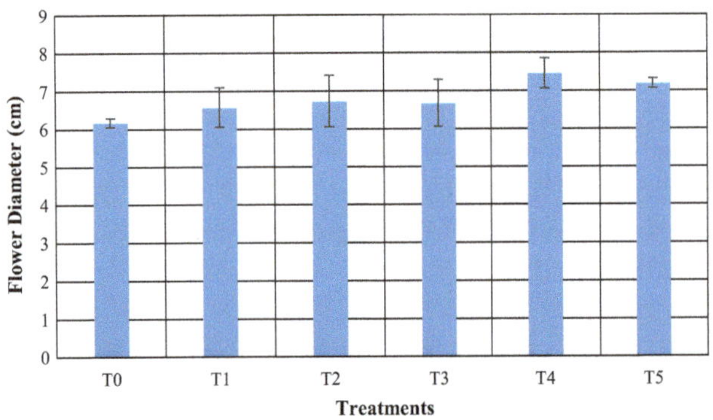

Fig. 1 Effect of different growing media on flower diameter of *Gerbera jamesonii*

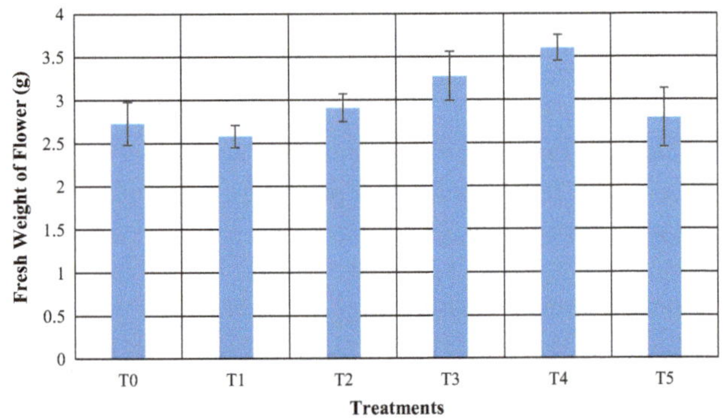

Fig. 3 Effect of different growing media on fresh weight of flower of *Gerbera jamesonii*

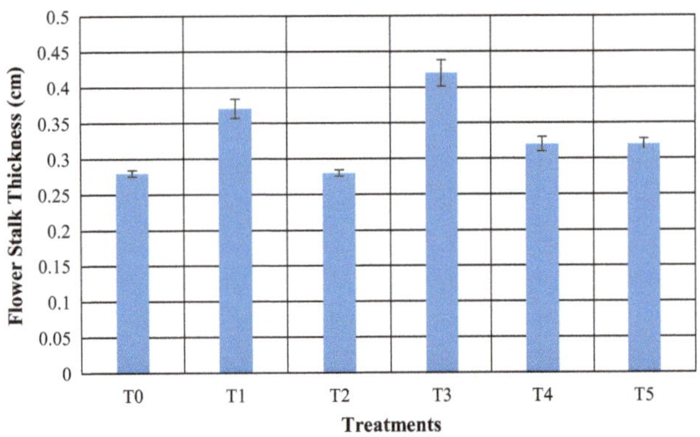

Fig. 2 Effect of different growing media on flower stalk thickness of *Gerbera jamesonii*

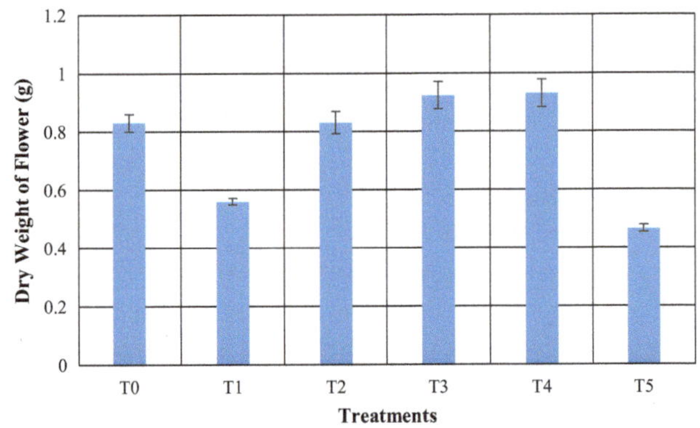

Fig. 4 Effect of different growing media on dry weight of flower of *Gerbera jamesonii*

(Fig. 6). Likewise, the mixture STL (T_5) again proved the superiority (34.85 %) regarding flowers per plant over control (GS) which produced the least flowers per plant (Fig. 7). Other mixtures including SCT (T_2), SST (T_4), and SFT (T_1) were statistically not different from each other with 23 to 30 % increase in flowers per plant compared to control (GS) (T_0).

Chemical characteristics of growing medium

The highly significant differences among organic matter contents, available potash (K), available phosphorus (P), pH and EC were observed in all the growth media used in this investigation. Growing media with SFT (T_1) combination provided better P (27 %) and K (72 %) contents

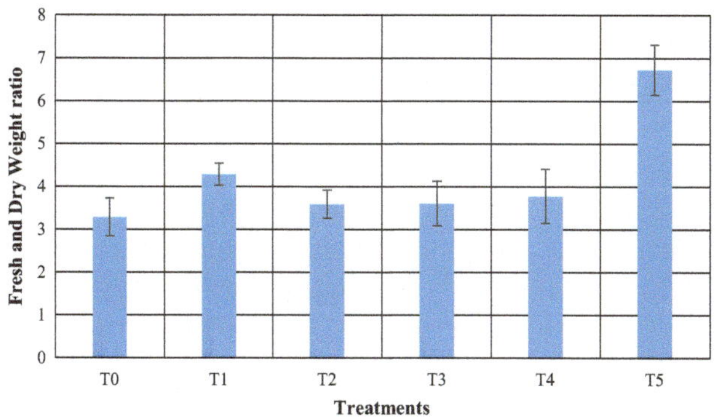

Fig. 5 Effect of different growing media on fresh and dry weight ratio of flower of *Gerbera jamesonii*

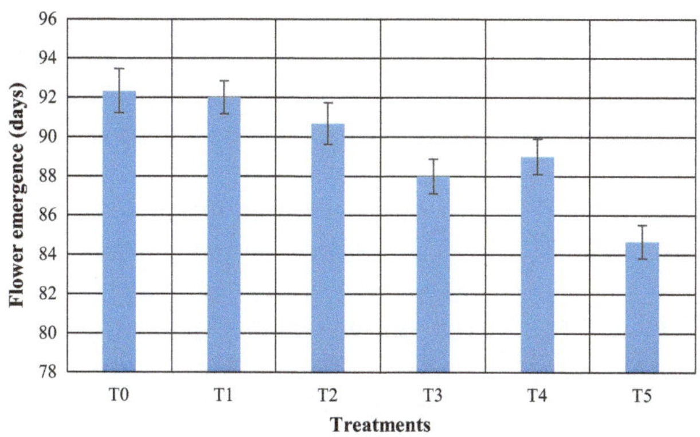

Fig. 6 Effect of different growing media on flower emergence of *Gerbera jamesonii*

compared to other growing media. While, maximum increase (up to 86.54 % over control) in organic matter contents was found again in SFT (T_1) and STL (T_5) mixtures without any significant difference. The SCT potting media gave the least percentage of available organic matter compared to other growing media treatments. Results regarding pH level of growing media exhibited that GS (T_0) gave the maximum value (9.54) while SST (T_4) and SFT (T_1) occupied 2nd position. The minimum level regarding pH (29 % less than control) was recorded in treatment containing SCT (T_2). The highest EC value (0.73) with 58.70 % increase over control was noticed with SFT (T_1) potting media whereas, GS (T_0) and SST (T_4) gave the lowest EC values (0.36) (see Table 3).

Discussion

Improved quality characters and growth traits of gerbera plants observed in this study with different growing substrates having garden soil, farm yard manure, coconut coir dust, a local made organic compost, and leaf compost are related to the previous findings (Saijeen et al. 2009; Ahmad et al. 2012). Appropriate concentration of nutrients in conjunction with growing medium has advancing effects on best flower growth and yield.

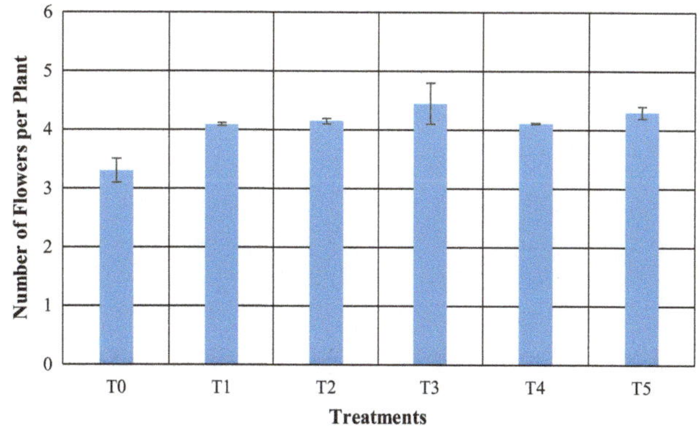

Fig. 7 Effect of different growing media on flowers per plant of *Gerbera jamesonii*

Results exhibited in previous section showed that balanced rooting media using agricultural waste can greatly affect the plant height and are essential for attaining maximum of it. These results are in accordance with Treder (2008) and Dubsky and Sramek (2008) who reported excellent growth of perennials in leaf mold and peat substrate with the addition of rock wool and observed maximum plant height. Though, number of leaves is influenced mainly by environmental conditions, nutrients present in growth media are also one of the factors which have prime importance in this regard. More number of leaves in plants reflect good vigor and their suitability to environment and growth media. Since nitrogen in growing media significantly affects the plant growth, increase in number of leaves of plants can also be due to adequate availability of nitrogen content in growing substrate (Benito et al. 2005; Khayyat et al. 2007; Riaz et al. 2008). More number of leaves could not be considered as factor contributing toward leaf area but it is mainly due to adequate supply of nutrients. Therefore, the nutrient rich substrate can be used to achieve significant results as maximum increase in size of leaves shows adaptability of plants to soil or growth media (Cardens et al. 2006; Riaz et al. 2014). Increase in number of shoots per plant highlighted the vigorous vegetative plant growth and only nutrient-rich growing media promote such growth in plants. Above mentioned results help to conclude that growing media in a combination of STL presented the best results for the number of shoots (Bashir et al. 2007).

Results regarding number of roots reflect that treatments containing mainly leaf compost represented superiority in promoting roots number. Similar trend has also been noticed by Lopez (2006) who found that root growth of geranium, petunia, and coleus was increased as the proportion of sewage sludge plus straw compost increased up to 50 % in peat. Maximum flowering as well as increase in flower bud

Table 3 Effect of growing substrates on organic matter, available potash, available phosphorus, pH and EC of *Gerbera jamesonii*

Treatments	Organic matter %	Available potash (ppm)	Available phosphorus (ppm)	pH	EC of media (dSm^{-1})
T_0	52 ± 1.07d	290 ± 11.99b	21.3 ± 0.76b	9.54 ± 0.39e	0.46 ± 0.048c
T_1	97 ± 1.93a	500 ± 8.91a	27 ± 1.36a	7.36 ± 0.49c	0.73 ± 0.02a
T_2	41 ± 8.02e	230 ± 8.49c	16.8 ± 1.42c	6.75 ± 0.67a	0.59 ± 0.05b
T_3	82 ± 3.21b	500 ± 13.89a	15.2 ± 1.19c	6.81 ± 0.22b	0.49 ± 0.004c
T_4	62 ± 2.13c	220 ± 6.94c	16.5 ± 2.72c	7.94 ± 0.71d	0.36 ± 0.01d
T_5	93 ± 7.01a	500 ± 10.02a	25 ± 2.02a	6.82 ± 0.1b	0.59 ± 0.08b
F value	1597**	952**	18.1**	154**	123**
p value	1.67×10^{-16}	3.70×10^{-15}	3.22×10^{-05}	1.88×10^{-10}	6.86×10^{-10}

* Significant ($P < 0.05$)

** Highly significant ($P < 0.01$), ±SE

Letters (a–f) exhibit significant difference among means at $P < 0.05$

diameter and subsequently increase in flower diameter indicates suitability of that growing medium, where organic matter in growing substrate and optimum amount of NPK manipulate plant growth and flower size (Kiran et al. 2007). Growing media containing silt and Lahore compost exhibited their superiority for stalk thickness, stem diameter, flower fresh and dry weights, and its ratio contrast to conventional media which proves that this kind of media can be used for gerbera production effectively. These results are in consistent with Grassotti et al. (2003) and Caballero et al. (2009) who observed that growing media having combination of silt + compost and manure can produce the maximum flower stalk thickness, fresh, and dry weights of gladiolus and lily. Similarly, the combination of sand, mushroom compost, coconut coir, and peat produced the best quality of plants with considerable increase in most of the growth parameters (Younis et al. 2013). Khalaj et al. (2011) also observed that maximum stem diameter of Gerbera plants attained with media in combination of perlite and peat. Rich supply of nutrients like K, N, and P in growing substrate is proved to be sufficient for production of good quality flowers (Younis et al. 2014) and adequate availability of these nutrients in growing media gave significant results regarding growth and flowering indices in this experiment. Organic matter and P content in growing substrate affected the quality of flowers as well. Growing media with excellent nutrient quality influences plant growth as availability of P and K contents in growth media has positive relationship with flowering indices. In current research, excellent nutrient quality with maximum phosphorus and potassium contents (27.00 and 500 ppm, respectively) was achieved in SFT growing media. Good quality flowers obtained in this combination is a confirmatory with Grassotti et al. (2003). Treder (2008) reported that plants had long stem, and maximum flowering in growing media with optimum P and K contents, where potting media comprising farm yard manure can give maximum amount of P and K contents (Strojny and Nowak 2004; Younis et al. 2007). Organic matter in growing media can individually promotes the plant growth due to high availability of nutrients (Fascella and Zizzo 2005; Younis et al. 2010), where maximum organic matter was found in farm yard manure containing substrates while other treatments showed no significant results. Physio-chemical attributes of high organic matter encourage the plant growth and improves substrate quality (Abad et al. 2001) which has been seen in this research work.

Various substrates having farm yard manure and compost are responsible to reduce pH and EC due to high nutrient supply (Bi et al. 2009; Diacono and Montemurro 2010). Actually, accessibility of nutrients to plants in growing media is coupled with the changes in media pH as increase or decrease in pH alters the solubility of nutrients and has direct effect on plant growth and development. Changes in pH above or lower than the optimum range adversely affected plants by damaging roots and decreasing nutrients availability (Altland 2006; Awang et al. 2009). The best range of dissolved salt contents i.e., EC in soil supports the better plant growth (Awang et al. 2009). Higher EC levels were not found to inhibit plant growth however; it decreased the biological activity of soil than the threshold value which negatively impacts plant growth (Miller 2001; Ribeiro et al. 2002). Suitable use of substrates for nutrient accessibility by plants not only depends upon nutrient composition but also on pH, nutrient form, organic compounds, and adsorption capacity of the growing media due to the presence of high organic matter content, P and K.

Conclusion

Results of this study indicated that application of coconut coir dust in combination with silt produced maximum plant height while addition of Lahore compost to substrate gave maximum number of leaves per plant and maximum flower stalk thickness. The addition of farm yard manure to silt and top soil produced maximum leaf area, highest value for number of roots, maximum fresh to dry weight ratio, and maximum flowers per plant, whereas maximum flower diameter, maximum fresh and dry weights of flower, resulted with mixture of sand with silt and top soil. All these organic waste substrates showed positive attribute to growth and flowering of gerbera with readily availability of nutrients for uptake. Therefore, it is concluded that suggested agricultural waste material can be used as growing media for better *Gerbera jamesonii* growth and can be explored further for other combinations.

Acknowledgments The Higher Education Commission (HEC) is highly acknowledged for providing funds for this research project.

References

Abad M, Noguera P, Bures S (2001) National inventory of organic wastes for use as growing media for ornamental potted plant production: case study in Spain. Bioresour Technol 77:197–200

Abad M, Noguera P, Puchades R, Maquieira A, Noguera V (2002) Physico-chemical and chemical properties of some coconut dusts for use as a peat substitute for containerized ornamental plants. Bioresour Technol 82:241–245

Ahmad I, Ahmad T, Gulfam A, Saleem M (2012) Growth and flowering of gerbera as influenced by various horticultural substrates. Pak J Bot 44:291–299

Altland JE (2006) Substrate pH, a tricky topic. Digger 50:42–47

Awang Y, Shaharom AS, Mohamad RB, Selamat A (2009) Chemical and physical characteristics of cocopeat based media mixtures and their effects on the growth and development of *Celosia cristata*. Am J Agric Biol Sci 4:63–71

Bashir MA, Ahmad M, Anjum MA (2007) Effect of various potting media on growth of rooted jojoba (*Simmondsia chinensis*) cuttings. Int J Agri Biol 9:147–151

Benito M, Masaguer A, Antonio RD, Moliner A (2005) Use of pruning waste compost as a component in soil less growing media. Bioresour Technol 96:597–603

Bi L, Zhang B, Liu G, Li Z, Liu Y, Ye C, Yu X, Lai T, Zhang J, Yin J, Liang Y (2009) Long-term effects of organic amendments on the rice yields for double rice cropping systems in subtropical China. Agri Ecosyst Environ 129(4):534–541

Caballero R, Pajuelo P, Ordovas J, Carmona E, Delgado A (2009) Evaluation and correction of nutrient availability to *Gerbera jamesonii* H. Bolus in various compost-based growing media. Sci Horti 122:244–250

Cardens MCA, Gomez IFR, Ronacancio VJF, Cordoba BC, Canola PW (2006) Growth analysis of standard carnation cv. 'Nelson' in different substrate. Acta Hort 718:623–630

Diacono M, Montemurro F (2010) Long-term effects of organic amendments on soil fertility. A review. Agron Sustain Dev 30:401–422

Dubey RK, Simrat-Singh Kukal SS, Kalsi HS (2013) Evaluation of different organic growing media for growth and flowering of petunia. Commun Soil Sci Plant Anal 44:1777–1785

Dubsky M, Sramek F (2008) Crushed rockwool as a component of growing substrates. Acta Hort 779:491–495

Fascella G, Zizzo GV (2005) Effect of growing media on yield and quality of soilless cultivated rose. Acta Hort 697:133–138

Grassotti A, Nesi B, Malrtta M, Magnani G (2003) Effects of growing media and planting time on lily hybrids in soilless culture. Acta Hort 603:395–399

Guerrero F, Gasco JM, Hernandez-Apaolaza L (2002) Use of pinebark and sewage sludge compost as components of substrates for *Pinus pinea* and *Cupressus arizonica* production. J Plant Nutr 25(1):129–141

Gupta R, Yadav A, Garg VK (2014) Influence of vermicompost application in potting media on growth and flowering of marigold crop. Int J Recycl Organ Waste Agri 3(1):1–7

Hernandez-Apaolaza L, Gascó AM, Gascó JM, Guerrero F (2005) Reuse of waste materials as growing media for ornamental plants. Bioresour Technol 96(1):125–131

Khalaj MA, Amiri M, Sindhu SS (2011) Study on the effect of different growing media on the growth and yield of Gerbera (*Gerbera jamesonii* L.). J Ornam Hortic Plants 1(3):185–189

Khayyat M, Nazari F, Salehi H (2007) Effects of different pot mixtures on pothos (*Epipremnum aureum* Lindl. and Andre 'Golden Pothos') growth and development. Am-Eurasian J Agric Environ Sci 2:341–348

Kiran M, Din J, Waseem K, Jilani MS, Khan MQ (2007) Effect of different growing media on the growth and development of Dahlia (*Dahlia pinnata*) under the agro-climatic condition of Dera Ismail Khan. Pak J Biol Sci 10:4140–4143

Lang HJ, Pannuk TR (1998) Effects of fertilizer concentration and minimum-leach drip irrigation on the growth of New Guinea Impatiens. Hort Sci 33:83–688

Lopez A (2006) Media influence on post harvest container plant quality in a retail nursery setting. Dissertation, University of Agriculture and Mechanical College, USA

McLean EO (1982) Soil pH and lime requirement. In: Page AL, Miller RH, Keeney DR (eds) Methods of soil analyses, part 2, chemical and microbiological properties, 2nd edn. American Society Agronomy, Madison, pp 199–224

Miller M (2001) Fulfilling special needs of nurseries. Biocycle 42:65–67

Nowak JS, Strojny Z (2003) Effect of different container media on the growth of Gerbera. Acta Hort 608:59–63

Olsen SAR, Sant MD, Gislerod HR, Solbraa K (1954) Nitrogen balance in bark composts used as growing media. Acta Hort 150:202–203

Rader JS (1998) Petunia. In: Ball V (ed) Ball Redbook. Ball Publishing, Batavia, IL, USA, pp 682–686

Riaz A, Arshad M, Younis A, Raza A, Hameed H (2008) Effects of different growing media on growth and flowering of *Zinnia elegans* cv. Blue point. Pak J Bot 40:1579–1585

Riaz A, Farooq U, Younis A, Karim A, Taj AR (2014) Growth responses of Zinnia to different organic media. Acta Hort 1018:565–572

Ribeiro HM, Vasconcelos E, Santos JQ (2002) Fertilization of potted geranium with municipal solid waste compost. Bioresour Technol 73:247–249

Saijeen S, Kaewman O, Suksawat M (2009) Evaluation of media, organic and chemical fertilizer applications on growth of pot gerbera (*Gerbera jamesonii*). Asian J Food Agro-Ind 2:S51–S56

Sindhu SS, Gholap DB, Singh MC, Dhiman MR (2010) Effect of medium amendments on growth and flowering in gerbera. Indian J Hort 67:391–394

Steel RGD, Torrie JH, Dicky DA (1997) Principles and procedures of statistics: a biometric approach. Mc Graw Hill Inc, New York

Strojny Z, Nowak JS (2004) Effect of different growing media on the growth of some bedding plants. Acta Hort 644:157–162

Tariq U, Rehman S, Khan MA, Younis A, Yaseen M, Ahsan M (2012) Agricultural and municipal waste as potting media components for the growth and flowering of *Dahlia hortensis* 'Figaro'. Turk J Bot 36:378–385

Torrey JG (1951) Cambial formation in isolated pea roots following decapitation. Am J Bot 38:596–604

Treder J (2008) The effects of cocopeat and fertilization on the growth and flowering of oriental lily 'star gazer'. J Fruit Ornam Plant Res 16:361–370

U.S. Salinity Lab Staff (1954) Diagnosis and improvement of saline and alkali soils. USDA Handbook 60, Washington DC, USA

Walkley A (1947) A critical examination of a rapid method for determining organic carbon in soils. Effect of variations in digestion conditions and of organic soil constituents. Soil Sci 63:251–263

Weeraratne TP, Daundasekera WAM, Wijesundara DSA (2012) Field survey of postharvest handling of cut flowers produced in the Up country of Sri Lanka for the local market. Ceylon J Sci 41:67–70

Wilson SB, Muller KL, Wilson PC, Incer MR, Stoffella PJ, Graetz DA (2009) Evaluation of new container media for Aglaonema production. Commun Soil Sci Plant Anal 40:2673–2687

Younis A, Ahmad M, Riaz A, Khan MA (2007) Effect of different potting media on the growth and flowering of *Dahlia coccinia* cv. Mignon. Acta Horti 804:191–196

Younis A, Riaz A, Wasim M, Khan MA, Nadeem M (2010) Production of quality croton (*Codiaeum variegatum*) plants by using different growing media. Am-Eurasian J Agric Environ Sci 7:232–237

Younis A, Riaz A, Siddique MI, Lim KB, Hwang YJ, Khan MA (2013) Anatomical and morphological variation in *Dracaena reflexa* 'Variegata' grown in different organic potting substrates. Flower Res J 21:162–171

Younis A, Anjum S, Riaz A, Hameed M, Tariq U, Ahsan M (2014) Production of quality dahlia (*Dahlia variabilis* cv. Redskin) flowers by efficient nutrients management. Am-Eurasian J Agric Environ Sci 14:137–142

Permissions

All chapters in this book were first published in IJROWA, by Springer; hereby published with permission under the Creative Commons Attribution License or equivalent. Every chapter published in this book has been scrutinized by our experts. Their significance has been extensively debated. The topics covered herein carry significant findings which will fuel the growth of the discipline. They may even be implemented as practical applications or may be referred to as a beginning point for another development.

The contributors of this book come from diverse backgrounds, making this book a truly international effort. This book will bring forth new frontiers with its revolutionizing research information and detailed analysis of the nascent developments around the world.

We would like to thank all the contributing authors for lending their expertise to make the book truly unique. They have played a crucial role in the development of this book. Without their invaluable contributions this book wouldn't have been possible. They have made vital efforts to compile up to date information on the varied aspects of this subject to make this book a valuable addition to the collection of many professionals and students.

This book was conceptualized with the vision of imparting up-to-date information and advanced data in this field. To ensure the same, a matchless editorial board was set up. Every individual on the board went through rigorous rounds of assessment to prove their worth. After which they invested a large part of their time researching and compiling the most relevant data for our readers.

The editorial board has been involved in producing this book since its inception. They have spent rigorous hours researching and exploring the diverse topics which have resulted in the successful publishing of this book. They have passed on their knowledge of decades through this book. To expedite this challenging task, the publisher supported the team at every step. A small team of assistant editors was also appointed to further simplify the editing procedure and attain best results for the readers.

Apart from the editorial board, the designing team has also invested a significant amount of their time in understanding the subject and creating the most relevant covers. They scrutinized every image to scout for the most suitable representation of the subject and create an appropriate cover for the book.

The publishing team has been an ardent support to the editorial, designing and production team. Their endless efforts to recruit the best for this project, has resulted in the accomplishment of this book. They are a veteran in the field of academics and their pool of knowledge is as vast as their experience in printing. Their expertise and guidance has proved useful at every step. Their uncompromising quality standards have made this book an exceptional effort. Their encouragement from time to time has been an inspiration for everyone.

The publisher and the editorial board hope that this book will prove to be a valuable piece of knowledge for researchers, students, practitioners and scholars across the globe.

List of Contributors

Jeremiah David Bala, Japareng Lalung and Norli Ismail
Environmental Technology Division, School of Industrial Technology, Universiti Sains Malaysia (USM), 11800 Pulau, Pinang, Malaysia

Farid Safari, Ahmad Tavasoli and Abtin Ataei
Department of Energy Engineering, Science and Research Branch, Islamic Azad University, Tehran, Iran

Jun-Ki Choi
Department of Mechanical and Aerospace Engineering, Renewable and Clean Energy, University of Dayton, Dayton, OH, USA

Payam Najafi
Department of Water Engineering, Agricultural Faculty, Isfahan (Khorasgan) Branch, Islamic Azad University, Isfahan, Iran

Jahangir Shams and Azita Shams
Young Researchers Club, Isfahan (Khorasgan) Branch, Islamic Azad University, Isfahan, Iran

I. A. Najar
Hydrobiology Research Laboratory, S. P. College,Srinagar 190001, Jammu and Kashmir, India

A. B. Khan
Department of Ecology and Environmental Sciences, Pondicherry Central University, Puducherry 605014, India

Loubna El Fels and Mohamed Hafidi
Laboratory of Ecology and Environment (L2E) (Unit Associated with the CNRST, URAC32), Faculty of Science Semlalia, Cadi Ayyad University, BP: 2390, Marrakech, Morocco

Yedir Ouhdouch
Laboratory of Biology and Biotechnology of Microorganisms, Faculty of Science Semlalia, Cadi Ayyad University, BP: 2390, Marrakech, Morocco

F. Parvin and Z. Ferdaus
Department of Environmental Sciences, Jahangirnagar University, Savar, Dhaka, Bangladesh

S. M. Tareq
School of Biosciences, The University of Nottingham, Jalan Broga, 43500 Semenyih, Selangor Darul Ehsan, Malaysia

T. R. Choudhury, J. M. M. Islam and M. A. Khan
Institute of Radiation and Polymer Technology, Bangladesh Atomic Energy Commission, Dhaka, Bangladesh

Majid Rostami
Department of Agronomy, University of Malayer, 657719-95863 Malayer, Iran

Stefano Monaco, Dario Sacco and Carlo Grignani
Department of Agricultural, Forest and Food Sciences,University of Turin, Lgo Braccini 2, 10095 Grugliasco, Italy

Elio Dinuccio
Mechanics Section, Department of Agricultural, Forest and Food Sciences, University of Turin, Lgo Braccini 2, 10095 Grugliasco, Italy

A. Mohammadi Torkashvand and M. Alidoust
Department of Horticulture, Faculty of Agriculture, Islamic Azad University, Rasht Branch, Rasht, Iran

A. Mahboub Khomami
Ornamental Plants and Flowers Research Station, Lahijan, Iran

Marc A. Zucco
USDA, Natural Resources Conservation Service, Grayling, MI 49738, USA

S. Alan Walters, She-Kong Chong and Brian P. Klubek
Department of Plant, Soil, and Agricultural Systems, Southern Illinois University, MC 4415, 1205 Lincoln Dr., Carbondale, IL 62901, USA

Joseph G. Masabni
Department of Horticultural Sciences, Texas A&M University, 2134 TAMU, College Station, TX 77843-2134, USA

S. K. Karn
Key Laboratory of Marine Environmental Corrosion and Biofouling, Institute of Oceanology,
Chinese Academy of Sciences, No. 7 Nanhai Road, Qingdao 266071, Shandong, China

S. K. Chakrabarti
Environment Management Division, Thapar Centre for Industrial Research and Development,
Yamunanagar 135001, India

M. Mirzaei Aminiyan, A. A. Safari Sinegani and M. Sheklabadi
Soil Science Department, College of Agriculture, Bu-Ali Sina University, Azadegan Street, 6517833131 Hamedan, Iran

W. Roshan Singh
Manipur Pollution Control Board, Imphal West DC Office Complex, Lamphelpat, Imphal 795004, Manipur, India

Shashi Kumar Panka and Ajay S. Kalamdhad
Department of Civil Engineering, Indian Institute of Technology Guwahati (IITG), Guwahati 781039, Assam, India

A. Koutarou, U. Seigo, H. Harada and Y. Mitoma
Department of Life and Environmental Science,Prefectural University of Hiroshima, 562 Nanatuka-cho, Shoubara, Hiroshima 727-0023, Japan

S. Oura and T. Watari
Department of Applied Chemistry, Saga University, 1 Honjo Saga, Saga 840-8502, Japan

B. K. Biswas
Department of Chemical Engineering, Jessore University of Science and Technology, Jessore 7408, Bangladesh

P. Bhaskar and S. K. Pyne
Department of ASEPAN P.S.B, Visva-Bharati, Sriniketan, Birbhum, West Bengal, India

A. K. Ray
Department of Zoology, Visva-Bharati, Santiniketan, Birbhum, West Bengal, India

Adel Ali Saeed Al-Gheethi
Faculty of Medical Science, Yemeni Jordanian University, Sana'a, Yemen

R. U. Shah, M. Abid and M. F. Qayyum
Department of Soil Science, Faculty of agricultural sciences and Technology, Bahaud din Zakariya University, Multan, Pakistan

R. Ullah
Soil and Water Testing Laboratory, Rajanpur, Punjab, Pakistan

K. M. Sangamithirai, J. Jayapriya and Ravi Manoj
Department of Chemical Engineering, A. C. Tech, Anna University, Chennai 600 025, India

J. Hema
Department of Biotechnology, PSG College of Technology, Coimbatore 641604, India

Rangasamy Anandham
Department of Agricultural Microbiology, Agricultural College and Research Institute, Tamil Nadu Agricultural University, Madurai 625 104, Tamil Nadu, India
Organic Agriculture Division, National Academy of Agricultural Science, Rural Development Administration, Suwon 441-707, Republic of Korea

Nagaiah Premalatha, Nellaiappan Olaganathan Gopal and Ramasamy Krishnamoorthy
Department of Agricultural Microbiology, Agricultural College and Research Institute, Tamil Nadu Agricultural University, Madurai 625 104, Tamil Nadu, India

Hyeong Jin Jee and Yong Ki Kim
Organic Agriculture Division, National Academy of Agricultural Science, Rural Development Administration, Suwon 441-707, Republic of Korea

Hang Yeon Weon and Soon Wo won
Korean Agricultural Culture Collection, National Agro Biodiversity Center, National Academy of Agricultural Science, Rural Development Administration, Suwon 441-707, Republic of Korea

Pandiyan Indira Gandhi
Regional Research Station, Tamil Nadu Agricultural University, Vridhachalam 606 001, Tamil Nadu, India

A. Mohammadi-Ghehsareh
Department of Soil Science, Faculty of Agriculture, Islamic Azad University of Isfahan, P.O. Box 81595-158, Isfahan, Iran

Amita Chattopadhyay
Department of Science & Technology, ICAR-Central Institute of Freshwater Aquaculture, Kausalyaganga, Bhubaneswar, Odisha 751002, India

Mohamed Fadel and Dina Helmy El-Ghonemy
Department of Microbial Chemistry, Genetic Engineering and Biotechnology Division, National Research Centre, 33 El-Bohouth St., Dokki, Giza 12622, Egypt

Jiwan Singh
Department of Civil Engineering, Indian Institute of Technology Guwahati (IITG), Guwahati 781039, Assam, India
Department of Environmental Engineering, Kwangwoon University, Seoul 139-701, Republic of Korea

Ajay S. Kalamdhad
Department of Civil Engineering, Indian Institute of Technology Guwahati (IITG), Guwahati 781039, Assam, India

Emmanuel Ibukunoluwa Moyin-Jesu
Agronomy Department, Federal College of Agriculture, Akure, Ondo, Nigeria

Ashish Kumar Nayak and Ajay S. Kalamdhad
Department of Civil Engineering, Indian Institute of Technology Guwahati (IITG), Guwahati 781039, India

Anoop Yadav and V. K. Garg
Department of Civil Engineering, Indian Institute of Technology Guwahati (IITG), Guwahati 781039, India

Raul Xavier Lopes, Romildo da Silva, Emerson Tokuda Martos and Eustá quio Souza Dias
Departamento de Biologia, Universidade Federal de Lavras (UFLA), CP 3037, Lavras, MG 37200-000, Brazil

Diego Cunha Zied
Universidade Estadual Paulista (UNESP), Câmpus de Dracena, Rod. Cmte Joaõ Ribeiro de Barros, km 651, Bairro das Antas, Dracena, SP 17900-000, Brazil

Rovilson José de Souza
Departamento de Agricultura, Universidade Federal de Lavras (UFLA), CP 3037, Lavras, MG 37200-000, Brazil

S. Sivasangari Ramya
Department of Biochemistry, Sri Sarada Niketan College for Women Amaravathipudur, Karaikudi, India

N. Vijayanand
Department of Botany, Arumugam Pillai Seethai Ammal College, Tiruppattur, India

S. Rathinavel
Centre for Research in Botany, Saraswathi Narayanan College (Autonomous), Madurai, India

Milad Mirzaei Aminiyan, Ali Akbar Safari Sinegani and Mohsen Sheklabadi
Soil Science Department, College of Agriculture, Bu-Ali Sina University, Azadegan Street, Hamedan 6517833131, Iran

Carla Bellera and Pere Rovira
CEMFOR, Forest Science Centre of Catalonia (CTFC), Carretera de Sant Llorenc¸ de Morunys, km 2, 25280 Solsona, Spain

Issa Abaalkheel and Jamal Alrefai
Petrochemicals Research Institute, KACST (King Abdulaziz City for Science & Technology), Riyadh, Saudi Arabia

Alireza Mesdaghinia and Simin Nasseri
Center for Water Quality Research (CWQR), Institute for Environmental Research (IER), Tehran University of Medical Sciences, Tehran, Iran
Department of Environmental Health, School of Public Health, Tehran University of Medical Sciences, Tehran, Iran

Amir Hossein Mahvi
Department of Environmental Health, School of Public Health, Tehran University of Medical Sciences, Tehran, Iran
Center for Solid Waste Research (CSWR), Institute for Environmental Research (IER), Tehran University of Medical Sciences, Tehran, Iran

Ramin Nabizadeh Nodehi
Department of Environmental Health, School of Public Health, Tehran University of Medical Sciences, Tehran, Iran

Mahdi Hadi
Center for Water Quality Research (CWQR), Institute for Environmental Research (IER), Tehran University of Medical Sciences, Tehran, Iran

Fernando Martínez San Pedro, Ignacio A. Domínguez Vara, José Luis Bórquez and Manuel Gonzalez-Ronquillo
Departamento de Nutrición Animal, Facultad de Medicina Veterinaria y Zootecnia, Universidad Autó noma del Estado de México, CP 50000 Toluca, Edo de México, México

Atif Riaz, Adnan Younis, Imran Ghani, Usman Tariq and Muhammad Ahsan
Institute of Horticultural Sciences, University of Agriculture, Faisalabad 38040, Pakistan